Micro Economic Analysis in Agriculture

Micro Economic Analysis in Agriculture

— Volume 1 —

K. Nirmal Ravi Kumar
Associate Professor & Head
Department of Agricultural Economics
Acharya N.G. Ranga Agricultural University
Agricultural College, Mahanandi
Nandyal, Kurnool District – 518 502

2012
DAYA PUBLISHING HOUSE®
New Delhi – 110 002

ISBN 978-81-7035-795-7 (Set)
ISBN 978-81-7035-796-4 (Volume 1)
ISBN 978-81-7035-797-1 (Volume 2)

Published by : **Daya Publishing House®**
4760-61/23, Ansari Road, Darya Ganj,
New Delhi - 110 002
Phone: 23245578, 23244987
Fax: (011) 23260116
e-mail : dayabooks@vsnl.com
website : www.dayabooks.com

Laser Typesetting : **Classic Computer Services**
Delhi - 110 035

Printed at : **Chawla Offset Printers**
Delhi - 110 052

PRINTED IN INDIA

– Dedicated to –

Late. K. RAMA CHANDRA RAO

ACHARYA N.G. RANGA AGRICULTURAL UNIVERSITY
ADMINISTRATIVE OFFICE, RAJENDRANAGAR, HYDERABAD-500030

V. NAGI REDDY, I.A.S
Vice-Chancellor

Phone : +91-40-24015035(O)
+91-40-24015031(F)
Email: angrau_vc@yahoo.com
GRAMS: "AGRIVARSITY"

Foreword

The subject 'Economics' has become a choice of study for many of the students in different Universities in India. The analytical aspects of this subject arouse interest among the student community and this motivates and guides them to pursue their post-graduate and doctoral programmes in the field of Economics. This text book entitled 'Micro-Economic analysis in Agriculture' is of no exception in analyzing the economic aspects in agriculture at the micro-level. The meritorious aspect of this text book is that, it presented all the topics in a more analytical manner that too considering simpler examples that we witness in our vicinity. Further, the author has quoted several examples from the field of agricultural sector and in this context, it is certainly a contributing one for the students specializing in Agricultural Economics. I complement the diligent, sincere and painstaking efforts of Dr. K. Nirmal Ravi Kumar in bringing out this important and timely publication, as it motivates and guides the students to pursue their career in the field of Economics. Further, it contains an up-dated and standard material on the various aspects of Micro-Economics duly covering the latest syllabi as prescribed by the several universities in India, thereby, it helps the Academicians to understand and explain the concepts in a more analytical manner. I hope this volume will also be very useful to the students, who desire to be competent enough in various competitive examinations.

V. Nagi Reddy

ACHARYA N.G. RANGA AGRICULTURAL UNIVERSITY

ADMINISTRATIVE OFFICE, RAJENDRANAGAR, HYDERABAD-500030

Dr. Y. ESWARA PRASAD
Ph.D
Professor & University Head &
Hon. Director
Department of Agricultural Economics
& CCS
College of Agriculture,
Rajendranagar – 500 030, Hyderabad
and
Member, Board of Management,
ANGRAU

Phone : +91-40-24015042
+91-40-24011561
Extn: 374
Mobile: 9441609517
Res: 040-23341406

Preface

In the modern era, economic aspects of agri-business are assuming paramount significance. In this context, understanding the basic concepts of micro-economic analysis with special reference to agricultural sector is essential. Moreover, in the recent period, there have been many new developments in the field of micro economic theory. This text book, '*Micro Economic Analysis in Agriculture*' covers both basic and advanced concepts of Micro Economics. Specifically, it dealt with the concepts of Micro Economics considering examples from the agricultural sector to the extent possible and the style of presentation of the concepts is very simple, analytical and easy to understand with attractive graphical representations. I complement the sincere efforts of Dr. K. Nirmal Ravi Kumar in bringing out this text book duly considering the latest syllabi, as prescribed by the IV Deans' Committee recommendations appointed by the Indian Council of Agricultural Research. This text book will be very useful and instructive for the graduates, post-graduates and doctoral students, who wish to pursue their career in the field of Economics in general and in Agricultural Economics in particular. It will also be of immense use for the Academicians, as it harbours standard and updated material pertaining to the concepts of Micro Economics. I hope this volume will even serve the students, who are aspiring high in the field of Economics and to be competent enough in various competitive examinations.

(Y. ESWARA PRASAD)

ACHARYA N.G. RANGA AGRICULTURAL UNIVERSITY
AGRICULTURAL COLLEGE, MAHANANDI
M.C. FARM – 518 502, MAHANANDI, KURNOOL DISTRICT

Dr. K. NIRMAL RAVI KUMAR
Ph.D
Associate Professor & Head
Dept. of Agricultural Economics

Phone : +91-08179696983
Email: drknrk@gmail.com

Author's Note

I consider this as a bold attempt to present a comprehensive text material on the fundamental aspects of Micro-Economics. With a limited experience in the field of teaching (5 years) in the Acharya N.G. Ranga Agricultural University, I will not dare enough to make a tall claim regarding the originality of the material incorporated in this text book entitled, '*Micro Economic Analysis in Agriculture*', but I certainly admit that, it is presented in a more analytical form to enable the students and Academicians to learn and conceptualize the concepts. As a teacher in the field of Agricultural Economics, I quoted examples from the agricultural sector to the extent possible, so as to benefit the students to understand the concepts more clearly. The present volume can serve as a standardized text book for the graduates, post-graduates and doctoral students, who wish to pursue their career in the field of Economics in different Indian Universities.

This task of writing and publishing the text book on the advanced versions of 'Micro-Economic Analysis in Agriculture' has been one of the most challenging and satisfying experiences of my life. I am especially thankful for the friendship and unfailing support of Mr.V. Sita Ram Babu, who practically demonstrated the true meaning of friendship by standing with me during the pursuit of this work. Finally, last but by no means least, a grateful kiss goes to my wife, Meena, my son, Sri Ram and my mother Janaki for their unwavering love and support. Their devotion, patience and understanding were indispensable. I gratefully dedicate this piece of work to my late father, Sri K. Rama Chandra Rao, who instilled in me at an early age the importance of education and the value of hard work. I only wish you could have been here to see me finish the journey I started with your efforts and initiatives. I know you would have been so proud.

K. Nirmal Ravi Kumar

Contents

Part I
Basic Concepts

1

Definitions and Scope of Economics

Economics is a social science, which is concerned with all phenomena incident to man's efforts to get what he wants, especially in so far as those efforts lead him to hire the services of others or to buy and sell goods. In simple sense, it deals with wants, efforts and satisfaction as the central idea. These three components constitute the subject matter of Economics and they are cyclical in nature and thus, never ending. To satisfy a human want, efforts will be made and if one want is satisfied another want will recur in the human mind. So, efforts will continue to satisfy his unlimited wants. The efforts will be in terms of different economic activities *viz.*, primary activities (agriculture, forestry, dairying etc), secondary activities (manufacturing industries) and activities in the services sector (banking, insurance, transportation etc). The existence of human wants is the starting point of performing all economic activities in the economy. Hence, Economics is considered as the *Science of wants*. But, human wants are unlimited and the means (resources) to satisfy them are limited or scarce. In fact, unlimited wants alone is not a problem, but certainly the problem exists when unlimited wants are combined with a limited means of satisfying those wants. 'Means', in the sense that, land, labour, capital, entrepreneur, time etc., are scarce and are not sufficient enough to satisfy unlimited human wants. So, Economics is otherwise termed as *Science of scarcity*. Moreover, wants are not of equal importance and since means are limited, it calls for prioritization of unlimited wants. On the other hand, if there is unlimited supply of means, then the problem of prioritization of wants will not arise. Since, means are limited, prioritization among wants is indispensable. However, some goods like air and sunshine are very important for human survival, but assume less significance in economic study, as they are unlimited in supply. So, Economics mainly deals with using only scarce resources in relation to satisfaction of prioritized human wants through performing various economic activities. The scarce resources will have alternative uses and thus, they can be employed among the prioritized wants to derive maximum satisfaction. This scarcity of resources or means will impose a limitation on the quantity of output that society can produce in the economy. So, the produced output in the economy will not satisfy all the human wants. So, it calls for choosing among the prioritized wants and allocation of resources to meet urgent wants. Hence, Economics is otherwise termed as *Science of choice making*. Since, there is choice making among human wants for the allocation of limited resources, it leads to opportunity cost. This concept of opportunity cost is inherent to the definition of Economics. Opportunity costs are everywhere, due to scarcity and the necessity of choosing the resources for their allocation to meet the unlimited human wants. This opportunity cost is not what we choose when we make a choice rather, it is what we did not choose in making a choice. So, it is the value of the forgone alternative *i.e.*, what we gave up when we got something. So, scarcity among resources results in choice and choice results in opportunity cost. Choice among human wants may also impose opportunity cost over time. The use of resources now, means that, those resources will not be available for future use. So, while taking a choice or decision regarding the employment of scarce resources over time, the opportunity cost is encountered, as to whether to allocate the resources for the present needs or future needs. There is yet more to discuss about the concept of opportunity cost, but this is the logical place to introduce this term.

The term Economics comes from the ancient Greek word *'oikonomika'*, means management of a household (*oikos* means 'house' + *'Nemein'* means 'to manage'). In modern usage, it refers to the efficient allocation of scarce resources in the production, distribution, exchange and consumption of goods and services to satisfy endless wants. As a branch of knowledge, Economics is the study of allocation and use of limited resources efficiently *viz.*, land, labour, capital and entrepreneur, so as to achieve maximum satisfaction in fulfilling the wants. Compared to other social sciences, Economics studies the human behaviour in the context of economic aspects only, so as to derive maximum satisfaction through achieving minimum cost in the production, distribution, exchange and consumption of goods and services. Hence, Economics deals with the decision making and management of the economy or economic systems. The decision makers or economic units of the economic system are households, businesses and Government. The fact that, Economics is in several ways an abstract science means that, various Economists have differing interpretations of models and the way the world works. Furthermore, it is difficult to run experiments on social situations, especially on a scale as large as an entire nation. Therefore, Economists rely on extensive mathematical tools to analyze real world situations that have occurred and use that to forecast future situations. This helps in shaping the trade and industrial policies of a nation, which finally aims at maximizing the welfare of the people. In this context, the so-called Father of political economy, Adam Smith, emphasized the advantages of free trade that England in 1846 abandoned its historic policy of protection. So, he quoted his famous book, *'The Wealth of Nations'*, with the subject matter of application of production and trade policies to enrich a nation. Because of this intimate relationship between the policies of Economics and the welfare of a nation, this science was commonly known as 'Political economy', till the end of 19th century. However, this name 'Political economy' has been less used in recent years for the reason that, Modern Economists are not politicians or statesmen and probably do not wish to have their efforts even seem to be coloured. They simply aim at enriching the body of knowledge like as in other sciences like Physics, Chemistry etc. The Economists will simply aim at formulating laws or principles or methods to address various economic issues like poverty, unemployment, inflation etc., in the economy, so as to promote the economic welfare of the whole nation. Hence, in this context, the Germans call it as 'National economy', instead of Political economy. The importance and utility of the subject of Economics can be judged from the fact that, it is now considered to be one of the most important and useful subjects as compared to any other branch of knowledge. The reasons for gaining its importance are that, it makes human welfare its direct and primary concern. In the words of Durbin, *'Economics is the intellectual religion of the day'*. About the importance of Economics, Malthus stated that, *'Political economy is perhaps the only science of which it may be said that the ignorance of it is not merely a deviation of good but produces great positive evil'*. In the words of Edmund Burks, *'the age of chivalry has gone, that of sophistry, economists and calculators has succeeded'*. The advantages of the study of Economics are as under:

- The subject is very informative and deals about man's behaviour regarding the allocation of limited resources among endless wants, when he is engaged in an economic activity.
- It conceptualizes various business principles, which can be applied to various sectors of the economy, so as to make the sectors both domestically and internationally competitive.
- It helps in understanding the present economic system based on which it guides to formulate developmental programmes and policies.
- It is applicable in various professions like agriculture, banking, marketing, industry etc., and thereby, guides the stakeholders of various sectors of the economy.
- It is useful in addressing the complex problems of economy like poverty, inflation, unemployment etc.
- Helps to understand the concepts of national income, employment, consumption, savings, capital formation, investment, price mechanism, demand and supply etc., and thereby, promotes logical thinking towards solving various economic problems.
- Economics is significant for the consumers, in the sense that, it guides them to allocate their scarce resources efficiently among unlimited ends, thereby, helps them in maximizing their satisfaction. The concept of choice making in Economics guides the consumer to be rational in his behaviouristic approach towards various economic problems.
- Economics helps the producers in determining the remuneration for different factors of production employed in the business. It enables the producers to enjoy the economies of large scale through making them understand about different cost concepts. It guides the producers in the fixation of competitive price for their output, and thereby, enable them to sustain in the business.
- Economics guides the workers or labourers to understand the concept of wages. It deals with various labour issues and their management and thereby, guide them not to get exploited in the business.

- Economics helps the politicians to understand and analyze about the causal factors of different economic problems in the society such as unemployment, inflation, poverty etc. It, therefore, guide them in planning the requisite strategies to address the same.
- Economics as a science provides scientific tools and techniques and this helps the academicians in developing the science in a systematic way. Also guides the researchers to conduct research on the prioritized issues in the economy, analyze cause and effect relationships, and thereby, helps in formulating need based strategies to address various economic problems.
- Helps the administrators in implementing various economic policies and programmes effectively in the society. Also guide them to plan various sources of raising revenue to the State and in mitigating the debt problems.
- Particularly in under developed and developing countries, under employment and unemployment are the major problems. So, understanding Economics, helps the policy makers to address various human resource development issues effectively.
- Helps to understand how the countries of the world are inter-dependent. Therefore, this helps the Governments of different countries to formulate and establish linkages between trade policies, environmental policies and sustainable development, so as to promote the living standards of people and to ensure full use of resources of the world reciprocally and mutually advantageous to all.
- Helps to forecast the future economic events.

The importance of the study of Economics can also be judged from the fact that, the daily newspapers cannot be understood without some knowledge of Economics. The newspapers often describe complicated economic problems such as inflation, balance of payment, balance of trade, imperfect markets, dumping, co-operative farming, sub-division and fragmentation of holdings, mechanization of agriculture etc., and in the absence of working knowledge of Economics, it is difficult to understand these diverse problems. This discussion infers that, the knowledge of Economics is very useful, as such it is necessary that every citizen, worker, administrator, consumer, etc., should have at least working knowledge of it. In the words of Sir Henry Clay, *'some study of Economics is at one a practical necessity and a normal obligation'*. Considering the importance of study of Economics, Wooten has rightly remarked that *'you cannot be in real sense a citizen unless you are also in some degree an economist'*.

1.1. Definitions

Studying the definitions of Economics will guide to understand about its nature and scope. There are differences of opinions among the Economists in giving the definitions about Economics. But, for the practical man, the simplest definition of Economics is 'the science which seeks to explain all business phenomena'. So, it implies, Economics is the science of business. Since, business embraces all forms of human activity, which mainly aims at increasing the wealth, the above definition can be refined as 'Economics is a science which seeks to explain the phenomena related to the production, consumption, exchange and distribution of wealth'. This concept was propounded by John Stuart Mill in his work, *'Principles of Political Economy'*. The older term for Economics is 'Political economy' and is still often used instead of Economics. The word *'Polis'* in Greek means a 'State'. So, the earlier Economists used the term 'Political Economy' for the management of the State. A person who runs a family is expected to make the best use of his income for meeting his household needs. Similarly, the State is expected to utilize its limited resources efficiently, so as to get the maximum benefit for the society. Hence the term 'Political Economy' is used instead of Economics. This definition is sometimes condensed as *'Economics is the Science of wealth'*. Other Economists seldom use this definition for the reason that, the word 'wealth' is a popular term, which cannot be defined in a clear-cut and scientific fashion. Adam Smith and J.B. Say too defined Economics on the same lines. However, other Economists re-defined the definition of Economics based on several criteria like based on welfare, scarcity, growth etc. So, there are several definitions about Economics based on various aspects or criteria, as perceived by the Economists of different times. The study of definitions of Economics is essential for the following reasons:

- Though Economics is studied by several Economists in the past, there lies no uniformity regarding the subject matter. So, by studying the definitions given by different Economists of different periods helps to trace out the uniformity of their approaches and thus helps to bring the unity in the science of Economics.
- Helps to study the scope of Economics.
- By thorough understanding the definitions of Economics, it helps to apply the principles in practice.
- Helps to analyze the growth of Economics.
- Helps to focus and prioritize the issues of Economics.

Like any science, Economics or Political economy, as it is often called, may be defined in different ways. Briefly, it is the science of business. This definition, however, takes it for granted that the reader knows exactly what is meant by business. Since business embraces all forms of human activities, such as production, consumption, exchange and distribution of wealth, we can conclude from this angle that, Economics primarily emphasizes on wealth aspects. However, different Economists defined Economics based on various criteria as they are discussed below.

1.1.1. Wealth Definition

The first person who introduced 'Economics' as a subject was Adam Smith (1723-1790). He is known as the Father of Political Economy and Father of Economics because, he was the first person who put all the economic ideas in a systematic way. It was only after Adam Smith, we study Economics as a systematic science. He defined Economics as *'An inquiry into the Nature and Causes of Wealth of Nations.'* in his famous book *'Wealth of Nations'* in 1776. In this book, he considered *'Economics as the science of wealth'*. That means, Economics as a subject, deals with the production of wealth, consumption of wealth, exchange of wealth and distribution of wealth. Production of wealth means the production of goods and services by combining four factors of production *viz.*, land (natural resources such as sea, minerals, live stock, forest etc), labour (mental or physical work which is done on reward basis), capital (man made resources which help to produce goods and services) and entrepreneur (act of combing other three factors of production to produce goods and services with a view to make profit). After the completion of production process, the generated wealth is distributed among the four factors of production for their performance. Rent is given to land, wages to labour, interest to capital and profit to organization. With these reward payments, people will satisfy their wants by spending their income to purchase goods and services.

The study of Adam Smith helped England to formulate and design the production, trade and industrial policies, as the book, *'Wealth of Nations'*, greatly emphasized the advantages of having free trade *vis-a-vis* protection trade. Because of this intimate relation between the conclusions of the Economist and the welfare of a nation, this science was commonly known as Political economy for a hundred years after Adam Smith. He used the famous term 'Invisible Hand' in his book, *'Wealth of Nations'* and he used this term only once in the following quotation:

'By directing that industry in such a manner as its produce may be of the greatest value, he intends only his own gain, and he is in this, as in many other cases, led by an invisible hand to promote an end which was no part of his intention. Nor is it always the worse for the society that it was not part of it'.

After observing the society as a whole, Adam Smith noted that, there was an 'invisible hand' in the form of market forces that turn the wheels of the economy and keeps it functioning. According to him, the allocation and efficient use of resources in the economy is guided by these market forces, which constitute the 'invisible hand'. He opined that, the participants in the economy are motivated by self-interest and that the 'invisible hand' of the market guides this self-interest into promoting general economic well-being.

The term 'wealth' has a special meaning in Economics. In the ordinary language, 'wealth' implies money, but in Economics sense, it refers to the goods and services that satisfy human wants. It is important to note that, all goods which satisfy human wants will not fall under 'wealth'. For example, free goods like air and sunshine are essential for human beings, but they are not regarded as wealth, as they available in abundance. So, only scarce goods having money value fall under the category of 'wealth' in Economics. Adam Smith was of the view that, Economics was concerned with the problems arising from wealth-getting and wealth-using activities of the people. He was interested mainly in studying the ways by which the wealth of all nations could be increased. But, the term 'wealth' was mistaken to be a substitute for 'money'. Hence, this classical definition became the target of attack and criticism by literary men like Carlyle, Ruskin and Russel. The following are the definitions given by other Classical Economists, emphasizing on the concept of 'wealth':

J.B.Say, J.S.Mill, J.E.Cairns, Senior, David Ricardo, Walker etc., supported Adam smith's view. J.B.Say defined Economics as *'the aim of political economy is to show the way in which wealth is produced, distributed and consumed'*. J.S.Mill stated that, *'Political economy professes to teach or investigate the nature of wealth and the laws of its production and its distribution'*. J. E. Cairns stated that, *'Economics deals with the phenomenon of wealth'*. The American Economist F.A.Walker says that, *'Economics is that body of knowledge, which relates to wealth'*. Thus, all these definitions relate to wealth.

Features of Wealth Definition

The following are the features of Adam Smith's definition:

- ☆ Economics deals with production, exchange, distribution and consumption of material goods.
- ☆ Only material goods, which are scarce constitute wealth and non-material goods like services and free goods like air and water etc., do not constitute wealth.
- ☆ Adam Smith and other Classical Economists defined Economics assuming an Economic man, who earns for fulfilling his economic self interest and to get maximum wealth. So, if every individual in the economy works for these objectives, the wealth of nation will automatically increase.

- According to Adam Smith, there is an 'Invisible Hand' in the economy in the form of market forces, which turns the self-interest of the Economic man into national interest.
- Economics deals with the causes of wealth changes. To promote wealth in the economy, production of material goods should be increased and this leads to division of labour and expansion of markets.

Criticisms of Adam Smith's Definition

Adam Smith's definition was criticized by Alfred Marshall and some other Neo-Classical Economists on the following grounds:

- According to Adam Smith and other Classical Economists, Economics is primarily the study of wealth. This definition was mainly criticized on the aspect that, it laid primary emphasis on the generation and distribution of wealth and secondary importance to man. In this way, the human being was degraded and ignored. As Marshall puts it, '*man occupies a primary place and wealth only a secondary one. Economics, on the one side, a study of wealth and on the other side and more important side, a part of the study of Man*'.
- Adam Smith considered only material goods in Economics that leads to wealth and excluded services like labours, doctors, teachers and lawyers services, which also leads to accumulation of wealth. These services are equally important in the economy as goods. So, Adam Smith's definition has restricted the meaning of wealth.
- Adam Smith's definition simply highlights the importance of earning of wealth. It has not studied about the means to earn the wealth.
- Adam Smith ignored the human welfare as compared to wealth. According to Classical Economists, wealth is more important than human welfare.
- The word 'wealth' itself is controversial and majority of the Economists have not accepted the definition emphasizing on wealth because, it is quite difficult to define wealth exactly.
- This definition does not analyze human behaviour in the context of multiplicity of human wants and scarce resources.
- This definition is incomplete because, it implies the human wants are satisfied only through earning and spending the money. However, a man can get satisfaction not only through earning and spending the money, but also through other activities like dancing, singing, playing games etc. These activities also involve economic aspects like choice making and satisfaction of human wants.
- When Adam Smith gave definition of Economics in 1776, British society was a religious society. So, religious minded people criticized the subject of Economics on the ground that, it will make the society as well as the individual materialistic and will take them away from spiritual values like friendship, brotherhood, love, sacrifice, patriotism etc. Instead, it will make them selfish, greedy, hypocrite etc., due to pursuit for materialism. The two main critics of Economics *viz.*, Carlyle and Ruskin stated that, 'Economics' as a subject teaches selfishness among the people and hence, criticized it as '*dismal science*' and '*dark science*' highlighting *Mammon* worship. Some other people said ironically that Economics, a '*science of materialism*' and as a '*science of bread and butter*', as it promotes selfishness and greed. They further opined that, if Economics was taught, the science of materialism will take mankind away from spiritualism. Hence, Carlyle even went to the extent of saying that, Economics is a *Pig Philosophy* and therefore, it should not be studied. But, this criticism is unfair because, it is based on a misunderstanding about the nature and scope of Economics. As the definition mainly emphasized 'wealth', they thought it is all about money. So, they concluded that, Economics will teach men and women how to make money. So, they called it a 'Selfish science', as in their opinion it emphasized on 'the means to get rich'. Russel, Edgeworth, and Jevons also criticized the wealth concept of Economics.
- The assumption of Invisible Hand in the economy, which turns self-interest of the people to national interest is unreal and illusory.
- Wealth by itself does not lead to economic development, unless it is invested for productive purposes.
- Wealth is only a means to an end and it is not an end in itself. So, this definition neglects man and gave more importance to wealth.
- The concept of 'Economic man' received more criticism from the Modern Economists, as they opined that 'Welfare Economics' assumes more significance rather than self-interest concept.
- Adam Smith's definition has two aspects *viz.*, to provide plentiful income to the people and to provide sufficient revenue to the State. So, Walras criticized this definition as unscientific because, these two aspects conflict with each other.

- ☆ Adam Smith's definition is based on assumptions like every man works more to satisfy his self-interests and there is no much difference between personal and social interests. But, the Neo-Classical Economists disregarded this view and opined that personal interests are different from social interests. So, they further stated that, *'Economics studies a common man and not Economic man'*.
- ☆ Adam Smith's definition emphasizes on the accumulation of wealth and it considered earning of wealth an end in itself. But, it pays no attention towards equitable distribution of wealth and its use for the welfare of the society.

Despite these criticisms, it is an acceptable fact from Adam Smith's definition that, wealth is the biggest reality of life and there is no life without wealth, particularly food. Wealth itself is not bad. Its use could be good or bad. So, in Economics sense, wealth does not refer to money alone. It refers to the scarce goods which satisfy our wants. Moreover, Classical Economists used the term 'wealth' in the sense of welfare. Here, the term 'wealth' should be considered as 'Aggregate wealth' and not 'Net wealth'. According to Adam Smith, 'wealth' gives real meaning to human wants in the sense of 'effective demand'. If the definition is correctly interpreted, it indicates, it has not neglected 'man', as wealth earning and wealth spending is done only for the sake of man and in the absence of 'Man', the objectives of economic science are not fulfilled. The critics might have misunderstood Smith's definition without knowing the context in which the term 'wealth' was used.

But, the greatest demerit of Adam Smith's definition is that, there lies overemphasis on wealth. No doubt, we deal with wealth concept in Economics, but it is only a part or component of the entire study. There is the other side. In fact, it is a more important side and that is the study of man. Economics is a social science. Hence, the prime concern of Economics is the study of man and not wealth alone. However, while looking into the definition given by Adam Smith, it is important to note the time when he gave the definition. This is so because, at the time of defining Economics by Adam Smith, England was on the eve of Industrial Revolution. The large investments of capital and use of large scale machinery enabled England to produce wealth on a large scale. So, this prompted Adam Smith to define Economics emphasizing on 'wealth' and rightly considered Economics as *'an inquiry into the nature and causes of the wealth of nations'*.

1.1.2. Welfare Definition

Alfred Marshall, a pioneer Neo-Classical Economist, re-oriented Economics towards the study of humanity and provided economic science with a more comprehensive definition. Marshall, in his famous book *'Principles of Economics'* published in 1890, defined Economics as *'Political Economy or Economics is a study of mankind in the ordinary business of life. It examines that part of individual and social action which is most closely connected with the attainment and with the use of material requisites of well-being'*. Thus, Marshall shifted the focus from wealth to welfare. He opined that, wealth is only a means to increase welfare. According to Marshall, Economics not only analyze the aspect of how to acquire wealth, but also how to utilize this wealth for obtaining material gains of human life. In fact, wealth has no meaning in itself, unless it is used to purchase all those things, which are required for our sustenance as well as for the comforts required for life. Marshall, thus, opined that, wealth is a means to achieve certain ends.

Features of Welfare Definition

The following are the features of Marshall's definition:

- ☆ Economics is a study of mankind *i.e.*, more recognition is given to man compared wealth.
- ☆ Human life has several aspects like social, religious, economic and political. But, Economics is concerned only with the economic aspect of life and hence, deals with economic activities only. Marshall clearly explains that, economic activity is different from other activities. For example,
  - ❖ If a student visits a friend who is ill, it is a social activity,
  - ❖ If a person gives his vote in an election, it is a political activity.
  - ❖ If a person goes to church or temple, it is a religious activity.
  - ❖ Marshall says that, economic activity is different from above mentioned activities. A farmer going to the field or a worker going to the sugar factory is an economic activity *i.e.*, they are working to earn money. With this money, they will buy goods and services to satisfy their wants. In other words, Economics deals with wants, efforts and satisfaction. In the words of Marshall, *'man earns money to get material welfare'*. So, Marshall gave importance to material welfare and man. Hence, this definition was termed as 'welfare' definition, because of its emphasis on material welfare.
- ☆ Promotion of welfare is the ultimate goal.
- ☆ Economics studies mankind in the *'ordinary business'* of life *i.e.*, engaged in earning activities. This is the activity of an ordinary man. So, Economics studies an ordinary man and not abnormal persons like misers, mentally imbalanced, saints etc.

- ☆ Main emphasis is on material welfare. It deals with the causes influencing material welfare. Hence, it is considered as a social science. The study of non-material welfare is outside the scope of Economics.
- ☆ He rejected the concept of 'economic man' and opined that, Economics deals with 'ordinary man', who tries to maximize the welfare through the wealth he earns from the society and is also influenced by feelings like love, affection etc.
- ☆ Wealth is sought only to promote welfare. So, Marshall gave more importance to welfare compared to wealth. But, it is interesting to note that, unlike Classical Economists, he gave emphasis to both human welfare and wealth. He regarded material welfare is the end and wealth is regarded only as means.
- ☆ Economics is not a pure science. It is only an art. It is one among the social sciences.
- ☆ Economics deals with individuals on one hands and social organizations of material activities on the other.

Marshall has made the following improvements in defining the Economics when compared to Adam Smith's definition:

- ☆ Classical Economists gave more emphasis to wealth and neglected the concept of human welfare. On the other hand, Marshall has emphasized both wealth and welfare, and further, he accorded prime place to human welfare compared to wealth.
- ☆ By shifting the emphasis from wealth to welfare, Marshall enlarged the scope of Economics.
- ☆ By giving emphasis to welfare concept, Marshall made Economics a Normative science, which implies welfare ought to be maximized.

Marshall agreed that, Economics studies about wealth. But, man is the center of study. According to him, the study of man is more important than the study of wealth. He emphasized that, Economics studies about man as 'buyer and seller, producer and consumer, saver and investor, employer and worker etc. It studies about how people earn their income, how they allocate the money, how efficiently they use the money etc. So, it helps, how people try 'to increase the material means of well-being'. This clearly implies that, Economics is the study of the causes of material welfare.

Edwin Cannon, Pigou and Beveridge followed Marshall and endorsed the welfare definition. Cannon stated that, '*the aim of Political economy is the explanation of the general causes on which the material welfare of human being depends*'. Pigou defined Economics as '*Economic welfare is the subject matter of economic science. Economic welfare is that part of social welfare that can be brought directly or indirectly into relation with the measuring rod of money*'. Beveridge defined Economics as '*the study of the general methods by which men cooperate to meet their material needs*'.

Merits of Marshall's Definition

The following are the merits of Marshall's definition of Economics:

- ☆ Marshall's definition has given due place both for man and wealth.
- ☆ Marshall's definition enlarged the scope of Economics by shifting the emphasis from wealth to welfare
- ☆ By prescribing the goal *i.e.*, welfare, Marshall has made Economics subject 'Normative'. Since, his definition links wealth and welfare, it implies that, welfare ought to be maximized.
- ☆ Marshall's definition lifted Economics to the status of a Science and made Economics free from all political influences.

Criticisms of Marshall's Definition

No doubt that, Marshall's definition of Economics is a great improvement over the definition of Adam Smith, as it laid more emphasis on social problems. Moreover, it tells us about the link between wealth and welfare. But, Lionel Robbins criticized Marshall's definition of Economics on the following grounds:

- ☆ Robbins criticized Marshall's definition as 'classificatory' and not 'analytical'. This is because, it is concerned with 'group of activities', rather than the 'aspect of activities'. This definition gave scope to classify the activities into 'activities which yield material welfare' and 'activities which yield non-material welfare' and also as 'economic activities' and 'non-economic activities'. This is purely non-scientific. This is because, it considers the production of material goods (*e.g.* rice, wheat, chairs, tables etc.) alone as economic activity and the services of a doctors, lawyers, judges etc., are not considered as economic activities, as they do not produce material goods. But, we make use of their services. Moreover, their services are paid and hence, their services have economic value. So, it is clearly misleading to say that, Economics is the study of the causes of material welfare only. As Lionel Robbins says, '*we do not say the production of potatoes is economic activity and the production of philosophy is not*'. So, the term welfare is used in a narrow sense to meet material welfare only. Thus, Marshall's definition has limited the scope of Economics.

- ☆ According to Neo-Classical Economists, welfare is a measurable concept. Robbins criticized this and stated that, welfare is not a measurable concept. According to him, it cannot be calculated that, how much welfare a good will give to an individual or a society, as it is a relative concept. The concept of welfare changes according to place, time, circumstances, individual etc. For example, a non-smoker considers smoking is injurious to health. But, a smoker considers, it is a need, as it given him satisfaction.
- ☆ If welfare is a measurable concept, it will create problems in policy making. This is because, Marshall's definition of Economics creates a problem of liking and disliking on the basis of welfare. For example, wine should not be taken and bread can be taken. This retards human freedom for the satisfaction of wants.
- ☆ As per Marshall's definition, material goods which do not promote welfare are excluded. For example, drugs, cigarettes, alcoholic drinks etc., are excluded from the subject matter. Hence, Robbins criticized this and stated that, the goods mentioned above, though, do not yield welfare to mankind, but satisfies human wants and production and distribution of these products involve economic activity. Hence, they should be included in the subject matter of Economics. Similarly, war does not promote material welfare. But, we have *'Economics of war'* and it is an important branch of Economics. There are many economic problems with regard to war. Sometimes, the economic causes of war are more important than the political and social causes. So, it is not right to say that, Economics studies material welfare only.
- ☆ Since, Marshall's definition introduced ethical concepts like welfare, Economics now turn as an inexact science. Moreover, it is rather difficult to measure welfare. Some economic policies may promote welfare for some people and may affect the welfare of others.
- ☆ According to Robbins, Economics should be called as 'Human science' rather than Social science. According to Marshall, Economics is a social science because, it deals with the people living in organized societies. But, Robbins opined that, Economics as a science deals with isolated men also. This is because, the generalizations of the theory of value are applicable to the behaviour of isolated men like Robinson Crusoe*. So, Economics should be considered as 'Human science'.
- ☆ Marshall's definition does not make Economics universal and restricts the scope of Economics to men living in organized societies.
- ☆ As per the Marshall's definition, Economics gives prime importance to human welfare and this implies Economics is a Normative science. So, as a Normative science, Economics has to pass moral judgments *i.e.*, what is good and what is bad in the economy. So, it becomes more an Ethical science. Robbins opined that, Economics should remain neutral between the ends.
- ☆ Marshall's definition does not explain the central problems of an economy
- ☆ Marshall's definition studies about ordinary man in his normal business activities of life. But, according to Robbins, all men will have economic problems pertaining limited resources and unlimited wants. So, all men faces the problem of scarcity of resources and not only an ordinary men.

1.1.3. Scarcity Definition

Lionel Robbins sought to define more precisely Economics as a science and to derive substantive implications. He defined Economics as a *'Science which studies human behaviour as a relationship between ends and scarce means which have alternative uses'* in his popular book, *'An Essay on the Nature and Significance of Economic Science'*. This definition fits neatly with the theoretical and applied versions of Economics and it includes the following features:

(a) Ends

It refers to human wants and they are unlimited. Wants recur. When a particular want is satisfied, another want will crop up in human mind. This multiplicity of wants makes the people to work ceaselessly to fulfill them, but they are unable to satisfy all the wants, as they are unlimited. Wants are of two types *viz.*, Natural wants and Artificial wants. Natural wants are those, which are satisfied by free gifts of nature or free goods like air, water, sunlight etc. As no economic activities are performed to fulfill these wants, they are not studied in Economics. Artificial wants are the wants, which are satisfied by economic goods like food, clothes etc., and they are scarce in the economy. Hence, they constitute the subject matter of Economics.

(b) Scarce Means

It refers to time, money and resources and they are limited. This scarcity aspect of resources forms the basis of Robbins's definition. Hence, Economics is also called as *'Science of scarcity'*. Moreover, the scarce resources at the disposal of human beings are not 'specific'. If the resources are specific, there would not have been much problem to

* The novel "Robinson Crusoe", describes a familiar story of a man, who is shipwrecked on an uninhabited island, and who is forced to make a life for himself equipped initially with nothing but a few basic tools.

satisfy a particular want. That means, non-specificity of resources is posing choice making problems in the economy. Thus, the definition of Robbins emphasizes that, the economic activity arises only, so long as the means are scarce. This scarcity of means also leads to choice making among the wants to be fulfilled. Had the resources were not scarce, there would not have been much problem regarding choice making among unlimited wants. A resource is said to be scarce, if its supply is less than its demand. Moreover, the term 'scarce' in the definition should be viewed in relative sense and not in absolute sense depending on the demand situation. For example, rotten eggs, though smaller in quantity, should not be considered scarce in economic sense and have no significance, as there is no demand for them in the market. Food grains, though supplied in larger quantities in the market, can be considered as scarce because, the demand is greater than supply. Hence, scarcity is considered in relation to demand only. This feature of scarcity of resources has given rise to the concept of 'scale of preferences', which has significant importance in economic analysis. In a practical situation, the individual faces with unlimited wants to be satisfied with limited resources. So, in such situation, the individual will act rationally. He will, then, prioritizes the wants and allocate the resources on the top most prioritized want, which gives him the maximum satisfaction. Then, he allocates the resources on the want, which will yield less satisfaction, and again on the want, which yields him least satisfaction. So, each individual has a scale of preferences or wants and he arranges them in the order of preference of importance to him. This does not mean that, every individual will have same scale of preferences. But, the priority order definitely changes across the individuals. So, this again highlights the importance of allocation of scarce resources among prioritized wants in accordance with the opportunity cost, which is discussed below.

(c) Alternative Uses of Resources

It refers to use of resources among different purposes. Economic problems arise due to non-specificity of resources. That means, the resources can be put into various uses. For example, land resource is capable of being used for cultivating paddy, wheat, maize, cotton, sugarcane, for constructing a house etc. Likewise, sugar cane can be used as a raw material for jaggery making, sugar, khandsari products etc. Such alternative uses of scarce resources give rise to the concept of Opportunity cost, which is of great significance in economic analysis. Since, all the wants cannot be fulfilled with limited or scarce resources, choice making is inevitable. So, if resources are allocated towards fulfilling one want, it leads to sacrifisation of other wants. That means, the choosing of one is the loosing of the other. Suppose, a farmer is having Rs. 10000 and he can invest the same in a crop production programme or else he can save the money in a bank (as he cannot do both). So, the opportunity cost of spending the money on the crop enterprise is the interest he misses from the money, if saved in the bank. This implies, opportunity cost is the cost equivalent to the returns from the next best alternative foregone or it is the cost equivalent to the returns from the missed alternative. According to Benham, *'every economic decision is a choice between alternatives. The cost of a thing, in the last resort, is not the money spent on it or the labour and materials employed on it, but the thing which was most nearly chosen instead, the alternative which was foregone. The Economists calls it as Opportunity cost'*. Because of opportunity cost, the aggregate demand for the resource becomes so large than its existing supply and thereby, acquires economic significance.

(d) Relative Importance of Wants

Wants differ in their importance. Since, resources are scarce, wants are prioritized and this leads to choice making. So, scarcity and choice making are central problems in Economics. Hence, Economics is defined as the *Science of choice*. It is important to note that, some resources like air and sunshine are not scarce and they are called free goods. But, many resources are scarce in the economy in relation to satisfy unlimited wants. Since, wants vary in their importance, they are prioritized and urgent wants are fulfilled first followed by others. For example, if the farmer is in requirement of seed, fertilizers, pesticides, weedicides, herbicides etc., for starting a crop production programme, he first purchases seed followed by other resources in the order of priority. So, Economics studies the human behaviour as a relationship between unlimited ends (wants) and scarce means. As means are limited, we have to pay a price for them. So, in Economics, the determination of prices of scarce goods will be studied. This makes Economics essentially a 'Valuation process or Pricing process'.

It is important to note that, mere multiplicity of wants or scarce resources or alternative uses of resources or relative importance among wants will not create an economic problem, if taken alone. But, if the four arises simultaneously, it will lead to economic problems and choice arises in relation to unlimited ends and scarce resources. For example, a farmer with his limited money income has to make choices among various alternatives *viz.*, expenditure on paddy, expenditure on livestock rearing, family living expenses like food, clothing, children's education, entertainment etc.

Merits of Robbins Definition

The following are the merits of Robbins's definition of Economics:

- ✰ The earlier definitions given by Adam Smith and Alfred Marshall include vague expressions *viz.*, 'wealth' and 'welfare' respectively. However, Robbins definition does not have such vague expressions and it is more towards reality. As this definition approaches the reality of what is seen in the economy, it elevated the Economics to the status of a Pure science.

- Unlike earlier definitions, Robbins definition is more analytical than classificatory. This is so because, it deals with any human activity, which falls in the scope of employing scarce resources among unlimited wants. This is because, all activities *viz.*, political, social, religious etc., have economic aspect and involves choice making among alternatives. It also considers any activity, which is influenced by scarcity of resources is an economic activity. So, it abolished the distinction between economic activity and non-economic activity. It does not confine Economics in the narrow sense of wealth or welfare, but extends to all activities of mankind, if they are related to the phenomena of scarcity and choice making. This indicates that, if there is scarcity of a thing in relation to the demand for it, it becomes the subject–matter of Economics. Hence, the services of farmers, lawyers, doctors etc., are taken for study in Economics. This makes the Robbins definition more scientific in its approach.
- Robbins concept of *'Science of scarcity'* has universal application. It is applicable to any form of economy like capitalism, socialism, mixed economy or even to an isolated economy like that of Robinson Crusoe. For example, even a sage in Himalayas has the problem of limited time. So, every individual in the society faces the problem of scarcity in terms of resources and this leads to choice making among human wants. So, this definition made Economics universal.
- Robbins definition made Economics a Pure or Positive science. According to Robbins, it is not the duty of an Economist to judge the rightness or wrongness of the things in the economy. The subject of Economics should remain neutral between the ends.
- The relation between scarce resources and unlimited ends calls for economizing the resource usage and this made the subject of Economics as Valuation or Pricing process.
- As there are alternative uses for the scarce resources, this definition gives rise to the concept of opportunity cost.
- Since the resources are scarce and every individual tries to maximize the resource use efficiency, he first tries to prioritize the wants, as they differ in their relative importance. This leads to arranging the wants in the order of their importance. So, this concept of definition leads to identifying the 'scale of preferences', such that, individuals will give prime importance to most urgent wants.
- Robbins definition clearly explained the differences between a 'technical problem' and 'economic problem' in an economy. According to him, Economics is not concerned with technical problems and deals only with economic problems. For example, if a farmer wants to construct a cattle shed, he can take up the construction work on a temporary basis with thatched roof or else he can take permanent construction with bricks and cement. This is a technical problem because, for one end (cattle shed) there are different means. On the other hand, for an economic problem, the ends (wants) are unlimited, but the means to achieve them are limited.

Robbins definition is certainly very convincing and scientific without any ambiguity. Eric Roll, Alec Macfie, Stigler, Caincross, Lerner etc., are the followers of Robbins. Prof. Macfie is much impressed with the definition of Robbins and he opined that, *'what Robbins has said cannot be resaid. To me, it appears final within its scope'*. They all opined that, Economics as a science aims at economizing the scarce means in relation to unlimited ends.

Criticisms of Robbins's Definition

Robbins definition was criticized on the following grounds:

- Robbins definition over-emphasized Economics as a positive science, as it deals exclusively with the valuation problem and neglected the normative aspect. As per his definition, Economics only says 'how man behaves' and will not suggest 'how he should behave' in the economy. Economics should remain neutral between the ends. But, Economists like Souter, Woofton and Macfie opined Economics should be both positive and normative in its approach. They further stated that, Economics being a social science, should deal with all the economic problems in the society and also provide solutions to overcome the same. So, Economists cannot disassociate themselves from ethics or welfare concepts.
- Robbins views are contradictory in terms of considering two aspects *i.e.*, scarce means and Economics is neutral between the ends. When Economics is neutral between the ends, effective allocation of scarce resources is of less importance.
- Robbins lays too much importance on the scarcity aspect of resources and it is the main reason for all economic problems. But, economic problems may arise even out of abundance. For example, the great depression of 1930s was caused not due to scarcity, but by plenty of goods (over production). That is why, the world depression was described as *'Poverty in the midst of plenty'*.
- According to Robbins definition, the scope of Economics is imprecise *i.e.*, either too wide or too narrow. In the wider sense, Economics embraces almost the whole of man's life and in the narrow sense, Economics would be a science of price theory or market equilibrium.

- Robbins definition lacks human touch. This is because, it is concerned only with the adjustment of scarce resources to meet unlimited wants and not concerned with human welfare. It does not try to establish a link between Economics and welfare. It looks at Economics only as the science of pricing process, but the scope of Economics is beyond the theory of value or resource allocation. Economics as a social science should aim for the promotion of human welfare.
- Robbins definition takes a static view of the economic problems and therefore, does not help us to solve the problems of development. It is inadequate to analyze modern economic problems like unemployment, underemployment, economic depression etc., as the definition deals with only the adjustment of limited resources among unlimited wants and does not consider the problem of how to increase the level of resources in the economy. So, the definition of Robbins is static and not dynamic, as it does not deal with the problems associated with growth and development of the economy.
- Robbins definition mainly emphasize that, Economics is micro in its approach, as it deals with individual's behaviour in relation to scarce means and unlimited wants. But, the economic problems are mostly related to the society rather than individual's behaviour.
- In the real world, it is difficult to separate ends and means. This is because, ends become means for further ends. For example, for a farmer, production of mangoes may be the end. But, for a processing firm, preparation of mango juice is the end. So, what is the 'end' from farmer point of view is the 'means' for processing firm. So, if ends and means are inseparable, the idea of 'economizing' becomes indefinable. Moreover, Robbins gave more importance to means rather than ends. But, neglecting ends implies eliminating the purpose in human actions. So, both ends and means should be given equal importance.
- Robbins conception about Economics will not offer solutions to the problems of under developed countries, as in these countries, economic problems are not due to scarcity of resources, but due to unused resources. Similarly, it is not applicable to rich countries like USA as well. This is because, in such countries, the people are constantly worried about the use of their expanding resources.
- The concept of scarcity was also emphasized by earlier Economists like Wicksteed, Meyer, Carl Menger etc., in their definitions. So, the concept of Robbins definition is regarded as *'Old wine in new bottle'*.
- Robbins definition has no universal application, as for example, in Communism, all the economic decisions are taken by the Government, thereby, the individuals have no freedom of choice.

In spite of the above criticisms, most of the Economists have accepted the definition of Robbins because, it emphasized scarcity and choice which are two important facts of life under all economic, political and legal systems. It is true that, there have been improvements in the methods of production because of technological advancements, but we experience scarcities in all aspects of resource allocations. Some Modern Economists like Stonier, Hague, Scitovsky, Ferguson, Samuelson Cairncross etc., have reformulated the definition of Robbins and among them Samuelson's definition has gained wide appreciation.

Similarities between Marshall's and Robbins's Definitions

On analyzing the definitions given by Alfred Marshall and Lionel Robbins for Economics, it can be concluded that, both the definitions differ completely. But, a close examination reveals the following similarities:

- Marshall gave emphasis for 'welfare' in his definition, whereas, Robbins emphasized on 'scarce means'. But, both the terms are similar indirectly. Robbins studied the concept of welfare indirectly. According to him, an individual tries to maximize the satisfaction through allocation of scarce resources among unlimited wants through choice making. This concept of maximum satisfaction implies 'welfare'.
- Both Marshall and Robbins gave primary place to man. Human being's welfare was the primary concern, but Marshall mentioned the same in his definition directly, unlike Robbins.
- Both the definitions conclude man is 'rational' in his behaviour. According to Marshall, man is rational in seeking material welfare, whereas, according to Robbins, man is rational in allocation of his scarce resources among unlimited wants.

Dissimilarities between Marshall's and Robbins's Definitions

There are lots of differences between the two definitions and they are discussed through Table 1.1.

1.1.4. Growth Definition

Scarcity concept explains the presence of economic problems. It is concerned with the positive aspect of the subject. But, Modern Economists are of the opinion that, how these (scarce) means should be further increased in the future to satisfy more wants and attain good living. So, this concept is called as Growth concept of definition of Economics. This conceptual definition of Economics given by Paul A Samuelson is also regarded as Modern definition of Economics. He

Table 1.1: Differences between Marshall's and Robbins's definitions.

Alfred Marshall's Definition	*Lionel Robbins's Definition*
Marshall's definition is classificatory and not analytical. It classified the activities into 'activities which yield material welfare' and 'activities which yield non-material welfare', economic activities and non-economic activities.	It is more analytical and not classificatory. It considered any activity which is influenced by scarcity of resources is an economic activity.
Marshall's definition considers Economics as a normative science.	Robbins definition overemphasized Economics as a positive science and neglected the normative aspect.
According to Marshall, Economics is a Social science because, it deals with the people living in organized societies.	According to Robbins, Economics should be called as 'Human science' rather than Social science.
Marshall's definition is nearer reality and forms basis for economic policies	Robbins definition is of an abstract nature and, therefore, moved away from reality.
Marshall's definition has human touch, as it laid more emphasis on human welfare	Robbins definition lack human touch, as it laid more emphasis on scarce means.
Marshall's definition is more relevant to capitalistic economies	Robbins definition is applicable to different types of economic systems, as scarcity of resources is the common phenomenon everywhere. Even a Robinson Crusoe is faced with the problem of scarcity of means.
Marshalls's definition is altogether distinct from its earlier definitions based on wealth.	The concept of scarcity was emphasized by earlier Economists like Wicksteed, Meyer, Carl Menger etc., in their definitions. So, the concept of Robbins definition is regarded as 'old wine in new bottle'.
Robbins definition made the scope of Economics too narrow, as it deals with the economic activities, which yield material welfare.	Robbins definition made the scope of Economics too wide and even includes non-economic problems.

was the first American to receive the Nobel Prize in Economics. According to Samuelson, '*Economics is the study of how man and society choose, with or without the use of money to employ scarce productive resources, which could have alternative uses to produce various commodities over time and distribute them for consumption now and in the future among various people and groups in society*'. The important features of this definition may be summarized as under:

- ☆ The relation between unlimited wants and scarce means leads to the problems of choice making. So, it is essential to prioritize the human wants and resource allocation should be made accordingly.
- ☆ Economics deals with the employment of scarce resources, which have alternative uses, to produce goods and services for present and future consumption.
- ☆ The available scarce resources should be used very judiciously and cautiously, as in the modern dynamic economic system, wants show increasing tendency. So, to satisfy as many wants as possible, best possible efforts should be made to increase the level of resources in the economy.
- ☆ Economics is not only concerned with the identification of economic problems but it should also suggest ways and means to solve these problems.
- ☆ Economics should analyze both positive and negative aspects of the economy, and as a science, it should suggest how best the resources are distributed among different individuals and groups.
- ☆ Economics should analyze the cost and benefits of improving patterns of resource allocation

Samuelson's definition is similar to Robbins's definition, but it has improved the definition and made it dynamic. This is because, it laid emphasis on the element of time, and thereby, allowed the possibilities of changes in scarce means and ends. It considers the problems of choice making in a dynamic setting. So, it applies to the past, present and future economic systems. Further, the interesting aspect of Samuelson's definition is that, the society may or may not make use of money. So, while allowing the growth aspect, Samuelson has made the definition applicable even to the barter economy. The phrases in the definition like *'with or without the use of money'* and *'now and in the future'* gave wider scope to the Economics. So, Samuelson goes a step ahead over Robbins definition and discussed the growth aspects, such that, the resources level should be increased, so as to keep pace with the dynamic nature of wants.

1.1.5. Employment Stability Centered Definition

John Maynard Keynes (1883-1946), whose landmark work, *The General Theory of Employment, Interest and Money*, was published in 1935. He was the Father of Modern Economics. According to Keynes, the problems in the economy are not due to scarcity of resources only, but even the scarce resources are not fully utilized. So, this leads to unemployment and underemployment problems in the economy. This calls for active Government intervention in the market place, so as to ensure economic growth and stability. He further stressed that, insufficient demand causes unemployment and the excessive demand results in inflation. Government should, therefore, manipulate the level of aggregate demand by adjusting levels of government expenditure and taxation. For example, to avoid depression, Keynes advocated increased government spending and easy money, resulting in more investment, higher employment and increased consumer spending. Keynes economic theory was based on a circular flow of money. According to Keynes, one person's spending goes towards another's earnings, and when that person spends his earnings he is, in effect, supporting another's earnings. This circle continues on and helps support a normal functioning economy. In his words, '*Economics is the study of the administration of scarce resources and also the determinants of income and employment*'. He further advocated that, economic theory does not provide readymade formulae to attend various practical problems in the economy because, economic problems are very complex in nature and ever changing, economic laws are not definite and exact and no practical problems can be solved solely by Economics only. For example, the problem of poverty should be viewed from social, economical and political angles and an Economist alone cannot solve the problem. But, Economics, as a science, simply gives us a technique of thinking, which enables to arrive at conclusions in a given situation more scientifically. However, Modern Economists criticized this version and opined that, Economics as a science should offer solutions to the economic problems, and hence, it should be both positive and normative in its approach.

On the whole, the study about definitions of Economics reveals that, the subject 'Economics' is a relatively new science, it came into being a little over two centuries ago. So far, it has developed into three main stages: the Classical, the Neo-Classical and Modern Keynesian schools. Corresponding to these, there are three distinct definitions of the subject. Initially, it was considered as a science of wealth through its fourfold activity of consumption, production, exchange and distribution. Marshall related the subject to economic welfare, most closely connected with the attainment and the use of material requisites of well being. However, Lionel Robbins gave the subject a positive scientific basis and this definition was widely acknowledged. Samuelson's definition though similar to Robbins definition, but it is more dynamic both in the context of scarce resources and unlimited wants. The definition given by Keynes has more applicability to the modern system of economy, which implies that, economic problems are not due to scarcity of resources, but due to the problems of bridging the gap between the existing level of resources and the resources which the economy is actually capable of using. So, the capacity of the economy to make use of available resources is more important rather than allocation of resources. Hence, 'making resources work' becomes a primary problem and 'allocation of resources' becomes only secondary. So, each definition puts forward its own contributing factor to the importance of the study of the subject. It should also be realized that, through logical thinking, the subject matter underlying each definition is more or less same and there is an underlying unity in these definitions. The above discussed concepts *viz.*, wealth, welfare, scarcity, growth with stability and employment are inter-related concepts. This is because, to promote welfare in the economy, producing more wealth is the aim. But, every society faces the problems of scarce resources and these should be effectively utilized among unlimited wants. Not only dealing with the existing resources, new resources should be developed to overcome scarcity and to achieve higher levels of production, employment and growth in the economy. The study about different definitions of Economics based on several criteria also highlights that, Economics is a growing science and it would not be possible to give a short and unalterable definition. In brief, the outline of different definitions of Economics is presented through the Table 1.2.

1.2. Scope of Economics

Scope refers to the area or boundary of the study. Regarding Economics, the scope can be studied with reference to the following aspects:

1.2.1. Subject-matter of Economics.

1.2.2. Economics is a social science.

1.2.3. Whether Economics is a science or an art?

1.2.4. Whether Economics is a positive science or normative science?

1.2.5. Macro and Microeconomic spheres.

1.2.1. Subject-matter of Economics

Economics as a science deals with man actions in the form of economic activities to fulfill his unlimited wants. It will not deal all the aspects of human life, rather only one aspect of it *i.e.*, economic aspect. It will not look into the physical, physiological, psychological, biological, ethical, social and political aspects of human life. It simply tells us how a man utilizes his limited resources towards the satisfaction of his unlimited wants and how best he can maximize the resource use efficiency. This is so because, as explained earlier, human wants are unlimited, but the means to satisfy them are limited. In an economy, we see the farmer working in the field, a worker working in a sugar factory, a doctor attending the patients, a teacher teaching the students and so on. They are all engaged in what is called as 'Economic

Table 1.2: Outline of different definitions of Economics.

Economists	*Emphasis on*	*Definition*	*What Economics is*
Adam Smith	Wealth	*'An inquiry into the nature and causes of the wealth of nations'.*	Economics as a subject deals with the production of wealth, consumption of wealth, exchange of wealth and distribution of wealth.
Alfred Marshall	Welfare	*'Political Economy or Economics is a study of mankind in the ordinary business of life. It examines that part of individual and social action which is most closely connected with the attainment and with the use of material requisites of well-being'.*	Economics studies mankind in the 'ordinary business' of life *i.e.*, engaged in earning activities with main emphasis on material welfare. He regarded material welfare is the end and wealth is regarded only as means.
Lionel Robbins	Scarcity of resources	*'Science which studies human behaviour as a relationship between ends and scarce means which have alternative uses'*	Emphasizes that, the economic activity arises only so long as the means are scarce and this also leads to choice making among the wants to be fulfilled.
Paul A Samuelson	Growth of scarce resources	*'Economics is the study of how man and society choose, with or without the use of money to employ scarce productive resources, which could have alternative uses to produce various commodities over time and distribute them for consumption now and in the future among various people and groups of society'.*	The available scarce resources should be used very judiciously and cautiously, as in the modern dynamic economic system, wants show increasing tendency and to satisfy as many wants as possible, efforts should be made to increase the level of resources in the economy.
John Maynard Keynes	Employment	*Economics is the study of the administration of scarce resources and also the determinants of income and employment'.*	Economic problems are not due to scarcity of resources, but due to the problems of bridging the gap between the existing level of resources and the resources which the economy is actually capable of using.

activity'. So, through performing economic activities, they earn money, purchase goods and services and fulfill their wants. This implies, earning money and purchase of goods is not the end, rather they are performed to satisfy human wants and this promotes human welfare. So, in brief, to fulfill the wants, a man has to make efforts through performing economic activities and these efforts lead to satisfaction. So, *Wants–Efforts-Satisfaction* constitutes the subject matter of Economics (Figure 1.1). This subject matter when examined involves two important aspects *viz.*, production and consumption. While performing an economic activity, the individual will put all his efforts in the business (production), so as to earn income for fulfilling the wants (consumption). When one want is satisfied, another want will crop up in the human mind and therefore, he has to continue his efforts in the business and this process is never ending, thereby, leading to circular flow of an economic activity. However, this is a traditional approach, as in the primitive society, the connection between wants efforts and satisfaction is close and direct. But, in a modern society, there is no such direct or straight relationship. This is because, a man produces what he does not consume and consumes what he does not produce. Even if he produces for himself and if the production is more, he has to sell the excess quantity. Similarly, he has to buy a product which is not produced by him. Thus, the process of buying and selling, which is called as 'Exchange' comes in between wants, efforts and satisfaction. In the modern days, most of the products are produced in factories. To manufacture these products, all the four factors of production *viz.*, land, labour capital and management or entrepreneur will contribute. They all get reward in terms of money, as land gets rent, labourer earns wages, capitalist earns interest and the management (entrepreneur) gets profit.

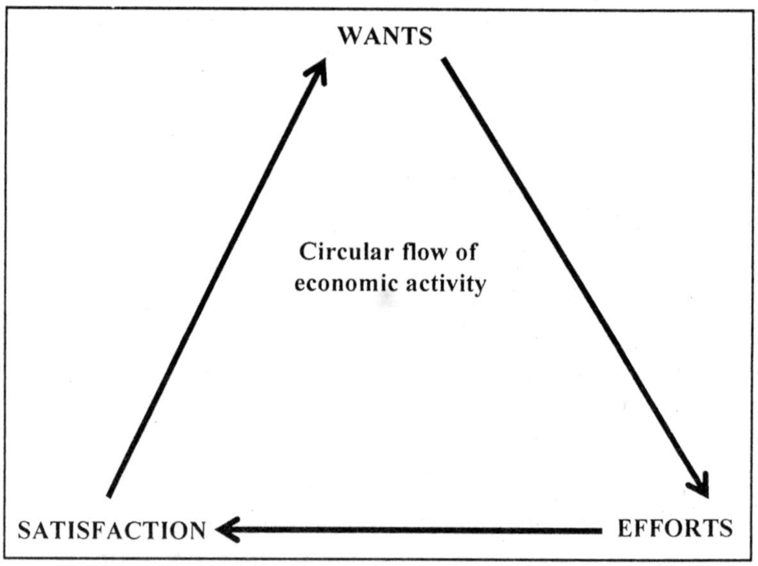

Figure 1.1: Subject-matter of Economics (Traditional approach).

Economics studies how these incomes *i.e.*, rent, wages, interest and profits are determined in the economy and this concept is called 'Distribution'. This also comes in between wants, efforts and satisfaction. Thus, according to traditional approach, the subject matter of Economics comprises of four basic components *viz.*, production, consumption, exchange and distribution.

Consumption

Consumption deals with the satisfaction of human wants. To satisfy human wants, economic activities are to be performed to earn money. When a particular want is fulfilled, the process is known as consumption. In Economics language, 'consumption' has a special meaning. The consumption of the services of an agricultural labour is used in the same sense as the consumption of food. This division in Economics has a special significance, as it deals with important concepts like nature of wants, classification of wants, Law of Diminishing Marginal Utility (LDMU), Law of Equi-Marginal Utility (LEMU), Engel's law of family expenditure, Law of demand etc.

Production

Production refers to the creation of wealth. Strictly speaking, it refers to the creation of utilities. Utility refers to the want satisfying capacity of a good. A good may create utility to a person in four different ways *viz.*, form utility (created by processing function), place utility (created by transportation function), time utility (created by storage function) and possession utility (created by buying and selling function). For production of goods, rather creating utility, all the four factors of production need to be employed and these efforts will produce the goods that satisfy human wants. This division of Economics deals with important concepts like *Laws* of returns, *Theories* of population, different forms of business organizations etc.

Exchange

The concept of exchange arises mainly because, no person or no country is self-sufficient in producing the goods. As mentioned earlier, a person produces something which he does not consume and consumes something which he does not produce. So, this calls for exchange of goods among the people through buying and selling mechanism and this process is called Exchange. In the past, goods are exchanged for other goods and it refers to barter system of trade. In modern times, goods are exchanged for money and with this money we buy requisite goods from the market. As goods are exchanged for money, in this branch, the important concepts like functions of money, role of banks, concepts of international trade etc., are studied.

Distribution

Distribution deals with payment of remunerations to all the four factors of production. The four factors of production *viz.*, land, labour, capital and management (entrepreneur) are rewarded in the form of rent, wages, interest and profits respectively. In Economics, functional distribution is accorded more importance over personal distribution, as these four factors perform different functions in the production of goods. This division mainly deals with the determination of payments for factors of production.

Besides above four basic components, in modern times, Economists added one more division to the subject matter of Economics *i.e.*, Public finance. It deals with the income and expenditure patterns of the Government, as the economic functions of the modern State have increased significantly. This division mainly deals with different aspects relating to taxation, public expenditure, public debt etc.

The division of subject matter of Economics into consumption, production, exchange, distribution and public finance is purely for the sake of convenience. This classification should not be viewed in a rigid way. This is so because, they are closely related to each other and they are inter-dependent also. For example, it is not possible to have consumption without production. So, production is means and consumption is the end. The concepts of buying and selling leads to exchange and it serves as a connecting link between production and consumption. Since the ultimate objective of exchange is to derive satisfaction or welfare and it is possible through earning factor payments. So, the economic welfare of people in a nation depends on how wealth is produced, exchanged and distributed. If there is no proper distribution, it will result in inequalities of income and wealth. This, in turn, will affect consumption and production. So, the divisions of subject matter of Economics are both inter-related and interdependent.

1.2.2. Economics is a Social Science

Economics is a social science, as it deals with human behaviour in the society regarding the allocation of limited resources to satisfy his unlimited wants. These efforts and human activities are carried on in groups or in societies. Economics as a subject deals with a social phenomenon like division of labor, cooperation of people in various economic activities, existence of money and its management, capital and technological knowledge etc. It further emphasizes on the creation of wealth from scarce resources, the production and distribution of goods and services for consumption, the behaviour, interaction and well-being of the groups involved in the above activities and the fact that, there is a trade-off involved in production and in consumption.

1.2.3. Whether Economics is a Science or an Art

Philosopher Karl Popper's widely accepted definition of Science says that, *'a statement is scientific only, if it is open to the logical possibility of being found false'*. This definition means that, scientific statements will be evaluated by testing them. A statement is non-scientific, if it takes no risk of being found false *i.e.*, if there can be no way to test the statement against observable facts or events. Popper called this distinction as the *'line of demarcation'*.

Economics as a Science

A Science is a systematized body.of knowledge. A branch of knowledge becomes systematized when relevant facts have been collected, analyzed and generalizations can be made after testing the results. Just as Physics and Chemistry are sciences, the same scientific approach is followed in Economics also. So Economics, like other sciences deals with cause and effect relationships and formulate useful laws and principles. Hence, Economics is treated as a science, as it studies the complex human behaviour with respect to the relationship between ends and scarce means. Robbins, Briggs, Jordan Robertson, Knight, Walras, Cournot and Senior are of the opinion that, Economics is only a Science. But, Economics is not as precise and exact as the Physical sciences. But, Economics has a greater right to be considered as a science than other social sciences like Politics or History because, in Economics, money is used as a measuring rod of utility. Though, it is only a rough measure, still it enables to give a concrete shape to the Laws of Economics. But, the Laws of Economics may not be true in some applied aspects. Such situations are common even in other sciences as well. For example, the weather report given by a Meteorological Department (Science of Weather) may go wrong in some cases. But, Meteorology is not disregarded as science. Similarly, even if some economic laws do not come true, we cannot say that, Economics is not a science.

Arguments Favouring Economics as a Science

Economics is considered as a science because of the following reasons:

- ☆ It is a systematized body of knowledge *i.e.*, there is systematic collection of data, compilation, analysis and interpretation of the findings.
- ☆ It has its own principles and laws, so that, various economic facts can be studied, analyzed and useful conclusions can be drawn. This further aids in policy formulations. These laws or principles have universal application. For example, the Law of demand, LDMU, Law of Diminishing Returns (LDR) etc., are applicable in all forms of economic systems.
- ☆ Economics as a science is self-corrective in nature. This implies that, it goes on revising its conclusions in the light of new facts based on observations.
- ☆ For deriving economic principles or laws, both deductive and inductive approaches are followed. The use of mathematical and statistical techniques for economic data makes the conclusions more realistic and scientific.
- ☆ 'Money' serves as a measuring rod for quantifying the economic welfare. For example, to measure the intensity of desire of an individual for having a commodity, we use 'price' as an indicator. Similarly, to study the extent of welfare yielded by a sugar factory in an area, incomes of the workers serves as the measuring rod. This makes Economics more exact than other social sciences like History, Ethics, Politics etc.
- ☆ The economic data facilitates conductance of research on the prioritized issues in the economy and the policy guidelines so formulated yield welfare measures to the public.
- ☆ Experimentation is possible regarding the verification of laws and principles of Economics. For example, the LDMU can be experimented. Similarly, devaluation of Indian rupee in 1966 was an experiment. Infact, Capitalism, Socialism, Mixed Economy etc., are the experiments of Economics.
- ☆ The economic laws or principles can be tested for their validity in a given practical situation.
- ☆ Economic predictions for the future may go wrong. The trends so projected for certain economic variables may deviate from the actual values. But, based on this, it is not justifiable to say Economics is not a science. This difficulty of predicting the future is common in other sciences also. Say, for example, in case of Meteorology, the predictions generally go wrong.

Arguments not Considering Economics as a Science

Based on the above positive features, it is easy to conclude that, Economics is a 'Science'. However, this concept is subjected to criticism on the following grounds:

- ☆ Human behaviour is uncertain and they behave differently under similar situations at different points of time. This leads to complexity in understanding the economic phenomena and thereby, it is difficult to build a science on uncertain issues.
- ☆ Controlled or laboratory experimentation is not possible and hence, the conclusions drawn from the economic laws and principles are not exact. This is because, Economics as a social science deals with the behaviour of

human beings, which is very complex. This makes the subject of Economics less exact compared to physical sciences. However, it is important to note that, laboratory experimentation is not needed to consider a subject to be called as Science.

- ☆ Even the application of mathematical and statistical methods to the economic data, it may not bring exactness to the validity of economic laws and principles. So, exact quantitative prediction is not possible in Economics. For example, if the price of the commodity is decreased, the demand may not increase, rather it may decrease, if the commodity is an inferior good.
- ☆ The Laws of Economics are not universal and the validity of the economic laws are subjected to certain assumed conditions. For example, regarding the applicability of the Law of demand, it needs some other factors like income of the individual, prices of substitutes and complementary goods, tastes and habits of the individual etc., to be held constant. This concept of assuming 'other things remaining equal or constant' refers to a Latin phrase, *'Ceteris Paribus'*.
- ☆ In Economics, future predictions cannot be made exactly compared to physical sciences.

To conclude, the positive views to consider Economics as a Science outweigh the negative views and moreover, the criticism have no force or truth in it. So, it can be justified that, Economics is a Science. But, there are certain limitations to call it as an exact science because,

- ☆ It deals with complex human behaviour.
- ☆ Money, which is used as measuring rod serves only a rough measure.
- ☆ Complex economic phenomena.
- ☆ Economic laws are based on several assumptions.

Economics as an Art

An art is the practical application of scientific principles for achieving the definite ends. In Economics, various principles are applied to different situations to address the problems in an economy. A Science teaches us to know a phenomenon and an Art teaches us to do a thing. For example, the study about causes and effects of inflation problem in the economy falls under Science. But, the measures implemented to control inflationary rise in prices in the economy falls under Art. Marshall, Pigou, J.S. Mill, J.M. Keynes etc., supported Economics as an Art. According to Modern Economists, the economic laws and principles should be applied to address various economic problems. If not applied, Economics as a Science, will lose its usefulness. So, Economics is not only a science, but also an Art. But, the strong criticism against considering Economics as an Art is that, Economics alone cannot solve all the practical problems in the economy and it should get integrated with other social sciences for addressing the problems on satisfactory grounds. So, even if Economics is developed as an Art, it becomes an imperfect art.

Arguments favouring Economics as an Art

Economics is considered as an Art because of the following reasons:

- ☆ Economics through its laws and principles can offer requisite solutions to various economic problems.
- ☆ Modern Economists aim at providing welfare measures to the people through addressing various economic issues.
- ☆ Practical application of economic laws and principles to justify their validity makes Economics as an Art.

Thus, Economics is both a Science and an Art. As a Science, it teaches us to know a phenomenon and as an Art, it teaches us to do a thing. After arriving at this conclusion, here arises another controversy: 'Is economics a Positive science or a Normative science?'

1.2.4. Whether Economics is a Positive Science or Normative Science

Production, consumption, exchange and distribution of goods and services are the basic economic activities of life. For executing these economic activities, every society has to face scarcity of resources and it is the scarcity of resources that gives rise to the problem of choice-making. The scarce resources should be efficiently allocated towards the production and distribution of goods and services in the economy. But, every society faces the problems of allocating these scarce resources in the production of different goods and services and in their distribution, which often leads to various problems like what is produced and in what quantities?, how are these goods produced?, for whom are these goods produced? etc., and these constitute the central problems of any economy. In principle, there are more than one ways of solving the central problems of an economy. These different mechanisms, in general, are likely to give rise to different solutions to those problems, thereby, resulting in different allocations of the resources and also different distributions of the final mix of goods and services produced in the economy. Therefore, it is important to understand which of these alternative mechanisms is more desirable for the economy as a whole. In Economics, we try to analyze and evaluate the different mechanisms by studying how desirable the outcomes resulting from them are. Often a distinction is made between positive economic analysis and normative economic analysis depending on whether we are trying to figure out how a particular mechanism functions or we are trying to evaluate it. In positive economic analysis, we study how the different mechanisms function and in normative analysis, we try to understand whether these mechanisms are desirable or not. Thus, there is difference of opinions among the Economists whether Economics is a positive or normative science. There is no need to pose this question whether Economics falls under a positive science or normative science. This is because, the distinction between positive and normative economic analysis is not a very sharp one. The positive and the normative issues involved in the study of the central economic problems are very closely related to each other and a proper understanding of one is not possible in isolation to the other. However, it is essential to study the positive and normative aspects of economic analysis and the distinctions between the two.

Economics as a Positive Science

Economics as a positive science refers to the body of systematized knowledge concerning 'what is', 'why', 'wherefore' and 'what will be'. According to R.T. Bye, *'positive science confines itself to accurate description of a phenomenon, it explains what is, how it works and what are its effects'*. So, Positive Economics is mainly concerned with the description of economic events and it tries to formulate theories to explain them. So, Positive economics is also called as Descriptive Economics. It deals with the actual happenings or explain the things as they are. It explains causes and effect relationship, but will not pass any moral judgements. So, it remains neutral between the ends. It presents the real picture of facts without any comments and suggestions. For example, Economics as a positive science studies what is the wage rate, how it is determined and why wage rate in India is low. Positive science is also called Pure science, as it deals only with the theoretical aspects. Lionel Robbins, J.S.Mill, Brown, Senior and Friedman have described Economics as a positive science. They opined that, Economics is based on logic. It is a value theory only. The following are the other examples of positive statements:

- ☆ Prices of fertilizers and pesticides are increasing in India
- ☆ There is unemployment problem in India
- ☆ There is inflationary rise in prices in India
- ☆ India adopted mixed economy.
- ☆ India is a developing country.
- ☆ There is inequality of income distribution in Indian economy.

It is evident from the above examples, a positive statement is a statement about *'what is'* and that contains no indication of approval or disapproval. Notice that, a positive statement can be wrong. For example, 'all skilled labour are equally efficient in executing the works in a sugar factory' is an incorrect statement because, efficiency varies from person to person. But, it is a positive statement because, it is a statement about what exists in a sugar factory.

Arguments Supporting Economics as a Positive Science

Economics is considered as a positive science because of the following reasons:

- ☆ Economics as a social science deals with the human behaviour in relation to scarce means and unlimited ends. The Economist is not concerned whether the ends are good and bad. So, Economics should remain neutral between the ends. This implies, *'the functions of the Economist are merely to explore and explain and not to advocate or condemn'*.
- ☆ In Economics, if normative aspect is included, it leads to judgement of economic policies and the Economists may differ in their opinions. This will affect the progress of Economics subject. Further, it gives scope for misunderstanding among the Economists, if the policies do not yield desired results in a given practical situation.
- ☆ Economics involves the application of logic in ascertaining a particular aspect. For example, through logical approach, we study about the merits and demerits of direct taxes and indirect taxes. But, Economics, as a positive science will not judge, which of these two will be better from consumer point of view.
- ☆ Economics should remain neutral between the ends, as it alone cannot address all the problems in the economy without integration with other social sciences like politics, ethics etc.

Economics as a Normative Science

Normative Economics is also called as Policy Economics or Regulative science. It deals with norms of facts and suggests, 'what ought to be' and 'what ought to have been'. According to R.T. Bye, it is regarded as *'Economic ethics'*. He opined that, Economics cannot be separated from ethics. It promotes social and economic values. It discusses whether a thing is good or bad and lays down principles or laws to address the issues and suggests measures for economic betterment of the people. So, in normative analysis, we give more importance to ethical judgements and concerned with the ideal rather than the actual situations. This indicates a normative statement expresses a judgment about whether a situation is desirable or undesirable. For example, Economics as a normative science studies 'what should be the wage rate?'. Similarly, 'the skilled labour will perform better in a sugar factory, if training is rendered to them in their relevant job works' is a normative statement because, it expresses a judgement about *'what ought to be'*. Notice that, there is no way of disproving this statement. So, a normative science judges the rightness or wrongness of the things. The following are the other examples of normative statements:

- ✰ India should formulate effective monetary and fiscal policies to check inflationary rise in prices in the economy.
- ✰ India should make productive investments in the economy to increase the level of employment.
- ✰ The fundamental principle of economic development should be the development of rural India.
- ✰ Agricultural income should also be taxed.
- ✰ There should not be multi-national companies in consumer goods industries in our country.
- ✰ Private sector should be encouraged for accelerating the pace of our industrialization.

Marshall, Pigou, Howtrey, Keynes and many other Economists regarded Economics as a normative science. According to them, the real function of the science is to increase the human welfare. So, they rendered useful suggestions in their works towards promotion of human welfare. For example, Malthus has given suggestions to check the rise in population. J.M.Keynes has suggested useful measures to remove unemployment and under employment problems in the economy.

Arguments Supporting Economics as a Normative Science

Fraser, Howtrey, Wolfe, Strecton etc., support the view that, Economics is a normative science. The following are their supporting views:

- ✰ If Economists do not deal with ethical aspects, they will be far away from realities and cannot understand about practical problems in the economy. Economists should give mankind some practical benefits by suggesting solutions for various economic problems like unemployment, under employment, poverty, inflation, unequal distribution of income etc. So, Economics cannot be separated from ethics. It should be both light bearing and fruit bearing and this is achieved when Economists deal with moral judgements.
- ✰ Present economies are open economies. This calls for addressing the economic issues through State intervention effectively. Hence, Economists has to deal with ethical aspects. Especially, at times of economic reforms, the Economists have to play a crucial role in offering solutions to various economic problems.
- ✰ Man is not only logical, but he is sentimental too. He is concerned more about his economic welfare. If the concept of welfare is not considered in Economics, it is no longer regarded as a social science.
- ✰ The ethical aspect in Economics forms the basis for economic planning towards the welfare of the society. In the modern days, with increased Government intervention in various developmental and welfare activities in different economic systems like capitalism, socialism, communism, mixed economy etc., the concept of ethics is gaining more significance.

Differences between Positive Economic Science and Normative Economic Science

The differences between Positive economic science and Normative economic science are presented in Table 1.3.

Economics is both Positive and Normative Science: Modern Approach

Lionel Robbins opined that, in Economics, positive statements should deserve special attention compared to normative statements, as normative questions (statements) fall under the purview of political and moral philosophy. He further opined that, an Economist need not pass any judgements regarding ethical issues of economic aspects, and if done so, Economics cannot be an exact science. But, many Modern Economists criticized this view and stated that, Economics being a social science, it has to consider ethical issues, as it aims at promoting human welfare. This led to the development of an important branch of Economics, known as 'Welfare Economics'. So, an Economist has the responsibility of giving useful solutions to the problems faced by the people in the economy like depression, unemployment, inflation, poverty etc., so as to promote their welfare. So, Economics has to deal theoretical aspects scientifically and at the same time offer relevant practical solutions to address various economic problems. Hence, according to Modern Economists, Economics is both positive science and normative science and hence, it is regarded as Prescriptive Economics (combination of positive and normative economic analyses).

If Economists confine to positive statements alone, it limits their views about various issues of Government policy. So, both positive and normative statements must be combined to make a policy statement. An Economist, while formulating a policy, must make a judgment about what goals are desirable (the normative part) and decide on a way of attaining those goals (the positive part). Hence, both positive and normative views play vital role in policy analysis. So, Economics is both positive science and normative science. It not only tells us why certain things happen, it also

Table 1.3: Differences between positive and normative economics.

Positive Economics	Normative Economics
Classical Economists viewed Economics as a 'positive science'	Modern Economists viewed Economics as a 'normative science'
It is body of systematized knowledge relating to 'what is', 'what was', 'why', 'wherefore' and 'what will be'. This implies, it explains causes and effect relationship.	It is a body of systematized knowledge relating to 'what ought be' and 'what ought to have been'. This implies, it judges the rightness or wrongness of things.
It considers Economics as a pure science and hence, it remains neutral between the ends	It considers Economics as an ethical science and hence, it starts with the assumption of good or bad.
In Positive Economics, the facts are merely stated.	In Normative Economics, the situation is analyzed and proclaims if it is desirable or undesirable.
Positive Economics is objective	Normative Economics is subjective
Positive economics mainly focuses on statistics, factual information and indulges on the scientific formula for determining what an economy should look like	Normative economics mainly deals with value judgments of economy.
It is based upon facts, and therefore, not suggestive.	It is based upon individual opinion and therefore, it is suggestive in nature.
It describes the things as they are and wont pass moral judgements	It evaluates the things and passes moral judgements
Positive statements can be empirically verified or tested.	Normative statements can't be verified or tested empirically.
The purpose of positive science is to make real description of an economic activity.	The purpose of normative science is to determine the ideals.
Examples: 1.There is inflationary rise in general price level in India 2. Price rises as demand increases.	Examples: 1. India should formulate effective monetary and fiscal policies to check inflationary rise in prices in the economy. 2. Rising prices is a social evil

says, whether it is right or wrong the thing to happen. For example, in the world, few people are very rich, while the masses are very poor. Economics should and can explain not only the causes of this unequal distribution of wealth, but it should also say whether this is good or bad. It might well say that, wealth ought to be fairly distributed. Further, it should suggest the methods of doing it.

According to Keynes, Economics as a science does not provide ready made formulae to address the economic problems. But, it simply gives us a technique of thinking, which enables us to arrive at a conclusion under a given situation. This is because, economic problems are highly complex, the economic laws are not as exact as physical sciences, no practical problem can be solved by Economics alone etc. But, Modern Economists disregarded this view and stated that, Economics should not move away from offering solutions to economic problems. This is because, a common man in the society expects solutions for his economic problems, and here, the Economists have to play a crucial role. Moreover, in the recent period, the concept of Welfare Economics is gaining more importance. This indicates the economic analysis proceed with a specific objective *i.e.*, promotion of economic welfare. So, the normative aspect 'welfare' is included in the Economics. But, Robbins criticized this view, as welfare cannot be exactly measured, being a relative concept. However, Economists should address this concept of welfare, as modern states are 'welfare states'.

Samuelson has introduced the concept of 'Net Economic Welfare (NEW)' and he stated that, *'NEW is an adjusted measure of total national output that includes only consumption and investment items that contribute directly to economic welfare'*. The word 'welfare' is more important in the above definition. This is because, if income of the people rises in the economy, they prefer leisure to work (income). When more time is devoted for leisure, Gross National Product (GNP) may come down. But, welfare goes up. So, the satisfaction derived from leisure should be given importance and it should be included by giving it a value in computation of NEW. Similarly, women do a lot of work at home and it is not taken into account while calculating GNP. But, it should be included while estimating GNP. This leads to correct estimation of NEW. Conversely, the negative aspects in the economy such as drugs trade, unaccounted money (black money), environmental pollution etc., should be deducted from the GNP to arrive at true picture of NEW. This concept of NEW has become very important in the study of National Income.

1.2.5. Macro and Microeconomic Spheres

Classical Economists like Adam Smith, Malthus, Ricardo etc., studied economic parameters from the view point of the economy as a whole. However, Neo-Classical Economists like Marshall, Pigou etc., attached greater significance to the study of micro parameters and the trend continued till the world's Great Depression in 1930s. So, to eliminate this dispersion, Prof. Ragnar Frisch of Oslo University divided the study of Economics into two branches and he coined the terms Macroeconomics and Microeconomics in 1933. Macro and Microeconomics and their wide array of underlying concepts have been the subject of a great deal of writings. Their field of study is vast and they are the two vantage points from which the economy is observed. Thus, according to Modern approach, the subject matter of Economics includes two dimensions *viz.*, Macroeconomics and Microeconomics.

(*a*) Macroeconomics

Macroeconomics studies the economic system in aggregate and hence, it gives the complete picture of the economy. 'Macro' term is derived from the Greek word '*Makros*', which means 'large'. It deals with large aggregates of the economy as a whole and hence, it is also called Aggregative Economics. This implies, it is the study of total economic activity. It looks at the total output of a nation and the way the nation allocates its limited resources in an attempt to maximize production levels and promote trade and growth for future generations. It is a vast field that concentrates on two areas, economic growth and changes in the national income. It is a study of the relations between broad economic aggregates such as total employment, saving and investment. Thus, national income, output, employment, general price level economic growth etc., constitute the subject matter of Macroeconomics. It is also called as Income and Employment theory, as the study mainly concerns with economy's income and employment. In Macroeconomics, we study the macroeconomic variables by splitting the economy into different sectors and hence, this method is also called 'Method of lumping'. In Macroeconomics, we take aggregate variables into considerations like aggregate demand, aggregate supply, total savings, total consumption, total investment etc and their inter-relationships are also studied. So, this method of study is called as General equilibrium analysis.

With the experience of Great Depression in 1990s due to the failure of free economy, macro-study gained significance, as the weakness of 'micro' was fully exposed. Keynes, Walras, Wicksell, Fisher etc., contributed towards the development of Macroeconomics and it was Keynes, who made Macroeconomics popular as a complete and important branch of Economics. With the increased Government intervention in different economic systems like capitalism, socialism, mixed economy etc., towards promoting welfare for the large masses of the people, the concept of Macroeconomics has gained momentum. So, the Government must have adequate knowledge about the aggregate variables like national income and its concepts, aggregate demand, aggregate supply, total savings, total investment, total consumption etc., and reliable data on the these aggregate variables are important for formulating macroeconomic policies. In this context, the study of Macroeconomics is essential. The following are the popular definitions about Macroeconomics:

'Macroeconomics concerns with such variables as the aggregate volume of the output of an economy, with the extent to which its resources are employed, with the size of national income and with the general price level'.

—(*Ackley Gardener*)

'Macroeconomics does not deal with individual quantities but with aggregate of these quantities, not with the individual incomes but with national income, not with the individual prices but with price levels, not with individual outputs but with the national output'.

—(*K.E. Boulding*)

'Macroeconomics deals with the functioning of the economy as whole'.

—(*Shapiro*)

Subject Matter of Macroeconomics

Macroeconomics deals with the following aspects (Figure 1.2):

- ☆ Deals with the theories of income and employment. The factors that determine them like investment function, consumption function, aggregate demand, aggregate supply etc., are analyzed.
- ☆ Deals with economic problems like inflation, deflation, poverty etc.
- ☆ Studies the factors that determine National Income and its concepts.
- ☆ Studies the nature and causes of trade cycles, effects, remedial measures etc.
- ☆ Macro Economists develop different types of models to study the relationships between factors such as output, national income, unemployment, consumption, savings, inflation, international trade, investment and international finances.
- ☆ Deals with relative shares of factors payments in National Income.

Importance of Macroeconomics

Macroeconomics is important considering the following grounds:

- ☆ Gives the complete picture about the working of an economy.

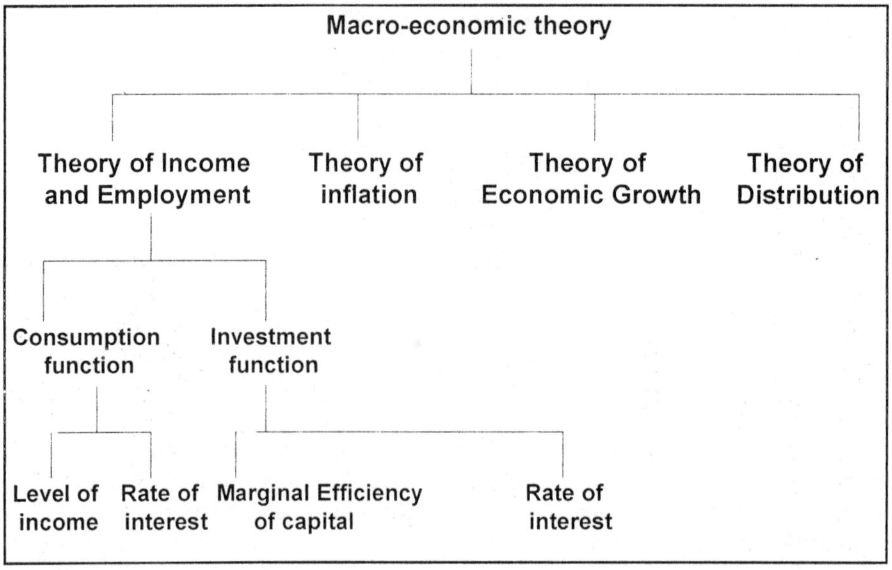

Figure 1.2: Subject-matter of Macroeconomics.

- The changes at aggregate level will lead to economic problems at micro level. For example, the increase in aggregate savings, will lead to low investments in the economy and this will affect the employment of workers in a factory.
- Since, Microeconomics addresses the economic issues only at micro level and not at macro level, the study of Macroeconomics fills the gap in understanding the economy as a whole and in making generalizations about economic principles.
- For formulation of economic policies at aggregate or economy level, the study and understanding about Macroeconomics is essential. This is because, the Government considers aggregate factors like aggregate demand, aggregate supply, national income, level of employment, general price level in the economy etc., to address various macroeconomic problems. Especially, in under developed and developing economies, there are innumerable problems like poverty, unemployment, under employment, inflation, over population, monetary problems, fiscal problems, fluctuations in business cycles, balance of payments etc., and to address these problems, Macroeconomic analysis is indispensable, as analyzing the individual behaviour to address these problems is practically impossible.
- Macroeconomics studies the economy in its dynamic concept. For example, it studies the inflationary rise in prices in the economy over a period of time and thereby, helps formulate the strategies to overcome inflation.
- Helps in building the data base on National Income and its related concepts, which guide the Government in planning the economic activities and to analyze the distribution of income among different group of people in the country.
- The study of Macroeconomics helps to understand the behaviour of individual units in the economy. For example, a fall in aggregated demand leads to fall in individual demand. So, the analysis pertaining to fall in aggregate demand helps to analyze the factors responsible for fall in individual demand. In fact, no principle of Microeconomics can be derived without the studying the relevant aggregates.

Limitations of Macroeconomics

However, the concept of Macroeconomics suffers from the following limitations:

- Macroeconomics ignores the individual and concentrates on the aggregates in the economy. But, the aim of Economics as a science is to study about individual's welfare.
- Macroeconomics considers aggregate behaviour is the sum total of individual activities. But, what is true at micro level, may not be true at the macro level and *vice versa.* This has been aptly characterized by Prof. Samuelson as '*Fallacy of composition*'. In his words, '*very definitely, in the field of Economics, it turns out what seems to be true for individuals is not always true for a society as a whole, and conversely, what seems to be true for all may be quite false for any one individual. For everybody to stand on tiptoe to watch a parade does no good, even though one person may gain a better view by so doing*'. For example, if an individual saves money in the bank, he may become rich. But, if all the individuals save the money in the banks, it will lead to fall in aggregate demand,

over production and hence, unemployment problem in the economy. Similarly, if an individual withdraw his deposits from a bank, it does not matter much for the bank. But, if all the people withdraw deposits from all the banks, the banking system will collapse. If there is rise in price of the commodity in the market, it will benefit the entrepreneurs producing that commodity through large profits. But, if the general price level in the economy rises leading to inflation, it leads to less demand from all the consumers, thereby, the entrepreneurs are adversely affected. Such paradoxes are rightly called as Macroeconomic Paradoxes, by Boulding. He points out that, *'it is these paradoxes more than any other factor, which justify the separate study of the system as a whole, not merely as an inventory or list of particular items, but as a complex of aggregates'.*

- ☆ Macro analysis, since deals with aggregates and totalities and considers aggregates as homogenous, is likely to ignore the differences existing within the aggregate groups. For example, different commodities have different units of measurement. So, it is not possible to measures all the commodities with the same measuring rod. For example, to study the aggregate output in the economy, it is not possible to combine the outputs of apples, milk, cotton, eggs etc.
- ☆ Sometimes, the study of aggregates is misleading. So, the aggregate variables may not have much significance at the grass root level. For example, if the National Income is increased, it does not mean that, the incomes of all the people have raised. This is because, the increase in income of few rich people might have offset the fall in the income of other people, thereby, leading to rise in National Income. So, this rise in National Income in the country will have little significance from the point of view of poor community. Hence, we can say, micro level changes are more important than macro level changes. For example, the general price level in the economy may remain constant, but the prices of cotton are falling. This means, the cotton farmers are adversely affected in the economy.

(*b*) Microeconomics

From the definitions of Macroeconomics, we can infer what would be Microeconomics. The concept of 'micro' in English language is derived from the Greek word *'micros'*, which means 'small' or 'millionth part'. So, Microeconomics is the microscopic study of the economy. Instead of studying the entire economy, as in Macroeconomics, it deals with individual parts or segments of the economy. As the name Microeconomics indicates, it is not aggregative, but selective in its approach. That means, it deals with the analysis of small units of the economy such as individual consumers, producers or firms and small aggregates like industries or markets. It is, otherwise, called as Price theory, since the major part of Microeconomics deals with the subject of pricing mechanism. Since it takes consideration of smaller units, this method is also referred as Slicing method. Microeconomic analysis involves the *'ceteris paribus'* concept *i.e.*, assumes certain conditions given or constant in the determination of a particular aspect. For example, the Law of Demand is applicable only under assumptions like income of consumer should remain same, no change in prices of substitute goods, no changes in tastes and habits of the consumer etc. So, this method of study is known as Partial equilibrium analysis. Microeconomics tends to be more scientific in its approach and studies the parts that make up the whole economy. Micro analysis makes an important tool of 'Marginal analysis'. The LDMU, LEMU, Consumers' surplus etc., are the outcome of marginal analysis in Microeconomics.

Adam Smith showed the way for microeconomic analysis assuming full employment equilibrium. The Classical Economists followed micro approach in determining how the goods are produced through proper combination of factors, how the value of final product is determined and how the distribution takes place among the factors. Though this micro approach was criticized by Malthus and Karl Max, the Neo-Classical Economists led by Alfred Marshall developed the Microeconomics to a level of almost perfection and made it popular. The following are the popular definitions about Microeconomics:

'Micro-Economics is the study of particular firms, particular households, individual prices, wages, incomes, individual industries, particular commodities'.

—(K.E.Boulding)

'Looking at the economy through a microscope to see how the millions of cells in the body of economic, the individuals or firms, play their part in the working of the whole economic system'.

—(Lerner)

'Micro-Economics deals with the division of total output among industries, products and firms and the collection of resources among competing uses. It considers problems of income distribution. Its interest is in relative prices of particular goods and services'.

—(Gardner Ackley)

'Micro-economic theory provides the framework within which the economist describes and analyses the behaviour patterns and inter-relationships of the elementary economic units like consumers, firms, industries, commodities and markets. The main objectives of micro-economic theory is to explain and predict how production, exchange and distribution of goods and services responds to the incentive structure operating in a given society'.

—***(Quirk)***

'Micro-economic theory explains the composition or allocation of total production, why more of some things are produce than of others'.

—***(Watson)***

'Microeconomics we examine among other things how individual prices are set, consider what determines the price of land and capital and enquire into the strength and weaknesses of market mechanics'.

—***(Samuelson)***

'Microeconomic theory or price theory deals with the economic behaviour of individual decision making units such as consumers, resources owners, business firms as well as individuals who are too small to have an impact on the national economy'.

—***(Leftwitch)***

Subject Matter of Microeconomics

The concept of Microeconomics basically deals with two markets *viz.*, factor market and product market and they are not independent. This is because, the factor incomes derived through the production process are spent in the product market. So, the changes in the product market reflect changes in the factor market. But, there are several differences between the two markets and this justify the need for having a separate theory of distribution dealing with factors markets. Microeconomics also deals with other aspects like output of a single firm or industry, methods of determining the supply and demand of the market, price determination of a commodity, price determination of factors, consumer behaviour, welfare theories, public finance, the gains from international trade and the distribution of gains among participant countries, determination of foreign exchange rate etc. For example, to be more specific, it focuses on the market's supply and demand factors that determine the economy's price levels. In other words, it concentrates on the 'ups' and 'downs' of the markets for goods and services and how the price affects the growth of these markets. So, it examines the behaviour of individual economic entities *i.e.*, firms and consumers. Incidence of tax on producers and consumers is determined only through microeconomic analysis. An important aspect of Microeconomics, is also to examine market failure, *i.e.*, when the markets do not provide effectual results.

Microeconomic analysis assumes the existence of full employment equilibrium in the economy. With this assumption, we proceed to know, how the producer and consumer try to attain equilibrium and how the resources of the community are allocated. It deals with the conditions that lead to economic efficiency of firms and industries. This helps to extend the analysis for the entire economy. This forms the basis for 'Welfare Economics', where we analyze the way the resources are allocated optimally, so as to ensure benefit to maximum number of people in the society (Figure 1.3).

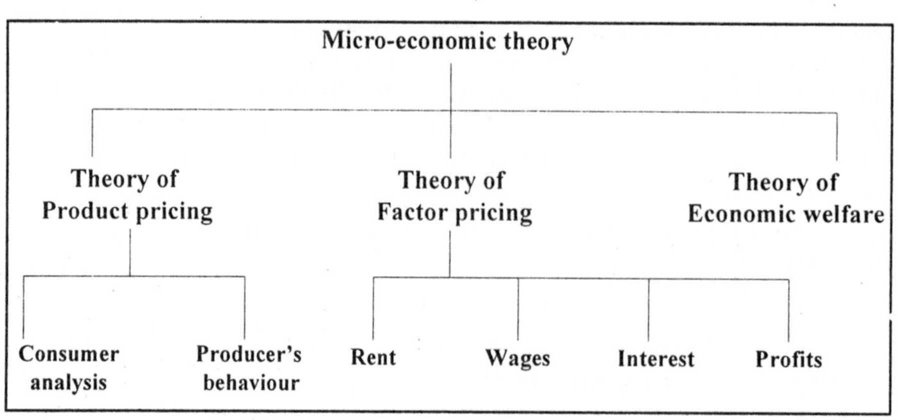

Figure 1.3: Subject-matter of microeconomics.

Importance of Microeconomics

Before Keynesian revolution, the body of Economics mainly consisted of Microeconomics. The Classical Economists as well as the Neo-Classical Economists belonged to the domain of Microeconomics. Microeconomics is important considering the following grounds:

- Helps to study the working of individual units of the economy like producers or firms, rational behaviour of consumers etc.
- It has a normative role in maximizing the satisfaction of the people and thereby, the welfare of them.
- Explains the determination of prices of both factors and products.
- Helps to analyze the conditions for optimality in the production programme, so as to ensure both cost minimization and profit maximization.
- Guides the State in the formulation of various economic policies duly understanding the phenomena at the grass root level. For example, if the Government wants to impose new taxes, the effects of taxes on social welfare, the distribution of burden of tax between sellers and buyers, analyzing the impact of direct tax vis-à-vis indirect tax on the consumers etc., can be studied through microeconomic analysis and this guides the Government to plan the policy mechanism.
- It forms the basis for conditional predictions. For example, it guides the Government in the allocation of resources, if the prices of commodities increased in the market.
- Helps the businessmen in addressing various issues like allocation of resources, efficiency of resources, cost structure, estimation of demand etc. So, it helps the businessmen in the attainment of maximum productivity with existing resources.
- It analyzes various concepts like gains from international trade, balance of payments, conditions of economic welfare, determination of foreign exchange rate of a country's currency etc.
- It forms the basis for Macroeconomic analysis, as aggregates are the totals of individual units.
- It provides the base for 'Welfare Economics'. The whole structure of Welfare Economics is built entirely on the price theory of perfect competition.

Limitations of Microeconomics

The concept of Microeconomics suffers from the following limitations:

- Microeconomic analysis is based on unrealistic assumptions like existence of full employment equilibrium, existence of complete free enterprise economy, prices, technology, tastes and habits of the individuals remains same in the short run etc.
- It is based on the assumption of laissez-faire. But, the policy of laissez-faire is no longer practiced., as it ended with the Great Depression of 1930s.
- When we deal with individual parts of the economy and neglecting the aggregates, it will not give the complete picture about the economy.
- The truths established through microeconomic analysis may be misleading, as no two individuals in the economy behave in an identical way and moreover, even the same person may act differently under similar situations at different periods of time. This is because, human behaviour is very complex.

Thus, Microeconomics is the study of the decisions of people and businessmen and the interactions of those decisions in the market. It analyzes the 'trees' of the economy as distinct from the 'forest'.

(*c*) Differences between Macro and Microeconomics

The fundamental difference between the macroeconomic theory and microeconomic theory is the macroscopic versus the microscopic views of the economy they take. However, there are several other differences between these two and they are presented in Table 1.4.

(*d*) Interdependence of Macro and Microeconomics

The differences between Macro and Microeconomics as presented in Table 1.4 is for only a matter of theoretical convenience. These distinctions are relative in nature. The problems of farmers in a district are macro in nature as compared to those of individual farmers, but a district unit is micro as compared to the state, and the state unit is micro as compared to the nation and the national unit can be considered micro in the context of the global economy. So, in practice, what is Macroeconomics in one situation becomes Microeconomics in another and one is dependent on the other. Again, all economic problems and activities, whether macro or micro, are ultimately connected with choice making and optimization. They emerge out of and are concerned with human behaviour. By looking at the above differences between Macro and Microeconomics, we can say that, both these approaches occupy a valuable niche in the field of Economics today. Though there are differences between Macroeconomics and Microeconomics, although, at times, it may be hard to separate the functions of the two. They are actually inter-dependent and complement to one another, since there are many overlapping issues between these two fields. Though Economists emphasize the need of any one of these two analyses depending upon the nature of the problem under study, one analysis cannot be exclusively

Table 1.4: Differences between Macroeconomics and Microeconomics.

Item	*Macroeconomics*	*Microeconomics*
Aggregates	Macro aggregates are large compared to micro aggregates. For example, the economy comprises of several industries and in Macroeconomics, we deal with all the industries in the economy.	Microeconomics also deal with aggregates, but micro aggregates are small compared to macro aggregates. For example, an industry is a small aggregate, which includes firms producing the same product.
Objectives	Macroeconomics studies the principles, problems and policies relating to full employment of resources and growth of resources.	Microeconomics aims to study the principles, problems and policies concerning the optimum allocation of resources at firm or industry level.
Scope	Macroeconomics is the study of total economic activity. It deals with aggregate economic behaviour of the people in general. It studies the immortal society *i.e.*, men may come and men may go, but the society remains for ever.	Microeconomics is the study of individual segments of the economy. It deals with an individual's economic behaviour. It deals with individuals and individuals are mortal
Subject matter	It deals with the general price level in the economy, National Income concepts, overall market structure, overall tax aspects of the country etc.	It deals with the pricing of a particular commodity or factor in an industry, income of a particular set of people, consumer behaviour, individual's tax aspects etc.
Approach	Macroeconomics takes a top-down approach to explain the economy	Microeconomics takes a bottoms-up approach in analyzing the economy
Assumptions	In Macroeconomics, we assume the proper distribution of factors of production and based on this assumption, it studies the achievement of full employment equilibrium in the economy.	In Microeconomics, the laws are studied by assuming full employment equilibrium, constant production, prices, technology, income etc. On the basis of these assumptions, it analyzes how factors of production and production are distributed among different uses.
Method of study	In Macroeconomics, we take aggregate variables into consideration like aggregate demand, aggregate supply, total savings, total consumption, total investment etc. Their inter-relationships are also studied in macro analysis. So, this method of study is called General equilibrium analysis.	Laws of Microeconomics are formulated by making certain assumptions and under such assumptions only, the laws are valid. For example, the Law of Demand is applicable only under assumptions like income of consumer remains same, no change in prices of substitute goods, no changes in tastes and habits of the individual etc. This method of study is known as Partial equilibrium analysis
Demand concept	The demand concept in Macroeconomics depends upon the organization's expectations.	The demand concept in Microeconomics directly depends upon the individual's expectations.
Relation among concepts	The concepts of Macroeconomics are interdependent on one another	The concepts of Microeconomics are independent concepts
Popularized by	The concepts were popularized by J.M. Keynes.	The concepts were popularized by Alfred Marshall.
Economic decisions	Macroeconomics deals with large-scale economic decisions. It focuses on countries or continents and large regions, and it generally has applications for Government policy makers.	Microeconomics focuses on small-scale economic decisions of individuals and firms. It examines how businesses can be the most successful and why individuals make the economic decisions that they do.
influence	Macroeconomics studies how the economy affects the world as a whole.	Microeconomics is the study of Economics from the perspective of the individual. It studies the effect of economy on the people in their day-to-day lives
Origin of activities	Problems of long-term growth depend upon the supply of productive resources	Micro activities emerge on the demand side of consumer's choices.
Conditions required	This approach is functional under dynamic conditions and complex long run changes.	This approach is functional under static conditions and small time intervals.
Methods	It deals with complex and dynamic changes inviting the use of advanced mathematical techniques.	It is concerned with small adjustments, for which the application of a marginal method is suitable.
Levels	Macro approach attempts to find the conditions of long-term expansions in output as a whole, assuming relative prices as constant.	Micro adjustments in resource allocation are made in response to changes in relative prices of goods and services. The aggregate level of income or total economic activities is considered to be constant.

Contd...

Table 1.4–*Contd...*

Item	*Macroeconomics*	*Microeconomics*
Time lag	In Macroeconomics, we consider the same people as buyers and sellers. For example, the labourers in the economy earn money wages as they sell their labour. With the help of wages, the same labourers will purchase the goods. So, from the point of view of economy as a whole, the total income forms the total expenditure. But, there will be considerable time lag to convert total income into total expenditure, as we deal with aggregates at macro level. So, the concept of time lag is considered in Macroeconomics.	In Microeconomics, we deal buyers and sellers as separate set of people. The time lag between income and expenditure is not considered at micro level.
Focus	Macroeconomics is a vast field, which concentrates on two major areas, increasing economic growth and changes in the national income.	Microeconomics focuses on the market's supply and demand factors and determines the economic price levels. So, the focus is mostly on optimization and equilibrium analysis.
Forces	Macroeconomics deals three forces *viz.*, aggregate demand, aggregate supply and income. This is because, aggregate demand and aggregate supply are influenced by income. Hence, income enters as a third force.	Microeconomics is limited to two forces *viz.*, demand and supply.
Importance	Study of Macroeconomics is important for formulation of economic policy of the whole nation. Macroeconomic concepts have more practical value.	Study of Microeconomics is important for resource utilization and for taking business decisions at individual level. Microeconomic concepts have more theoretical value.
Examples of comparison of macro activity with complementary micro activity	Inflation rate Economic growth Total employment	A product's price Pollution Number of labour employed in a sugar factory

carried out without the help of the other. Microeconomic problem leads to macro analysis and macroeconomic analysis necessarily involves micro analysis. Both Macro and Microeconomics provide fundamental tools for any professional and should be studied together in order to fully understand how the economic system is administered and sustained. The interdependency of Macro and Microeconomics are explained with the help of following examples:

Dependency of Macroeconomics on Microeconomics

- ☆ In Macroeconomics, the aggregate income at the national level will be studied. But, this aggregate income at national level is the sum of incomes of individuals in the economy.
- ☆ Keynes, who made Macroeconomics popular, used microeconomic theory to explain the rise in prices due to increase in money supply in the economy. According to Keynes, with increase in money supply in the economy, it leads to rise in aggregate demand and consequently more output is produced. But, with increase in output, the cost of production will increase due to operation of LDR in the business. This leads to rise in prices in the economy. The concepts of cost of production, operation of LDR etc., fall under the purview of Microeconomics and these have to be studied to analyze the general rise in prices in the economy, which falls under the purview of Macroeconomics.
- ☆ Unless we analyze the production pattern and cost structure of individual firms in the economy, it is not possible to assess the total production at national level.
- ☆ The determination of general price level in the economy belongs to macro study. But, the general price level in the economy is the outcome of the study of relative prices of various products produced and remunerations paid to different factors of production in different sectors. So, the theory of prices of products and factors which comes under 'micro' have to be necessarily related to 'macro' study of determination of general price level in the economy.

Dependency of Microeconomics on Macroeconomics

- ☆ The total income at national level determines the individuals income. If the national income is more, the individuals income will be more and *vice versa.* So, it is not possible to assess the changes in individuals income without analyzing the national income.

☆ Inflation (macro effect) would cause the prices of raw materials to increase for firms (micro effect) and, in turn, affect the end product's price charged from the public.

☆ The determination of wages of labour in a firm is a micro problem. The wages are determined based on the demand and supply of labour for that firm. But, this depends on demand and supply of labour in that region or locality, the industry to which the firm belongs to and also in the whole economy. So, the wage determination in a firm, though a micro problem, depends on the study of macro problem.

In stressing the importance of 'macro' study for solutions of 'micro' problems, Augustin Cournot stated that, *'for a complete and rigorous solution of the problems relative to some parts of the economic system, it was indispensable to take the entire system into consideration'*. Thus, the 'total' affects the 'parts' and the 'parts' affect the 'total'. So, it is difficult to separate Micro and Macroeconomic analysis. If they are treated separately, it may lead to wrong conclusions, and thereby, lead to wrong and harmful policy making.

1.3. Agricultural Economics

The word 'agriculture' comes from the Latin words *'agre'*, referring to the soil and *'cultura'*, to its cultivation. Agriculture, in its widest sense, can be defined as the cultivation of crop plants and production of products of allied enterprises. Until recently, this was a fairly accurate definition about agriculture. But todays' agriculture is radically different. This is because, the definition of agriculture had to be expanded to include more than production. Farmers rely on the input industries to purchase requisite inputs to conduct production programmes and they also rely on commodity processors, food manufactures, food distributors and retailers to market their products. So, it has evolved into agribusiness and has become a vast and complex system including all the stakeholders, who are involved in bringing food and fiber to the consumers. This food and fiber system is increasingly being referred to as 'Agri-business'. It is the sum total of all the operations or activities involved in the business of production and marketing of farm supplies and products for achieving the targeted objectives. So, an agri-business system has undergone a rapid transformation, as new industries have evolved and traditional farming operations have grown larger and more specialized. The transformation did not happen over night, but came slowly as a response to a variety of both domestic and international forces. The term agri-business was first introduced by Davis and Goldberg in 1957. It involves three important sectors *viz.*, agricultural input sector, production sector and the processing-manufacturing sector. To capture the full meaning of the term 'agri-business', it is important to visualize these there sectors as inter-related parts of a system in which the success of each part depends heavily on the proper functioning of the other two. Economics is a science, a social science that studies how society chooses to allocate its scarce resources, which have alternative uses to provide goods and services for the present and future consumption. Having discussed about the definitions and meaning of Economics, it is quite convenient to define Agricultural Economics. According to Prof. Goodwin, *'Agricultural Economics as a social science is concerned with human behaviour during the process of producing, processing, distributing and consuming the products on farms and ranches'*. According to Prof. Black, *'Economic principles imply economizing. The objective of the science of Economics in agriculture as anywhere else, is to provide a basis for the economizing of resources'*. Prof. Jouzier stated that, *'Agricultural Economics is that branch of Agricultural science, which treats the manner of regulating the relations of different elements comprising the resources of the farmer, whether it be the relations of each other or to human beings in order to secure the greatest degree of prosperity'*. In the words of Taylor, *'Agricultural Economics treats of the selection of land, labour and equipment for a farm, the choice of crops to be grown, the selection of live stock enterprises to be carried on and the whole question of the proportions in which all these agencies should be combined. This definition clearly emphasized the agriculture from the stand point of farm management'*. Prof. Hubbard defined Agricultural Economics as *'the study of relationships arising from wealth-getting and wealth-using activity of man in agriculture'*. Prof. Gray defined Agricultural Economics as the *'science in which the principals and methods of Economics are applied to the special conditions of agricultural industry'*.

Penson, Capps, and Rosson defined Agricultural Economics as *'an applied social science that deals with how producers, consumers and societies use scarce resources in the production, processing, marketing, and consumption of food and fiber products'*. Casavant, Infanger and Bridges defined Agricultural Economics as *'Economics applied to agriculture and rural areas'*.

Through critically reviewing the above definitions given by the earlier Economists, we can generalize the definition of Agricultural Economics as, the 'study of the allocation, distribution and utilization of the resources used along with the commodities produced by farming'. It is otherwise defined as 'the branch of Economics in which the basic principles of Economics are applied to address complex agricultural problems, so as to maximize the economic welfare of the people concerned with that sector'. This clearly implies, Agricultural Economics views the farming as a business and as an industry.

1.3.1 Nature and Scope of Agricultural Economics

The study of Agricultural Economics is of recent origin. The business principles of Economics are applied to agricultural sector with a view to minimize the costs and maximize the profits in a production programme. This application of principles of general Economics to agricultural sector is called as Agricultural Economics. So, Agricultural Economics is an applied science of Economics to the field of agriculture. Though, enormous literature has been generated on the subject of Agricultural Economics, still there is lot of scope to address various topics, especially in addressing the situation specific problems, evaluation of projects, issues of international trade etc. Initially, earnest efforts were made by the Economists to apply the principles of Economics to address various agricultural production related problems. This earlier concept of Agricultural Economics is 'Agronomics' that specifically dealt with land usage. Agronomics mainly focused on maximizing the crop yields giving due emphasis on maintaining a good soil ecosystem. With the expansion of scientific body of knowledge, presently, the scope of this discipline is much widened and rightly recognized as Agricultural Economics. Agricultural Economists have made many striking contributions to the Economics field such as the Cobweb model, Pricing models, Multi-factor productivity concept, Random coefficients regression etc. This agricultural sector is taken as a good example for explaining perfect competition economic paradigm. In the recent period, the subject of Agricultural Economics is expanded both horizontally and vertically taking the relevant tools particularly from Mathematics and Statistics. This applied science helps the policy makers in tracing out cause and effect-relationship between various economic variables in the field of agriculture by using sophisticated techniques like production function and programming models. This application of Economic principles to the field of agriculture gained immense significance especially with the agricultural depression occurred in the last quarter of 19th century and middle of 20th century, so as to find out the plausible causes and solutions for the said depression. Here, the contributions made by Agronomists, Agricultural Economists, Plant Breeders, Horticulturists etc., are noteworthy. Since, agriculture is an integral part of the world food system having the foundation links between crops and animal production systems, Agricultural Economists have to play a major role here, in understanding the intricacies involved in the foundation systems. Even, the students of Agricultural Economics should have a clear insight about the employment of resources in the production programmes, understanding the cost structure, enterprise combinations, profitability criteria, distribution of factor payments, finance, marketing aspects of factors and products, sophisticated econometric programmes for analyzing the relationships between various economic variables etc., so as to design solutions to address micro-level economic problems.

Agricultural Economics began as a way to study the allocation of scarce resources in a farming context. Over time, however, the discipline grew in scope to encompass important issues like natural resource management, livelihood security, international trade, Environmental Economics etc. The roots of the discipline, however, can be found in the writings of the Classical Economists of the 1700s and early 1800s. Some renowned Economists like Henry Moore of Columbia University and Henry Schultz of Chicago University has made significant contributions in the field of Agricultural Economics by applying quantitative methods. Moore conducted a number of demand and supply studies in 1920, which were being strongly influenced by Walras and Pareto. Another major contribution was brought by Black in 1926 in Production Economics in agriculture. The works of Adam Smith, Thomas Malthus and David Ricardo discussed land as a factor of production and issues of human population versus their ability to produce food. In the 1960s, the scope of Agricultural Economics was expanded beyond issues of farm and ranch management and agricultural production and this development gave the subject more of an international focus.

The subject of Agricultural Economics deals with both micro and macroeconomic aspects of agriculture. When the principles of Economics deal with an individual farmer or a single firm or small aggregates of firms, it implies Microeconomic analysis of agriculture. Conversely, when the subject of Agricultural Economics deals with the entire agricultural sector, it implies Macroeconomic analysis of agriculture. The microeconomic aspects of agriculture include important concepts like enterprises relationship or enterprises combination that aim at profit maximization *i.e.*, making a choice between different crops and allied enterprises and select the best option that leads to profit maximization in the business, studying the relationship between the factors, as it ensures scientific factor substitution or combination (complements) that leads to cost minimization in the production programme, studying the resource use efficiency in agriculture, application of cost principles, diminishing returns, farm inventory analysis, different types and systems of farming, pricing of a product and factor at firm level etc. However, the macroeconomic concepts form the basis or background for the microeconomic aspects of the agriculture.

Like Pure Economics, Agricultural Economics includes various branches of study such as Agricultural Production Economics, Farm Management, Agricultural Finance, Agricultural Price Analysis etc. The new concept of Agricultural Economics is the Agri-Business Management, where in the business principles of Economics are applied to the field of agriculture. In the recent period, as the funds allocated to the field of agriculture by the Public sector is on declining trend, there lies immense need to prioritize the issues or problems in agriculture. Even the fundings from the international donors is also encouraged to address various issues in the field of agriculture through the execution of collaborative projects. This highlights the importance of Agricultural Economics as a subject in prioritizing the (researchable) issues in the field of agriculture, studying the technical and economic feasibility of different technological interventions, formulation of strategies to improve the livelihood security of the people associated with the field of agriculture. However, the scope of the subject of Agricultural Economics can be better highlighted through assessing the roles of Agricultural Economists and they are listed here under.

- ☆ The role of Agricultural Economists in policy research is very vital, as indicated by their contributions in economic evaluation and dissemination of generated or refined technologies to the farmers-beneficiaries. Their role found to be more significant in the modern era of agriculture, as the capital investments both from public and private sources declined in the recent period. This calls for efficient utilization of resources both in production and marketing programmes, where the active role of Agricultural Economists is indispensable.
- ☆ The policy guidelines formulated by the Government either at National or State level involves the suggestions offered by the Agricultural Economists, as they look into the livelihood security, poverty alleviation, employment generation and sustainability perspectives of the farming community.
- ☆ Agricultural Economists play dual role in conducting the policy research, as a 'researcher' in generating the technologies on one hand (in collaborative research) and as an 'evaluator' in studying the economic viability, technical feasibility and implications of generated technologies at the grass-root level on the other hand. This economic evaluation of technologies is very important, as it forms the basis for recommending the policy suggestions to the farmers-beneficiaries. In the recent period, their role was further extended towards dissemination of generated and evaluated technologies to all the stakeholders in close association with Agricultural Extension specialists. This enabled the Economists to conduct impact assessment studies of the generated technologies and suggest the cost-effective technologies to the farmers. This active sharing of information, ideas and knowledge with the farmers and other stakeholders facilitate congenial environment towards quick dissemination of generated or refined technologies and also helps to identify the new researchable issues for the future. The research and the policy suggestions that are going to be executed in the future in the identified prioritized areas will drive the economy in the right direction suiting to the global trends.
- ☆ With the advent of trade liberalization, the role of Agricultural Economists is much widened, so as to make the domestic farmers really dynamic in the competitive world situation. Their approach was also changed towards addressing the farmers' livelihood security and to improve upon the same by alleviating the poverty rather than mere evaluation of the generated technologies. This approach has brought down some desirable changes in the economies of developing countries like India, where the policy parameters were mainly planned to alleviate poverty situation in the country. These policies were effective in two ways *viz.*, adoption of cost-effective production technologies and efficient marketing strategies. Their role was found more crucial now, with the advent of modern technologies and increased foreign direct investments in agricultural research. The research findings of the Economists, if found reliable by the donors, will have a strong bearing on the access for the future research projects and this also facilitates to win their confidence in better execution of the projects. During the fifties and sixties, the contributions of Economists in the country were mainly related to the dissemination of production technologies to the farmers to increase their agricultural production. During seventies and eighties, *i.e.*, after realizing the benefits of Green Revolution, their approach was changed towards economic evaluation of generated technologies in the context of achieving food security to the mounting population. From early nineties (in the liberalized regime), their execution of research was totally re-oriented and they are playing the central role in executing, evaluating, suggesting and disseminating the cost-effective production and marketing technologies. Moreover, they are functioning in collaborative research projects in close association with other Scientists (of different disciplines) such that, policy measures can be suggested for the selected project area as a whole. With the increased disbursement of financial resources to the farmers-beneficiaries through financial institutions, their role assumed greater significance in shifting the farming community towards taking up profitable enterprises in the priority areas. Further, their accountability was also increased with the participation in project mode of activities.

1.3.2 Agricultural Economics: Applied Science or Pure Science

Many of the Economists opined that, Agricultural Economics is an applied science, as it involves the practical application of basic economic principles to the field of agriculture. Economists like Frostern, Leoger, Gray, Snodgrass, Wallance, Ashby etc., supported the view of applied science of Agricultural Economics. They all opined that, Agricultural Economics basically aims at solving the problems associated with the field of agriculture. This applied science will guide the farmer in the pursuit of better living. However, Prof. Black critically commented that, Agricultural Economics is not an applied science. This is because, an applied science, in general, involves the application of basic principles to a chosen field with no modifications of basic principles. Same is the case with reference to the field of Engineering, where the basic principles of Physics are applied as they are without modifications or adjustments. But, in case of Agricultural Economics, the basic principles of Economics are applied to the field of agriculture through making relevant modifications suiting to the practical situations under consideration. So, According to Prof. Black, Agricultural Economics can be regarded as a specialized form of pure science rather than applied science.

1.3.3 Agricultural Economics: Science or an Art

Agricultural Economics is considered as a Science, as it involves the study of cause and effect relationship between the economic variables of agriculture. It deals with various scientific principles of basic Economics and the information of these principles has to be viewed from the economic angle with reference to the field of agriculture. For example in basic Economics, we study that, a firm will maximize the profits when Marginal Costs (MC)=Marginal Returns (MR). The same principle is applied to a firm in the field of agriculture to analyze the optimality situation. This method of applying basic economic principles to the field of agriculture constitutes art. So Agricultural Economics is both a science and an art.

1.3.4 Why a Separate Branch of Agricultural Economics

As mentioned earlier, the concepts of Agricultural Economics include the application of general principles of Economics to the agricultural sector. But, there lies requisite modifications in the principles of Economics when applied to the agricultural sector, as we have to deal with a particular practical situation and agricultural sector differ from other sectors on several aspects. So, this application of modified general Economics principles to the agricultural sector gives a complete justification to the branch of Agricultural Economics as a separate branch of knowledge. With the gradual shift in orientation of practicing agriculture from subsistence to market oriented and gaining popularity towards agri-business managerial concepts, the subject matter of Agricultural Economics has broadened and this led to consideration of Agricultural Economics as an important and popular field of applied science. With the gaining importance for agriculture as an industry in the concept of agri-business, presently, the subject of Agricultural Economics is gaining wider recognition. This agri-business concept has further widened the scope of Agricultural Economics and also highlights the significance of pursuing Agricultural Economics as a separate field.

1.4. Methods of Economic Analysis

Economics is the study of economic behaviour of human beings. As both human behaviour and economic phenomena are complex in nature, the laws or principles or theories that we use in Economics are only the statements of economic tendencies of the people. While propounding the useful laws or generalizations in economic theory, the Economists generally follow two methods *viz.*, Deductive method and Inductive method. These two ways of deriving economic generalizations are explained here under.

1.4.1. Deductive Method

The deductive method is otherwise termed as Analytical or Abstract or Scientific method or Method of logical reasoning. Mill characterized it as *'a priori'* method. Francis Bacon described it as a *'Descending process'*. This method of economic analysis proceeds from 'general to particular on the basis of general truth'. So, deduction in Economics starts with a generally accepted principle and proceeds to the specific. That means, it aims at deriving conclusions at micro level from established general truths. According to Wilson Gee, *'by deductive method is meant the reasoning from general to particular or from universal to individual'*. In this method, we accept certain general truths or facts and apply them in specific cases to prove the accepted truths. For example, the accepted general truth is, a firm gets maximum profits when MC=MR. So, this general truth is applied to a particular firm and if the firm fulfills this principle, we accept the firm is getting maximum profits. Similarly, it is an established fact that, 'man is entirely motivated by self-interest'. The same will be applied at the micro level and its reliability will be established. So, this method involves the application of general principles to a particular situation and draw useful conclusions. This method assumes that, the behaviour of the general public will also be the behaviour of individual person. This method is based on *a priori* reasoning and conclusions are drawn from certain fundamental assumptions. In general, Macroeconomic theories are based upon deductive method. Studies of national income, employment, general price level and international trade are made on the basis of deductive method. The Classical and Neo-Classical school of Economists notably, Ricardo, Senior, Cairnes, J.S. Mill, Malthus, Marshall, Pigou, Robbins, Frank Knight, Karl Marx, etc., advocated this method. The main steps involved in deductive logic are discussed here under (Figure 1.4).

(a) Selecting the Problem

Identification of problem is the basic step. The problem selected for the investigation should be stated very clearly. The problem under consideration may have wide scope such as addressing issues like poverty, inflation, unemployment etc or a narrow scope. But, if the focus or scope of the problem is too narrow, it facilitates the researcher to get accurate results. The problem must be one of significance for the actual world and should have economic significance. While selecting the problem, it is important on the part of investigator to be clear about the objective(s) of the research activity.

(b) Defining Technical Terms and Variables in the Problem

After finalizing the problem for study, the investigator should get well versed with the technical terms involved in the study. This helps him to understand about various concepts, identify variables and their inter-relationships.

(c) Framing Meaningful Assumptions

The study about technical terms and inter-relationship among variables helps the investigator to frame meaningful assumptions, which further guide him to formulate hypothesis. This method is also known as Hypothetical method, as the assumptions formulated may not correspond to actual facts, but are very near to actual facts and these assumptions are used as premise for starting, reasoning and drawing conclusions. The assumptions may be behavioural relating to the behaviour of variables or technological relating to the state of technology or factors employed in the study. Sometimes, crucial assumptions are also made on the basis of observations or introspection like, farmers try to maximize the profits in the production programme, consumer try to attain maximum satisfaction etc.

(d) Process of Logical Deduction

Generally, a hypothesis is formulated from the assumptions framed through logical deduction. This logical deduction may be carried out verbally or symbolically. In the recent days, mathematics is mostly used in the method of logical deduction.

(e) Formulation of Hypothesis

Now, the research investigator will formulate the hypothesis duly considering the variables for the study, their inter-relationships, assumptions framed, *a priori* knowledge etc.

(f) Making Predictions and Testing the Hypothesis

After formulating the hypothesis, predictions can be made on its basis and they are tested. If the predictions are in agreement with the facts, then the theory is accepted. On the other hand, if the predictions are in conflict with the facts, the theory is discarded in favour of superior alternative and again the process is to be initiated duly modifying the assumptions.

(g) Verification of Hypothesis

It is an important step in the deductive method. Verification refers to confirming whether the hypothesis formulated earlier is in agreement with the facts or not. Two possible cases may arise regarding the verification of hypothesis.

- ☆ ***Hypothesis is accepted/confirmed:*** For example, a hypothesis is formulated stating that, 'the firm will try to maximize the profits'. This is said be accepted, when the firm under consideration for study is found behave in the same way.
- ☆ ***Hypothesis is rejected/not verified:*** If the hypothesis is not verified, it implies two important aspects *viz.*, the hypothesis formulated is correct, but the results are contradictory, as the variables under study behaved in a different way due to special circumstances. Or, the investigator might have formulated a wrong hypothesis without having thorough knowledge about the variables, their inter-relationships, faulty assumptions made, lack of *a priori* knowledge etc.

The above second case *i.e.*, contradictory results not confirming hypothesis is more common in economic sciences, as both human behaviour and economic phenomena are very complex, unlike physical sciences. Moreover, laboratory or controlled experimentation is not possible in Economics.

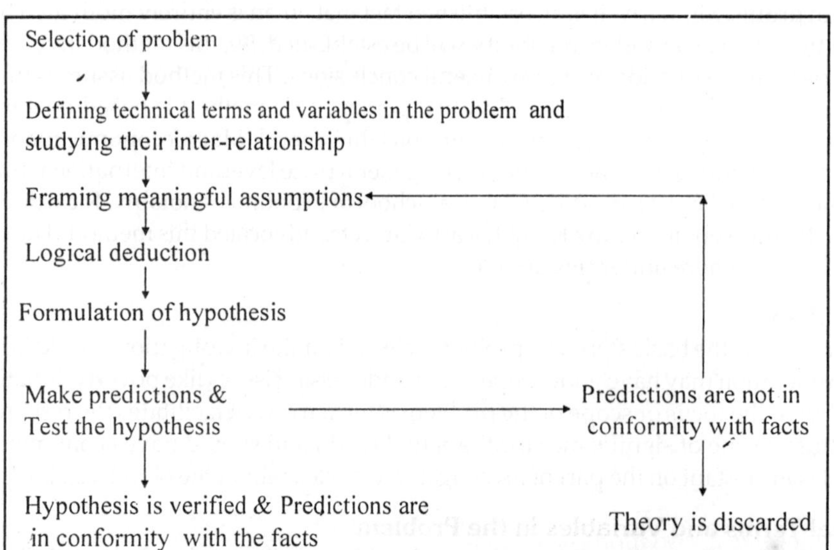

Figure 1.4: Steps involved in deductive method of economic analysis.

Merits of Deductive Method

The merits of deductive method are as under.

- ☆ This method is nearer to reality. It consumes less time and less expensive compared to inductive method, as useful economic theories can be verified with less data.
- ☆ The use of mathematical techniques and logical reasoning brings exactness and clarity in economic analysis.
- ☆ This method is indispensable, as there is limited scope for experimentation in Economics. So, this method serves as a substitute for experimentation. Since experimentation is not possible in Economics, this method is based on logical reasoning. As Boulding puts it, *'it is the method of intellectual experiment'*.
- ☆ This method is simple because it is analytical and it has easy application. It involves abstraction and simplifies a complex problem by dividing it into component parts. We know, the Marginal Utility (MU) derived by a consumer goes on diminishing with every successive consumption of a commodity is a self-evident truth. From this, we can formulate several logical conclusions. For example, rich people have lesser MU for money compared to poor people. This will guide the Government to impose progressive taxation policy. So, the Principle of Progressive taxation is derived from the LDMU through deductive reasoning.
- ☆ The inferences drawn through deductive method have universal validity.
- ☆ We try to derive useful conclusions from the well known generalizations, through collecting elaborate statistical information or data and establish the law.
- ☆ This method helps to explain complex economic phenomena through making simple assumptions and draws conclusions.
- ☆ If the premises or hypotheses are true, the conclusions will be true and exact.

Demerits of Deductive Method

No doubt, deductive method is simple and precise only, if the underlying assumptions are valid. There is big 'IF' in the statement. Criticisms were raised against this method by the Historical School, which flourished in Germany. The shortcomings of the deductive approach are as under.

- ☆ More often, the assumptions turn out to be based on half truths or have no relation to reality. In such case, the conclusions drawn from such assumptions will, therefore, be misleading. That means, if the assumptions framed are not true, the conclusions will be wrong and the generalizations cannot be accepted as a law. If such generalizations are used in framing economic models and policies, the results would be disastrous. For example, J. B. Say, claimed universal validity for his law, *'Supply creates its own demand'*, so that, there will not be any over production in the market. But, this law was discarded, when Keynes proved that, Say's law was wrong and over production would be possible. Such laws or models, which are built with little empirical content, remain as 'intellectual toys' and they have no operational significance.
- ☆ Professor A.P. Lerner describes the deductive method as 'Arm Chair' analysis. According to him, the premises from which inferences are drawn may not hold good at all times and places. As such, deductive reasoning is not applicable universally. For example, Keynes theory of employment is not applicable to under developed economies.
- ☆ This method is highly abstract and requires great skill or competence in drawing inferences from various premises.
- ☆ This method is based on the assumption that, economic conditions remain constant, But, in practical situation, economic conditions are continuously changing.
- ☆ It is important that, the verification of theories or laws or principles is Economics is mainly based on observation. A right observation depends on the accuracy of data. If faulty and inadequate data are employed in the research, it leads to faulty conclusions and thereby, there is a chance of refuting the law/principle under consideration.
- ☆ It is not easy to test the results obtained through deductive method. So, it should be supplemented by induction.

As the deductive method employed by the Classical and Neo-classical Economists led to many facile conclusions due to reliance on imperfect and incorrect assumptions, the German Historical School of Economists criticized this method and advocated a more realistic method for economic analysis known as Inductive method.

1.4.2. Inductive Method

According to Wilson Gee, *'inductive method is the process of reasoning from particular to general or from individual to universal'*. This method is also known as Historical or Concrete or Posterior or Realistic or Empirical method. It was adopted by the Historical School of Economists of Germany. It was Francis Bacon, who advocated inductive method in scientific enquiry and described it as an *'Ascending process'*. In this method, first the data are collected about a certain economic phenomenon, they are compiled, systematically arranged and statistical techniques are employed and from these findings general conclusions are drawn. Thus, inductive reasoning in Economics does the reverse of deductive reasoning *i.e.*, it begins with an individual problem or question and proceeds to form a general principle based on the evidence observed in the real world of economic activities. This method involves conductance of micro level study first and this study will be replicated. If same conclusions were drawn from such studies, we generalize the statement as an observed truth and the theory is propounded. The inductive method can be conducted in two ways *viz.*, through experimentation and through statistical means. The former is generally used in biological and physical sciences, and the latter is used in Economics. This is because, in Economics, there is limited scope for experimentation and both economic phenomena and human behaviour are complex in nature. However, some experimentation is possible to validate some economic laws like laws of production, cost principles, price discrimination practiced by monopolist etc. So, the Economists had no other option to rely upon statistical means in building the economic laws or principles. However, some *a priori* knowledge is also essential, as it gives the researcher some direction in which way he has to proceed. For example, farmer 'A' purchased less quantity of a particular brand of fertilizer when its price was increased in the market. Farmers B, C and D also showed the same pattern of demand due to rise in price of that fertilizer. So, finally, we can generalize their behaviour and an economic theory can be propounded stating that, 'quantity demanded varies inversely with the price of the commodity'. Similarly, assume there are 500 farmers in a village. The data analysis reveals that, nearly 490 farmers are purchasing chemical pesticides manufactured by Multi National Companies, which are highly effective. Out of the ten which remains, nine farmers are purchasing local products just to patronize their own products, while the remaining one farmer is not adopting plant protection measures due to lack of awareness. From this analysis, we can easily conclude that, farmers like to buy the products from Multi National Companies on the criteria of effectiveness, unless they are guided by patriotism or lack of awareness about application of an effective chemical pesticide.

The LDR, LDMU, LEMU, Consumer's surplus etc., have been developed on the basis of inductive method. Malthus formulated the theory of population after analyzing the facts about population trends in different countries. Similarly, Engel formulated the propositions of family expenditure after a factual study of family budgets. Theories of rent, wages, interest and profits are also based upon inductive method. This implies, microeconomic theories are formulated according to inductive method. However, some of the concepts of Macroeconomics *viz.*, consumption function, the development and use of accelerator principle have been obtained through inductive analysis. Economists like Frederick List, Hilder-brand, Roscher Hutchinson etc., advocated this method. The main steps involved in the application of inductive method are given here under.

(a) Selection of Problem

In order to arrive at generalization of an economic phenomenon, the problem should be properly selected and clearly stated.

(b) Defining Technical Terms and Variables in the Problem

The investigator should study about different technical terms related to the study. This helps him to identify different variables and their inter-relationships.

(c) Collection of Data

This inductive method can be applied in two distinct ways *viz.*, experimentation and statistical form. The former is used in natural and physical sciences and the latter is used in Economics. For the problem under consideration, the relevant data should be collected about the variables and a preliminary thinking will be done to study the possible functional relationships between the selected variables.

(d) Processing of Data

The collected data for the selected variables should be compiled and analyzed using appropriate statistical techniques.

(e) Development of Theory

From the analytical findings, a theory can be developed, which can be further refined and tested using statistical tests.

(f) Making Predictions

Once the theory has been developed, predictions can be made on its basis. If predictions of theory are in agreement with the facts and actual behaviour of the economy, then a new reliable theory has been developed. If a new theory explains the concepts in a better way than the existing ones, it replaces them. On the other hand, if the predictions are in conflict with the facts and actual behaviour of the economy, the theory is discarded and again data collection is to be initiated with much accuracy to make further predictions.

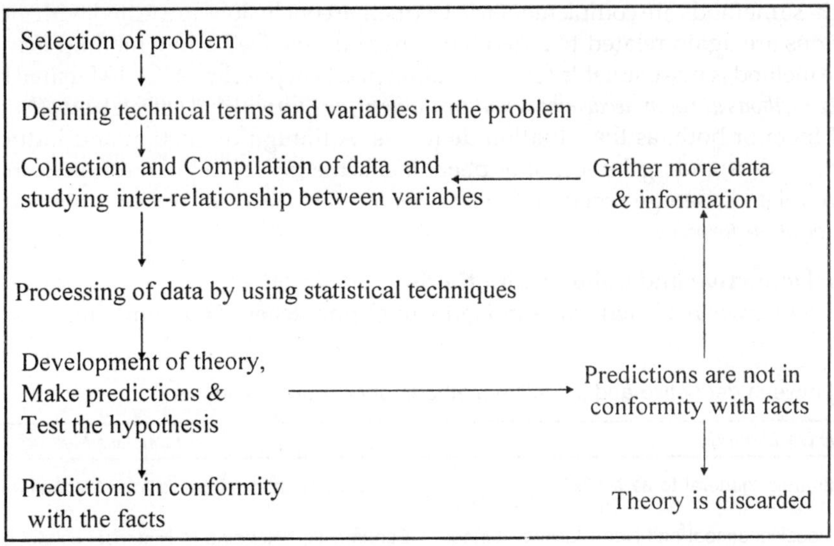

Figure 1.5: Steps involved in inductive method of economic analysis.

Merits of Inductive Method

The following are the merits of inductive method:

- This method is more realistic, as it is based on facts.
- It has more practical approach for formulating economic generalizations
- This method employs statistical techniques to test various economic principles, thereby, it brings exactness, clarity and reliability for the findings.
- Inductive method is dynamic. The changing economic phenomenon is analyzed and on the basis of collected data, solutions and conclusions are drawn from them.
- This method is more useful because, its prepositions can be tested and verified easily.
- This method considers the complex relationships found in actual life and examine them carefully.
- The generalizations drawn through inductive analysis are *historico-relative* in Economics. Therefore, they are relative to time and place and hence, highly practical.
- Inductive method discovers and proves general principles and this helps in future investigations.
- The conclusions drawn through deductive method can be verified by inductive method.

Demerits of Inductive Method

The following are the demerits of inductive method:

- If insufficient data are used for the analytical study, the conclusions drawn from the study will lead to faulty economic generalizations.
- The collection of data consumes more time and cost and the sources and methods employed in the collection of data differ from investigator to investigator. The results, therefore, may differ even with the same problem. So, there is a danger of investigator bias entering into propositions and this affect the concreteness.
- Since the propositions obtained through this method are based on few facts, they may not have universal applicability and hence, there is limited scope for verification.
- This method has limited scope in Economics unlike physical sciences because, controlled experimentation is not possible. This is because, in Economics, we deal with complex human behaviour and even economic phenomena are very complex. Every economic event is unique in the sense that, it will favour one segment of the population and disfavour another segment of the population. Further, one economic problem is related to another economic problem and thereby, deduction has its own role.
- In this method, there is a danger of giving hasty and hurried conclusions from insufficient number of facts.

Inspite of the above demerits, inductive method is widely used in Economics. The modern era has been called as the Inductive era.

Thus, none of the above two methods provides satisfactory system for solutions of problems. Since, both the methods have their own merits and demerits, we cannot rely exclusively on any one of them. Modern Economists are of the view that, both these methods are complementary. Deductive conclusions are tested with inductive methods and inductive generalizations are again related to a deductive hypothesis. Deductive method is more suitable to pure theory, while inductive method is more suitable for formulating practical policies. Alfred Marshall has rightly remarked, *'Inductive and Deductive methods are both needed for scientific thought, as the right and left foot are both needed for walking'*. We can apply any one of them or both, as the situation demands. Although deduction and induction represents two differing approaches for understanding the economic phenomena, the 19th century American Economist, Henry George observed that, they are related. George noted that, *'induction involves the use of human reason to investigate facts, while deduction is the derivative of the former'*.

Differences between Deductive and Inductive Methods

The major differences between the deductive and inductive approaches of economic analysis are depicted in Table 1.5.

Table 1.5: Differences between deductive and inductive methods of economic analysis.

Deductive Method	*Inductive Method*
Deductive method proceeds from 'general to particular'.	Inductive method proceeds from 'particular to general'.
Economists belonging to Classical school and Neo-Classical school advocated this method	Economists belonging to Historical school advocated this method
Aims at deriving conclusions at micro level from established general truths.	Examines sample data to make inferences about a population.
Mainly aims at verifying the established truths or principles or laws or methods	Mainly aims at establishing the truths or principles or laws or methods as generalizations.
Deductive systems are usually normative in nature.	Inductive systems are usually descriptive in nature.
It consumes less cost and less time.	It consumes more cost and more time, as it is based on extensive collection of data.
This method is indispensable, as there is limited scope for experimentation in Economics.	This method helps to make generalizations about laws or principles for which there is scope for experimentation.
In general, macro economic theories are based upon deductive method.	In general, micro economic theories are based upon deductive method.
Limited use to-date.	More frequently used to-date.
It does not give any new knowledge, as it is a method of verification	It gives new knowledge, as it is a method of discovery
It is a downward process of thought and leads to useful results.	It is an upward process of thought and leads to formulation of economic principles.

1.5. Economics in Relation to Other Social Sciences

Economics as a social science deals with the human behaviour in relation to satisfaction of unlimited wants with limited means. Being a social science, it is related to other social sciences like sociology, politics, history, ethics, jurisprudence and psychology in various aspects. We know, the economic development of a nation depends not only on the economic factors, but also on historical, political and sociological factors. If all these factors are congenial, it promotes the economic development of a country at a steady pace. So, it is essential to discuss in-detail the relationship of Economics with other sciences (Figure 1.6).

Economics and Ethics

Economics is very closely related to ethics. The early economic thinkers (Marshall) described Economics as the 'Handmaid of ethics'. Earlier, in the university teachings, Economics was a part of moral sciences. Ethics is a branch of philosophy, which emphasizes how people ought to conduct themselves in the society and balances of rights and duties. So, it describes the moral code of behaviour and is much more related to customs and traditions. It discusses the rules that govern right conduct and morality. It deals with the questions of right and wrong and aims at promoting good life. Economic laws or principles assume *'Ceteris Paribus'* concept, referring to moral or social factors that are supposedly held equivalent for all choices that one might make. Ethics is the science of moral conduct. It asks the question 'what ought to be'. Earlier, Economists like Robbins, J. S. Mill, Brown, Senior and Friedman are of the opinion that, Economics is only a positive science dealing with 'what is', and not concerned with the question 'what ought to be' (normative aspect). But to-day, practically all Economists hold that, Economics is a Positive science as well as an Ethical science *i.e.*, it is concerned with both 'what is' and 'what ought to be' in the economic sphere. So, there is close

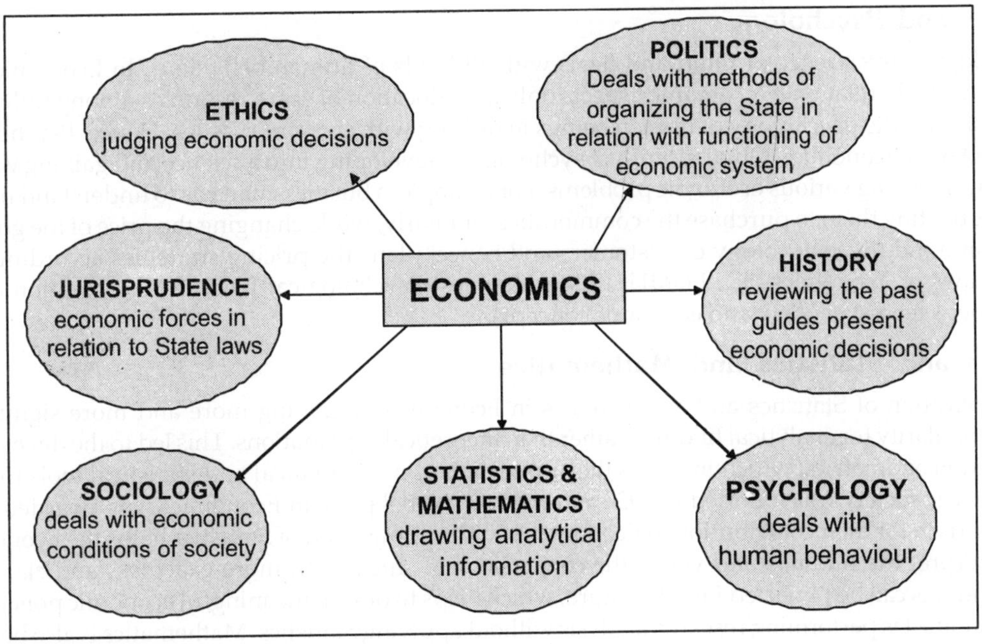

Figure 1.6: Economics: Its relationship with other sciences

connection between Economics and Ethics. Economics describes, while ethics prescribes. Economics is amoral, while ethics is moral. Economics aims at promoting material welfare, while Ethics aims at promoting moral welfare. Ethics is concerned with ends as such, while Economics is concerned with arranging means to ends and it has nothing to do with ends as such. Ethics is concerned with the mode of behaviour and the conduct of man, while Economics takes for granted all these things. Ethics is concerned with all immaterial, eternal and sublime things, while Economics, many times, is concerned with all material things in life. While dealing with economic problems, ethical issues are also considered. For example, the Government of Andhra Pradesh introduced prohibition of liquor for ethical reasons, though there was heavy loss of revenue from it. Taking this close relationship between Economics and Ethics, the modern Economists believed that, Economics cannot be dissociated from Ethics.

Economics and Politics

These two sciences overlap each other. Politics is the science of the State or political society and deals with the methods of organizing the State. It studies about man in relation to the State. The functional aspects and efficiency of different economic organizations in the State depend very much on the fundamental philosophy of the State. Economics comes into closer relation with Politics especially with the methods of organizing a State. In matters of legislations, like company law and nationalization or in matters of public finance, Economics plays a very important role. The economic planning in a country depends upon the political stability. The issues like Nationalization, Privatization, Globalization, Prohibition etc., will be viewed both from economic angle as well as political angle. In the modern days, the Government is performing several functions such as, education, medical care, support of old people, etc., which were formerly looked after by the church or family or group organizations. While formulating economic policies, Government emphasizes on maintenance of full employment equillibrium, optimum allocation of economic resources, equitable distribution of national income etc., be it a socialist state or a capitalist state. So, taking the vital role of Government in addressing various economic issues, Economics is otherwise regarded as Political economy. Sometimes, the economic conditions may influence the political ideas, institutions and even economic systems. For example, socialism was born because of wide economic inequalities and severe exploitation of working class in England during the industrial revolution.

Economics and History

History is a record of the past events. In History, the economic, political and social conditions of the people will be reviewed. For an Economist, it is important to study and analyze various economic problems prevailed in the past and how best they are controlled. The aspects like taxation and other sources of raising revenue, trading policies etc., will be reviewed from the history and accordingly they are modified and planned suiting to the current period. This also helps us to draw the comparative picture with the past. This comparative analysis of economic policies has special importance, as controlled experimentation is not possible in Economics. Realizing the importance of linkage of Economics with History, a new branch of Economics has gained significance in the recent period known as 'Economic History'. So, we may say Economics is the fruit of History and History in the root of Economics. This implies that, *'Economics without History has no root and History without Economics has no fruit.*

Economics and Psychology

Psychology is the science of mind and deals with all kinds of human behaviour. In Economics, we deal with human behaviour in relation to economic aspects only *i.e.,* allocation of scare resources among unlimited wants. So, Economics studies the psychology of human beings in dealing with economic issues. Hence, Psychology comes into closer contact with Economics. Quite recently, Psychology is developing into a science and gaining wider recognition especially, in analyzing various economic problems. For example, a businessman has to understand the psychology of buyers in motivating them to purchase the commodities. Similarly, while changing the price of the good, the business man keeps in mind the psychology of customers and hence, plans the pricing strategies accordingly. Some of the important Laws of Economics *viz.,* LDMU, LEMU, Keynes' Psychological Law of Consumption etc., involve the association between psychology and economic concepts.

Economics and Statistics and Mathematics

The application of Statistics and Mathematics in Economics is gaining more and more significance with the increase in popularity for analytical findings rather than theoretical explanations. This led to the development of a new science called Econometrics, which involves the application of mathematical and statistical tools to economic data. The Econometric society was founded in 1930 and the First Nobel prize in Economics was awarded to Jan Tinberen and Ragnar Frisch for their contribution to Econometrics. The application of Statistics helps the Economist to analyze the data, interpret the findings and verify the conclusions of study with more exactness and clarity. Further, the analytical findings can be projected into the future, which helps to design meaningful economic policies. Infact, now a days, no Economist is performing research analysis without applying Statistics. Mathematics is also being increasingly used in Economics with a view to sharpen the economic analysis.

Economics and Sociology

Sociology is a general social science, which deals with all aspects of society. It attempts to discover the facts and laws of society as a whole. But, Economics deals only with the economic aspects of a society. It deals with the study of human behaviour in relation to scarce means and unlimited wants. For a student of sociology, social institutions like marriage, religion, political institutions and economic conditions are all important subjects for study. But in Economics, the above aspects are studied only to the extent of their influence on the economic life of a society. So, the economic conditions of a society can be best studied through the proper interaction between Economics and Sociology concepts and hence, they are closely connected. However, though Economics is a branch of Sociology, it should be treated as a separate and distinct branch because, the concepts in Economics are more exact and concrete compared to Sociology.

Economics and Jurisprudence

Jurisprudence is the Science of Law. Just as Economics is closely related to the Science of Politics, so it is also closely related to the Science of Law. Government and Law form a framework in which the economic forces act. The economic progress of a country depends to a great extent on its legal system. Good laws promote economic progress and bad laws act as an impediment to growth. For example, to encourage foreign investments in a country, the laws of taxation and labour legislations should be congenial, simple and clear, so as to promote economic progress.

1.6. Economic Laws

The term 'law' means a general proposition or statement of tendencies, more or less certain, more or less definite. Many such statements are made in every science, but we do not, indeed we cannot, give to all of them a formal character and name them as 'laws'. We must select and the selection is directed less by purely scientific considerations than by practical convenience. So, a law is a statement of what must happen under given certain conditions. The laws of all sciences are hypothetical or conditional. Like other sciences, Economics has its own laws. For example, the Law of demand explains that, when there is fall in the price of the commodity, the demand for the commodity extends. It implies that, there is a tendency among the people to buy more when there is decline in price of the commodity and *vice versa.* According to Alfred Marshall '*economic laws or statements of economic tendencies, are those social laws, which relate to branches of conduct in which the strength of the motives chiefly concerned can be measured by a money price*'. This definition indicates the following aspects:

- ☆ Economic laws are the statements of economic tendencies
- ☆ Economic laws are like social laws
- ☆ Economic laws are related human behaviour
- ☆ The human behaviour can be measured by money

Lionel Robbins defined '*economic laws are statements of uniformities about human behaviour concerning the disposal of scarce means with alternative uses for the achievement of ends that are unlimited*'. This definition has the following interesting aspects:

- Economic laws deals with human behaviour
- Economic laws guide the individuals in using the scarce resources efficiently
- Economic laws guide the individuals in allocating the scarce resources which are having alternative uses
- Economic laws guide the individuals in fulfilling his wants

From the above two definitions given by Marshall and Robbins, it is clear that, the economic laws are the statements of tendencies in relation to human behaviour.

Scientific Nature of Economic Laws

All scientific laws studies cause and effect relationship. Economic laws also establish the same about economic behaviour of man and economic phenomena. Since human wants are unlimited, but the means to derive satisfaction are limited, every individual will act in peculiar way. So, to analyze this behavioural pattern, Economists have formulated several economic laws. But, it is important to note that, these economic laws are hypothetical or conditional. This is because, human behaviour is so complex and a person acts in different ways for the similar situations he faces in his life. Moreover, the economic phenomena are very complex. As a matter of fact, all scientific laws are conditional. For example, in Physics, the law of gravitation states, any object, if thrown upward, tends to move towards earth due to the influence of gravity, unless no opposing forces act on the object. We see birds, aero planes etc., will not fall to ground, because of influence of some opposing forces. So, the law of gravitation operates only when opposing forces do not act on the objects. Similarly, the law of demand operates only, when the assumed conditions holds good. This implies, all scientific laws hold good only under certain conditions and Economics is not an exception. But, the economic laws are less exact compared to physical sciences like Physics and Chemistry, as it deals with complex human behaviour. But, economic laws are more exact compared to other social sciences like ethics, sociology etc., as it involves the measuring rod in terms of money to measure satisfaction. The economic laws are formulated generally by the use of introspective or psychological method. That means, by knowing the reactions of an individual to a certain economic phenomenon, the Economists assume that, others also tend to show the same behaviour in the similar situation. The following Table 1.6 shows the ways, how an economic conclusion and a physical one differ:

Table 1.6: Ways an economic conclusion and a physical one differ.

Item	*Physics*	*Economics*
Environment	Fully controlled	Slightly controlled
Primary strategy	Mathematical induction and deduction	Observation
Correctness	Always true	Not at times
Unanimity	Yes	Not at times

Features of Economic Laws

The following are the features of economic laws:

- ***Economic laws are statements of economic tendency:*** Economic laws, like other laws establish relationship between causes and their effect. As these laws are not certain and they express a tendency, they explain the likelihood of human behaviour under specific conditions. As explained earlier, the law of demand is not applicable when the assumed conditions are not constant. So, it implies, the economic laws are not as exact as the laws of physical sciences.
- ***Economic laws are hypothetical:*** According to Prof. Seligman, economic laws are 'essentially hypothetical' *i.e.*, they simply indicate the general tendency of human behaviour under given conditions. These laws are not applicable in every situation. So, every economic law is associated with the phrase, *'Ceteris Paribus'*, which means 'other things being equal' or 'other things remaining the same'. For example, the LDMU explains that, additional utility derived from the consumption of successive units of a commodity goes on falling. But, this law is applicable, when units are uniform, sufficient, there is continuity in the consumption pattern, MU of money remains constant etc.
- ***Economic laws are relative:*** Economic laws are closely related to time, place and situations. If these changes, economics laws must be re-stated. So, they are not universal. Change of time, place and situations make them ineffective. So, economic laws are considered as *'historico-relative'*.
- ***Economic laws are human laws:*** Economic laws may be termed as human laws because, they explain the behaviour of human beings in the society regarding economic aspects. Since, laboratory experiments cannot be conducted regarding human behaviour, the economic laws are said to be less exact. Moreover, the economic phenomena are very complex and even the individual behaviour cannot be exactly predicted under similar situations. So, they are not so rigidly stated.

- ***Some economic laws have universal application:*** Certain economic laws have universal application such as LDMU, LDR, Law of Demand etc. However, most of the laws in Economics are not universal and everlasting, as they good only under certain assumed conditions. That means, some economic laws are valid for developed countries and they may not be applicable for developing or under developed countries. For example, the Keynes theory of employment is applicable to capitalistic countries and not relevant to India.

Assumptions of Economic Laws

Since economic laws are less exact than laws of physical sciences because of the reasons explained earlier, the economic laws will assume the following aspects to ensure wide applicability:

(a) '*Ceteris Paribus*' Concept

All economic laws carry the Latin phrase, *'Ceteris Paribus'*, which indicates, 'other things being equal' or 'other things remaining the same'. It shows that, an economic law is applicable only under certain given conditions. As explained earlier, for the validity of law of demand, it assumes other factors like income of consumers, their tastes, preferences, prices of substitutes etc., do not change. If the assumed conditions are not fulfilled, the law is not applicable. So, the Economist has to use the phrase, *'Ceteris Paribus'*. To make economic laws more exact and valid, attempts should be made to make the assumptions of the laws as realistic as possible.

(*b*) Rationality of Human Behaviour

Economic laws presume that, the consumers are 'rational'. That means, they will spend their limited resources, so as to derive maximum satisfaction.

Economic Laws vis-à-vis Laws of Other Sciences

The following points highlights the comparative picture between economic laws and laws of other sciences:

- Economic laws are not as exact as physical sciences like Physics and Chemistry, as economic laws operate only under certain assumed conditions and show probable trends. If the conditions change, the economic laws are not valid. This is because, Economics deals with human beings, whose behaviour is very complex, governed by several external forces, which are not under his control and moreover, experimentation is not possible in Economics. Hence, Alfred Marshall has compared the Laws of Economics as 'Laws of tides', as tides rise and fall on the influences of the Sun and the Moon, the validity of economic laws depends upon the existence of assumed conditions. For example, if price falls, the quantity demanded for a commodity will extend. But by how much? It again depends upon two factors *i.e.,* when assumed conditions are valid and when assumed conditions are not valid. That means, if the assumed conditions are static, *i.e.,* if the income of the consumer, tastes and habits of the consumer, prices of substitutes or complements etc., remains same, the Law of demand is valid, otherwise it is not valid. So, the per cent increase in demand for the commodity due to fall in price show probability trends depending upon the variations in other factors. However, due to the use of statistical analysis and mathematical methods, economic laws are approaching towards exactness.
- When economic laws are compared to the laws of other social sciences, economic laws are more exact, precise and accurate, as they are less hypothetical. This is because, in Economics, there is a measuring rod of money for studying the consumer's satisfaction, which is not available in other social sciences like Ethics, Sociology etc. This gives some concrete shape to the validity of economic laws. However, economic laws are less certain like social laws, as money value fluctuates frequently.
- Accurate predictions are not possible through economic laws. Even in sciences like Biology and Meteorology, accurate predictions cannot be done. However, through laws of physical sciences (physics and chemistry), there is scope for accurate predictions.
- Most economic laws are behaviourist, such as the LDMU, LEMU, Law of demand etc., which depend upon human behaviour. So, they are less exact compared to physical sciences.
- Unlike scientific laws, economic laws are not assertive. Rather, they are indicative. For example, the Law of demand states that, as price falls, the demand for the commodity extends. This law is applicable only under the phrase *'Ceteris Paribus'*. But, it does not assert that, demand must extend, when price of the commodity falls.
- There are certain generalizations in Economics, which may be considered as truism. They are like axioms and they do not need empirical data to verify them and they are universally valid. For example, 'human wants are unlimited', 'consumption is a function of income', 'human behaviour is rational' etc. On this ground, economic laws are superior to scientific laws. But, all economic laws are not like axioms.

Reasons for Uncertainity of Economic Laws

From the above discussion, it is evident that, economic laws are not exact as Physical sciences, but more exact compared to other Social sciences. In this context, Alfred Marshall quoted that, *'the laws of Economics are to be compared with the laws of tides rather than with the simple and exact laws of gravitation'*. This implies, economic laws are less certain and therefore, based on economic laws, accurate predictions are not possible for the future. The following are the important reasons regarding the uncertainity of economic laws:

- ✰ Human behaviour is complex and rational. It is influenced by various social, economical, political, religious, cultural and many other factors.
- ✰ Even though, money serves as a measuring rod in Economics, but the value of money fluctuates frequently. So, the measuring rod is defective in exactly quantifying the human welfare.
- ✰ Less scope for experimentation in Economics.
- ✰ Assumptions are more and therefore, the validity of laws are subjective to the assumed conditions.
- ✰ It is very difficult to identify all the factors influencing the economic events. In the words of Durbin, *'the true determinants of economic events are not adequately discovered'*.

1.7. Schools of Economic Thought

Schools of economic thought describe the variety of approaches in the history of economic theory noteworthy enough to be described as a 'school of thought'. While all Economists do not always fit into a particular school, particularly in modern times, classifying them into different schools of thought is common.

1. Classical Economics

The term 'Classical Economics' was coined by Karl Marx. It is widely regarded as the first modern school of economic thought. Its major developers include Adam Smith, Jean Baptiste Say, David Ricardo, Thomas Malthus and John Stuart Mill. It is 'classical' in the sense that, it based on the belief that, competition leads to an efficient allocation of resources and regulates economic activity that establishes equilibrium between demand and supply through the operation of market forces.

2. Neoclassical Economics or 'Marginalism'

This body of theory formed from about 1870 to 1910. The term was originally introduced by Thorstein Veblen in 1900. The term 'Economics' was popularized by such Neoclassical Economists like Alfred Marshall. It was later used by John Hicks, George Stigler, Carl Menger, William Stanley Jevons, John Bates Clark. This school of thought increased the use of mathematical equations in the study of various aspects of the economy. This approach was developed in the late-nineteenth century, based on the contributions from William Stanley Jevons's Theory of Political Economy (1871), Carl Menger's Principles of Economics (1871), and Leon Walras's Elements of Pure Economics (1874–1877). These three Economists have been said to have begun 'the Marginal Revolution'. It is 'neo' in the sense that, it departs sharply from the classical viewpoint in its analytic approach that places great emphasis on mathematical techniques. In opposition to Keynesian Economics, this school states that, savings determine investment (not the other way round) and is concerned primarily with market equilibrium and growth at full employment instead of with the under-employment of resources. Since its inception, Neoclassical Economics has grown to become the primary take on modern-day economics. An important change in Neoclassical Economics occurred around 1933. Joan Robinson and Edward H. Chamberlin, with the near simultaneous publication of their respective books, *The Economics of Imperfect Competition* (1933) and *The Theory of Monopolistic Competition* (1933), introduced models of imperfect competition. Theories of market forms and industrial organization grew out of this work. They also emphasized certain tools, such as the marginal revenue curve. Joan Robinson's work on imperfect competition was a response to certain problems of Marshallian partial equilibrium theory. Anglo-American Economists also responded to these problems by turning towards general equilibrium theory, developed on the European continent by Walras and Vilfredo Pareto.

However, Neoclassical economics is sometimes criticized for having a normative bias and in formulating many unfounded and unrealistic assumptions that do not represent real situations. Therefore, many critics believe that, this approach cannot be used to describe actual economies.

3. Keynesian Economics

A school of economic thought founded by the UK Economist John Maynard Keynes (1883-1946) and developed by his followers. It was his simple explanation for the cause of the Great Depression for which he is most well-known. Keynes' economic theory was based on an circular flow of money. In Keynes' theory, one person's spendings goes towards anothers earnings and when that person spends his earnings he is, in effect, supporting anothers earnings.

This circle continues on and helps support a normal functioning economy. When the Great Depression hit, people's natural reaction was to hoard their money. Under Keynes' theory this stopped the circular flow of money, keeping the economy at a standstill. In 1936, at the height of the great depression, Keynes' landmark book, *The General Theory of Employment, Interest and Money* caused a paradigm shift for Economics *i.e.*, it suddenly replaced their emphasis on study of the economic behaviour of individuals and companies (Microeconomics) to the study of the behaviour of the economy as a whole (Macroeconomics). The main plank of Keynesian theory is the assertion that, the aggregate demand created by households, businesses and the government and not the dynamics of free markets is the most important driving force in an economy. This theory further asserts that, free markets have no self-balancing mechanisms that lead to full employment. Keynesian Economists urge and justify a government's intervention in the economy through public policies that aim to achieve full employment and price stability. Their ideas have greatly influenced governments the world-over in accepting their responsibility to provide full or near-full employment through measures (such as deficit spending) that stimulate aggregate demand. Thus, Keynesian Economics advocates a mixed economy, predominantly private sector, but with a large role of Government and public sector and served as the economic model during the later part of the Great Depression, World War II, and the post-war economic expansion (1945–1973), though it lost some influence following the stagflation of the 1970s. Keynesian Economics argues that, private sector decisions sometimes lead to inefficient macroeconomic outcomes and therefore, advocates active policy responses by the public sector, including monetary policy actions by the central bank and fiscal policy actions by the government to stabilize output over the business cycle. The advent of the global financial crisis in 2007 has caused a resurgence in Keynesian thought. Keynesian Economics has the following successors:

(*a*) Post-Keynesian Economics

The term 'Post Keynesian' was first used to refer to a distinct school of economic thought by Eichner and Kregel (1975) and by the establishment of the Journal of Post Keynesian Economics in 1978. Prior to 1975 and occasionally in more recent work, Post Keynesian could simply mean Economics carried out after 1936, the date of Keynes's *'The General Theory of Employment, Interest and Money'*. Post-Keynesian Economics can be seen as an attempt to rebuild economic theory in the light of Keynes's ideas and insights. The theoretical foundation of Post Keynesian Economics is the principle of effective demand, that demand matters in the long as well as the short run, so that a competitive market economy has no natural or automatic tendency towards full employment. The positive contribution of Post Keynesian Economics has extended beyond the theory of aggregate employment to theories of income distribution, growth, trade and development in which demand plays a key role, whereas in Neoclassical Economics these are determined by general equilibrium forces. In the field of monetary theory, Post-Keynesian Economists were among the first to emphasize that the money supply responds to the demand for bank credit, so that, the central bank can choose either the quantity of money or the interest rate, but not both at the same time. This view has largely been incorporated into monetary policy, which now targets the interest rate as an instrument, rather than the quantity of money. Major Post Keynesian Economists of the first and second generation after Keynes include, Victoria Chick, Paul Davidson, Shiela Dow, Alfred Eichner, Geoff Harcourt, Nicholas Kaldor, Micha Kalecki, Steve Keen, Jan Kregel, Hyman Minsky, Bill Mitchell, Basil Moore, Luigi Pasinetti, Joan Robinson, George Shackle, Anthony Thirlwall, Eric Tymoigne, William Vickrey, Sidney Weintraub, L. Randall Wray etc.

(*b*) Neo-Keynesian Economics

Neo-Keynesian Economics is a school of macroeconomic thought that was developed in the 1930s and 1940s of post-war period from the writings of John Maynard Keynes. A group of Economists notably, John Hicks, Franco Modigliani and Paul Samuelson, attempted to interpret and formalize Keynes' writings in terms of formal mathematical models. In a process termed 'the neoclassical synthesis', they combined Keynesian analysis with neo-classical economics to produce Neo-Keynesian economics, which came to dominate mainstream macroeconomic thought in the 1950s, 60s and 70s.

However, in the 1970s, a series of developments occurred that shook Neo-Keynesian theory. The advent of stagflation and the work of monetarists like Milton Friedman, cast doubt on Neo-Keynesian theories. The result would be a series of new ideas to bring tools to Keynesian analysis that would be capable of explaining the economic events of the 1970s. The next great wave of Keynesian thinking began with the attempt to give Keynesian macroeconomic reasoning a microeconomic basis. The new Keynesians helped create a 'new neo-classical synthesis' that currently forms the mainstream of macroeconomic theory. Following the emergence of the new Keynesian school, neo-Keynesians have sometimes been referred to as Old-Keynesians.

(*c*) New Keynesian Economics

This modern Macroeconomic School of thought that evolved from classical Keynesian Economics. This revised theory differs from Classical Keynesian thinking in terms of how quickly prices and wages adjust. New Keynesian advocates maintain that prices and wages are 'sticky', meaning that, they adjust more slowly to short-term economic fluctuations. New Keynesian Economics as a philosophy took root in the 1980s in response to the criticisms of many of Keynes' original precepts as espoused by Classical Economists in the previous decade. Within this group, researchers tend to share with other Economists the emphasis on models employing micro foundations and optimizing behaviour but with a narrower focus on standard Keynesian themes such as price and wage rigidity. Its major developers include, Alan Blinder, J. Bradford DeLong, Jacques Drèze, Jason Furman, Raj Chetty, Carmen Reinhart, Michael Woodford, David Romer, Julio Rotemberg, Nouriel Roubini, Jordi Gali, Mark Gertler, Robert J. Gordon.

2

Basic Economic Terms and Concepts

Like in any other science, in Economics also, there are certain basic terms and concepts relating to various economic phenomena. They are used in a different sense relating to the subject of Economics and hence, the knowledge pertaining to them is essential for the readers. This will enable the readers to understand and analyze the economic principles in a scientific way. Following are some of the important terms and concepts, which are frequently used in this book and they have been explained in what sense they are used in Economics.

2.1. Goods

In Economics, good refers to any commodity or service for which there is a demand, irrespective of whether it is any sense 'good' or 'bad'. So, any thing that can satisfy a human want is called a 'good' in Economics. In narrow sense, 'good' includes a tangible physical product, unlike a service which is intangible. But, in Economics, 'good' includes both commodity and service because, both can satisfy human wants. For example, 'rice' is a good and similarly the service rendered by a labour in a sugar factory also falls under 'good' in Economics. All those economic actions which are done for money rewards are known as services. There are two types of services *viz.*, physical and mental services. The service, which is done physically by a person is known as Physical service such as, work rendered by agricultural labour. That action, which is done mentally for satisfying needs is called Mental service. For example, the services of Agricultural Economist, Plant Breeders, Agronomists etc. However, there exist slight differences between 'goods' and 'services'. Goods are tangible and transferable, while the services are intangible and non transferable. Goods are separable and non perishable, while services are perishable. Goods may be homogeneous or heterogeneous, while services are heterogeneous. The good may yield the satisfaction to the consumer either directly or indirectly. For example, consumption of food gives the satisfaction directly. On the other hand, machinery produces goods and these goods will satisfy human wants, thereby, machinery yields satisfaction indirectly.

2.1.1. Classification of Goods

Goods can be classified based on several criteria (Figure 2.1), discussed here under.

1. Based on Supply

The goods can be classified as Free goods and Economic goods.

(a) Free Goods

Free goods are the free gifts of nature. So, they exist in plenty and they can be used as much as we like. In Economics, we use the term 'free good', in the sense that, it is not scarce. The supply of these goods in the economy is

Goods

- Based on supply – Free goods and Economic goods
- Based on consumption – Consumption goods and Capital goods
- Based on locational existence – Personal goods and Impersonal goods
- Based on concreteness – Material goods and Non-material goods
- Based on transferability – Transferable goods and Non-transferable goods
- Based on tangibility – Tangible (material) goods and Intangible (non-material) goods
- Based on the ownership – Public goods and Private goods
- Based on rivalrous nature – Rivalrous goods and Non-rivalrous goods
- Based on excludability – Excludable goods and Non-excludable goods
- Based on urgency of fulfilling the wants – Necessaries, Comforts and Luxuries
- Based on income elasticity of demand – Normal, Inferior, Neutral and Superior goods
- Based on price elasticity of demand – Ordinary goods and Giffen goods
- Based on relationship between the goods – Complementary goods and Competitive goods
- Based on durability – Durable goods, Non-durable goods and Services
- Based on shopping habits – Convenience, Shopping, Specialty and Unsought goods
- Based on quality information – Search, Experience and Credence goods
- Based on welfare implications to society – Merit goods and Demerit goods

Figure 2.1: Classification of goods based on several criteria.

more than demand. So, these goods will have value-in-use only and no value-in-exchange, as no price is paid to avail the use of these goods. This is because, no efforts are needed in the economy to produce free goods. Their usefulness to mankind is more or less same across the individuals. As the free good is available in a great quantity, its opportunity cost is zero. It is important that, if the good is offered at zero price to the customers, it does not necessarily a free good. For example, a pesticide firm may offer free samples of pesticide bottles to the farmers for promoting the products among the farming community. So, such offering of pesticide on a free sample basis to the farmers would not be a free good in an economic sense. Examples of free goods include, air, sunshine, rainfall etc. There is another form of free goods, which are produced as by-products. For example, while processing sugar cane in a sugar factory, molasses is obtained as a by-product and it serves as a free good obtained for use in wine industry (distillery).

(b) Economic Goods

These are the goods produced in the economy through man made efforts. In general, their supply is limited and usually less than demand and hence, these goods will have both value-in-use and value-in-exchange. So, price must be paid to avail the services of these goods. Their usefulness to the mankind differs from person to person, as these good are priced in the market in accordance with their quality, features, attractiveness etc. In Economics, we are concerned with economic goods only and they are often considered as 'wealth'. There would have been no science of Economics, if all goods had been free goods or in the absence of economic goods. Examples include, fertilizers, pesticides, tractor etc.

However, the distinction between free goods and economic goods is not permanent. For example, water in nature is a free good, but the farmers have to pay water cess to the Government for utilizing canal water for irrigating the crops. Similarly, households have to pay tax to the Government for using municipal water for drinking purposes. In this connection, we have to remember that, what is a free good in one place can become an economic good in another place. It all depends on the supply of a good and the demand for it. For example, in some villages firewood is a free good. But in a city and town, where we have to pay a price for it, thereby, it becomes an economic good.

2. Based on Consumption

On the basis of this criterion, goods are classified into Consumption goods and Capital goods.

(a) Consumption Goods

These are the goods which yield satisfaction directly to the individual or consumer. So, these goods help the individuals to satisfy their wants directly. Hence, these goods are also called as Goods of First order. These goods are cheaper compared to capital goods. These goods can further sub-classified into two major categories according to the degree of durability *i.e.,* Perishable goods (fruits, milk, vegetables etc) and Durable goods (pen, radio, house, car etc).

(b) Capital Goods

These goods are used to produce consumption goods. So, these goods yield satisfaction to the individuals indirectly. Hence, these goods are also called as Producer goods or Goods of Second Order. These goods generally involve more investment compared to consumption goods. The strengthening of these goods in the economy implies capital formation. For example, sugar is a consumer good. But, there must be some processing machine to process sugar cane to sugar and that machine is known as capital good. Other examples for capital goods include tools, equipments etc.

But what is a consumers' good in one place can become a producers' good in another place. For example, when electricity is used for lighting purposes at home, it is a consumers' good. But the same electricity when used for farming (say, for irrigation purpose) it becomes a producers' good.

Besides these two goods, some goods are Intermediate goods. For example, the raw material used in sugar factory *i.e.*, sugar cane is an intermediate good. This is because, it will not yield satisfaction to the consumers directly. It has to be used as a raw material in sugar factory to produce sugar, which then yields satisfaction to the consumer. So, it serves an intermediate good *i.e.*, as a raw material only and capital goods process it to produce final consumption good *i.e.*, sugar. To avoid confusion, in general, capital goods and intermediate goods are termed as Producers' goods, whereas, consumption goods are referred as Consumers' goods.

3. Based on Locational Existence

Based on this criterion, goods are classified into Personal goods and Impersonal goods.

(a) Personal Goods

These goods are internal to the individual and hence, they are related to personal qualities of an individual. So, these goods are also called as Internal goods. These goods are non-materialistic and non-transferable in nature. In short, personal goods indicate 'what he is'. Examples include, skill, intelligence etc.

(b) Impersonal Goods

These goods are external to the individual. So, these goods are materialistic and transferable in nature. These goods are also called as External goods. In short, impersonal goods indicate 'what he has'. Examples include, land, tractor etc.

4. Based on Concreteness

This criterion classifies the goods into Material goods and Non-material goods.

(a) Material Goods

These goods are concrete in nature. They are generally produced through capital goods. They are comprised all useful material things. Examples include, products of agriculture, tractor, machinery, table, building, furniture, books etc.

(b) Non-material Goods

These goods are non-materialistic in nature. In general, non-material goods fall under two classes *viz.*, internal and external. Internal non-material goods lie within the individual such as personal skills, ability, dynamism etc. External non-material goods includes all kinds of services such as agricultural labour service, bullock labour service, services rendered by a worker in a food processing firm etc.

5. Based on Transferability

This criterion classifies the goods into Transferable goods and Non-transferable goods.

(a) Transferable Goods

These goods are mostly materialistic in nature, so that they can be easily transferred from one person to another. Here, transferability implies change in ownership. Examples include, land, tractor, machinery etc. However, some materialistic goods like conduct certificate, fitness certificate etc., are not transferable.

(b) Non-transferable Goods

These goods are generally non-materialistic in nature and hence, they cannot be transferred from one person to another person. Example include, skill, intelligence, attitude etc. However, some non-material goods are also transferable like goodwill of business.

The detailed classification of goods based on locational existence, concreteness and transferability is shown below along with examples in the Table 2.1.

6. Based on Tangibility

This criterion classifies the goods into Tangible goods and Intangible goods.

Table 2.1: Classification of goods based on locational existence, concreteness and transferability.

Category Based on 'Locational Existence'	*Category Based on 'Concreteness'*	*Category Based on 'Transferability'*	*Example*
External	Material	Transferable	Table, chair, fertilizer bag etc
External	Non-material	Transferable	Goodwill of a business, Patent-rights, Copyrights
External	Material	Non-transferable	Conduct certificate, Degree certificate, Driving license etc.
External	Non-material	Non-transferable	Friendship, Courage, etc.
Internal	Non-material	Non-transferable	Skill, Ability, Dynamism etc

(a) Tangible Goods

These are the goods which can be measured or quantified. A tangible good is something which a firm produces that the customer can see or hold, as against something like advice. Examples include, table, chair, fertilizer bag etc.

(b) Intangible Goods

An intangible good is a good, which is incapable of being touched. In economic sense, intangible goods include 'services'. The major distinguishing features between a tangible good and an intangible good is that, an intangible good is not quantifiable, can be instantaneously delivered to the buyer and it can be easily replicated with low cost. Examples include, agricultural labour service, doctor's service etc.

7. Based on the Ownership

This criterion classifies the goods as Public goods and Private goods.

(a) Public Goods

These goods are owned by the State or Government. That means, these goods are owned and used on collective basis. In general, these goods are meant for public-welfare motive. In general, a public good is a good that is non-rivalrous and non-excludable. Non-rivalry, in the sense that, the consumption of public good by one individual does not reduce its availability to others in the economy. Non-excludability means no individual in the economy can be effectively excluded from using the good at the same time. Examples include, railways, roads, dams etc. There are two sub-types of public goods *viz.*, Pure public goods and Impure public goods.

- ☆ *Pure public goods:* These goods ensure equi-distribution of benefits among the public. *e.g.* National defense, public parks etc.
- ☆ *Impure public goods:* These goods are offered at particular locations and hence, do not ensure equi-distribution of benefits among the public. For example, establishment of an agro-based industry by the Government in a particular area helps the farmers of that locality comparatively more than the farmers of distant areas.

(b) Private Goods

These are the goods which are possessed and owned by the individuals. The individuals have complete economic freedom in using these goods in the economy. These goods are mainly produced with a profit-motive. In general, a private good is a good that is rivalrous and excludable. Examples include, tractor, fertilizers, pesticides etc.

8. Based on Rivalrous Nature

According to this criterion, there are two types of goods *viz.*, Rivalrous goods and Non-rivalrous goods.

(a) Rivalrous Goods

These are the goods, whose consumption by one consumer prevents simultaneous consumption by other consumers. These rivalrous goods include both durable and non-durable goods. For example, tractor is a durable rival good. This is because, if one farmer uses the tractor, it prevents other farmers in using the same at the same time. However, the first farmer does not 'use up' the tractor in his usage. This indicates that, durable rival goods can be shared through time. However, an apple is a non-durable rival good. This is because, once if an apple is eaten, it is 'used up' and can no longer be shared by others. Non-tangible goods can also be rivalrous. For example, a labour service engaged by the farm manager is not accessible to other farm managers in that area at the same time. It is important that, almost all private goods are rivalrous.

(b) Non-rivalrous Goods

In contrast, non-rival goods may be consumed by one consumer without preventing simultaneous consumption by others. Most examples of non-rival goods are intangible. For example, when a farmer is gathering useful information from the internet, it will not prevent other farmers in accessing the internet services. Similarly, radio broadcasting, telecasting, telecommunications, National defense, clean air, intellectual property rights etc., are other classical examples for non-rivalrous goods. It is important that, the good may be rivalrous, but the services may be non-rivalrous. For example, the radio itself is a rival good, but radio broadcasts are non-rival goods. These examples clearly indicate that, non-rivalrous goods can be enjoyed simultaneously by an unlimited number of consumers. So, goods that are both non-rival and non-excludable are called Public goods.

9. Based on Excludability

This criterion classifies the goods into Excludable goods and Non-excludable goods.

(a) Excludable Goods

A good is said to be excludable, when it is possible to prevent people from having access to it, who have not paid for it at the same time. *e.g.* Watching a movie in a theatre.

(b) Non-excludable Goods

A good is said to be non-excludable, when it is not possible to prevent other people from having access to it at the same time. *e.g.* Downloading free e-books from the internet.

However, based on rivalry and excludability concepts, there are some other goods *viz.*, Common goods and Club goods. The goods, which are rivalrous but non-excludable are called Common goods or Common pool resources. On the other hand, the goods, which are non-rivalrous but excludable are called Club goods. For easy understanding, the same concept was presented in the Table 2.2.

Tablɘ 2.2: Classification of goods based on rivalrous nature and excludability.

Item	*Excludable*	*Non-excludable*
Rivalrous	**Private goods** *e.g.* land, fertilizers, pesticides	**Common goods** *e.g.* timber, coal
Non-rivalrous	**Club goods.** *e.g.* private parks	**Public goods.** *e.g.* National defence

10. Based on the Urgency of Fulfilling the wants

Based on this criterion, the goods are classified into Necessaries, Comforts and Luxuries.

(a) Necessaries

These goods satisfy most urgent wants. This is because, without these goods, life of human beings is totally unimaginable. These goods are less expensive compared to comforts and luxuries. For these goods, the demand remains almost constant irrespective of the level of income of the individual. So, they are income inelastic. Examples include, food, clothing, water etc.

(b) Comforts

These goods satisfy less urgent wants compared to necessaries. These goods are more expensive compared to necessaries, but less expensive compared to luxuries. For these goods, the demand raises quickly compared to necessaries, but rather slowly compared to luxuries with increase in the level of income of the individual. So, they are more income elastic compared to necessaries, but less income elastic compared to luxuries. For example, a cushioned chair is a comfort to increase the efficiency of a student.

(c) Luxuries

These goods are meant for superfluous consumption. After getting necessaries and comforts, human beings desire for luxuries, as they aim to lead fuller life. So, luxurious goods satisfy least urgent wants. These goods are more expensive compared to necessaries and comforts. For these goods, the demand raises quickly compared to both necessaries and comforts with increase in income of the individual. So, these goods are more income elastic compared to necessaries and comforts. Examples include, gold, silver, costly furniture etc. There is a thin line difference between comforts and luxuries. For a student, a cushioned chair add comfort to increase his efficiency, but an air-conditioned room constitute luxury.

11. Based on the Income Elasticity of Demand (E_Y)

This criterion classifies the goods into Normal goods, Inferior goods Neutral goods and Superior goods.

(a) Normal Goods

A normal good is anything that the consumer is willing and able to buy more when his income increases. That means, for these goods, the demand varies directly with the income of the consumer. That means, if the income of the consumer increases, more quantity of (normal) good is demanded (price remains constant) and *vice versa*. So, normal goods have a positive E_Y. The term 'normal' does not necessarily refer to the quality of the good. Examples include, new clothing, new car, new computer, new brand of pesticide, new brand of fertilizer, latest models of television, DVD player, mobile phones etc.

(b) Inferior Goods

These are the goods, whose demand varies inversely with the income of the consumer. That means, if the income of the consumer increases, less quantity of (inferior) good is demanded and *vice versa*. Normally, it would be expected that, a rise in income would lead to a rise in the quantity demanded of a good, but an inferior good is an exception. An inferior good is a cheap good, but an inferior substitute to another good. So, when income of the consumer increases, consumers can afford to buy the superior product and so demand for the inferior good decreases. So, for an inferior good, there will be negative E_Y. Inferiority, in the sense that, it is related to affordability rather than quality of the goods. In general, goods and services used by poor people for which richer people have alternatives exemplify inferior goods. Examples include, low quality food like coarse cereals, used clothing, used car, used computer, used books, instant noodles, old brand of pesticide, old brand of fertilizer, a second hand/used television etc. Another classical example is inter-city bus service. This form of transportation is cheaper than air or rail travel, but it is more time-consuming. When the consumer possesses less money, he prefers traveling by bus service. But, when money is more abundant, he prefers rapid transport *i.e.*, air or rail.

The concept of 'inferiority' of goods is inconsistent across geographic regions. For example, potato is an inferior goods in the Andean region (Peru), where the crop is originated. But, in several countries like China, India and Bangladesh, potato is not an inferior goods because of its rich source of calories and considered it as a prestige food, especially when eaten in the form of 'French fries' by urban elites.

(c) Neutral Goods

These are the goods, whose demand is independent of income. The demand for basic necessaries is independent of income. For example, quantity of food consumed per day remains more or less same irrespective of the level of income.

(d) Superior Goods

These goods are a special type of normal goods. The income elasticity of a superior goods is above one. That means, as income increases, the expenditure on superior goods will increase more than proportionately. For a good to be considered as 'superior good', it must possess two important characteristics *viz.*, scarcity and high price. The public in the economy should consider that, this good is distinguishably better compared to other related goods and the possession such good will add some prestige to them in the society. If the price of the superior goods declines, consumers will purchase less of the goods. A superior goods may be a luxury good, as it is not purchased below a certain level of income. Examples include, vaccum cleaners, diamonds etc. In general, the 'superior goods' are considered as the antonym of 'inferior goods'. These superior goods are often confused with normal goods. If the consumption expenditure on a goods increase with increase in income in absolute terms, but not in terms of the percentage share of budget allocated on it or less than the percentage increase in income, then the goods is said to be a normal good. For example, if the income of the individual is increased from Rs. 5000 to Rs. 7500 *i.e.*, by 50 per cent, and if the consumption expenditure on the good increases from Rs 1000 to Rs.1500, *i.e.*, by 50 per cent, then the good is said to be a normal goods. Sometimes, even the percentage increase in consumption expenditure on normal goods is less than 50 per cent. Conversely, for a superior good, the percentage increase in consumption expenditure is more than the percentage increase in income. Say, in the earlier example, for a superior goods, the percentage increase in consumption expenditure will be more than 50 per cent.

12. Based on Price Elasticity of Demand (E_P)

This criterion helps to distinguish Ordinary goods and Giffen goods.

(a) Ordinary Goods

An ordinary good is a good, which is price elastic, unlike Giffen good. A normal good is always ordinary, while an ordinary good can be either normal or inferior. *e.g.* Television, cars, fertilizers, pesticides etc.

(b) Giffen Goods

For most products, E_P is negative. That means, price of the commodity and quantity demanded move in opposite directions. So, if price goes up, then quantity demanded goes down or *vice versa*. Giffen goods are an exception to this. Their E_P is positive and therefore, demand curve slopes upward. That means, when price goes up, the quantity demanded also goes up and *vice versa*. In order to be a true Giffen goods, price must be the only thing that changes to get a change in quantity demanded and conspicuous consumption does not enter the picture (such a situation would indicate a Veblen good). So, a Giffen goods is an extreme type of inferior good, which disobey the 'law of demand'. This concept implies that, potential consumers of the goods believe that, if the price is higher, the quality of those goods are higher and *vice versa*. To be a Giffen goods, the item must lack easy subsitutes and it must be an inferior good *i.e.*, a goods for which demand declines as the level of income of the consumer increases. Giffen goods are named after its conceptualizer, Sir Robert Giffen. Ex. bread, potatoes, households in the Hunan province of China were shown to buy more rice at high price and less when the price was subsidised. This is because, even at high price, rice was still considered as the cheapest source of calories available. So, when the price of rice was subsidized, the households had more money left over after buying rice. This left over money was spent on buying more expensive foods such as meat, vegetables and fruits, which reduced their need for rice.

It is important that, Normal goods and Inferior goods are distinguished mainly based on E_Y, whereas, Ordinary goods and Giffen goods are distinguished based on E_P. To summarize, the differences are shown through Table 2.3.

Table 2.3: Different types of goods based on E_P and E_Y.

Income Change			*Price Change*		
Item	*Normal Goods*	*Inferior Goods*	*Item*	*Ordinary Goods*	*Giffen Goods*
Income up	Consumption up	Consumption down	Price up	Consumption down	Consumption up
Income down	Consumption down	Consumption up	Price down	Consumption up	Consumption down

13. Based on Relationship between the Goods

Based on this criterion, the goods are classified into Complementary goods, Competitive goods and Independent goods.

(a) Complementary Goods

Complementary goods are the goods, which have negative cross elasticity of demand (E_C). For example, pen and ink are complementary goods. If the price of ink rises in the market, the demand for pen falls. So, if price of one good is increased, the demand for other goods will decrease in the economy. Conversely, if the price of one goods is decreased, demand of other good will increase. Other examples include, tea and sugar, shoes and socks, printers and ink cartridges, DVD players and DVDs etc.

(b) Competitive Goods

These are the goods which substitute each other and have positive E_C. These goods are also called as Substitute goods. For example, coffee and tea are good substitutes. If the price of coffee in the market increases, the demand for tea rises. So, if price of one good is increased, the demand for other goods will increase in the market. Conversely, if the price of one good is decreased, demand for other goods will decrease. Other examples include rice and wheat, human labour and machinery usage, organic manures and chemical fertilizers etc.

(c) Independent Goods

Two goods are said to be independent, if they have a zero E_C. That means, if the price of one goods is increased or decreased, there is no change in quantity demanded of other goods. So, changes in the price of one goods will have no affect on the demand of other good. For example, if price of wheat is increased, it will not affect the demand for mangoes in the market.

14. Based on Durability

This criterion classifies the goods into Durable goods, Non-durable goods and Services.

(a) Durable Goods

These goods are tangible and survive for many number of years. It refers the producers goods such as machinery tools, implements etc.

(b) Non-durable Goods

These goods are also tangible like durable goods, but survive for less period of time compared to durable goods. They get consumed in normal consumption process. They are purchased frequently and consumed quickly. Examples include, salt, soap, chocolates, fertilizers, pesticides etc. such goods being known as non-durable consumer goods to distinguish them from durable consumer goods like furniture, cars etc.

(c) Services

These goods are intangible, perishable and inseparable. They need more quality control, credibility and more adaptability. Examples include, repairs of machinery, skilled labour works etc.

15. Based on Shopping Habits

According to this criterion, goods are classified into Convenience goods, Shopping goods, Specialized goods and Unsought goods.

(a) Convenience Goods

These goods are purchased very frequently, immediately with a minimum of effort. These goods are again sub-classified into three types *viz.*, Staples, Impulse goods and Emergency goods.

- ☆ ***Staples:*** These are the convenience goods, which are purchased on regular basis. Examples include, news paper, cigarettes etc.
- ☆ ***Impulse goods:*** These are the convenience goods purchased without planning and effort. Until the customer locate or spot the goods, he won't think of purchasing them. *e.g.* purchasing magazines while travelling etc.
- ☆ ***Emergency goods:*** These are the goods purchased at times of emergency only. Examples include, purchase of umbrella during rainy season, swetter during winter season etc.

(b) Shopping Goods

These are the goods purchased by the customers based on differences in quality, features, attractiveness, price differences, tastes and habits. They are characterized by pre-planning, information search and price comparisons. Examples include, clothes, furniture, fridge etc. These goods are again sub-classified into two types *viz.*, Homogenous shopping goods and Heterogenous shopping goods.

- ☆ ***Homogenous shopping goods:*** These goods are more or less similar in quality, but price differences enable the customers to purchase the goods. So, for these goods, price difference is more prominent than quality difference. *e.g.* Cotton clothes
- ☆ ***Heterogenous shopping goods:*** These goods differ greatly in quality or features. So, quality difference is more prominent than price difference. These goods require more advertisement and propaganda for promoting sales. *e.g.* Fridge, Televisions etc.

(c) Specialty Goods

These goods have special or unique features and brand identification. So, the customers move to these shops, where they are located to purchase the goods. Examples include, computers, DVD players, cars etc.

(d) Unsought Goods

These goods are not normally known to the customers and generally they do not think of purchasing them. *e.g.* Vaccum cleaners, Smoke detectors, Encyclopedias etc.

16. Based on Quality Information

This criterion classifies the goods into Search goods, Experience goods and Credence goods.

(a) Search Goods

These are the goods, where the buyer can pre-determine its features and characteristics like quality and price before purchasing it from the seller. So, these goods are subjected to substitution and price competition. For example, pre-shipment physical examination of apples, mangoes, grapes etc., by the buyer to ascertain the quality of the good.

(b) Experience Goods

Unlike search goods, the experience goods cannot be evaluated in terms of its features and characteristics like quality and price in advance. But, the product features and characteristics can be ascertained upon its consumption. This concept of experience good is originally due to Philip Nelson and he compared an experience good with a search good. Branding and detailed product specifications make the experienced goods to transform into a search good. Due to lack of knowledge about the product features and characteristics in advance, the consumers generally experiences difficulties in making consumption choices accurately. Due to this, these experience goods are less elastic compared to search goods, as consumers generally fear that, lower prices may be due to poor quality or any other unobservable problems. Examples include, jams, jellies, butter etc.

(c) Credence Goods

These goods are also called as Post-experience goods. For these goods, it is very difficult on the part of the consumers to ascertain the quality even after they have consumed them. However, the sellers may or may not know about the quality of the product. That means, for consumers, it is difficult to ascertain about the utility or harmful affects of these goods even after consuming them, unlike experience goods. For example, consumption of groundnut may lead to afflatoxin infestation, which is not known to the consumer even after consuming the good. But, the harmful affects of afflatoxin on the health of the individual will be realized in a long run and not in the short run. Other examples include, vitamin supplements, car repairs, pesticide application on the crop, medical treatment, bio-fertilizers application on the crops etc. The demand for credence goods show direct relationship with price, as price is the only possible indicator of quality. The least expensive products might be avoided in order to avoid suspected fraud and poor quality.

17. Based on Welfare Implications to the Society

This criterion classifies the goods into Merit goods and Demerit goods.

(a) Merit Goods

The concept of a merit goods was introduced by Richard Musgrave. A merit good is a good, whose consumption is considered healthy and socially desirable due to the exertion of positive effects on the consumers and the society. Examples include, education, medical services, crop insurance etc.

(b) Demerit Goods

A demerit good is a good, whose consumption is considered unhealthy and socially undesirable due to the perceived negative effects on the consumers themselves and the society. Such goods are over-consumed, if left to market forces. Hence, Government often levy taxes on these goods to regulate the consumption or banning the use of these goods. Examples include, consumption of tobacco, alcoholic beverages, drugs, etc.

Besides different types of goods discussed above based on several criteria, there are other important types of goods, which are relevant to understand the concepts explained in the ensuing pages. They include:

1. Composite Goods

A composite good is a good, which represents all goods in common, besides the one in question. This composite good concept is used in Indifference Curve (IDC) analysis, where the goods under consideration is taken on one axis and all other goods (together considered as composite good) are taken on another axis represented by consumers' income. Thus, this concept of composite goods helps to analyze what is given up along consumer's budget constraint to consume more of the first good. This concept is introduced because, if single good is taken into consideration for analyzing the price effects, the cost of a good cannot be related to any other opportunities. So, opportunity cost cannot be calculated. Similarly, if two goods are taken into consideration, it allows to compute opportunity cost, but with reference to the second good only. Therefore, this concept ignores all other choices of budget allocation. But, if we go on adding more number of goods, it makes the analysis more confusing. So, to avoid this problem, the concept of composite good is introduced, as this goods represents all other goods as a single good except the goods under consideration for analysis. So, the analysis is between two goods *viz.*, goods under consideration and composite good representing all other goods. For example, if we want to analyze the price effect on rice, all other goods except rice constitutes composite goods. The important aspect is regarding representation of unit of composite good. As different goods have different units of measurement, to avoid confusion, the income of the individual is represented on one axis (to represent composite goods) and the goods under consideration is taken on another axis.

2. Club Goods

These goods are sub-types of public goods and commonly considered as artificially-scarce goods. These goods are excludable but non-rivalrous. So, they are often provided by a natural monopoly. Examples include, cable television, access to copyrighted works, services provided by social or religious clubs to their members.

3. Veblen Goods

These goods like Giffen goods disobey the law of demand. The purchases of these goods increase with increase in prices of the goods. This is because, purchasing a commodity at a higher price confers rich status to the individual. If the price of such commodity decreases in the market, people's preference for buying these goods also decreases, as they are no longer perceived as exclusive or high status products This is called Veblen effect, named after the Economist, Thorstein Veblen, who first pointed out the concepts of conspicuous consumption and status-seeking. Examples include, luxury car (Audi), diamonds, selling Sir Donald Bradman's Cricket kit at huge price etc.

2.2. Utility

As explained above, there are different types of goods and these goods will satisfy human wants upon their consumption. This wants satisfying power or capacity of a good is called 'utility' in Economics. So, we can say, 'utility' is the inherent quality in a good by virtue of which it is capable of satisfying a human want. Stanley Jevons was the first Economist, who conceived the concept of 'utility'. It is a mental phenomena because, physical or physiological characteristics differ from person to person. Utility is an abstract concept rather than a concrete and observable quantity. The units to which we assign an 'amount' of utility, therefore, are arbitrary, representing a relative value. Human wants can be satisfied through air, water, sunshine, products of agriculture, tools, equipments, machinery etc. So, all types of goods *i.e.*, both free goods and economic goods will satisfy human wants and hence, we can say, they all possess utility. However, the term 'utility' has a different meaning in Economics when compared to its usage in everyday life. This is exemplified below.

- ☆ Utility and Usefulness are not synonyms. So, utility does not mean usefulness. For example, manufacturing cigarettes, alcoholic drinks etc., involve economic activities and these goods, if consumed, will satisfy human wants. This implies they possess utility. But, these goods are harmful to human health and not useful to mankind in any sense. So, a good may possess utility, but may not be useful. Thus, utility and usefulness are not the same. This implies, the concept of utility in Economics has no ethical significance. This is because, even if a good is harmful to an individual, but if it satisfies his want, it is said to possess utility.

- Utility is not synonymous with pleasure. That means, a good may possess utility, but it may not yield pleasure to the person when consumed. For example, a medicine prescribed for a patient is useful to him for curing his illness. But, its consumption will not yield pleasure, as it is distasteful. Thus, pleasure is different and utility is different.
- Utility is subjective or psychological concept. That means, the same commodity or service gives different utilities to different people. So, the concept of utility is independent of the consumer and it is the consumer's mind that gives utility to the good. For example, a literate farmer finds utility in going through the literature on agricultural aspects, as he is able to read and understand the information. On the contrary, an illiterate farmer never find any utility, even if he posses the literature. Similarly, for a vegetarian, mutton has no utility. Thus, utility depends on man's mind rather than on the things itself.
- Utility does not mean satisfaction. The former stands for 'expected satisfaction' and the latter stands for 'realized or derived satisfaction'. This indicates, utility is the inherent quality of the good and this quality will yield satisfaction to the individual. That means, 'utility' is the quality in the good and 'satisfaction' is what we derive after consuming the good. When a consumer wants to buy a commodity, he thinks about the utility of the commodity or how much of satisfaction the commodity is capable of giving. Only after purchasing and consuming the commodity, he realizes the 'satisfaction'. In practice, the expected utility may not be as realized. So, there be a gap between 'utility' and 'satisfaction'. When 'expected satisfaction' (*i.e.*, utility) is not realized by the consumer after consumption, it guides the consumer not to purchase that commodity in the future. Another important distinction between the two is that, 'utility' can be measured indirectly by means of the price paid for the commodity. But, satisfaction cannot be measured either by direct or indirect methods. In spite of these differences, these two terms are used interchangeably in Economics and this assumption has been made for purpose of convenience. For example, saying that, 'rice gives utility' is incorrect. To put it in a right way, 'rice possesses utility and by consuming it we derive satisfaction'.
- Utility varies across the situations. That means, the same goods will have different utilities, if used for different purposes. For example, water resource has different utilities when it is used for irrigation, drinking, bathing and washing purposes.
- Utility varies at different times. For example, if paddy seed is made available in the market just before the commencement of crop season, it will have utility, as the farmers purchase them for cultivation. On the hand, if the seed is made available to the farmers after the sowing period, it will not give utility to the farmers.

Types of Utility

Economic goods yield different types of utility to the individuals and they are studied here under.

- ***Form utility:*** The change in form of a commodity will bring change in utility to the consumer. For example, consumption of raw mango will give utility to the consumer. But, if the mango is processed into juice or jelly, its utility will change to the consumer. Similarly, paddy produced in its original form is not useful to the consumer directly. But, if it is processed into rice, it will yield utility to the consumer. Forest wood when transformed into furniture, utility will increase. The above examples imply that, 'processing function' adds form utility to the commodities.
- ***Place utility:*** The change in place of availability of the good will add utility to the goods. For example, farmers cultivate vegetables, fruits, food grains etc., in the villages and if they are made available in towns and cities, it adds utility to these commodities. So, 'transportation function' will add place utility to the commodities.
- ***Time utility:*** The change in time of availability of the good will add utility to the good. For example, farmers cultivate different crops like vegetables, fruits, food grains etc., in the villages and if they are made available to the consumers at the same time, due to increase in supply in the market, they fetch low prices. So, if the farmers stock the commodities in godowns and made available to the consumers at times of scarcity, more utility can be realized. So, 'storage function' adds time utility to the commodities.
- ***Possession utility:*** We know, utility refers to want satisfying power of a commodity. So, anything, which satisfy human want is said to possess utility. For getting access to the commodities, they have to be exchanged in the market economy between sellers and buyers. So, the 'buying and selling function' leads to transfer of ownership of commodity from seller to buyer and this adds possession utility to the commodity. This increased usefulness created by marketing enable the consumer to own, use and consume the commodity. So, it is also called as Ownership utility. Since the possession utility is due to exchange function, only economic goods will have possession utility. For example, paddy in the hands of farmers has less utility. But, once it reaches the consumers through exchange mechanism, it will add utility to the commodity and this increase in utility refers to possession utility.

2.3. Value

In ordinary speech, we use the term 'value' denoting the meritorious features of a particular aspect under consideration. For example, we generally say, professor's guidance has a great influence on the education career of the students, parents advises has a great value, fresh air is always valuable etc. However, in Economics, the term 'value' is used very frequently, but in a different sense. It is not used in that sense, as we use it in ordinary speech. In general, value of a goods is expressed under two different concepts *viz.*, value- in-use and value-in-exchange.

- ***Value-in-use:*** The goods which are freely available to man-kind without any role in exchange mechanism will have only value-in-use. For example, all free goods like air, rain, sunshine etc., have value-in-use. But, they don't have value-in-exchange because, they are not involved in economic exchanges.
- ***Value-in-exchange:*** In Economics, 'value' mainly refers to value-in-exchange. That means, goods and services must have value in the exchange mechanism. In Economics, we are interested only in those goods which have value-in-exchange. Since only economic goods possess value-in-exchange, they will be studied in the 'value' concept. So, for a good to have value or value-in-exchange, it must possess the following three important attributes:
    - It must possesses utility
    - It must be scarce, and
    - It must be transferable and marketable.

It is important that, for a good, to say, it is having 'value', it should possess all the three characteristics stated above. In the absence of any one of the three characteristics, the goods is considered invaluable in the exchange market. For example, infested seeds, fruits contaminated with pesticidal residues etc., may be scarce in the market. But, they do not possess utility and not marketable. So, these commodities do not have 'value' in the exchange market. Similarly, fresh air has utility, but it is not scarce and non-marketable. So, fresh air has no 'value' in the exchange mechanism. On the other hand, a fertilizer bag has 'value' in the exchange market, as it possess all the above three attributes. From the above discussion, it is quite easy now, to define the concept of 'value', as under.

- Value of a commodity refers to the commodities or services that can be obtained in exchange or return for it. So, the term 'value' refers to the exchange qualities of a good. According to Marshall, *'the term value is relative and expresses the relation between two things at a particular place and time'*. Most Economists use the word 'value' in the sense of purchasing power or exchange power. It is generally measured in terms of money and the value of a thing changes according to time and situation. For example, the value of an agricultural commodity is more during lean marketing season than peak marketing season. Similarly, ice has more value in summer than in winter.
- It is a purchasing power of a commodity in terms of other commodities and services.

According to Marshall, the concept of 'value' is purely relative in its approach and not absolute. That means, we use the term 'value' in terms of value in exchange only. So, to say a good is said to have 'value', it must have the capacity to earn something in exchange for itself in the market. In the earlier days of the barter system of economy, where the goods are exchanged for other goods, value of a good is measured in terms of other goods that it earns in exchange mechanism. Say, for example, if paddy is said to have value in the market, it implies that, by exchanging paddy the farmer gets something (say, jowar) from other farmers. If the farmer exchanges one bag of paddy for two bags of jowar, it indicates the value of a bag of paddy is equal to two bags of jowar. So, value of a commodity is expressed in terms of relationship with other commodity or commodities in an exchange mechanism. Say, in the above example, if one paddy bag fetches three bags of jowar, it implies, value of paddy in the market is increased or value of jowar in the market is decreased. That means, in valuation mechanism, if the value of one good is increased, the value of other good must necessarily be decreased and *vice versa*. So, we can conclude that, there cannot be a general rice or fall in value of goods in the market. Similarly, in the monetary market, if a quintal of paddy fetches Rs. 1000/- in one situation and Rs. 800/- in another situation, it indicates, the value of paddy is decreased in the market in terms of money income or the purchase value of a 'Rupee' in the market is increased. This expression of value of a good in terms of money is called Price, and it is summarized here under.

2.4. Price

As explained earlier, when we say, a commodity is having value in Economics, it implies value-in-exchange. In barter system of economy, the terms value and price are used synonymously. But later on, goods are exchanged for money and not for other goods. This exchange of goods between a seller and buyer will takes place at an equilibrium price. So, the concept 'price' measures on article in terms of other particles, price measures all goods and services in terms of money. This indicates that, price expresses or measures the value of a commodity in terms of money. So, when value of a commodity is expressed in terms of money, it is called Price. Prices in the open economy are determined by the interply of two forces of demand and supply. Like commodities, the payments we make like rent, wages and interest for the use of land, labour and capital respectively are also prices. It is the price system that regulates the economic activity of a society. But, unlike value, there can be general rise or fall of prices in the economy.

2.5.Wealth

In ordinary speech, when we use the word 'wealth', it refers to money, well-being, prosperity and abundance in its widest sense. But, in Economics, the word, 'wealth' has altogether different meaning and it refers to those goods, which have value-in-exchange in the economy. So, the concept of wealth is related to economic goods only, which are scarce in the economy. Wealth has been defined as 'stock of goods existing at a given time that have money value'. In Economics, wealth is most commonly defined as 'consisting of all useful and agreeable things which possess exchange value and this again is generally regarded as co-extensive with all desirable things except those which do not involve labour or sacrifice for their acquisition in the quantity desired'. In the words of J.M. Keynes, *'wealth consists of all potentially exchangeable means of satisfying human needs'*. In short, wealth means anything which has value. So, for a good to be considered as wealth it must include the following attributes:

- ☆ It must possesses utility. That means, the good must have the power to satisfy a want. As Marshall says *'they must be desirable'*.
- ☆ It must be scarce or limited in supply. In this context, free goods like air, sunshine, rain water etc., do not constitute wealth because, they are not scarce.
- ☆ It must be transferable. That means, it should be possible for us to transfer the ownership of good from one person to another.
- ☆ It must be external to man.

Among the above four attributes, first three are the attributes of 'value'. So, for a good to be considered as wealth, first it must have 'value' in the market economy. If the answers to the above four attributes are positive, then a good is considered as wealth. We can apply the above four attributes to the following categories, so as to assess whether they represent wealth or not.

- ☆ Material and Non-material goods like tractor, money, land, livestock, building, furniture, good will of business, human labour service, bullock labour service etc., they are all considered as wealth because, they obey all the four attributes of wealth.
- ☆ LIC bonds, Fixed deposit bonds, Bills of exchange, Documents of property etc., are all considered as wealth and they are popularly regarded as 'Representative wealth'.
- ☆ Air, sunshine, rainfall, are not considered as wealth because, they are not scarce and not marketable.
- ☆ Personal qualities like skill, ability, intelligence, dynamism etc., are not considered as wealth, as they are not transferable. Rather, they are the sources of wealth and hence, they enjoy the courtesy title, 'Personal wealth'.
- ☆ Oceans, the sun, the moon, deserts etc., are not considered as wealth because, they cannot be owned or bought and sold..
- ☆ Infested agricultural produce is not wealth because, it cannot be marketable.
- ☆ A stock of outdated equipment, machinery, tools etc., cannot constitute wealth because, they are not marketable.

Types of Wealth

Wealth can be classified as follows:

- ☆ ***Individual wealth:*** This wealth is conveniently categorized into Tangible wealth and Intangible wealth. Tangible wealth includes the properties such as land, tractors, machinery, buildings etc., which can be measured or quantified. Intangible wealth includes goodwill of business, Patent-rights, Copy-rights etc., which cannot be measured or quantified. It is important that, while assessing the individual wealth, the loans indebted by the individual to others should be deducted from the market value of all his possessions.
- ☆ ***Social wealth:*** Also called as Communal wealth or Collective wealth. It implies, the State or Government holds the ownership rights on the property. This wealth aims at equi-distribution of welfare across the beneficiaries in the society. All public properties like dams, railways, roads, museums, libraries etc., fall under this category of wealth.
- ☆ ***National wealth:*** It is the aggregate of both individual wealth and social wealth. In narrow sense, it includes the wealth of all the citizens in the country, excluding their debts. But, according to Marshall, National

wealth should also include free gifts of nature such as rivers, mountains, good climate etc. But, taking the attributes of wealth into consideration, they cannot be regarded as wealth. So, this indicates again, the concept of wealth is related to economic goods only.

- ☆ ***Cosmopolitan wealth:*** It is the extension of concept of National wealth to the entire globe. That means, it includes the sum total of wealth of individual nations. This concept of wealth includes the free gifts of nature such as forests, oceans etc., across the countries along with economic goods.
- ☆ ***Negative wealth:*** It refers to the debts owned by the individuals or Government to others. Any undesired output/result realized during the course of execution of economic activities constitutes Negative wealth. For example, if the exports of grapes from India into the international market are affected due to the presence of pesticidal residues in the produce, it implies Negative wealth. Similarly, crop damages due to rodents, termites, pigs, elephants etc., fall under the category of Negative wealth.

In general, the concept of 'wealth' is often confused with money, income and welfare. However, there exists lot of differences between wealth and the above three concepts and they are discussed below.

Wealth and Money

It is important to note that, all money is wealth because, it satisfies all the four attributes of wealth stated earlier. But, all wealth is not money because, even the capital goods like machinery, tractor, land, furniture, buildings etc., constitute wealth. So, wealth includes the goods in several forms and money is one form of wealth. So, all wealth is not money.

Wealth and Income

Wealth and income are quite different and should not be used interchangeably. Wealth represents a stock of goods, which yields income. So, we can say, wealth is a fund or stock of goods and income is a flow. This implies, income is what wealth yields. For example, if a farmer possess one acre of land worth of Rs. Two lakhs, it is his wealth. If the farmer gives that land for lease purpose outside, it may earn him Rs. 10000 per year, and it is his income. If the asset does not yield any income to the owner, it cannot be considered as wealth, as it implies, it cannot be marketed outside.

Wealth and Welfare

In general, both wealth and welfare are used in the same sense that, wealth always leads to welfare. But, in economic sense, it is not true. Say, for example, narcotic drugs, alcoholic drinks, cigarettes etc., fall under the category of wealth, as they obey all the attributes of wealth. But, on consumption of these goods, they affect the human health and hence, do not yield welfare. So, wealth does not always lead to welfare. This also indicates that, an increase in wealth in the economy does not necessarily mean welfare has increased because, harmful goods might have shown increment in the economy. However, most of the goods in the economy such as cars, furniture, machinery, tractor etc., constitute wealth and they lead to human welfare in the economy. So, we can say, wealth refers to stock of goods in the economy and serves as a powerful means of promoting human welfare. Professor A.C. Pigou is associated with this view of Economics. However, the following are the important differences between wealth and welfare:

- ☆ Wealth refers to stock of goods, whereas, welfare is the state of mind of the individual.
- ☆ Wealth is a means to end, whereas, welfare is the end.
- ☆ Only economic goods constitute wealth, whereas, both free and economic goods leads to welfare
- ☆ Wealth includes both useful and harmful goods, whereas, welfare includes only useful goods.

2.6. Income

According to Seligman, *'Income in the economic sense, is the flow of satisfactions from economic goods'*. We know that, all economic goods in the economy constitute wealth. Income is what this wealth yields. So, wealth is a pre-requisite for income. For example, if the farmer owns a land, it is his wealth. If the land is leased outside, it earns income. There are two kinds of income *viz.*, Money income and Real Income.

(*a*) Money Income

In Economics, when we refer to income, generally it refers to Money income. It is also known as Nominal income. It implies, the income earned by the person in terms of money through performing an economic activity.

(*b*) Real Income

It is the money income expressed in terms of goods and services. It refers to the command of a person over actual goods and services in the economy. So, real income depends upon the prices of goods and services and it is inversely related. If prices rise, real income will fall, as the individual purchases less quantity of goods and services. The standard of living of people in a country depends on their real income. This is because, real income depends upon the purchasing power of money and this, in turn, depends on the price level of commodities. If money income of the individuals is increased, it does not imply they are better off. This is because, the general price level in the economy may be increased. Suppose, if an individual's money income is Rs. 100 and price of rice in the open market is Rs 25/kg, and then he can purchase four kilograms of rice. This implies, real income of individual is worth of four kilograms of rice. In the next month, if the money income is raised to Rs. 125, but the price of rice is increased to Rs. 35/kg, then the consumer can purchase only 3.6 kg of rice. So, the real income worth is decreased to 3.6 kg from 4 kg . Therefore, in spite of increase in money income, the real income of individual has come down due to higher increase in price. In short, real income is price adjusted money income in terms of goods and services.

2.7. Savings

Savings refers to the excess of income over consumption. Out of the total income, consumer spends a portion on the consumption of goods and services and remaining amount is saved. In developing countries like India, savings are essential, as their investment leads to capital formation in the economy.

2.8. Investment

The savings made by the individuals will be invested to create capital goods in the economy. So, investment refers to making an addition to the stock of existing capital goods in the economy.

2.9. Human Wants

As discussed in the earlier chapter that, human wants are unlimited and to fulfill these wants on prioritization basis, people engage themselves in various economic activities. So, the existence of human wants is the basis for all economic activities. Human wants are not only unlimited, but they vary in their importance across the individuals in the society. That means, some wants are essential, in the sense that, they must be fulfilled. They are considered as necessaries and without fulfilling them, it is very difficult to imagine the human life. They include, food, clothing, shelter, medication at times of sickness etc. However, some wants are not so important, but their fulfillment makes the mankind to lead fuller life in the society. These wants also multiply with the civilization.

In general, all desires, tastes and motives of human beings are called wants in Economics. They arise in the society due to various causes like psychological (food, clothing and shelter), social causes (a particular type of dress and food habits), customs and habits (drinking coffee, tea etc) and advertisements (cars, bikes, using different brands of fertilizers, pesticides etc). In the early stages of civilization, human wants are comparatively few and simple. But, with the advancement of civilization, human wants have multiplied rapidly and they are complex in nature. So, an individual will perform economic activities and with his limited income he try to satisfy his prioritized wants. This concept of fulfilling the wants through performing economic activities is called consumption. It is also defined as 'the beginning as well as the end of all economic activity'. It is, otherwise, regarded as 'destruction of utility'. In Economics, the concept of consumption is related to the study of nature of wants, achieving consumer's equilibrium in satisfying the wants and the behaviour of demand. Before going into the details of analyzing the consumer's equilibrium in deriving maximum satisfaction in fulfilling the wants, it is essential to discuss about characteristics and different types of wants in the economy.

Characteristics of Human Wants

The following are the characteristics features of human wants:

- ***Human wants are unlimited:*** Man is a bundle of desires. He will face the peculiar situation or relation between unlimited wants and scarce resources. If one want is fulfilled by his limited resources, another want will immediately crop up in his mind and to fulfill this again, he has to engage himself in an economic activity. Even the richest man in the society will have a list of wants to be fulfilled. So, the fulfillment of human wants is thus, a never ending cycle.
- ***Any particular want is satiable:*** Though human wants are unlimited, but a particular want can be fulfilled, provided the individual possess adequate resources. This concept is explained by the LDMU.
- ***Wants are complementary:*** Some wants are complementary in nature *i.e.*, the individual prefers some goods together to fulfill a particular want. For example, if the farmer wants to plough the land, he should employ two resources *viz.*, tractor and labour (driver). So, these two resources combine in fixed proportion to satisfy a particular want. In the absence of one of the resources, human want cannot get fulfilled. The theory of joint demand is based on this feature of human wants.
- ***Wants are competitive:*** Since human wants are unlimited, it is not possible on the part of consumer to satisfy all his wants with his limited resources. So, he will study their relative importance and then allocate his

limited resources to the top most prioritized want to derive maximum satisfaction. This implies that, wants compete with each other and the want which is felt urgent will receive the immediate attention and other wants are rejected at that point of time. For example, machinery and human labour, hired labour and family labour compete with each other in agricultural production programmes. Similarly, tea and coffee, sugar and jaggery are good substitutes and hence, they compete each other. This competitive nature of human wants has given rise to the formulation of LEMU.

☆ ***Some wants are both complementary and competitive:*** In some cases, the fulfillment of wants involves both complementary and competitive relationships. For example, if an agro-based industry installs machinery, it displaces unskilled labour. This indicates competitive relationship. But, at the same time, the firm engages skilled labour to operate the machinery. This indicates complementarity relationship.

☆ ***Wants are alternative:*** There are several ways of satisfying a particular want. That means, there are several goods in the economy to satisfy a particular want. However, the final choice of want or selection of good depends on the availability of money resources and the relative prices. For example, in meeting the nutrient requirements of crops, the farmers can apply chemical fertilizers or organic manures or bio-fertilizers etc.

☆ ***Wants vary with time, place and person:*** Wants are not static in nature, as they change with time, place, persons, generation, culture, society, geographic location and the extent of economic development. With reference to time, the farmers' wishes to avail crop loan facility at the beginning of crop season rather than at the end of crop growth period because, he has to procure all the inputs well-in advance. In terms of place, people in Northern parts of India prefer to consume wheat, whereas, in Southern parts of India prefer rice. With reference to person, to get engaged in economic activities, the farmers wants are totally different compared to software professionals. Even the wants of our forefathers are different from the wants of the present generation.

☆ ***Wants vary in urgency and intensity:*** All wants are not equally important to the consumer at a point of time. For example, from the farmer point of view, the want of procuring quality seed at the beginning of crop season is most urgent when compared to other wants like purchasing fertilizers, pesticides, herbcides, rodenticides availability of labour etc. This is because, these operations will come into picture, only when the farmer takes up the desired crop with good quality seed.

☆ ***Wants multiply with civilization:*** With the advancement of civilization, the wants will multiply. This is because, the human beings get exposed to latest inventions and innovations and therefore, his desire to fulfill them will increase. In this context, the urban people will have more and diversified wants compared to rural people. For example, people living in urban or metropolitan areas will think about the use of internet, playing video games in the computer, purchase of sophisticated technologies like mobile phones, laptop computers, digital cameras etc. However, the farmers living in rural areas will have little orientation about these latest inventions and innovations due to lack of media and awareness.

☆ ***Wants recur:*** Some wants, once fulfilled, will enable the individual to realize complete satisfaction and they will not recur. For example, if an individual wishes to purchase a lap top computer, he can fulfill his want, if he is having adequate money resources. However, some wants are recurring in nature. For example, if an individual consumes food, his want gets satisfied. But again, the same want will recur after a few hours. That is why, we say wants are recurring in nature.

☆ ***Wants change into habits:*** If a particular want is repeatedly satisfied, it becomes a habit on the part of individual. For example, drinking coffee or tea, smoking, use of drugs etc., make the individuals addict to these wants regularly. So, wants become habits and habits are responsible for fulfilling the wants. This feature of wants helps in the formulation of Engel's law of family expenditure.

☆ ***Wants are influenced by income, salesmanship and advertisement:*** Through increase in income and getting good exposure about various inventions and innovations, human wants multiply and they are also divergent in nature. Effective advertisement and personal solicitation about the goods and services will create new wants and the existing wants gets modified. The brand loyalty and dealer loyalty enable the customers to fulfill divergent wants. For example, farmers use latest brands of chemical fertilizers and pesticides due to advertisements and salesmanship of products.

☆ ***Wants are the result of custom or convention:*** As a part of custom and convention, the individuals will fulfill certain wants. In true sense, they are not essential, but for the sake of society, the individuals will invest on certain goods and services. Examples include, the expenses made by the individuals during birthday parties, marriage functions, festivals etc.

☆ ***Present wants are more important than future wants:*** Since future is uncertain, human beings are more concerned to fulfill present wants rather than future wants. A rupee earned today is relatively more important compared to rupee earned in the future. In this context, the farmers generally prefer to save money in the bank or go for short term investments rather than going for long term investments with long gestation period. The time preference theory of interest is based on this feature of human wants.

- *Wants vary with age:* Human wants vary with age, as we need different goods and services at different ages. For example, a child needs toys for playing, whereas, a farmer needs seeds, fertilizers, pesticides in his crop production programme.
- *Wants vary with tastes:* We know tastes of different people are different. For example, a farmer may prefer to go for using organic manures, another farmer may apply chemical fertilizers and another farmer may go for using bio-fertilizers in their respective crop production programmes.
- *Wants vary with sex:* We know human needs or wants vary with sex. For example, boys' wants are different from girls' wants regarding their dressing pattern.
- *Wants vary with climate:* Depending upon the nature of climate, human wants vary across the people. For example, in Rajasthan, people use cotton clothes, in Chirrapunji, people prefer rain coats. Even cultural differences across the people also lead to differences in human wants. For example, the dressing pattern of the people in Andhra Pradesh is different from the people of Assam.
- *Wants vary with state of health:* Depending upon the state of health of the individuals, human wants vary. For example, a sick person needs medicines whereas, healthy person does not.

Classification of Wants

Having studied the characteristics of wants, it is essential to categorize the wants, as it enables the consumers to plan their resources allocation across the categories in view of deriving maximum satisfaction. Further, it enables the consumers to know, which category of wants should be accorded more importance over the other. So, taking into consideration the urgency in fulfilling the wants by the consumers, the wants are classified into three broad categories *viz.*, Necessaries, Comforts and Luxuries. Though these aspects are discussed in-brief in the earlier pages, it is the right place now to discuss them in-detail.

Necessaries

These are the goods which are essential for living. Human beings require certain basic goods and without these it is difficult to imagine human life. Examples include food, clothing and shelter. These can be sub-divided as:

- *Necessaries of existence:* These goods are most essential for human life and without these we cannot exist. Examples include water, food, clothing, shelter etc.
- *Necessaries of efficiency:* These goods help the individuals to execute the works with more efficiency. They may not be so essential, but their presence will add efficiency in fulfilling the given tasks. For example, a chair and a table are the necessaries of efficiency for a student. Similarly, tools and implements used by farmers fall under this category. Nutritious diet to the human beings, teaching aids in the class rooms etc., represents this category.
- *Conventional necessaries:* These are the necessaries, which arise due to social customs and traditions. Examples include, celebrating festivals, dressing habits etc.

Comforts

These are the goods and services, which are not essential for living, but required to lead a happy life. These goods promote efficiency of the individuals in using them. For example, for a student, a table and chair will increase the efficiency, but a cushioned costly chair is a comfort. Similarly, eating fruits, drinking milk etc., also fall under the category of 'comfort' goods. So, after satisfying necessaries, human beings desire to have comforts. This indicates, comforts satisfy less urgent wants and makes life more comfortable.

The main difference between a comfort and necessary good is that, the benefit the consumer derives from the former is less than the investment made on it, while on the latter, the benefit derived is much greater than the amount spent on it. Further, the consumer is desperate to pay any amount to have the necessary good in time. But, in case of having a comfort good, he can temporarily postpone the consumption. On the other hand, comforts compared to necessaries add more social recognition and importance to human life.

Luxuries

Luxury means superfluous consumption. Once after enjoying comforts, human beings aim for having luxuries, as they ensure fuller life to the individuals. So, luxurious goods are used to show one's higher status in life. These goods are not essential for mankind and without these goods, human beings can lead a comfortable life. But, every individual in the society aim at having luxury goods, as they ensure fullest satisfaction to them. Examples include, gold, diamonds, silver, costly furniture etc.

The important distinction between a comfort good and a luxury good is that, the investment made on the comfort good will have a significant return or compensation to the consumer, whereas the compensation is of less significance on the consumption of luxury good. The dividing line between comforts and luxuries is a very loosely defined one. The term comforts may be understood to include things that are not necessaries, but which under the pressure of public opinion or through habit have come to be looked upon as necessaries. This class of things is sometimes spoken of as conventional necessaries. Luxuries are those economic goods, which do not come within either of the classes of necessaries or comforts. Whether or not a good is to be considered as a luxury will often depend as much upon the circumstances of the consumer as upon the character of the goods.

Significance of Necessaries, Comforts and Luxuries

The classification of goods and services into necessaries, comforts and luxuries is purely relative in its approach and not absolute. That means, what is 'necessary' to one may be a 'comfort' to another and a 'luxury' to a third man. For example, car is a necessary good for a doctor. It is a 'comfort' good for a lecturer and a 'luxury' good for a student. Similarly, a good may be a necessity in town, but a comfort or luxury in a village. So, the categorization of goods into necessaries, comforts and luxuries is purely relative in its approach and it all depends on the income of the person, social status, tastes and preferences, urgency for the want, opportunity cost etc.

Justification of Comforts and Luxuries

From the above discussion, it is evident that, human beings after fulfilling necessaries aim for having comforts and luxurious goods in order to lead a fuller life. But, these goods are not so essential for the survival of human beings, unlike necessaries. But, still they are produced in the economy and the production of such goods can be justified on the following grounds:

- ☆ These goods may have trade advantage for the nation, as the categorization of goods is purely relative in its approach.
- ☆ The production of such goods promote employment opportunities.
- ☆ They make the life more comfortable.
- ☆ Taking the specificity of resources into consideration, the production of such goods ensure maximum resource use efficiency.
- ☆ Some goods may enjoy trade protection in the form of Geographical Indications, Industrial Designs, Trade Secrets etc.
- ☆ Production of these goods requires strengthening of capital goods and this enhances capital formation in the economy.
- ☆ The consumption of these goods promote quality life among the individuals.
- ☆ The earnings from these goods enable the economy to mobilize the resources towards producing more necessary goods and this ensures equi-distribution of income.
- ☆ These goods serve as a stimulant for further inventions and innovations and this positively influences the technical progress in the economy.
- ☆ The expenses on these goods transfers wealth from idle rich people to the working class in the society.

Besides above classification, human wants can be further categorized as given below.

1. *Economic wants:* These are the wants of the people that are satisfied by the goods and services brought and sold in the market. Examples include fertilizers, pesticides, television etc.
2. *Non-economic wants:* These are the wants that are required by the people in the society, but cannot be bought and sold in the market. Examples include, a farmer should have good relations with his neighbouring farmers, peace of mind to the family members, love and affection from the elders in the family etc.
3. *Personal or Individual wants:* These are the wants, which are felt required by an individual person in the society. For example, a farmer requires good quality seed in the crop production programme, an individual want for television, computer etc.
4. *Collective or Social wants:* These are wants, which are felt required collectively by group of people in the society. Examples include schools, hospitals, museums, libraries etc.

3

Economic Systems

An economic system is the system of production, distribution and consumption of goods and services in an economy. Alternatively, an economic system refers to the organizational arrangements and processes through which a society makes its production and consumption decisions. In creating and modifying its economic system, each society chooses among alternative objectives like efficiency, growth, liberty and equality. In choosing these alternative objectives, the economic system can be considered as a part of the social system and hierarchically equal to the law, political, cultural systems etc. It is the set of principles and techniques by which problems of economy are addressed through the allocation of economic resources to satisfy unlimited human wants. So, economy is an economic system by which people get a living. This implies, an economic system comprises of people and economic institutions, including their relationships regarding the allocation and use of productive resources. In general, the economic system is often confused with economic institution. From the above statement, it is clear that, an economic institution forms part and parcel of the economic system, which is never ending. For example, an economic institution like a joint family, private property etc., may survive for a period and die. But, an economic system comprises of several economic institutions and the working of the system does not come to an end or die, but only grows or changes its form as Capitalism, Socialism, Mixed economy etc. So, 'economic systems' is the economics category that includes the study of respective systems. At one extreme, there is a free enterprise economic system, in which all the means of production is privately owned. This system, following Adam Smith, is based on the belief that, the common goal is maximized, when all members of society are allowed to pursue their rational self-interest. At the other extreme, there is pure-communist system, in which all the resources are publicly owned with the intention of minimizing inequalities of wealth and achieve other agreed-upon social objectives among the people. This system was advocated by Karl Marx and Vladimir Lenin. No nation exemplifies these two extreme forms of economic systems. As one move from Capitalism through Socialism to Communism, a greater share of a nation's productive resources is publicly owned and a greater reliance is placed on economic planning. Across these different forms of economic systems, the individuals and economic institutions execute various economic activities in the context of fulfilling the goals and objectives of respective economic systems. In the course of execution of basic economic activities, every society will face certain economic problems and they are mentioned here as under.

- ☆ ***What is produced and in what quantities?*** - Every society must decide on how much of each of the many possible goods and services it should produce. For example, the society has to consider whether to produce more of basic necessaries or to have more of luxurious goods, whether to have more agricultural goods or to have industrial products and services, whether to spend more resources in the fields of education and health or in building military services, whether to have more of basic education or more of higher education, whether to have more of consumption goods or to have investment goods etc.

- ☆ ***How are these goods produced?*** - Every society has to decide on how much and which of the resources to be engaged in the production process of each of the different goods and services. For example, whether to use more labour or more machines, whether to go for organic farming or use chemical fertilizers in the production of crops etc.
- ☆ ***For whom are these goods produced?*** - Who gets how much of the goods that are produced in the economy? How best the goods produced in the economy will be distributed among the individuals in the economy? Who gets more and who gets less? Whether elementary education and basic health services are available or access to everyone in the society?

These problems are popularly regarded as central problems of the economy, as every economy faces the above problems. These three basic problems are inter-dependent. In the process of solving these problems, subsidiary problems like allocation of resources among alternative uses, determination of input and output, fixation of prices, distribution pattern etc., will also arise. The society must make proper choice about them in order to meet the development aspirations of people satisfactorily. To address these central problems, every economic system is composed of people, resources and institutions like banking institutions, educational institutions etc., and they should be properly planned and utilized in a coherent manner. The way of addressing these problems vary across different economic systems. So, an economic system is a way of answering these basic problems. Examples of contemporary economic systems include Capitalism, Socialism, Communism and Mixed economy. Currently, Capitalism is the world's most dominant form of economic system.

3.1. Capitalist Economy or Capitalism

A capitalist economy is an economic system in which the production and distribution of commodities take place through the mechanism of free trade. Any economy that derives most of its production from the employment of capital may be said to be capitalistic. According to Pigou *'Capitalism or capitalist system is one of the main part of whose productive resources is engaged in capitalist industries in which the material instruments of production are owned or hired by private persons and are operated at their orders with a view to selling at a profit the goods or services that they help to produce'*. This implies, Capitalism is an economic system stressing individual initiative with a central role for a market economy, freedom of enterprise, profit motive and ownership of means of production by private individuals and corporations. Thus, the working of a capitalistic economy is dependent upon self-interest and invisible hand. In Capitalism, each individual be it a producer, consumer or resource owner has considerable economic freedom. That means, every individual has right to plan his own business or enterprise for which he has the necessary means. So, Capitalism is also called as Free enterprise economy or Laissez faire economy or Market economy or Free trade economy or Natural economic system or Systems of price mechanism or System of invisible hand etc. The Government is not supposed to interfere or control such an economy and private markets are the primary vehicles used to allocate resources and generate incomes, as profit is the driving force behind all economic activities. All the resources of the economy like land, business and capital etc., are owned by the persons. Individuals can use the resources in whatever manner they choose, subject to common protective legal restrictions. They also have the freedom to buy and sell any number of goods and services and to choose any occupation. Thus, a market economy has no central coordinator guiding its operations. But, self-organization emerges amidst the functioning of market forces namely supply, demand and price. Since, there is no central planning authority to make the fundamental economic decisions, the market economy uses the price mechanism, which plays a vital role in the working of the economy. Any imbalances are solved and corrected automatically through the price mechanism and demand-supply interaction. There are adequate rewards for greater efficiency and hard work in the form of giving compensations. There is also incentive to save and invest and provide incomes for the present and future generations. Market mechanism also enables entrepreneurs to take risks for higher profits, undertake innovations giving rise to technological progress. The State or Government takes least part in the economic activities of the people. Its role is limited to the issues like defense, maintenance of law and order, provision of a legal framework for protecting rights of private property, foreign affairs, currency and coinage and some important civil works such as the construction of roads and bridges, which the private entrepreneurs may not find them as profitable ventures. Under this system, State undertaking would mean wastage of society's resources. The slogan of Capitalism is to observe three 'P's *i.e.*, Profits, Prices and Private property. There is no true 'opposite' of Capitalism, but Capitalism stands in most direct contrast with 'Labourism' or economies that derive most of their production from human labour. In practice, there are no economies that are purely capitalistic or purely labouristic. This economic system now exists in all countries of the world. The United States and Japan are examples of a Capitalist Economy. Under this system, there are two types of classes in the society *viz.*, Capitalist class and Working class. This class division is the essential feature of Capitalism.

Capitalist Class

The means of production and distribution of goods in the economy are owned by a small minority of people, called Capitalist class. They invest their capital in the production programmes and make decisions regarding production of the goods, making of expenditures, saving, investment etc.

Under Capitalism, the capitalists have the right to organize and produce the goods and services with the resources at their disposal. This class will produce the goods that yield maximum profits to them and not concerned with public welfare. With the profits earned from the business, the capitalists will reinvest them in the economy for the further accumulation of wealth.

Working Class

This class involves majority of the people, who sell their ability to work in return for a wage or salary in the factories or firms owned by the capitalists. The working class is paid for their services in the production programmes and the products are then sold for a profit. As the capitalist class aims at profit maximization, the working class is exploited by the capitalist class.

History offers various types of Capitalism over time. In 17th century, Capitalism refers to the total freedom of enterprises and hence, it is referred as Free Market Economy. Government only focused on defense and law and order. Capitalist, with a view to maximize his profits, exploited the labourers. This fact forced the Government to interfere in the business and designed labour protection laws. With the growth of monopolies, it further forced the Government to check the monopoly exploitation in the economy. To achieve the aim of social welfare, Government controlled and directed the system of price mechanism. Thus, the today's Capitalism is not a pure type of Capitalism or a free market economy. It is a controlled or regulated Capitalism. In fact, Capitalism is a relatively new social system. There is a difference between fixed capital and circulating capital. Fixed capital of a company or plant includes plants, stock of raw materials and semi furnisedh goods etc. which are used in the production of final goods. However, Capitalism is not a rigid, but an evolving and dynamic concept. It has successfully fought off many crises and emerged stronger.

Essentials of Capitalism

The main features of Capitalism are as follows:

- ☆ ***Private property:*** The various means of production are under the private ownership of individuals. The people have full right to use their own or private property in any manner as they like with profit motive and this right is protected by Government. For example, a farmer can make use of his land in any manner he likes for cultivating paddy or sugar cane or for grazing cattle. He may use the land himself or hire out for getting rent or construct a building. Like wise, material means of production or capital assets are allowed to be owned and managed by private individuals, groups, associations and companies. This private ownership of factors of production plays a dominant role in Capitalism and it helps to distinguish it from other economic systems. The ownership of means of production combined with free enterprise by private individuals enables them to channelize the flow of economic resources in such manner as they like. There is no limit and they can own any amount of property. Right of inheritance by the sons and daughters or other legal heirs is implicit in this right. This right in capitalistic system adds greater significance to private property and becomes an added incentive to put hard work for further accumulation of wealth. This system of private property in Capitalism encourages savings and accumulation of capital for the future. However, the State can place some minimum restrictions on the use of the property. There is some public property in the economy as well, such as, roads, railways, electricity, parks etc.
- ☆ ***Price mechanism:*** It is the automatic price mechanism that makes the capitalist economy function efficiently. In Capitalism, prices are determined by the forces of demand and supply. Equilibrium between demand and supply is brought through the process of price mechanism. These market forces are the signals to direct the system. Government does not interfere in the fixation of price. This automatic price mechanism directs consumption, production, exchange, distribution, investment, saving etc., without any central planning and control. Adam smith quoted this automatic price mechanism as 'Invisible hand', which operates the capitalistic economy.
- ☆ ***Economic freedom:*** In Capitalism, there is no restriction on the right of the individual to plan any business or enterprise for which he has the necessary means. He can take independent decisions with regard to what, where, how and when to produce. Similarly, consumers have the freedom to buy anything they want, workers are free to take employment wherever they get higher wages and investors are free to invest wherever they get higher money returns. The basic concepts of private property and profit motive have full meaning only, if everyone has the freedom to make use of his property and to take such economic activities that brings him maximum return or profit. This economic freedom resulted in the establishment of corporate form of enterprises called Joint Stock Companies.

☆ ***Competition:*** In Capitalism, freedom of enterprises leads to more and more competition, which ultimately results in efficiency of industrial units and production of goods at low costs. Competition exists among producers, sellers, buyers, job seekers, employers, investors etc. This is achieved through cost control, price cutting, advertisements etc. This form of competition among all groups without any restriction or hindrance would manifest into what may be called as 'Perfect competition'. Because of free and perfect competition, production tends to meet to the demands of the consumers and it leads to maximum efficiency. This perfect competition, theoretically, is the essence of free Capitalism.

☆ ***Profit Motive:*** In Capitalism, the chief motive of producing goods and services is to sell them for earning profits and not to satisfy people's needs. The main objective of producer is to produce the product(s) and to find a buyer for its sale with the main objective of profit-making. That means, the producers produces the goods not with a view what consumers actually want and not what consumers are prepared to pay for it, but with a profit-motive. Those goods may satisfy human needs, but those needs will not be met, if people do not have sufficient money. This profit motive is not just the result of greed on behalf of individual capitalists, rather they do not have a choice about it. This is because, they have to make a profit in the business for not losing their investments and also their position as capitalists in the economy. So, strong profit motive serves as inducement to invest in any economic activity. This profit-motive leads to cut throat competition among the capitalists and hence, the Capitalist class keeps their means and methods of production up to date. Generally, production decisions involving high risks are taken by the individuals only to earn large profits. This class division and profit motive of Capitalism are the root causes of most of the world's problems today, from starvation to war, to alienation and crime. This is because, people needs or welfare comes only second to profit making. So, profit-motive is at the heart of a capitalist and it guides the entrepreneur to invest in any economic activity. As profit becomes the driving force for the firms, there will be tremendous increase in production and Capitalism has the capacity to cope with any amount of increase in production. This profit motive channelizes the resources, determines the character of business and also leads to variations in production with improved technology. So, profit motive, private property and personal initiative under Capitalism are inter-related and inter-dependent, as one will have no meaning and significance without the other.

☆ ***Role of entrepreneur:*** The role of entrepreneurial class is very crucial in the success of the capitalist economy. This is because, if the entrepreneurs are more dynamic, it promotes healthy competition in the economy.

☆ ***Consumer's sovereignty:*** In a free market economy, wishes and preferences of the consumers direct the economic activities. The consumer buys only those things which give him maximum satisfaction. Utilization of resources, techniques of production, output in a capitalistic economy depend on the extent of consumer's demand. Hence, the consumer is considered as a king under Capitalism.

☆ ***Minimal role of Government:*** As most of the basic economic problems are expected to be solved by market forces, the Government has minimal role to play in the economy. There is no conflict between the individual interests and the society. Further, the role of the Government is not only minimum, but also neutral. The Government is not expected to interfere with the market mechanism nor take any steps to curb the economic freedom. The State operates only in the fields, where private enterprises will be inadequate and cannot afford huge investments, as in case of educational facilities, public health etc. Its role will be limited to some important functions like regulation of market, maintenance of law and order in the country, defense, foreign policy etc.

☆ ***Competition and Co-operation goes side by side:*** Due to strong profit motive, there will be severe market competition among the entrepreneurs to occupy due share in the market. Of course, to realize this, they have to employ modern techniques of production. So, the workers and other resources co-operate the entrepreneurs in the desired direction, so that the production pattern runs as per the planned schedule.

☆ ***Prevalence of two classes:*** As mentioned earlier, in Capitalism, the society is divided into two classes, *viz.*, the Capitalist class and the Working class. The former owns the factors of production, while the latter earn for living through their labour services. The capitalists not only utilize their capital but also hire labour. Hence, in addition to commodity market, there is 'labour market' in Capitalism. So, labour becomes a commodity and is sold just like any other commodity.

☆ ***Decentralization:*** All economic decisions are decentralized and made by individuals in a process of spontaneous coordination throughout the economy.

Merits of Capitalism

The following are the merits of capitalistic economy:

☆ ***Rich variety of goods and services:*** In Capitalism, there is diversified production of goods and services suiting to the needs and demands of the consumers.

- ☆ ***Best utilization of resources:*** Since profit making objective drives the Capitalism, the main aim of the producers of goods and services is the efficient use of resources in the production programmes. Capitalism ensures the best utilization of resources because, each individual have economic freedom, act as per his self-interest, profit motive, severe competition etc. So, every producer tries to make use of the resources efficiently and such producers will sustain in the industry and others will be automatically eliminated.
- ☆ ***Inducement to work***: Private ownership and Law of succession induces the people to work more under the Capitalism, so as to earn more income. This profit making motive makes the producers to invest more in the business.
- ☆ ***Efficient production and increase in the standard of living:*** In Capitalism, every firm tries to compete with one another in producing better quality products through cost effectiveness. Modern and advanced techniques of production will be employed to ensure efficient production. Moreover, as the producers have full freedom in selecting the enterprises, it further ensures increased production of the commodities having comparative price advantage in the market. So, from the producers' point of view, the profit making objective and from the consumers' point of view, purchasing the qualitative products at affordable prices ensures improvement in standard of living to a greater extent.
- ☆ ***Automatic market mechanism:*** In Capitalism, the balance between demand and supply is maintained automatically. In case of disequilibrium, say if demand exceeds the supply, new producers will enter the market and start producing that product and existing producers will increase its production, thereby, equality between demand and supply is attained. Similarly, if supply exceeds the demand, some firms which experience losses will leave the industry and again the equality of demand and supply forces is attained. Thus, shortages and surpluses in the economy are generally adjusted by the forces of demand and supply and the economy operates automatically through the price mechanism.
- ☆ ***Growth of entrepreneurship:*** Various features of Capitalism like private ownership, intense competition, profit motive etc., induce the entrepreneurs to become more efficient, competent, hard working and dynamic.
- ☆ ***Maximizes the welfare:*** The promotion of self-interest both in production of commodities by the producers and purchases of qualitative commodities by the consumers leads to welfare maximization. Due to profit-motive, there is general increase in the supply of goods and services. Wage rates are high and people work hard to increase their material welfare. So, per capita income, prosperity and standard of living of the masses will increase.
- ☆ ***Technological improvements:*** Goods and services can be produced with the help of two types of techniques *i.e.* capital intensive technique and other labour intensive technique. The element of competition under Capitalism and profit-making motive among the producers guide the researchers to develop new cost-effective technologies.
- ☆ ***Liberty and equal rights:*** Under Capitalism every one gets equal right to work, accumulate wealth and become rich. Even ordinary workers through hard work, build up their resources and improve their standards of living. In fact, American Economists believed that, USA has become rich and powerful only by giving equal rights and liberty. Liberty in the sense of political freedom, which limits the Government's interference within the framework of social justice and security.
- ☆ ***Flexibility:*** Another achievement of Capitalism is its flexibility or the adaptability to changing conditions. For example, at times of scarcity of food products, agro-based industries will be given priority and under normal conditions, the economy reverts back to normal functioning emphasizing all the industries. This economic system can be adopted in different countries, to different peoples, to different technological changes and even with different degrees of Government intervention.
- ☆ ***Increase in capital investment:*** In order to earn profits, businessmen invest more and adopt new technologies, which results in the capital formation and rapid economic growth.
- ☆ ***Freedom of choice of goods:*** Consumers are not dictated by any agency to purchase or not to purchase a particular type of good. They purchase the goods as per their will and choice and according to their purchasing power. Similarly, the labour class can opt for any career in accordance with their capabilities and skill. So, the sovereignty of consumer is given due importance.

Demerits of Capitalism

The following are the demerits of capitalistic economy:

- ☆ ***Unequal distribution of wealth:*** The greatest demerit of the system is unequal distribution of wealth and income. This inequality is due to private property, profit motive, less Government interference, freedom of enterprises etc. Though the system theoretically provides equal opportunity for all, the people who possess the resources are in an advantageous position to amass more wealth. Thus, those who own property under this system, can become richer and they can easily exploit the weak class. Thus, this system will make the rich richer and the poor poorer resulting in gross economic inequalities. So, the poor with limited means are unable to compete with the rich and this widens the gap between them. This inequality tends to be perpetuated by the system of inheritance. A person of a rich family has the initial advantage over a person of a poor family. Further, to become rich by means of hard work alone without the initial advantage of wealth or property is only a myth. Moreover, the Right of inheritance of private property makes possible the growth of large lazy propertied class, who gets their incomes not from hard work, but from ownership of property. This 'unearned income' gets more and more accumulated, further increasing the inequalities. When the society progresses, the value of the property rises and the benefit goes to the capitalists, who never contributed anything to the progress of society by means of their work. This 'unearned increment' is the bane of capitalistic societies. The distribution of income in almost all capitalistic economies shows that, there is concentration of income at the higher level, while large percentage of people get a poorer share of the national income. This economic inequality, in turn, lead to social and political inequalities. Socially, it denies equal opportunity for all. Politically, it causes discontent and instability.

- ***Consumer's sovereignty is a myth:*** In Capitalism, the whole of economic activity is guided by consumer's choice. The consumer makes his wishes prevail in the market and the producer should produce only those commodities, which are demanded in the market. But, in the modern days, there are several limitations from the consumers' side to execute his full powers and they include *viz.*, development of monopolies and oligopolies in the market economy, advertisement and propaganda, restrictions imposed by the Government, frequent changes in fashions, habits, tastes of individuals etc. So, consumer's sovereignty is only a myth in Capitalism and not a reality.
- ***Class struggle:*** On account of inequalities of wealth and income, there exist two classes in the capitalist society *viz.*, Capitalist class and the Working (Labour) class. Capitalism does not offer enough security to working class, thereby, leaving them at the mercy of the capitalists. This insecurity and the exploitation of workers were mainly due to inequalities of wealth and income among the two classes. This will lead to labour unrest and curtailment of production due to stoppage of work and other connected evils.
- ***Exploitation of labour:*** Another bane of Capitalism is the exploitation of labour. Under Capitalism, labourers get wages, which are less than their marginal productivity. Stringent labour laws are enacted favouring the profit-motive of capitalists. Such laws also aim at exploiting the labour by keeping their wage rate at its lowest minimum. The organisation of the workers formed with a view to improve their working conditions and to raised the level of wages.
- ***Wasteful competition:*** As profit-making is the heart of Capitalism, there lies cut throat competition among several firms in the economy for producing and transacting the products. So, the firms will invest huge amounts of money on the advertisement and salesmanship for increasing the sales of the product, which may be considered as wastage of productive resources. Moreover, due to lack of Central Planning Authority or directive to make use of the resources in the best and rational manner possible, there is every possibility of over-production and under-production in Capitalism because of competition. This will further lead to wastage of resources. Quick changes in fashions and tastes also misguide the producers, leading to under-use or over-use of resources. If there is any efficient and economical use of resources in Capitalism, it is purely accidental.
- ***Business fluctuations and unemployment:*** Since, Capitalism encourages mechanization and automation, it will result in unemployment particularly in labour surplus economies. Moreover, automatic character of capitalistic economy is responsible for causing business fluctuations. The economy faces with alternative cycles of boom and depression, leading to good deal of wastage of resources. Whenever trade depressions occur, it leads to unemployment and under employment in the economy and this causes greater uncertainty and suffering. So, the need of State intervention is essential at this juncture. The automatic price mechanism is not sufficiently capable enough to restore the economy to normal conditions. So, Keynes advocated substantial State intervention is essential in the form of fiscal and monetary measures to restore the normal conditions.
- ***Disregard of public welfare:*** In Capitalism, welfare motive comes only second to profit motive. So, high prices are quoted for the goods with a view of making profits in the business. Moreover, the entrepreneurs aim at producing luxurious goods rather than basic goods, as luxurious goods have more elastic demand. Thus, commodities which yield higher profits will be produced by the entrepreneurs without caring the actual requirements of the economy. When the economy is in shortage of foodstuffs and other basic goods necessary for life, the resources would be diverted towards the production of luxurious goods demanded by the few rich people in the market. So, the resources will be mis-used and mis-directed for profit sake and the welfare of workers, consumers and the community will not be considered at all.

☆ ***Basic social needs are ignored:*** Considering profit-motive as the main objective in capitalist economy, the basic social sectors like literacy, public health, poverty, drinking water, social welfare etc., are ignored. This is because, these sectors are less profitable from the entrepreneurs point of view. So, most of the vital human issues are ignored in a capitalist system.

☆ ***Lack of coordination:*** Due to lack of central planning, there is very little coordination among different firms and different industries regarding the total output in the economy. So, it leads to excess production of some goods more than the required quantity, while other goods fall short of the actual requirement.

☆ ***Leads to monopoly:*** Though perfect competition is supposed to exist in Capitalism, it is only in theory. The severe cut throat competition among the firms gradually leads to monopoly. The companies, who enjoy advanced techniques of production replace smaller firms, which produce goods with outdated technologies. It is common to see that, firms combine to form cartels and in this process brings about reduction in number of firms engaged in production. So, gradually the firms aim at monopoly gains from the market against the welfare of consumers. Exploitation of the consumer would become limitless and the monopolist would grow richer and richer leading to inequalities of income. This will adversely influence the self-generating and self-adjusting character of the economy.

☆ ***Depression:*** There is over-production of goods in the economy due to adoption of advanced technologies, severe competition, profit motive etc., and this leads to glut in the market and hence, depression.

☆ ***Social costs:*** As profit-motive drives the economic activities under Capitalism, the entrepreneurs run the enterprises with no consideration about the social costs on the economy. The pollution and health hazards are severe, as factory wastes are not properly disposed off.

☆ ***Increase in production only illusory:*** The contention that, in Capitalism, production will increase as resources are used efficiently is refuted. Since the business fluctuations are common in capitalistic economy, the production at any rate may not be higher than what it would be under Socialism.

☆ ***Poverty in the midst of plenty:*** Under Capitalism, the increased output cannot be absorbed in the economy because of the low purchasing power of the majority of the people. Moreover, in capitalist countries, the marginal propensity to consume is low among the consumers and therefore, over production is common. This over production of goods in the economy would lead to unemployment and this would still aggravate the problem of the poor. This leads to poverty in the midst of plenty. This would be the condition in all capitalistic economies particularly during the period of depression.

☆ ***Lack of social security:*** As the entrepreneurs are profit-motive, they exploit the working class and even they do not show any concern about their social security. The working class are not provided with pension, accident reliefs, relief to the family if a worker dies in employment etc.

From the above features it is clear that, capitalism in its pure form is not seen today. This is because, Government intervention is indispensable to correct some imperfections in the capitalistic economy. But, the extent of Government intervention depends upon the severity of the economic problems in the economy. Prof. Galbraith sums up the ineffectiveness of capitalism thus: *'there is much that the market can usefully encourage and accomplish - as it cannot put a man into space so it cannot bring quickly into existence a steel industry where there was little or no steel making capacity before. Nor can it quickly create an integrated industrial plant. Above all no one can be certain that it will do so in countries where development has lagged and where there is not only need for development but an urgent demand that it should occur promptly. To trust to the market is to take an unacceptable risk that nothing, or too little, will happen.'*

Modern Capitalism

In the present world, capitalism of pure form does not exist anywhere in the world. This is because, several limitations will be imposed on the private sector by the State, thereby, limiting its economic freedom. Similarly, from the consumers' side, the sovereignty is restricted. As in Capitalism, welfare motive comes only second to profit motive, the entrepreneurs aim at producing luxurious goods rather than basic goods. When the economy is in shortage of foodstuffs and other basic goods necessary for life, the resources would be diverted towards the production of luxurious goods demanded by the few rich people in the market. This leads to misallocation of resources in the economy. Similarly, other important social sectors like literacy, public health, poverty, drinking water, social welfare etc., are ignored. Because of lack of controlling mechanism from the State, some firms even aim at monopoly gains from the market at the cost of public welfare. Such activities will aggravate the problem of social costs in the economy. Further, all the businessmen aim for profit making through increased production of goods. But, the economy is not in a position to absorb all the output because of low purchasing power of the majority of the people and it is rightly quoted as *'Poverty in the midst of plenty'*. From the consumers' side, he cannot execute his full powers to enjoy his sovereign powers in the economy because of several limitations like development of monopolies and oligopolies in the market economy, advertisement and propaganda, restrictions imposed by the Government, frequent changes in fashions, habits, tastes of individuals etc. So, this calls for Government's intervention to reduce the Capitalism generated problems like economic inequalities, unemployment, trade depressions, the animosity between the 'have' and the 'have-nots' etc. So, particularly, after the Second World War, the State started assuming a positive role in capitalistic economies to overcome the short comings of uncontrolled Capitalism. So, this form of Capitalism with Government intervention is called Controlled capitalism. The various forms of Government's interventions include *viz.*, anti-monopoly legislation, minimum wages legislation, regulation of public utilities, credit controls, laws relating to industry, trade and commerce etc. Further, the Government also employs fiscal measures to check inflation and unemployment problems. So, Modern Capitalism is a mixture of private enterprise and Government control.

3.2. Socialism

Socialism is an economic system, in which the material means of production are owned by the whole community or State or Government and the production is meant for the welfare of the community and not for the profit of a few individuals. Defining Socialism in exact terms is very difficult, as it is a very elusive concept having different shades. Karl Marx defined '*Socialism has been an intermediate state in the inevitable transformation of capitalism into communism*'. Webb and Webb defined Socialism as '*the essential feature in Socialism is that, industries and services, with the instruments of production which they require, should not be owned by individuals and the industrial and social administration should not be organized for the purpose of obtaining private profit*'. This definition is quoted in the negative sense, as it indicates, what is not Socialism regarding the ownership and operationship of means of production. Dickinson stated that, '*Socialism as a economic organization of society in which the material means of production are owned by the whole community according to a general economic plan, all members being entitled to benefit from the results of such socialized planned production on the basis of equal rights*'. This definition is fairly comprehensive, as it indicates various aspects of Socialism *viz.*, means of production should be owned by the state or state owned organizations, production in the economy should be welfare-motive and not profit-motive and equi-distribution of benefits. This implies, Socialism is both an economic system and a belief system. The socialist state has been characterized by the dictatorship of the proletariat and existence of scarcity.

The above definitions show that, in socialistic economy, the means of production are owned and operated by the State and hence, the chief objective of production process is to meet the basic human needs. This clearly indicates, what community actually needs will be produced rather than what perceived by the few rich people with profit motive. The old slogan of '*from each according to ability, to each according to needs*' would apply in Socialism. In Socialism, as the system functions based on democratic approach, it is quite easier to identify what actually the human needs are. The democratic system will itself be the outcome of future democratic decisions. The economic decisions will be taken at a number of different levels *i.e.*, from micro to macro. Though, exact predictions are not possible for each and every individual, but the common needs can be well ascertained. In this way, it will help the Government to streamline the prioritized issues through democratic participation of every individual and target them effectively, so as to ensure mass benefits. All decisions regarding production and distribution are taken by the Government or Central planning Authority. Hence, the socialist economy is also called as Planned economy or Command economy. A planned economy may consist of state-owned enterprises, private enterprises directed by the state or a combination of both. Though 'planned economy' and 'command economy' are often used as synonyms, there is slight distinction between the two. Under a command economy, the means of production are publicly owned. But, a planned economy is an economic system in which the government controls and regulates production, distribution, prices etc. But, a command economy, while also having this type of regulation, necessarily has substantial public ownership of industry. Therefore, command economies are planned economies, but not necessarily the reverse.

Private property rights are strictly limited to small tools that an individual needs for his occupation. Land, factories and major machinery are not privately owned. Prices are set by the state rather than by the forces of demand and supply. Economic decisions about what, how, how much and for whom to produce are all made by the State through its Central Planning Authority and other administrative units. Hence, the Authority only decides for what purposes the profits should be used. The benefits from production would be used for the public welfare, improving their living conditions and attempting to give everyone an equal chance of comfortable living. This implies, Socialism is democratic in the pure sense of the word. In contrast, a capitalist system might insist that, it is democratic, but the power lies with the people possessing more wealth. Some of the most successful socialist or planned economies are Canada, Germany, Sweden, Australia, Great Britain, Saudi Arabia, Iran, Burma etc.

In Socialism, considerable room is given to the concept of equality and the focus is put on curing the ills of society, especially hunger and poverty. Many, who strongly believe in Capitalism would argue that, compassion and caring for others is not reserved for socialists. The key difference here is, how the two philosophies view the ills of a society like hunger, poverty etc. A capitalist might think that, hard work and investment will rectify these problems, though they are considered an unavoidable part of life. In socialist eyes, these social ills are not an acceptable part of life. Rather, they are created by the actions of men and women and can thus be eliminated with correct changes. The socialists believe that, in case, if problems arise in the economy due to miscalculation about the priorities, they can be addressed only through the fundamental change of attitude. This change in attitude must then be carried through to action. So, this leads to no class-discrimination between a wealthy owner and a production worker and it leads to the society, which will be free from hunger, poverty, unhealthy working environment etc.

It follows that, Socialism must be either forms of 'Capitalism' or forms of 'labourism'. There is no much difference regarding the basis of production in Socialism compared to Capitalism and labourism. It is true that, the economic system in Socialism is typically a Mixed economy model, in which the state owns and controls some means of production and the private sector others. But, the main difference between Socialism and both Capitalism and Communism lie in the aspect of dispersing the ownership and control of the means of production. Regarding Socialism, there is wide dispersion of the ownership and control of the means of production, so that the individual enjoys the fruits of his own labour, compared to Capitalism and Communism. To put this in another way, Socialism aims at promoting equity-building Capitalism or equity building 'labourism'. The opposite of Socialism is economic elitism (or degenerate economic conservatism), which is any economic system that seeks maximum good for a tiny elite at the expense of the majority. Socialists even believe that, if a true socialist system was there from the beginning at the global level, it would have prevented several economic problems. Now the existing economic problems at the global level can be dealt or addressed effectively only through the introduction of the socialist philosophy. So, this philosophy must be put into action, across state and national boundaries, so as to ensure a healthy society, especialiy in the context of eliminating rich and poor discrimination, which was the root cause of all the economic problems.

Essentials of Socialism

The following are the features of socialistic economy:

- ✰ ***Collective welfare:*** In Socialism, social or collective welfare is the chief motive. In general well being of pepole which depends on the extent to which social services are available to them. The decisions taken by the Government aim at promoting the public welfare. Social well-being of people is the ultimate goal of any developmental policy of the Government. There is no room for the profit motive in true Socialism, unlike in Capitalism.
- ✰ ***Abolition of private enterprise:*** There is no significant room for private enterprises, instead production is carried out by the State. The State also fixes the prices of finished goods and gives to the consumers a free choice of the goods available in the market at fixed prices.
- ✰ ***Economic equality:*** Socialism mainly aims at reducing economic inequality. This can be ensured due to the absence of private property and profit motive. The Planning Authority will take all possible steps in the execution of different policies and programmes to see that, there will not be any income disparity among the people. So, Socialism mainly aims at reducing economic inequality, if not completely achieving economic equality.
- ✰ ***Collective ownership:*** All properties of the country will be owned by the State, indicating that, the ownership is collective in nature. This social or collective ownership of means of production is the fundamental feature of Socialism. No individual can accumulate too much property, as in the case of Capitalism. This implies, the right to private property is limited. The management of resources will be directly under the control of the Government through a department of the ministry or may be entrusted to a specially created authority for the purpose such as Corporation, Board etc. The administration may be centralized or decentralized, but the power to take decisions rests with the public authorities, who work for the sake of general interest of the public rather than profit motive of few individuals. It is important to note that, there will be private sector in socialist economy, where individuals carry their activities on profit-basis. But, these private operations are very negligible, insignificant and not capable enough for playing an influential role in the economy. So, these private sector units have to necessarily depend upon the Government for executing their activities.
- ✰ ***Central economic planning:*** There is a Central Planning Authority in the socialist economy and it will take all the policy decisions of all the sectors for promoting the welfare of the society. Though planning may be available even in non-socialistic systems, it is a must in Socialism. In Socialism, planning carries utmost significance because of State control, public-welfare motive, absence of profit making, price mechanism is not automatic etc. In the words of Pigou, *'Socialism is nothing but central planning and the entire economic activity of the nation has to be decided on the basis of a pre-conceived and determined plan'.*
- ✰ ***Non-automatic market mechanism:*** Production, consumption, exchange and distribution of goods in the economy will be governed by Central Planning Authority, as social welfare is the prime motive. Market forces have limited role to play in directing these economic activities. So, unlike Capitalism, the economy in Socialism will not function automatically. Government does interfere in the fixation of prices of commodities and all the economic activities are centered on its decisions.

☆ ***Well-defined social and economic objectives:*** In Socialism, there are well defined social and economic objectives and to achieve them there is a Central Planning Authority. It aims at planning for better utilization of resources, production, distribution, fiscal measures, credit policies, social programmes etc., such that, the targeted goals and objectives like industrialization, economic growth, social justice etc., are attained, so as to ensure general public welfare.

☆ ***Formulation of long term plans:*** The Central Planning Authority takes the responsibility of formulating economic plans for the welfare of the people. It formulates long term plans, which are otherwise called as Perspective plans. For convenient sake, these perspective plans are divided into various short term plans and annual plans and the objectives of these plans fits into the broad framework of long term plans. For planning into the future, the State aims at conservation of natural resources, unlike capitalism. But, in capitalism, wastage of natural resources is more common, as private entrepreneurs does not care for the future and aim only for short-term benefits. So, a socialistic economy can plan for the future in a better way compared to a capitalistic economy.

☆ ***Equal opportunity:*** The Planning Authority aims at giving equal opportunities to all and shows no discrimination between 'haves' and 'have nots'. So, it ensures social upliftment of the people at a faster pace. Due to the absence of profit motive, the exploitation of weaker sections is greatly checked. Classless society and elimination of class struggle by giving equal opportunity is one of the basic features of Socialism.

☆ ***People's co-operation:*** As Socialism aims at maximization of public welfare, the people's co-operation is essential for the planned execution of various Governmental programmes and policies. The accountability of various stakeholders counts much in the success of execution of developmental programmes in the socialistic economy.

☆ ***Wasteful competition minimized:*** Unlike capitalism, the expenditure on non-price competition is very low in socialistic economy. This is because, the State aims to produce the goods meant for public welfare. If there is any advertisement and propaganda about the products, it is meant only for giving information to the customers.

☆ ***No over production:*** An advantage of a planned economy is that, it is not subjected to major pitfalls of market economies and marked-oriented mixed economies. A planned economy, in theory, does not suffer from fluctuations in business cycles, it does not experience crises of over production (the one that was believed to have contributed to the Great Depression) and massive misallocations of resources.

The above features clearly indicate that, Socialism is an economic phase, where everyone gets benefited from industrialization and the number of workers was more than the number of traders. Being the majority, more power was vested in the hands of the workers. Socialism was only an economic system and not a political system. It was more advanced and liberal, where everyone could contribute in the working of the economy.

Merits of Socialism

The following are the merits of socialistic economy:

☆ ***Efficient use of resources:*** Socialistic economy ensures efficient utilization of resources because of well defined social and economic objectives. Planning is done in the socialist economy considering the availability of resources to address the basic human needs. This avoids wastage or mis-utilization of resources, as resources are allocated in proportion with the severity of the basic problems in the economy. So, working in the frame work of well planned objectives ensures effective utilization of resources, quality output and its proper distribution among the people in the economy. Absence of profit-motive further ensures planning the production programmes in tune with the actual basic needs of the people rather than producing only remunerative goods or luxury goods. So, by giving due emphasis on social welfare, the resources are utilized efficiently to produce a wide variety of goods that are useful to the society. It further leads to rapid progress and development of the economy.

☆ ***Flexibility:*** The socialistic economy is much more flexible compared to capitalist economy. This is because, the State will allocate the resources as per the priorities of the economy keeping in view of the public welfare.

☆ ***Absence of monopoly:*** Since, State assumes pivotal role in regulating the economy and there is absence of private ownership, the element of monopoly is eliminated in Socialism. In fact, the State acts like a monopoly, but functions on welfare motive rather than profit motive. So, the concepts like exploitation of labour and class struggle will not arise in socialism.

☆ ***Economic stability:*** In Socialism, economic stability is ensured because, Government assumes the key role in regulating the production sector. Because of the meritorious planning by the Government, economy will be free from business fluctuations, thereby, there will be no over-production and unemployment. This system eliminates cyclical fluctuations in the economy and tries to attain stability. For example, during the 'Great

Depression' in the early thirties, when all countries of the world were under the grip of depression and unemployment, Russia was markedly unaffected by the worldwide depression. Further, this economic system is not affected by frictional unemployment, as the methods of production and different activities are well coordinated. All these contribute to rapid economic development compared to capitalist economy.

☆ ***Maximization of social welfare:*** The State concentrates on the production of basic necessaries instead of luxurious goods and they are effectively distributed among the people to ensure maximum social welfare. The State provides quality education, cheap and congenial housing, public health amenities, social insurance covering accidents, unemployment, sickness, old age pensions, medical and education facilities etc., for the people. So, in Socialism, commodities needed by the poor will be produced and hence, maximum number of people will enjoy maximum satisfaction. All the sectors *viz.*, agriculture, industry and service sectors will be given due attention. The slogan is *'to each according to his needs, from each according to his capacity'*. This clearly indicates, overall human development is the chief concern and not only their economic development.

☆ ***No extreme inequality:*** As the State aims at promoting social welfare, the inequality regarding distribution of wealth is effectively checked or rather prevented. So, welfare motive ensures equitable distribution of wealth in the socialistic society. Earnings vary according to the ability of the individual. Exploitation of man by man is not allowed. Taxation measures are used to redistribute national income This is important because, inequality, under any sense, cannot be justified *i.e.*, either morally, socially, politically and economically because, it widens the gap between rich and poor in the economy and further prevents the poor to rise to a full stature.

☆ ***Capital formation:*** It is a process of investment in fixed assets causing net addition to the stock of basic capital, gross fixed capital for nation includes depreciation, repair and maintenance expenditure, where as net capital formation, excludes them all. It is undertaken directly by the State through planning.

Demerits of Socialism

The following are the demerits of socialistic economy:

☆ ***Inefficiency:*** Due to absence of profit motive, the hope of gain and fear of loss are totally absent in the business. This leads to greater fall in efficiency and hence in output, thereby, adversely influencing the national income. As socialistic economy is managed by the Government officials, their work efficiency is not in proportion with the monetary resources they earn from the Government. Red tapism, nepotism and corruption are severe and this further leads to inefficiency of production system.

☆ ***Complexities of administration***: Since, Government intervenes in each and every activity of the welfare of the people, the burden of administration in Socialism is very high. This is mainly because, all the basic problems in the socialistic economy are not decided automatically through price-profit mechanism.

☆ ***Slow technological progress:*** As the works are performed monotonously, there is complete absence of creativity, unless there is complete change in the production process, as suggested by the Authority. So, it leads to absence of drastic changes in technological progress, as new innovations cannot be easily enforced in the production patterns.

☆ ***Loss of liberty:*** Under Socialism, all economic activities are directed by the Planning Authority. No significant role is given to private investment and initiative. The workers will have no choice of occupation and 'labour' just like any other economic resource under planned management. The workers should work in such places as decided by the authorities. But, the only consideration is, there is no unemployment problem in the economy and this security of employment is no compensation for loss of liberty. This curtailment of individual freedom and liberty would even go to the extent of exploitation of the individual.

☆ ***Loss of consumer's sovereignty:*** The commodities, which promote the general public welfare will be produced in the economy rather that what actually the consumers want. The State fixes the prices of the goods and services on its own will and not according to the influences of market forces. So, the consumers are forced to purchase the goods produced by the State at the announced fixed prices. So, from the consumers' point of view, their choices are limited, as they cannot get the goods as per their preferences.

☆ ***No automatic price mechanism:*** The greatest obstacle as well as the weakness of socialistic system is the absence of automatic pricing process. In the absence of price mechanism, there will be no basis for calculation of costs and prices. In Capitalism, the price mechanism directs the allocation of resources. In general, the price of the product must be high to cover the average total cost. If this is so, the entrepreneur can make profits in the business and this guides him to continue and expand the business. However, due to absence of profit-motive in Socialism, the resource allocation will be done arbitrarily and this will not lead to proper expansion of business. Further, the costs of resources usage cannot be determined exactly, as the factors of production are owned by the Government. So, it becomes difficult to estimate how to produce, what to produce and how much to produce in the economy accurately. So, both production and pricing under Socialism will be done arbitrarily without taking into consideration the actual needs of the people. This will lead to misallocation of resources and thereby, production will be inefficient. But, many of the socialists opined that, there lies no much difference in the pricing of the products particularly of large scale enterprises, as in both socialistic and capitalistic economies, the large scale enterprises are managed by the Government. However, this supportive statement by the socialists may be logically true, but in practicality, due to the absence of profit-motive in Socialism, the price mechanism of Socialism cannot be on par with the competitive and automatic price mechanism of Capitalism.

- ☆ *Lack of incentive:* Since the socialist economy functions under the directions of Central Planning Authority, bureaucratic administration is seen and this makes the people to work hard out of fear of higher authorities. So, labour does not take much interest in executing the production activities and this kills the scope of performing inventive and innovative works. As a result, the workers do not enjoy freedom in accomplishing the tasks, as both the jobs and the places of work are governed by the Authority and hence, they simply perform the works monotonously. From the workers point of view, the job growth does not depend on their efficiency and success, rather depends on seniority. So, the workers suffer from lack of initiative, inability to take quick decisions and become lethargic in executing the job duties. Since, every worker, irrespective of their quality output, gets the same reward and there lies no meaning in accomplishing quality tasks in the production programmes. Since, showing wage differentials is not permissible as per the basic principles of Socialism, in USSR, the Government introduced awarding non-monetary incentives like medals and certificates to the meritorious workers and imposing punishments and fines to the workers, who failed to achieve minimum standards in the quality of output.
- ☆ *Misallocation of resources:* Some goods are produced more than the actual demand therefore, they are wasted. Sometimes, there may be shortage of production of some goods resulting in the uncontrolled demand from the people. This is because, the Government alone cannot be able to utilize the whole country's resources to the maximum benefit.

The features of socialism reveal that, it has more applicability in the underdeveloped nations, as in such economies, achieving social equity or public welfare is the immediate concern. With the active planning from the Government, all the national problems can be addressed through the active participation of all the stakeholders.

3.3. Communism

According to Karl Marx, Communism is the final form of an economy towards the development of an egalitarian society. Here, all the resources are state-owned and it determines its distribution based on the needs in an effort to bring about equality. In other words, Communism is 'State Capitalism' in which, all or most means of production are owned and controlled by 'the State', and few or no means of production are owned and controlled by individuals. So, it is an extreme form of Socialism. The term is defined not only as public ownership of means of production but also the "withering away" of central government authority. It is conservative in its approach. By using State coercion to fulfill unmet demands, it restricts individual freedom. Communism necessarily takes the form of totalitarianism, since it is up to the State to decide who gets what. So, Communism is characterized by the absolute rule of the State, beneath whom everyone is equally subservient.

Karl Marx is known as the 'Father of Communism'. The people who follow his ideology are known as 'Marxists'. Karl Marx and his friend, Frederick Engels analyzed various economies of the world regarding their mode of functioning and working mechanism of different sectors of the economies. They researched the development of various economies and the reasons for it and developed the concept of Communism based on their findings. They opined that, Communism aims at the banishment of all the differences in the society and people will share all the things equally with equal status. There shouldn't be any kind of exploitation or social inequality like women being dominated by men, racial discrimination and backward ideology. Communist society is supposed to provide equal status to everyone with the opportunity to use the resources equitably. Communism also means an extensive outlook for the betterment of society. The symbol of Communism is the hammer and the chisel, wherein the hammer represents the workers and the chisel represents the farmers or the peasants. China, Cuba, Vietnam, North Korea, Soviet Russia etc., are the examples for communist countries.

Communism is a system in which all people share the things with each other without any scope for profit making. Since, there was sharing and all the things were in abundance, there was no need for any disciplinary measures. Though capitalists were still there, they could not exert any significant impact. The features, merits and demerits discussed under Socialism broadly covers Communism also, as Communism is one form of Socialism or to put it precisely, extreme Socialism. It is also called as Marxian Socialism. It was Karl Marx through his celebrated book 'Das Kapital' published in 1867 attempted to provide a scientific basis for the theory of Socialism. He extolled labour to dizzy heights and contented that goods are produced in the economy by labour alone. But, instead of labour getting the larger share of the value of the product, his services are getting exploited by the employer. So, he advocated that, in the economy, all the means of production including labour service should be treated as a property of the Government. He emphasized that, the economic differences between two classes of people *viz.*, 'haves' and 'have-nots' are solely responsible for all the economic problems in the economy. So, according to Marx, a classless society should emerge to address all such problems and such full-fledged Socialism gives rise to Communism.

Essentials of Communism

The following are the features of Communism:

- ☆ ***Equality:*** The State will dictate the terms in the economy and regulate few rich capitalists, so as to ensure social justice. By promoting equality among all the citizens, the rich people will not exploit the poor in the society and a classless egalitarian society will be formed, in which all people live by earning and no person lives by owning.
- ☆ ***Wages in proportionate to labour:*** For labour, the wages will be paid in proportionate *i.e.*, *'for an equal quantity of labour, an equal quantity of products'*. The labour laws are very stringent in the sense that, *'he who does not work shall not eat'*. So, in Marxian ideology, Socialism is the transitory stage between two extreme forms of economic systems *viz.*, Capitalism and Communism. According to Marxian Socialism, the economy's output is distributed by the formula, *'from each according to his ability, to each according to his labour'*.
- ☆ ***Collective ownership:*** Communism is a system in which, the State plans and controls the whole economy. Since, the Government owns all the property and due to complete absence of private property, the people would not take advantage of power and become greedy. But, the main difference between Communism and Socialism is that, the former believes and adopts violent revolutionary methods to capture the machinery of Government, while the latter believes in peaceful and parliamentary methods.
- ☆ ***Work culture:*** People must work more, but receive less in terms of wages and benefits. This is because, if the people earns more, it makes people more greedy and take more advantage of enjoying power. In order to overcome this, the Government distributes the goods and services.
- ☆ ***Employment:*** All citizens are the hired employees of the State so that, everyone is equal and cannot get better or more powerful than anybody else.
- ☆ ***No freedom of speech:*** In Communism, there is no freedom of speech and if anyone speak out of line, he appears to be greedy, wanted personal wealth and abused people. If any one goes against the State and did not believe in Communism, he will be re-educated about what is Communism and explain him where he went wrong and if he still criticizes the Communism, he will be killed.
- ☆ ***Management:*** In Communism, there is only one political party, the Communist party, which runs the Government. The control of the State can be secured only through revolutionary methods and measures.
- ☆ ***Communist party:*** Only one political party, the Communist party, runs the government. According to communists, the revolution should be conducted on a world-wide scale, as the interests of the labour or working class are more or less the same everywhere. Communist parties in different countries are, therefore, closely linked with each other. After acquiring control over the State, the communists must resist the capitalists against their exploitation and dictate the terms for the formation of egalitarian society. Then, for designing policy formulations for the State, a set of people *i.e.*, the Communist party, who represent the interests of the working class should be identified and are trained for that said purpose. The communist party aim at promoting equality among the people and they serve as the supervisors of the system.

Thus, revolution, dictatorship and totalitarian planning are the features of Communism. It is a Government by bullets and not by ballots. The advantage of totalitarian planning is witnessed in Soviet Russia in the form of rapid industrialization. Because of deliberate concentration of heavy industries and restriction of consumer goods, by collectivization of farms and the transfer of agricultural surpluses towards industries etc., within a short period, Soviet Russia has succeeded in realizing the benefits of industrialization. Of course, it is equally accompanied by much suffering of the people and at the loss of personal liberty. Thus, there are many benefits and dangers of communism. The experiences of various countries have shown its weaknesses and drawbacks. However, Communism has remained only as a dream for all the Socialist Republics of Europe and Asia. Even the Socialism in its totalitarian form has not yielded significant positive results in the USSR.

Merits of Communism

The following are the merits of Communism:

- ☆ ***Stable economic system***: As State plans and controls the whole economy, it ensures internally stable economy system.

- ☆ *Stronger social communities*: By promoting equality among the people in the country, it builds strong social communities, so that, they have common needs and goals. Of course, there is equality for everybody except for the Communist party, who has an advantage over the citizens.
- ☆ *Equality:* Both men and women are treated indifferently and women get equal opportunities like men.
- ☆ *Public welfare:* State controls the public welfare measures such as schools, health care, alleviation of poverty, housing etc.
- ☆ *Full employment:* There is virtually no unemployment in the economy, as all the citizens are the employees of the State.
- ☆ *Business fluctuations:* Because of totalitarian planning, business fluctuations in the economy can be eliminated.

Demerits of Communism

The following are the demerits of Communism:

- ☆ *Lack of inventiveness:* As the Government owns all the businesses and properties, leaving no scope for private ownership, it will adversely affect the inventiveness and innovativeness of the workers in the economy. So, it will hamper the production of creative or luxurious goods. Moreover, the entrepreneurial abilities and skills are taken for granted.
- ☆ *No freedom of speech:* This hampers the identification of prioritized issues or problems in the society. So, the resources allocated based on supply driven approach by the State, may not be utilized in an efficient manner.
- ☆ *Lack of common goals:* Large or geographically broad populations tend to be diverse, making it difficult to maintain a common goal or set of rules for shared efforts and resources.
- ☆ *Central planning is difficult:* Complete central planning is difficult to execute, as the needs and goals of large diversified populations greatly differ.
- ☆ *No consumer's sovereignty:* There is no scope for consumer's sovereignty, as the approach is totally supply driven from the State.
- ☆ *Low productivity:* The complete absence of profit motive hampers the productivity and efficiency of workers in the organizations.
- ☆ *Balance between demand and supply:* It is difficult to achieve internal balances between supply and demand without a price mechanism.
- ☆ *Corruption and Bureaucratization:* Since 'the state' owns the means of production, the individuals highly placed in the Communist party tends strongly toward corruption. Disassociated from his own self-interest and dominated by the corrupt and bureaucratic apparatus of the Communist party, the average individual found himself politically powerless, poorly motivated economically and unable to enjoy most of the fruits of his own labour.
- ☆ *Misallocation of resources:* Complete absence of automatic price mechanism leads to misallocation of resources at larger scale.

It can be said that, Communism is only the ideal imaginary world of Karl Marx. This is because, in the present day context of globalization and trade liberalization, it is very difficult to run the economic system with totalitarian planning from the Government.

3.4. Mixed Economy

Capitalism and Socialism, discussed earlier, have their own merits and demerits. An exclusive reliance on either of these systems would expose an economy to the peculiar drawbacks inherent in it. So, a mid-way between these two economic systems is important, as it leads to a judicious policy of having both State control and at the same time free enterprise economy. Mixed economy is used in context to the simultaneous working of the two sectors, public and private *i.e.* both economic system capitalism and socialism survives in same time. In this context, Mixed economy is an important outcome of having the meritorious features of both Capitalism and Socialism and it is regarded as an ideal compromise between Capitalism and Socialism. So, Mixed economy is a system in which both public sector and private sector co-exist and each of the sectors has respective roles to play towards the welfare of all the sections of society. The public sector functions as a socialistic economy with a welfare motive and the private sector, as a free enterprise economy with a profit motive. Thus, Mixed economy implies demarcation and harmonization of the public and private sectors. In it, free functioning of the market mechanism alone is not permitted and the Government intervenes or regulates the private sector in such a way that, the two sectors become mutually re-enforcing. So, a Mixed economy represents an achievable balance between individual initiative and social goals. Planning and market mechanisms are so adjusted that, each is used for realizing the objectives of the economy to which it is most suited. There is a commitment on the part of both the sectors towards national objectives and priorities. A Mixed economy possesses the freedom to hold private property, to earn profit, to consume, produce and distribute and to have any occupation. But, if these freedoms affect public welfare adversely, they are regulated and controlled by the State. This implies, the freedom in the economic activities is influenced by the Government's regulation and licensing policies. So, in a Mixed economy, there is the presence of private economic freedom with centralized planning and with a common goal of avoiding the problems associated with both Capitalism as well as Socialism. It is too difficult to define a country's economy as Capitalist, Socialist or Mixed. But, as the experiences tell, the role of the Government has increased significantly in all the economic systems, particularly after the world-wide depression. It is important to note that, Mixed economy is not merely an economic concept and the rights of the individual are respected and protected subject only to the requirements of public law and order and morality.

The origin of the concept of 'Mixed economy' dates back to early thirties of the 20th century, when the weaknesses of Capitalism got exposed in many parts of the world during the period of the Great Depression (1929-1933). Due to inherent weaknesses in Capitalism, many countries could not recover from the set backs of the depression, thereby, leading to adverse falls in production, prices and employment. This actually called for the intervention of Governments in different countries to restructure the economies in the form of increasing investments, announcing subsidies, meeting food security needs of the population, expanding credit facilities etc. So, the role of Government in different economic activities started assuming greater significance. Some of the policy makers in different countries even advocated the substitution of Socialism instead of Capitalism, as it helped to recover the set backs of depression. One of the greatest Economists of 20th century, J.M. Keynes, opined that, pure Capitalism with its inherent defects was not a permanent solution for various economic problems. At the same time, he further opined that, dictatorial Socialism would be harmful to individual freedom. So, he suggested a compromise between the two forms of economic systems *i.e.*, free Capitalism on the one side and Socialism on the other. So, it was from those ideas of Keynes and other Modern Economists that the concept of Mixed economy saw the light of the day. This mid-way concept is flexible and has its own means and methods of approaching economic, political and social issues. To achieve clarity in understanding the concept of Mixed economy, the essentials of Mixed economy are discussed here under.

Essentials of Mixed Economy

The following are the features of Mixed economy:

☆ *Co-existence of public and private sectors:* In a Mixed economy, both the State as well as the private ownership takes part in the means of production, distribution and other types of economic activities. Thus, it has the features of both Government or Planned economy and a Free market. So, it has two sectors *viz.*, Public and Private and there is a clear demarcation of the boundaries between these sectors. The core sectors, strategic sectors and the public services that private businesses cannot afford to finance *viz.*, public transportation, law and order, public health care, defense, judicial system etc., are invariably in the public sector, which is managed by the State or Government. Thus, the most strategically and nationally important sectors of the economy will be reserved for the public sector. The economic viability criteria for investment decisions in the public sector are based on social cost-benefit analysis. The Government intervenes in the economy to prevent undue concentration of economic power, monopolistic and restrictive trade practices, looks after the food security needs of the weaker sections through public distribution system, execution of poverty alleviation programmes etc., so as to achieve social welfare. The private sector is that part of the economy, which is owned and operated by private bodies, sole traders or partnership or joint stock companies. Profit motive influences the decision-making in the private sector. The private sector is regulated and controlled by the State against its excessively optimistic or pessimistic postures and moderated through fiscal and monetary measures. Government's overall economic policies and economic planning have a powerful influence on the economic decisions in private sector. So, Mixed economy is a free market economy with Government intervention thrown in and hence, it is regarded as a balance between the market economy and the planning mechanism. Since, both public and private sectors cannot function in isolation, the co-operation between the two sectors counts much. In India, public sector offer funds for various developmental projects in which private sector also execute the activities. In general, the funds available with the public sector are more compared to the private sector and hence, the healthy competition between these two sectors also help the private sector to expand its activities. Besides these two sectors, there are other important sectors in the Mixed economy *viz.*, joint sector (where both Government and private enterprises will execute the activities on shared basis) and Cooperative sector (where co-operative methods of production and distribution are encouraged). Thus, the important feature of the Mixed economy is the functioning of various sectors.

It is incorrect to regard every country as a Mixed economy just because some features of Capitalism or Socialism are present in that system. That means, mere existence and functioning of both public and private sectors in an economy will not make the economy a 'mixed' one. In practice, even capitalistic economies have 'public sector' and socialistic economies have 'private sector'. But, they are not regarded as 'Mixed economies'. Countries like Sweden, Norway, Austria, France, India, Pakistan, Israel etc., are mixed economies. A Mixed economy must have the structural characteristics and also profess the social democratic ideology. Through law, the different sectors in the Mixed economy should be well-defined and the scope of each sector should also be clearly spelt. Countries, that put greater stress on decentralized socialist market tend to approximate to or are approaching a Mixed economy in the allocative aspect. Capitalist countries that put more stress on an egalitarian distribution of property and incomes (Japan, South Korea, Taiwan and Singapore) are approaching the Mixed economy ideal from the other end. Thus, even though, Mixed economy is a mixed form of Capitalism and Socialism, it has an identity of its own. The evils of extreme economic systems of pure Capitalism and pure Socialism are avoided in a Mixed economy. Therefore, it presents a middle path between these two. For example, in India, the fluctuations in trade cycles have induced the Government to create some institutions that will help in maintaining economic stability and they include *viz.*, The Planning Commission, Tariff Commission, The Commission for Agricultural Costs and Prices, Foreign Investment Promotion Board, Monopoly Enquiry Commission, Finance Commission, Securities and Exchange Board of India etc. Also, the Apex bank of the country, the Reserve bank of India has wide power to intervene in both money and foreign exchange markets. The provision of powers to these institutions represents a broad public interest in reducing the instability that are associated with the free operations of market mechanism.

☆ ***Government intervention in market mechanism:*** Mixed economy envisages a role both for price system or market mechanism and Government directions or guidelines. In market mechanism, demand and supply forces determine the prices of commodities and services. Prices, in turn, direct the business activities in the economy. If prices rise in the economy, it catches the attention of new firms and they enter the industry. On the other hand, if prices fall, less competitive producers either leave the industry or start producing different goods. If inflationary or deflationary trends are observed in the economy, the Government imposes regulatory forces and they serve as a guide to the private sector in its functioning mechanism and at the same time, they check on the exploitative nature of private sector. Thus, when the automatic price system fails to reap benefits either to producers, consumers and other stakeholders, then the Government intervenes in the price mechanism, so as to safe guard their interests.

In factor market, Government enacts laws to check the exploitation of factors of production. It does not remain mere spectator, while market mechanism plays havoc with their interests. Government also addresses issues beyond the reach of market forces. It provides welfare and unemployment benefits to people, cost of medical care for the aged, subsidies to below poverty line people, regulates private industry to limit air and water pollution, provides low-cost loans to people who suffer losses as a result of natural disasters etc.

Government also intervenes in the economy to provide goods and services, which private businessmen are hesitant and unable to produce and provide them in sufficient quantities at a reasonable price. For example, production of socially desirable goods such as, equipments for the mentally retarded and handicapped people of society may be unprofitable for private businessmen and hence, Government produces such goods in State owned factories. Government puts ban on the imports and restricts the local production of those items, which may be harmful for the society such as drugs, narcotics etc.

☆ ***Role of co-operative sector:*** In the words of Horace Plunkett, *'Co-operation is self-help made effective by organization'*. In mixed economy, co-operative organizations play a vital role in different sectors like production, marketing, banking etc. These organization functions with a motto, 'each for all and all for each'. So, protecting the weak is the basic objective of cooperative organization. For example, in India, there are co-operative farming societies, co-operative marketing societies, co-operative finance institutions etc., and these organizations serve as a link between the Government and the public for promoting the welfare of the masses. They function with a non-profit motive.

☆ ***Dual pricing:*** It implies prevalence of two different prices for the same commodity. Dual pricing of essential commodities rice, wheat, sugar, edible oil etc. is the basic feature of mixed economy. In this economy, the prices are determined by two market forces *viz.*, demand and supply. Such prices are called open market prices and generally they are very high compared to the cost of production of commodity. These essential commodities at such high prices are not affordable by the people below poverty line. So, the Government offers these essential commodities to them at lower prices through fair price shops. Such prices are called Issue prices and they serve as a social security measure. This mechanism of having two different types of prices for the same commodity refers to dual pricing and it is an important feature of mixed economy.

☆ ***Planning the economy:*** The existence of different sectors in the mixed economy and fixation of achievable targets across the sectors calls for scientific planning at the national level. In India, there is a Central Planning Commission to draft Five Year Plans duly incorporating the nation's goals and the roles and targets of each sector towards the achievement of these goals. All the sectors in the economy *viz.*, public, private, joint and cooperative sectors will design their activities such that, they will fit into the broad framework of the designed plans. So, planning will direct the relative roles of different sectors and their respective jurisdictions. The ownership of means of production of different sectors is so decided such that, there is a balance between personal and social incentives and sectional and general interests. The forces of demand and supply and market prices influence the resource allocations in different sectors of the economy. In general, for public sector, there is 'Planning by direction' and for private sector, there is 'Planning by inducement'.

Merits of Mixed Economy

The following are the merits of Mixed economy:

- ***Efficient utilization of resources:*** The resources are utilized efficiently, as good features of both Capitalism and Socialism co-exist. If there is misallocation of resources, the State controls and regulates it.
- ***Consumer's sovereignty:*** In Mixed economy, the consumer's sovereignty is largely protected. Consumer priorities and preferences are duly reflected in production decision making in both public and private sectors. The consumer's role is so great, in fact, that the nation is sometimes characterized as having a 'consumer economy'.
- ***Consolidation of merits of Capitalism and Socialism:*** Mixed economy is expected to retain the merits of both Capitalism and Socialism. This implies both profit motive and social interest are blended. For example, the Government is expected to allow private investment, but at the same time, it controls monopolies. Private sector works with a profit-motive and aims at capital accumulation and rapid industrial expansion. Public sector functions with a welfare-motive and it executes activities on par with the private sector. Sometimes, if the private firms suffer from losses, they are nationalized and allowed to function in the interest of the public welfare. The Government allows the profit motive of private sector only to a certain extent and will not allow it to go for exploitative monopoly gains in the market against the welfare of the people.
- ***Rapid economic development:*** In mixed economy, rapid economic development is possible because of scientific planning by the public sector. It is a very useful instrument regarding the creation of savings, mobilization of savings and investment of savings in the economy and this leads to capital formation and hence, economic development.
- ***Incentives:*** In mixed economy, incentives will be offered to both producers and consumers in the economy. The incentives will be in the form of subsidies and grants. For example, in India, the Government is supplying fertilizers to the farmers at subsidized rates.
- ***Prices are administered:*** In general, the price determination is automatic, as influenced by demand and supply forces. But, in the case of goods which are scarce, the prices are administered by the Government and such goods are also rationed. For example, in India, The Commission for Agricultural Costs and Prices will recommend the Government about Minimum Support Prices of agricultural commodities.
- ***Social welfare:*** In a Mixed economy, planning is centralized and there is overall welfare. Workers are given incentives and reward for any inventions. There is social security provided to the workers. Inequalities of income and wealth are reduced.
- ***Healthy competition:*** The healthy competition between private sector and public sector will guarantee quality production.

Demerits of Mixed Economy

The following are the demerits of Mixed economy:

- ***No harmony between public and private sectors:*** Sometimes, the demerits of both Capitalism and Socialism may appear in the Mixed economy, thereby, adversely affecting the functioning of the economic system. The public sector, which dominates the scenario in the Mixed economy, may not allow the private sector to function up to its fullest efficiency by imposing larger restrictions. The public sector may expand to a greater extent, such that, it may devour the petty private sector and the economy may slip into complete Socialism, instead of Mixed economy. On the other hand, if the private industrialists are powerful, they may not cooperate with the public sector in various ways and this makes the economy more like capitalistic economy. Moreover, public sector spends huge capital resources in the economy and works with welfare motive, whereas, private sector uses these capital assets with a profit motive and lacks social responsibility. Further, the private sector firms also aim at monopoly gains from the market through formation of Cartels. Conversely, the public sector firms are often criticized for their poor management and inefficient production. So, the lack of harmony between these two sectors is the major set back in Mixed economy and this leads to inequalities of income, concentration of economic power, labour problems, misallocation of resources, bureaucratic administration, loss of efficiency and productivity, loss of liberty etc. This makes the economy less efficient to achieve its broader goals and objectives. Moreover, the disadvantages of a Mixed economy really depend on how 'mixed' it is. For example, if the economy is mixed more towards a free-market, it indicates there is little regulation and this may affect the public welfare motive, but if it is mixed more towards a command economy, it may affect the profit motive of the private sector.

- *Improper planning:* It is very difficult to plan and take fruitful economic decisions in Mixed economy. This is because, there is no guiding mechanism as seen in Capitalism or a full-fledged planning mechanism, as seen in Socialism. In the absence of these two guiding factors, the planning is not proper and efficient, thereby, the performances of different sectors in the economy are not satisfactory upto the level of expectations and this also hampers coordination among them.
- *Bureaucratization and red tapism:* Power is concentrated in the hands of public sector, which takes all decisions regarding investment, production, distribution and consumption with reference to all the sectors in the economy. This leads to bureaucratization, red tapism and a very cumbersome and expensive system of administration, which cannot deliver the goods.
- *No direct control on private sector:* As the mode of planning differs across the public and private sectors, the Government cannot exercise direct control on the private sector. But, the it can formulate indirect control measures on the private sector through the execution of monetary and fiscal policies. But, the indirect control measures are less effective compared to direct control by the Government.
- *Instability:* If the public sector dominates the scenario in the Mixed economy, it may affect the functioning of private sector. The private sector may lose its individual identity and simply remain as the dependent sector on the public sector. This will definitely show adverse influences on the allocation of resources, pricing, output, distribution etc., as everything is to be moved under the directions of the Government. But, if private sector remains aloof, the economic plans may not be fulfilled. This is because, in general, the private sector will be subjected to many ups and downs in its activities, resulting in periodic recession and unemployment.

However, when compared to Capitalism, the shortcomings of Mixed economy can be easily rectified. The major set back in Mixed economy will arise only due to lack of coordination between different sectors functioning in the economy. If this is fulfilled, the economy functions as per the planned expectations. So, the success in Mixed economy depends on the integrity of different sectors both at the administrative and execution levels. If the Government views all the other sectors in a harmonious manner through democratic concepts, the best balance of different sectors can be achieved and this leads to successful operation of the economic system.

3.4.1. Features of an Indian Economy: A Mixed Economy

Indian economy is a mixed economy in having different sectors like public, private, joint and co-operative sectors working in coherent manner. It is the eleventh largest economy in the world in terms of nominal GDP and the fourth largest by purchasing power parity. Between 1947 to 1991, India was under social democratic-based policies characterized by extensive regulation, protectionism, public ownership and slow growth. However, with the advent of liberalization policy since 1991, India transformed towards a market-based economy from self-regulated economy. This led to the formulation of need based economic policies and this witnessed in the form of significant economic growth during the post-liberalization period and by 2010, India had established itself as the world's second-fastest growing major economy. Economists predicted that, by 2020, India will be among the leading economies of the world, as India harbours sound natural resource base, a very large pool of human resources, a growing large pool of scientific manpower etc.

In India, there are three major occupational sectors *viz.*, agriculture, industry and service sectors. In terms of providing employment opportunities, agricultural sector is the predominant occupation, offering for about 52 per cent of total employment in the country. Service sector stood second followed by industry with 34 and 14 per cent shares respectively. In terms of contribution to country's GDP, service sector stood first with 55 per cent share followed by industrial and agricultural sector with 28 and 17 percents respectively. Comparing the present trends of various features of Indian economy (after trade liberalization) with the features prevailed at the time of Indian Independence, revealed the following interesting points:

- Indian economy still remains as an agrarian economy. India ranks second world-wide in farm output. Agriculture and allied sectors like forestry, logging and fishing accounted for 17 per cent of the GDP in 2009. Despite a steady decline in terms of share in the GDP, agriculture still remains as the largest economic sector and plays a significant role in the overall socio-economic development of the country. Yields of all crops have grown since 1950 due to steady improvements in irrigation, technology, application of modern agricultural practices, provision of agricultural credit and subsidies etc. However, international comparisons reveal that, the average yield in India is generally 30 to 50 per cent lower than the highest average yield in the world for majority of the crops.

- ☆ Regarding labour force, more than 50 per cent of the working population is engaged in agriculture. This percentage is very high compared to developed countries.
- ☆ Industry accounts for 28 per cent of the GDP and employs 14 per cent of the total work force. In absolute terms, India is 16th in the world in terms of nominal factory output. With the advent of trade liberalization, Indian industries had to face stiff competition with foreign firms and some firms are even privatized in order to revitalize them. Post-liberalization period increased the threat especially to the firms, which produce the products with no emphasis on cost-effectiveness and due to cheaper imports from China.
- ☆ Textile manufacturing is the second largest source for employment after agriculture and accounts for 26 per cent of manufacturing output.
- ☆ India ranks 15th in services output. It provides employment to 23 per cent of work force and it is growing fast at a rate of 7.5 per cent in 1991–2000 compared to 4.5 per cent in 1951–80. It has the largest share in the GDP accounting for 55 per cent in 2007, up from 15 per cent in 1950.
- ☆ The Indian Information Technology is spread across four key sectors *viz.*, information technology services, information technology enabled services, software and e-business. These sectors account for a revenue more than $85B in 2008. All these key sectors are fastest growing sectors contributing to one third of the total output of services. The present share of information technology sector in the country's GDP is around seven per cent in 2008 and this share is significant compared to the past.
- ☆ The accelerated economic growth in India particularly after trade liberalization was mainly due to significant contributions from the service sector. Its share in country's GDP is 55 per cent and hence, the success in this regard has been termed as 'India's Services Revolution'.
- ☆ Organized retail marketing is gaining more significance and growing at a faster pace.
- ☆ Unemployment in India is characterized by chronic or disguised unemployment. With the advent of globalization, private sector gained momentum in the country and the decreased role of the public sector opened up more employment opportunities in the private sector. It is estimated that, the labour force in India is growing at 2.5 per cent annually, but employment only at 2.3 per cent.
- ☆ India's per capita income has increased from Rs.255 in 1950-51 to Rs. 44,345 in 2009-10, but still it is one of the lowest figures in the world. Countries like USA and Japan have more per capita income than India.
- ☆ As on 2007, India's share in global trade is only 1.5 per cent. According to the World Trade Statistics of the World Trade Organization, in 2006, India's global trade covering both merchandise and services was of the order of $437 billion, up by a record 72 per cent from a level of $253 billion in 2004. India's trade has reached a still relatively moderate share 24 per cent of GDP in 2006, up from 6 per cent in 1985.
- ☆ In the field of banking, the public sector banks hold over 75 per cent of the total assets of the banking industry, while the private and foreign banks holds 18.2 and 6.5 percents respectively. Since liberalization policy (1991), the Government has implemented significant banking reforms. While some of these relate to nationalized banks (like encouraging mergers, reducing government interference and increasing profitability and competitiveness), other reforms have opened up the banking and insurance sectors to private and foreign players.
- ☆ Being the founder member of General Agreement on Tariffs and Trade in 1947 and its successor, the WTO in 1995, India actively participated in the negotiations at different Rounds of WTO. India has continued its opposition to the inclusion of issues like labour and environment and other non-tariff barriers into the trade framework. The value of India's international trade has risen to 63,080,109 crores in 2003–04 from 1,250 crores in 1950–51. The major trading partners of India include *viz.*, China, the USA, the UAE, the UK, Japan and the European Union. India enjoy export competitiveness with reference to pharmaceuticals, gems and jewellery, textiles and garments, agricultural products, iron ore and other minerals. Major import commodities included crude oil and related products such as machinery, electronic goods, gold, silver etc.
- ☆ The pubic sector plays a dominant role in Indian economy. The economic activities are carried out through planning process. Between 1951-52 to 2010-11, Indian Government has completed 11 Five Year Plans (FYP) and outlay during this period has increased from Rs. 2400 crore in I FYP to Rs. 3644719 crores in 11th FYP. Presently, 12th FYP is in progress.

Performances of Different Sectors

The Mixed economy in India comprises of a well-regulated private sector and the fast expanding public sector. Besides these two, there are other sectors *viz.*, joint and cooperative sectors. The analysis of the performances of these sectors highlights the performance of Indian economic system as a whole.

1. Public Sector

Enterprises owned and run by the Government are called public sector enterprises. In India, the public sector enterprises include *viz.*, road transport, railways, airlines, large factories like Indian Oil Corporation, banking, insurance, posts and telegraph etc. Since past three decades, the public sector's share in domestic investment has been nearly halved, but its output share has remained roughly constant at about a quarter of GDP, suggesting a sustained rise in productivity. With the advent of globalization and trade liberalization, the Indian Government realized that, a strong and growth oriented nation could be built only if India grows as a part of the world economy and not in isolation. So, one of the important policy measures taken by the Government is the reduction of investments in different public sector ventures and encourage privatization, as it ensures more accountability for the use of resources.

2. Private Sector

The enterprises owned and run by individuals or a group of individuals with an aim of profit making are called private sector enterprises. They include *viz.*, retail and wholesale shops, factories like Hindustan Lever, Reliance etc. As per the recent reports, in the decade 2000-2010, the private corporate sector overtook the public sector both in terms of net sales and net profits. The private sector's share in the net sales of manufacturing and services sectors output increased from 48.83 per cent in 2000-01 to 68.55 per cent in 2009-10, while the public sector's share consequently declined from 51.17 per cent to 31.45 per cent. Similarly, the private sector's share of net profit in the non-agricultural economy increased from 39.17 per cent to 63.86 per cent for the same period, with a decline in public sector share from 60.83 per cent to 36.14 per cent. This phenomenal growth of private sector contribution in India can be attributed to political will, financial reforms, rise of enterprises especially in the energy, telecommunications, civil aviation, manufacturing, finance and banking and information technology sectors, usage of more advanced technology, increase in foreign direct investments etc. On the other side, the decline in public investment has contributed to a decline in the share of the public sector.

Earlier, the Indian market was ruled by the Government enterprises, but the scene was changed towards privatization, when the markets were opened for investments in the era of globalization and economic liberalization. This prompted the private sector companies to produce the products suiting to the needs and wishes of the customers. This further fueled healthy competition between the public and private enterprises. This also went further and forged joint venture private Indian companies, especially in sectors like, telecommunication, petroleum, housing and infrastructure. This inculcated healthy competition and finally the end consumers are benefited, since the cost of service or products come down substantially. The following are the indicators highlighting the growth of private sector in India:

- ✰ India leads developing nations in private sector investment
- ✰ Manufacturing industries, hotels, financing, insurance, real estate and communication registered positive growth during the past two decades (since early nineties).
- ✰ The trade in food sector is estimated to be of US$ 200 billion and it is expected to grow to $310 billion by 2015
- ✰ Merchandise exports recorded significant growth during the past decade.
- ✰ Transport has also become an important sector attracting 34 per cent investment commitments in 2006.
- ✰ India attracted more investment commitments to infrastructure projects with private participation in 2006 than any other developing country. Indeed, commitments in India were nearly twice those in its nearest rival, Brazil and well ahead of those in China.

3. Joint Sector

Enterprises, which are jointly owned and executed by both the Government and individuals (private sector) are called Joint Sector Enterprises. An enterprise is called a joint sector enterprise, if the State Industrial Development Corporation or Government holds 26 percent equity or more, private promoter 25 per cent and the balance by the general public. The concept of a joint sector is basically an extension of the idea of a Mixed economy. In joint sector, the relationship between public and private sectors is much closer and more inter-reliant, as these two sectors executes the work as a team under unified management towards the desired objectives. Thus, the meritorious features of these two sectors will co-exist. This concept is very attractive and ideally suited to the mixed economies like India. In India, the joint sector concept became stronger in the 1980s, following its success in such countries as Japan, South Korea, Taiwan, Singapore and Hong Kong. The following are some of the examples for joint ventures in India:

- ✰ IFFCO and KRIBHCO in the fertilizer industry.
- ✰ Indian Railways.
- ✰ Industries like steel, automobiles, cement and banking.

Table 3.1: Comparative features of major economic systems.

Features	*Capitalism*	*Socialism*	*Communism*	*Mixed Economy*
Sectors involved	Private sector dominates the economic system. The role of public sector is limited to the issues like defense, maintenance of law and order, foreign affairs, currency and coinage and some important civil works such as the construction of roads and bridges, which the private entrepreneurs may not find them as profitable ventures.	Public sector dominates the economic system. There will be private sector in socialist economy but, the private operations are very negligible, insignificant and not capable enough for playing an influential role in the economy.	Only public sector prevails and there is complete absence of private sector.	Both public and private sectors co-exist. It is mid-way between Capitalism and Socialism.
Motive	Profit-motive is the only motive.	Social-welfare motive is the only motive.	Social -welfare motive is the only motive, so as to create classless egalitarian society.	Both profit-motive and welfare motives are seen in this system.
Means of production	The various means of production are under the private ownership of individuals. The people have full right to use their own property in any manner as they like and this right is protected by the Government.	The various means of production are owned by the State, indicating collective ownership. Private ownership is limited and no individual can accumulate too much property as in the case of Capitalism.	The entire means of production are owned by the State. No private property is allowed.	This system allows both public and private ownership of means of production.
Price mechanism	The price mechanism is automatic, as influenced by domand and supply forces in the economy.	Government does interfere in the fixation of prices of commodities and all the economic activities are centered upon its decisions. Market forces have limited role to play in directing these economic activities.	Completely dictated by the State.	Mixed economy envisages a role both for automatic price mechanism and Government directions or guidelines. When the automatic price mechanism fails to reap benefits either to producers, consumers and other stakeholders, then the Government intervenes in the price mechanism, so as to safe guard their interests.
Economic freedom	The individual can plan any business for which he has the necessary means. He can take independent decisions with regard to what, where, how and when to produce. Similarly, consumers have the freedom to buy anything they want, workers are free to take employment wherever they get higher wages and investors are free to invest wherever they get higher money returns.	The production is carried out mainly by the State. There is no significant room for private enterprises and these private operations are very negligible, insignificant and not capable enough for playing an influential role in the economy.	The production is carried out only by the State. There is no scope for private enterprise to operate in the economy.	Mixed economy is a free market economy with Government intervention thrown in. So, there is considerable economic freedom in the mixed economy, as private sector and public sector co-exist. The Government allows the profit motive of private sector only to a certain extent and not allows the private sector to go for exploitative monopoly gains.
Economic equality	It is not possible to achieve economic equality in capitalist economy due to private property, less Government interference, profit motive, freedom of enterprises etc.	Socialism aims at minimizing economic inequality, if not completely achieving economic equality. This is possible due to the absence of private property and profit motive.	Aims at ensuring classless egalitarian society. But, this is quite difficult and purely imaginary, especially in the era of globalization and economic liberalization.	Aims at economic equality, as public sector functions with welfare-motive. Though private sector aims at profit maximization, its activities are regulated by the Government, so as to ensure public welfare.

Contd...

Table 3.1–*Contd...*

Features	*Capitalism*	*Socialism*	*Communism*	*Mixed Economy*
Consumer's sovereignty	The consumer is considered as a king under Capitalism because, his wishes and preferences will direct the economic activity. But, in modern days, this consumer's sovereignty is only a myth and not a reality, because several limitations like development of monopolies and oligopolies in the market economy, advertisement and propaganda, restrictions imposed by the Government, frequent changes in fashions, habits, tastes of individuals etc., affect the consumers to execute his full powers.	Consumer's sovereignty is limited, as they cannot get the goods as per their preferences. Moreover, there is no significant room for private enterprises and private operations.	There is no scope for consumer's sovereignty, as the approach is totally supply driven from the State.	Consumer's sovereignty is largely protected here. Their priorities and preferences are duly reflected in production decision making in both public and private sectors.
Role of Government	The Government has minimal role to play in the economy, as the basic economic problems are expected to be solved by market forces. Further, the role of the Government is not only minimum, but also neutral. The State operates only in the fields, where private enterprises will be inadequate and cannot afford huge investments.	The Government in the socialist economy will take all the policy decisions of all the sectors for promoting the welfare of the society. Its role is crucial in Socialism, because of public-welfare motive, absence of profit making, price mechanism is not automatic etc. It regulates and directs the activities of private sector in the economy.	The State plans and controls the whole economy and it assumes greater significance compared to socialism due to complete absence of private sector. .	Government's role is very crucial in Mixed economy even compared to Socialism because, it has to plan its own sectors and at the same time it should impose regulatory forces to guide the private sector in its functioning mechanism. It should also check the private sector against its exploitative nature.
Class struggle	It is severe because of large economic inequalities.	Compared to capitalistic economy, it is lower in socialism, as the Government functions with the objective of promoting public welfare.	Aims at classless egalitarian society, which is purely imaginary.	Aims to check class struggle by reducing income inequalities through implementing various developmental programmes and policies.
Administration	Complete decentralization	Administration may be centralized or decentralized, but the power to take decisions rests with the public authorities.	Complete centralization.	Administration may be centralized or decentralized, but the power to take decisions rests with the public authorities.
Inducement to work	More, because the economy functions with profit-motive	Because of bureaucratic administration, the inducement to work is low. This kills the scope of performing innovative and inventive works. The workers suffer from lack of initiative, inability to take quick decisions and become lethargic in executing the job duties.	Since Government owns all the businesses and properties, there is absolutely no scope from the workers side for performing inventive and innovative works in the economy. This will hamper the production of creative or luxurious goods.	The workers capable of producing inventive and innovative goods will work in the private sector and reap higher benefits in proportion of their hard work. The Government also encourages such sectors by offering various subsidies and other benefits to the private companies.
Efficient production	Modern and cost-effective production techniques will be employed to ensure efficient production. Since, producers have full economic freedom in selecting the enterprises, it further ensures increased production of the commodities having comparative price	Since a socialistic economy does not suffer from fluctuations in business cycles, crises of over production and massive misallocations of resources, the production at any rate will be higher than what it would be under capitalism.	Because of totalitarian planning, business fluctuations in the economy can be eliminated and ensures higher production. But, this is purely imaginary, as the approach is completely supply driven.	It ensures efficient production in the economy, as both profit motive and welfare motive are to be attained.

Contd...

Table 3.1–*Contd...*

Features	*Capitalism*	*Socialism*	*Communism*	*Mixed Economy*
	advantage in the market. But, the fluctuations in production are common in capitalist economy, and thereby, increase in production is only illusory.			
Growth of entrepreneurship	Various features like private ownership, intense competition, profit motive etc., induce the entrepreneurs to become more efficient, competent, hard working and dynamic.	Bureaucratic administration will hamper the growth of entrepreneurship.	Complete centralization of powers will leave no scope to the workers for performing inventive and innovative works.	There is every possibility to excel in developing good entrepreneurial skills both in public and private sectors, as these sectors work in harmony towards the attainment of objectives.
Technological improvements	The element of competition and profit-making motive among the producers guide the researchers to develop new cost-effective technologies.	As the works' are performed monotonously, there is less scope of generating new technologies.	Complete absence of private ownership will check the technological developments.	As both public and private sectors work in harmony with a healthy competition, there is bright scope for generating and refining the technologies.
Liberty and equal rights	Everyone gets equal right to work, accumulate wealth and improve his standards of living.	Since, all economic activities are directed by the Planning Authority, the workers will have no choice of occupation. This curtailment of individual freedom and liberty would even go to the extent of exploitation of the individual.	There is complete absence of individual rights.	There is the presence of private economic freedom, but with centralized planning. But, if the private economic freedoms affect the public welfare adversely, they are regulated and controlled by the State.
Flexibility	It is more flexible, as can be adopted in different countries, to different peoples, to different technological changes and even with different degrees of Government intervention.	This is much more flexible compared to capitalist economy because, the State will allocate the resources as per the priorities of the economy keeping in view of the public welfare.	Absolutely not flexible in its approach.	More flexible compared to other economic systems, as it deals with twin objectives *viz.*, public-welfare and profit motive simultaneously.
Formation of monopolies	More common	Less scope of formation of monopolies, as there is no significant room for private enterprises.	Absolutely no chance.	Check the formation and exploitation by monopolies.
Social costs	More, as the entrepreneurs aims at profit making only.	Less	Minimum	Considerably high
Social security	Not concerned	Concerned	Concerned	Concerned
Approach	'Survival of the fittest'.	'You get your share no matter what your contribution is'.	'You put in your fair share, you get your fair share'.	Both 'Survival of the fittest' and 'Protection of the weak'.
Examples	USA, UK, Japan, Germany, Sweden.	Canada, Australia, Great Britain, Saudi Arabia, Iran, Burma.	China, Cuba, Vietnam, North Korea, Soviet Russia.	India, Pakistan.

- ☆ In a few cases, equity participation by foreign enterprises in the public sector enterprises was also allowed. Madras Fertilizers Ltd. for example, was established as a joint enterprise in participation with Amoco Inc. (USA) and National Iranian Oil Co.(Iran). The same foreign companies were partners in Madras Refineries Ltd too.
- ☆ Raasi Cement Ltd.
- ☆ Reliance Petrochemical Limited.
- ☆ Power Trading Corporation.

☆ Andhra Pradesh Fibres Ltd.

☆ Gujarat Drugs & Chemicals Ltd

The number of companies of the joint sector form is gaining increasing importance in India since last three decades. The formation of Tata Indian Oil Refineries (Tata Chemicals is the private party), Mangalore Refinery (Indian Rayon of the Birlas), Haldia Petro-chemicals (Goenkas) and Gujarat Heavy Chemicals (GTC industries of Dalmias) in the joint sector indicates a new phase in the process of Indian industrialisation. Nearly all States, irrespective of their relative levels of industrialization have utilized this mechanism to promote new corporations. The joint sector represents a wide spectrum of industries with the modern industries like electrical and electronics dominating the scene.

4. *Co-operative Sector*

A cooperative enterprise is a business organization owned and operated by a group of individuals for their mutual benefit. The members join the organization on voluntary basis, so as to meet their common economic, social and cultural needs. The era of trade liberalization gave equal opportunities to all the sectors in Indian economy to excel in the competitive environment. But, the disheartening aspect is that, the reform measures were mainly aimed at industrial and service sectors, comparatively neglecting agriculture and cooperative sectors. During post-liberalization period, this cooperative sector remained in the backdrop, inspite of its predominant position in various fields of economy. This is because, it was felt erroneously that, cooperatives would not be able to withstand the stiff competition posed by the dynamic private sector. But, there are certain enterprises in the cooperative sector such as Housing cooperative, Building cooperative, Agricultural cooperative, Co-operative wholesale society, Credit unions and Cooperative banking etc., are regarded as dynamic on par with the private enterprises.

The current status of co-operatives in India reflects both a threat and an opportunity. It is a threat, because cooperatives have failed, to a large extent, in delivering efficient goods and services unlike the private sector and an opportunity, because the new economic scenario will offer enough opportunities, which could be effectively utilized by the cooperatives to prove their case of continuation. The important strategies for the future include complete re-organization of cooperatives in the light of changing priorities and competitive environment without depending on any State support, enhancement of competitive strength through merging or by division, liquidation of non-viable societies, active involvement of members in day-to-day business aspects etc.

3.5. Differences Across the Economic Systems

The different economic systems studied earlier have their own merits and demerits. In the context of giving freedom to the entrepreneurs with a profit motive, capitalism weighs much over other economic systems. But, if social welfare is to be maximized, Socialism or extreme socialism *i.e.*, Communism should be given emphasis. However, to justify both *i.e.*, profit motive and social welfare, Mixed economy suits best. In the present day world, Mixed economy seems to offer better solution to overcome the defects of both capitalism and socialism. Most of the capital economies of the world have become more or less of 'mixed type', whether they are called mixed or not. Even in capitalist country like USA, State plays a very important role in the economic field. In this sense, all systems are mixed economies. Mixed economy is specially suited to developing economies like India, as in these economies both public and private sector investments are crucial for capital formation and to achieve overall economic development. Analyzing the features of different economic systems, it is felt appropriate by the author to differentiate the economic systems, as it helps the readers to compare the merits and demerits across the systems (Table 3.1).

4

Central Problems of an Economy

It is a known fact that, economic resources are limited, but human needs and wants are unlimited. This indicates, economic resources are not adequate enough to fulfill unlimited wants or to produce desired goods and services in sufficient quantities. This scarcity of resources in relation to unlimited wants leads to the problem of allocation of resources. So, every individual has to make a choice for a certain set of wants from among unlimited wants, which are to be satisfied by the available limited resources. In Economics, this very problem of scarce resources versus choice making among unlimited wants constitute economic problem. This economic problem is popularly regarded as 'basic', as it is commonly seen in all the sectors of different economic systems and gives rise to 'central or fundamental problems' in every economy. According to Robert Awh, *'economic problem is the problem relating to the necessity of the choosing what, how and for whom to produce and how to achieve economic growth'*. Explaining it, Prof. Friedman has also said that, *'whenever limited resources are used to satisfy unlimited ends, economic problems will arise'*. Prof. Eric Roll stated that, *'economic problem is basically a problem arising from the necessity of choice, choice of the manners in which limited resources with alternative uses are disposed off. It is the problem of husbandry of resources'*. So, the premise of the model of the economic problem is that, wants of all the members of society cannot be fully satisfied in a given time. In reality, they have been multiplying at rapid pace due to changes in incomes, tastes, habits, fashions etc. So, at any given point of time, there exists innumerable unsatisfied wants in a community. However, resources to meet these unlimited human wants are always scarce to the amount of natural and human resources available. So, it leads to competition among the use of scarce resources and this gives rise to pricing or valuation of the resources. This is because, if something is scarce, it will have a market value. Say, if the supply of a good or service is low, the market price will rise, provided there is sufficient demand from consumers. Conversely, if goods and services are in plentiful supply, they will have a lower market value because, supply can easily meet the demand from consumers. This clearly implies that, scarcity of resources is the root cause of all economic problems. Another peculiarity of use of scarce means is that, they have alternative uses. For example, a farmer can use the same piece of land for cultivating different crops like paddy, cotton chillies etc., taking up allied enterprises, constructing a house etc. Since, the resources are limited, he cannot fulfill all the wants simultaneously. So, he has to prioritize the wants depending upon their relative importance to him and to the extent of availability of resources. Similarly, a processing firm might wish to produce both fruit juices and khandsari products. But, due to scarcity of resources, the firm has to choose one among the two alternatives. Likewise, the resources of the different individuals in the economy are limited and they can be put into several alternative uses. So, both scarcity of means and their alternative uses are the main causes of origin of economic problems and this leads to choice making among the unlimited wants. 'Choice' may be defined as the range of options available to the individual, household, firm or Government, while making decisions to fulfill the prioritized wants. This concept of choice making is inevitable, while taking all economic decisions. Here comes the important role of an Economist to explain and analyze the nature of choice for addressing this basic economic problem towards the betterment of different sections of the society *viz.*, producers, consumers, Government etc. As this basic problem is common across different forms of economic systems, a net work of organizations will function efficiently in the society of each system to formulate different methods around the idea of efficient choice making to curb the economic problem. For example, in capitalistic economy, there is complete economic freedom and the individuals can plan their decisions or choice making on their own, as guided by the automatic price mechanism. In socialistic economy, major economic choices are best made for the societal welfare by the Government. However, in Communism, all economic choices will be made through Central Planning Authority of Government. In Mixed economy, the choice making mechanism lies with both State and the individuals, but the individual's choices should fit in the broad frame work of Government's programmes and policies and should not be exploitative in nature.

4.1. Causes of Economic Problems

Economic problems arise due to the following facts of economic life:

1. Unlimited Wants

Wants are the effective desires for a particular product or service, which can only be obtained by working for it. Human wants constitute the major driving force for the conductance of all economic activities. They stimulate the demand for goods and services and to produce them economic activities are executed. Since, human wants are unlimited, the individuals categorize their wants, prioritize them and allocate resources efficiently in an organized production to curb the economic problem. The basic assumption in taking economic decisions by the producers is that, humans are greedy and thus, they must produce as much as possible to satisfy them.

2. Scarce Resources

The scarce resources can be conveniently divided into four factors of production *viz.*, land, labour, capital and entrepreneur. Land includes all natural resources, which are free gifts of nature, labour includes all physical and mental efforts made by the individuals, capital includes man-made aids to production such as machines, tools etc., that go into further production and entrepreneurship is the art of combining the other three factors and bears uncertainity in the production programme. It is important to note that, the concept of scarcity does not mean 'few', rather it should be viewed from the demand side. So, the concept of scarcity is 'relative' in approach. For example, few mangoes in a basket are spoiled due to pest infestation. But, we won't pay price for them even though, they are scarce in nature. Farmers in India produced 233.38 m. tonnes of food grain in the year 2008-09. But, they are considered scarce because, there is lot of demand for food grains in the market and hence, we are paying price for them. This implies that, scarcity is an economic condition, where all the goods and services, though they may be plentiful, are not sufficient for all those who desire them. So, the concept of scarcity is also called as Relative scarcity.

The availability of factors of production differs from one economy to another economy and it directly influences the production of a wide range of goods and services in an economy. This is referred as Factor endowments and it clearly indicates that, those economies with a high factor endowment will lead to a better choice-making in the society. In Economics, we deal with only scarce resources, as we pay price for the use of them. Free resources are of less importance compared to scarce resources.

3. Alternative Uses of Resources

The peculiarity of scarce resources is that, they have alternative uses. For example, a farmer can allocate his limited resources among different enterprises like paddy, maize, wheat, cotton, chillies, dairy, sheep, goat etc. It is because of alternative uses of scarce resources, the concept of choice making arises. That means, in order to gain one alternative, the other alternative should be foregone and this leads to the concept of opportunity cost for gained alternative. In short, scarcity of means and their alternative uses are main causes of origin of economic problems.

4. Choice-making

Choice-making among unlimited wants is inevitable, while making economic decisions because of scarcity and alternative uses of resources. So, all economic problems fundamentally revolve around the idea of choice, which ultimately must answer the problem. This choice making guides the entrepreneurs in addressing the central problems in an economy. Consumers are considered as the biggest influences of this choice making, as their preferences, tastes, incomes etc., guide the entrepreneurs in producing the goods and services and moreover, the prices of the goods produced must be well in their reach of their money income. So, price mechanism will also guide the entrepreneurs in making choices regarding the production of goods and services. For example, the entrepreneurs tend to produce more of luxurious goods in the economy compared to basic necessaries, as they fetch high price in the market and have high elasticity of demand (E_D).

It is important to note that, choice making normally involves a trade-off *i.e.*, choosing more of one thing means giving up something else in an exchange mechanism. This is because, choice making is an unavoidable issue, while dealing with scarce resources. For example, trade-off is involved when the farmer faces the situation of purchasing a new tractor or hiring a tractor. There are both costs and benefits while taking the tractor on a rent basis or choosing to buy a new tractor with a mortgage. Both decisions involve a degree of risk.

4.2. Central or Fundamental Economic Problems

Resources scarcity leads to scarcity of goods and services. That means, as goods and services are the combinations of scarce resources, they themselves are scarce. So, the goods and services produced in the economy will not satisfy all human wants. This leads to choice making both from the producer's side and from the consumer's side in fulfilling their wants with their scarce resources. Since, the consumer's choice-making guide the producers in producing the goods and services, the scarcity of resources at the producer's level will lead to five inter-related problems, popularly regarded as Central or Fundamental problems of an economy. They include *viz.*, What to produce and How much to produce?, How to produce?, For whom to produce?, How to achieve fuller utilization of resources?, How to accelerate growth of resources?. That means, an economic system must decide the best allocation of resources, optimum enterprise combination, better distribution of output, understanding the resource relationships etc., irrespective of the nature and its level of development. The system of allocation of resources may differ across the economic systems, but these problems remain same across the systems. So, every economic system attempts to find ways to answer these central economic problems. The existence of various problems in the economy such as shortage, glut, starvation, famine, poverty etc., can be attributed to misallocation of scarce resources. The central problems of an economy are discussed here under.

1. What to Produce and How Much to Produce?

It implies, enterprise or enterprise combinations to be produced and how much quantity of each enterprise (commodity) should be produced in the economy. This is the first and foremost fundamental problem that every economy faces regarding what type of goods and services to be produced with the available scarce resources, such that, maximum wants of the people are satisfied. The major factor influencing this problem is that, the resources are scarce in relation to human wants. In fact, it is the problem of allocation of scarce resources among alternative uses and it is referred as resource allocation. Since, all the wants cannot be satisfied with the available scarce resources, it calls for categorization of wants, their prioritization and prioritized wants will be addressed through efficient allocation of resources. So, every economy has to make a choice about what wants can be satisfied and what can be foregone. In other words, for a rational allocation of resources, a society must set priorities among their unlimited wants. Thus, the wants which are decided to be satisfied calls for two more decisions *viz.*, what to produce and how much of it to produce.

(a) What to Produce?

It implies whether the society has to produce more of consumer goods or capital goods. Normally, a society tries to establish a 'balance' between producing consumer goods and capital goods, so that, it ensures good standard of living to majority of the people. If the economy aims at satisfying present wants compared to future wants, then production of consumer goods will be given more importance. On the contrary, if the economy aims at increasing the present production capacity, so as to satisfy future wants, then more importance will be given towards the production of capital goods in the present context. This clearly indicates that, if we give importance to one type of good in the production process, we have to sacrifice the resource usage on the other, thereby production level of the other good will be affected. For example, if the farmer diverts the resources towards paddy cultivation, prawn culture gets affected. This sacrifice made on the idea taking up prawn culture is known as Opportunity cost. So, opportunity cost is the cost equivalent to the returns from the next best alternative foregone. It is, otherwise, explained as the next best use to which the productive resources could have been devoted in the production process.

(b) How Much of it to Produce?

This implies, how much quantity of requisite goods should be produced in the economy, so as to fulfill the wants of more number of people in the economy. This problem arises when the economy decides what type of goods to be produced. This problem again depends upon the level of resources in the economy. That means, the level of resources guide the production possibilities of different types of goods *i.e.*, consumer goods and capital goods. We can ascertain such information through the Production Possibility Frontier (PPF), which was discussed in the ensuing pages.

2. How to Produce?

This fundamental problem arises from the first problem because, when the society decides what to produce and in what quantities, it has to allocate the resources now to produce such goods. So, the problem under consideration is 'how to produce' the planned goods and services in desired quantities. So, this problem is concerned with the choice of technique. The following aspects will be considered in dealing with this problem:

- ☆ Planning the resource usage in the production of desired goods and services and preventing their mis-usage on unwanted goods and services.
- ☆ Planning the different combinations of resources that can be used in the production of desired goods and services with out affecting the desired level of output. For example, organic manures and chemical fertilizers can be used in combinations to boost the yields of the crops with due emphasis on cost-effectiveness. Sometimes, the substitution between the resources is done, so as to ensure cost-effective production without affecting the level of output. For example, capital-intensive methods will be employed instead of labour-intensive production techniques.
- ☆ Identifying the firms or industries in the economy, which are committed towards the production of desired goods and services, so as to ensure maximum resource use efficiency. For example, if the economy decides to produce more of cotton and mangoes taking their export competitiveness into consideration, it must select such farmers who are willing to produce these commodities. So, these farmers, research agencies, export agencies and other relevant stakeholders must be provided with adequate resources, so as to ensure the desired level of output with cost-effectiveness.
- ☆ The economy has to take into consideration the least cost combination of resources to produce a given level of output of desired goods and services. In other words, it will have to decide upon the most appropriate technology to be adopted in the production programme. This concept of least cost combination of resources arises because of scarcity of resources. Had the resources available in plenty, this concept will not arise.

This concept also aids in normative aspect of Economics. We know, by using the Genetically Modified (GM) seeds of crops, output will increase significantly. But, many countries have banned their use. This is because, the adoption of GM technology will adversely affect the ecological balance. So, an Economist on ethical grounds should not recommend the adoption of GM technology.

3. For Whom to Produce?

This problem is related to distribution concept. That means, it deals with the question, how an economy distributes its total production among different economic units. In terms of income concept, it implies how the national income is distributed to each factor as factor incomes. That means, this problem deals with important questions like 'who gets how much of the goods that are produced in the economy? Who gets more and who gets less? How should the produce of the economy be distributed among the individuals in the economy? What is the best method of distributing the products to ensure the maximum level of wants and needs are met? Will some people get a bigger share of resources than other? Will some people get so few resources that they cannot survive, while others live in luxury? Should there be a minimum wage? If so, at what level should it be? etc. Thus, every economy faces the problem of allocating the scarce resources to the production of different possible goods and services and of distributing them among the individuals within the economy. In some countries, deliberate attempts have been made to reduce income inequalities by re-distributing wealth and income from the richer members of the community to the poorer members through execution of sound taxation policies. However, in capitalist countries, no policies of re-distribution exist and individuals are free to earn whatever they can. To address this problem, it is worth noting that, the Economists have to be aware of philosophy, politics, sociology and other subjects that form part of the decision-making process.

4. How to Achieve Fuller Utilization of Resources?

This implies, every economy has to decide about the scale of utilization of its resources. There should not be any wastage of these resources. This is because, economic development will get affected if some resources are wasted, under-utilized or kept idle. For example, if the economy makes large-scale exploitation of resources like coal, oil, natural gas etc., in the present use, then their availability in the future will go down. This planning of scale of resource usage determines the level of employment of other resources in the economy. This concept of fuller utilization of limited resources gained momentum particularly after experiencing Great Depression in 1930s and they are the important components under Welfare Economics. This problem also relates to achieving allocative efficiency and productive efficiency of resources.

(a) Allocative Efficiency

It implies, if the economy is producing a combination of commodities, it maximizes the overall level of satisfaction or welfare of the population. To understand this, it is important to realize that, the factors of production a firm employs are scarce and thus have an opportunity cost. If a firm is able to produce one more unit of a commodity by utilizing more factors of production, somewhere else in the economy a commodity must be foregone. This is because, the resources are scarce and not sufficient enough to produce everything we want. So, allocative efficiency occurs, when an economy allocates the resources in such as way that, it is impossible to reallocate them and further increase welfare. So, if all the resources are used efficiently in the production programme, the only way that someone can be made better off through the production and consumption of an extra good is, if someone is worse off, because they have to sacrifice a good. This condition, where it is impossible to make one person better off without making someone else correspondingly worse off is called as Pareto optimal allocation of resources. If it is possible to reorganize the production programme and make someone better off without making someone equally worse off, then clearly the welfare of society was not being maximized.

(b) Productive Efficiency

It is also known as Technical efficiency. It occurs when the economy is utilizing all of its resources efficiently *i.e.,* producing most output from least input. Thus, a productively efficient firm organizes its factor of production in such a way that, the average cost of production is at lowest point. This concept is explained through PPF, as any point on the frontier indicate the point of maximum productive efficiency. That means, a point on the frontier implies, the given level of resources can produce that maximum level of output of the two commodities under consideration and more output beyond the frontier is not possible. In simple words, when production of one good is achieved at the lowest possible cost, given the production of the other good(s), it indicates productive efficiency.

5. How to Accelerate Growth of Resources?

This problem is related to Developmental Economics. We know, increase in population is the common feature of the economy. So, it becomes necessary that, the rate of economic development must be faster than the rate of increase in the population, so that, it ensures sufficient standard of living to the people in the economy. To accelerate the economic development, the economy has to increase the capital formation. Through creation, mobilization and investment of savings, capital assets in the economy will be created and this, in turn, creates employment for other resources. To increase the rate of economic growth in a country, its social and institutional factors like, ownership of means of production, land reforms and techniques of production must undergo a radical change.

The above central problems are inter-related. Irrespective of the nature of economic system, every economy has limited means and unlimited ends. So, the economy has to take the decisions about the type of goods to be produced, techniques of production to be employed and for whom to produce, so as to satisfy maximum number of people in the economy. It is also required to make the best possible use of resources, so that, economic development takes place at a faster pace. To study and address these economic problems, the economic theory has been divided into Micro Economics and Macro Economics. The economic problems like, what to produce and how much to produce?, how to produce? and for whom to produce? are decided on the basis of price mechanism and hence, these problems are addressed through the concepts of Micro Economics. The other associated problems like how to achieve fuller utilization of resources? and how to accelerate growth of resources? are decided on the basis of aggregate concepts and hence, they are addressed through the concepts of Macro Economics.

4.3. Solutions to Central Economic Problems

So far, we discussed about the causal factors and the types and nature of central economic problems. Since these problems are inevitable for any economic system, it is essential to formulate the strategies to address the same effectively. In earlier period, people lived in traditional economies. In such an economy, the problem 'what to produce' is answered simply by producing what has always or traditionally been produced. The problem 'how to produce' is again determined by traditions *i.e.,* following the same pattern of resources allocation as done earlier. Regarding 'for whom to produce', the goods are distributed in the customary manner with shares being determined by the traditional values and rankings of the various members of the society. Typically, a traditional economy tends to be predominantly an agrarian economy, where agriculture is mainly subsistence oriented and not market oriented. However, with the modernization of economies, this traditional concept of addressing the central problems will not hold good and they are addressed mainly through two important methods *viz.,* Market or Price mechanism method and Central planning method.

1. Market or Price Mechanism Method

In this method, all the central economic problems are addressed by the free play of market forces *i.e.,* demand and supply in the economy. This mechanism of addressing the central economic problems is seen in economies, where the entrepreneurs have complete economic freedom regarding what and how much to produce, how to produce, how to distribute etc. From the consumer's side, there is also complete freedom regarding what type of goods they have to purchase and in what quantities? Since, we cannot produce everything that we desire in the economy because of scarcity of resources, the PPF shows the limits of producing the commodities (Figure 4.2). So, prices prevailing for the commodities, resources and services in the economy direct the production, consumption, exchange and distribution pattern. For example, if the demand for paddy is more in the market, it encourages production of paddy and *vice versa.* Besides commodities, prices of factors (*i.e.,* remunerations paid to different factors of production) are also determined based on their market forces. Since these factor prices determine the incomes of the factor owners and they again influence the prices of the commodities in the economy. At this equilibrium level of prices of factors, the entrepreneurs will decide the choice of technology or the combinations of resources to be employed in the production programme. This method of addressing the central problems through automatic price mechanism is commonly seen in free enterprise capitalistic economies.

2. Economic Planning Method

When the mechanism of addressing central problems is left to price mechanism alone as in free enterprise capitalistic economies, it leads to various complications such as, the entrepreneurs aim at monopoly gains at the cost of social welfare, tend to produce more of luxurious goods rather than basic necessaries, consumer's sovereignty gets limited and basic social issues like literacy, public health, poverty, drinking water, social welfare etc., are ignored. All these complications advocate the intervention of Government in addressing the central problems of an economy. So, in this method, the Central Planning Authority of the Government addresses the central problems of the economy and not through automatic price mechanism. That means, the problems *viz.*, what and how much to produce, how to produce, how to distribute etc., are all planned and decided through State intervention. This clearly implies, what community actually needs will be produced and distributed in the economy, rather than what perceived by the few people with profit motive. So, the economy addresses the central problems through democratic approach and this helps to identify what actually the human needs are. The economic decisions to address the central problems will be taken at a number of different levels *i.e.*, from micro to macro. In this way, the Government will streamline the prioritized issues through democratic participation of every individual and target them effectively, so as to ensure mass benefits. This method of solving the central problems through formulating the economic plans by the State is seen in socialistic and communistic economies. In such economies, the private entrepreneurs have no role in dictating the economic activities on their own and everything is planned by the State. This State intervention in different economic activities is gaining more and more significance in the recent period, due to inherent weaknesses in the free enterprise capitalistic economies. Moreover, significant time period was consumed for restructuring the capitalistic economies particularly after the Great Depression. But, this method of State intervention to address the central problems becomes difficult, if the economy faces more complex issues and increases in choice making among the individuals. It is notoriously difficult to devise incentives and rules that get people to do what the rule maker wants. For example, if the planning authority decides to cultivate wheat alone, then the consumers have to adjust for wheat consumption only and not to other food grains.

The causes and methodology of solving the central economic problems in different economic systems is depicted in Figure 4.1.

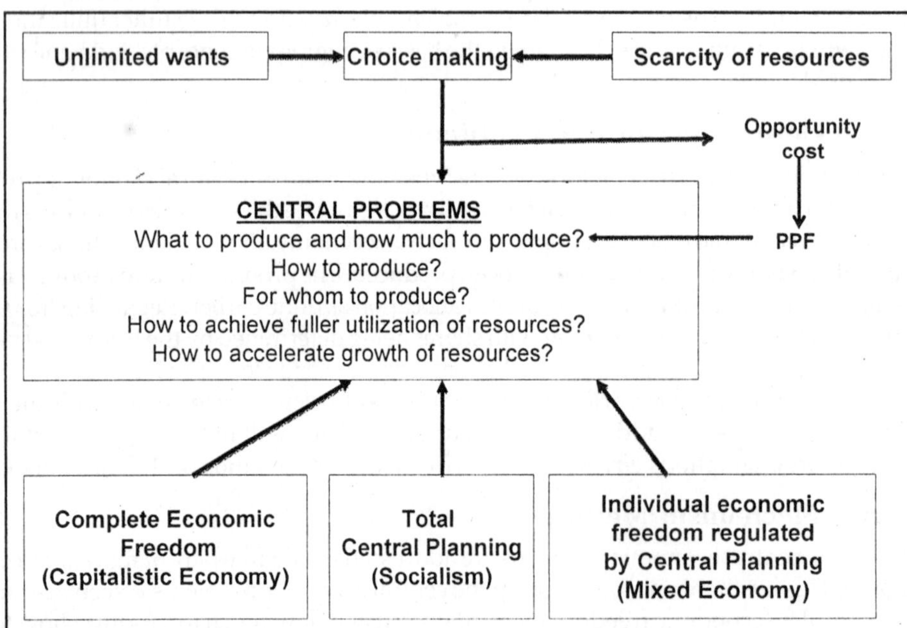

Figure 4.1: Causes and methodology of solving the central economic problems.

4.4. Production Possibility Frontier (PPF)

The starting point in an economic analysis is to address the first and foremost central problem *i.e.*, 'What to produce and How much to produce? Since the resources are scarce in an economy, there is a limit for production of output in any sector of the economy. These scarce resources have alternative uses and hence, every society has to decide on how much level of the scarce resources should be employed in the production of different goods and services. In other words, every society has to determine the resource allocation across the production of different goods and services. An allocation of a particular set (level) of scarce resources of the economy gives rise to a particular combination of different goods and services. Given the total amount of resources, it is possible to allocate the resources in many different ways, so that different mixes of all possible goods and services can be produced in the economy. This collection of all possible combinations of the goods and services that can be produced from a given amount of resources and a given stock of technological knowledge is called the Production Possibility Set of the economy. To decide what to produce and in what quantities, it is first necessary to know what is obtainable from the given level of resources. The limits of production of different commodities in the economy for a given level of resources can be ascertained through PPF. Thus, PPF answers the first central problem. It is otherwise called as Production Possibility Curve or Production Possibility Line. It is also known as a Product Transformation Curve, as the resources are transformed into varying units of the two goods.

A PPF is a graphical representation showing all possible combinations of two commodities that an economy produces efficiently during a specified period of time with the given level of limited resources and technology. It assumes *ceteris paribus* concept to explain the production possibilities of two commodities for the given level of resources. This curve is a 'frontier' indicating the limits of production of two commodities under consideration for a given level of resources and technology. Further, it indicates that, there are certain maximum limits of production of commodities and to achieve efficiency, the society must decide what combination of commodities and services can be produced. Along this frontier, there is productive efficiency. That means, if an economy is fully utilizing its resources efficiently, then it will be producing the commodities on the PPF. It shows the maximum of each commodity that can be produced using all the resources at maximum efficiency. In dealing with two commodities, PPF represents how much of the production of latter commodity must be sacrificed for a given increase in production of the former. Any point on the PPF implies the efficient use of given level of resources in producing the two commodities under consideration. If the economy is not producing the quantities of the commodities on the PPF, it indicates the resources are being managed inefficiently and the production of society will dwindle. So, PPF is a hypothetical representation of the quantities of two different commodities that can be obtained by shifting resources from the production of one to the production of the other. This curve, therefore, is used to describe a society's choice between two different commodities.

4.4.1. Assumptions

The following are the assumptions formulated under PPF analysis:

- ☆ The quantities of factors of production and the technology available to the economy are fixed.
- ☆ The economy can produce only two goods with the available scarce resources. However, considering only two goods for analysis is a case of theoretical simplification and convenient for graphical analysis. However, if one good is of primary interest, all other goods can be represented as a composite good and we can draw their production possibilities.
- ☆ The resources are fully employed.
- ☆ The resources are efficiently employed.
- ☆ The resources are not equally efficient in production of two commodities. Thus, if resources are transferred from production of one commodity to another, the marginal opportunity cost increases and it determines the shape of the PPF. If this assumption changes, the shape of PPF changes. If marginal opportunity cost is decreasing, PPF will have convex shape, If marginal opportunity cost is increasing, PPF will have concave shape and if marginal opportunity cost is constant, PPF is a straight line.
- ☆ Each of the goods can be produced using changing ratios of the factors of production. This is called 'variable factor proportions'. That means, the same resources can be used to produce either or both of the commodities and can be shifted freely between them.
- ☆ PPF concept is a short period phenomenon because, factors of production and techniques of production are constant in the short run and change in the long run.

4.4.2. Construction of a PPF

The PPF shows the maximum output of two commodities that can be produced in an economy with a given level of resources and technology and at a given point of time. If an economy is fully utilizing its resources, then the outputs of two commodities lie on the PPF. To make analysis simple, assume the economy is producing two commodities A and B (Table 4.1 and Figure 4.2). If all the resources and technology of an economy are devoted to commodity A, then Q_A output of commodity A will be produced and none of commodity B would be produced. Similarly, if all the resources and technology of an economy are devoted to commodity B, then Q_B output of commodity B will be produced and none of commodity A would be produced. But, if resources and technology were divided between the two commodities, then a range of combinations of these two commodities will be produced. At point X on the PPF, the economy produces OQ_{A1}

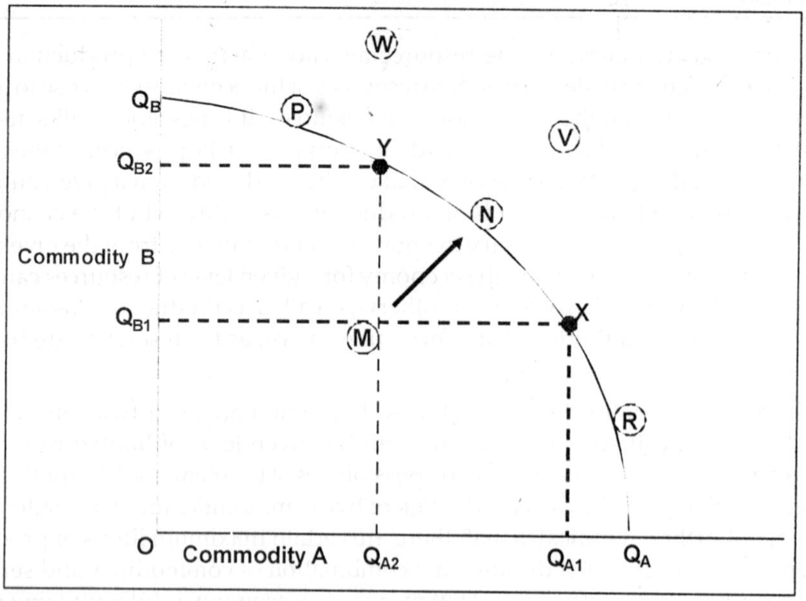

Figure 4.2: Production Possibility Frontier.

quantity of commodity A and OQ_{B1} quantity of commodity B. Similarly at point Y on the PPF, the economy produces OQ_{A2} quantity of commodity A and OQ_{B2} quantity of commodity B. All the points on the frontier, such as X, Y, P, N and R are the points of maximum productive efficiency, *i.e.*, no more output can be achieved from the given resources, as the commodities are fully utilizing the allocated resources. This indicates two important aspects *viz.*, the resources are not wasted in producing the two commodities and when an economy is producing the commodities efficiently (*i.e.*, on PPF), it can only produce more of one commodity by producing less of another. This is because, the resources have to be shifted from one commodity to another, as the resource level for a given PPF is constant. So, in order to increase the production of one commodity the production of other commodity should be decreased. The PPF, therefore, illustrates the concept of opportunity cost. This is because, more output of commodity A can be produced through sacrificing the output level of commodity B and *vice versa*. It is evident from the Figure 4.2 that, the opportunity cost of producing extra $Q_{A2}Q_{A1}$ output of commodity A is possible only by sacrificing $Q_{B2}Q_{B1}$ output of commodity B.

Table 4.1: Production possibility schedule of two commodities.

Output of Commodity A (Units)	*ΔA*	*Output of Commodity B (Units)*	*ΔB*	$MRPS_{AB}$
0		15		
5	5	14	–1	0.20
9	4	12	–2	0.50
12	3	9	–3	1.00
14	2	5	–4	2.00
15	1	0	–5	5.00

When the economy is not fully utilizing its resources or not using the resources efficiently, *i.e.*, when there is unemployment or under employment of resources, the economy operates inside the frontier, say at M. No doubt, production is feasible at point M, but the resources are not efficiently utilized. This implies Productive inefficiency, as the economy operates at less than full capacity and there is still scope for the economy to increase the output of the two commodities through allocating more resources. So, no economy will operate below the PPF because, it is wasting its resources. If all the resources in the economy are allocated and used efficiently, then the economy will move from M to P or Y or N or X or R and makes the production efficient. But, we generally observe in the economy that, some factories or shops closed down, high levels of unemployment, some shops with very few customers in them etc., and all these are due to productive inefficiency. This can happen, if resources are not re-allocated effectively among the commodities, when conditions in the economy alter. For example, if the demand for commodity A is increased in the market, it leads to movement of resources from commodity B to commodity A, resulting in close down of firms producing commodity B. But, this is purely theoretical because, if the employees and managers of the firms producing commodity B do not have necessary skills and expertise in producing commodity A, there will be no shift of resources from B to A in the short run.

So, the economy will struck at 'M' and this leads to inefficient production of both the commodities. However, in the long run, if the employees and managers of firms producing commodity B are provided with adequate trainings regarding the production aspects of commodity A, they acquire the skills and this leads to shift in resources from commodity B to commodity A. Then the economy now operates at point 'R', indicating two important aspects *viz.*, more of commodity A is produced compared to commodity B and the production turns efficient, as the economy now operates on PPF again. This implies, the economy may be productive inefficient in the short run, but can be turned productive efficient in the long run. Alternatively, if there is lack of demand for both the commodities *i.e.* B and A, though the economy is capable of producing both the commodities at point 'N', it can only afford the combination of commodities at point 'M'. Again, over time, the demand will hopefully increase and the economy will end up on the frontier.

All points outside the frontier (say, W and V) are infeasible, as the output level would exceed the given level of resources and capacity of the economy and thus unattainable in the short run. So, a point on the PPF indicates the allocative efficiency, also called as Pareto efficiency, as for a given preferences and distribution of income, any movement along the curve will positively influence the someone duly affecting the other. The economy would require an increase in the level of factor resources or an increase in the efficiency (or productivity) of factor resources or an improvement in technology to achieve the combination of goods at W or V or Y.

4.4.3. Shapes of PPF

PPF will assume different shapes like concave, straight line and convex and they are discussed here under.

1. Concave Shape PPF

We normally draw a PPF on a diagram as concave to the origin. This is because, as shown in the Table 4.1, if we go on increasing the output of commodity A (Added commodity) by allocating more resources, it will replace more and more units of commodity B (Replaced commodity). That means, the rate of replacement of commodity B will increase, as the added commodity A is increased in its magnitude. In other words, as we move down the PPF, more and more resources are diverted from commodity B towards commodity A, thereby, the additional or marginal or extra output of commodity A will gradually decrease or total output will increase at diminishing rate. So, in terms of commodity B, more and more of it gets replaced or given up in order to produce the extra output of commodity A in each successive case. That means, if more of commodity A is produced, increasingly larger amounts (output) of commodity B must be given up. This is because, experienced producers of commodity B are not efficient in producing commodity A (and similarly, experienced producers of commodity A are not efficient in producing commodity B). That means, if we move down the PPF, it implies, more and more resources are diverted towards commodity A from commodity B, but the diverted resources from commodity B are not equally well suited for commodity A. Thus, more and more resources well suited for the production of commodity B are diverted towards commodity A and this leads to increasing rate of sacrifisation of output of commodity B, if production of commodity A is increased. This implies, increasing rate of product substitution of commodity B by increasing the output of commodity A and this gives a concave shape to PPF, as the diverted resources from commodity B are not equally suited to produce commodity A. Since, more and more resources are devoted towards commodity A, it leads to diminishing MRs or increasing costs in the production of commodity A at the expense of commodity B. Thus, this is in accordance with the famous law given by Ricardo *i.e.*, the diminishing MRs or increasing costs principle. The above schedule (Table 4.1) can be inverted by reading it upwards, in which case, one can assume that, more and more units of commodity B can be produced at an increasing rate of sacrificing the production of commodity A units.

Here comes the important concept of opportunity cost to explain the concave shape of PPF. As mentioned earlier, for a given level of resource usage, if the production of one commodity is increased, it leads to sacrifisation of the other because, the resources must be transferred to the preferred commodity away from the other. So, the points along the PPF describe the trade-off between the commodities. The sacrifice we make in the production of the commodity B due to increase in production of commodity A is called the Opportunity cost (because, increasing production of the commodity A entails losing the opportunity to produce some amount of commodity B). So, opportunity cost is measured in terms of the number of units of the commodity sacrificed or forgone due to increase in production of the other. So, this opportunity cost will directly determine the shape of PPF. In the Figure 4.3, producing 20 more units of commodity A (movement from Y to X on the PPF), where the economy is already close to its maximum potential, the opportunity cost is 30 units of commodity B. Similarly, the opportunity cost of producing extra 5 units of commodity B (movement from Q to P on the PPF) is 10 units of commodity A. The ratio of opportunity costs is determined by the Marginal Rate of Transformation, which is nothing but Marginal Rate of Product Substitution (MRPS) or Marginal Opportunity Cost, given by:

$$MRPS_{AB} = \frac{\text{Change in number of units of replaced commodity B}}{\text{Change in number of units of added commodity A}} = \frac{\Delta B}{\Delta A}$$

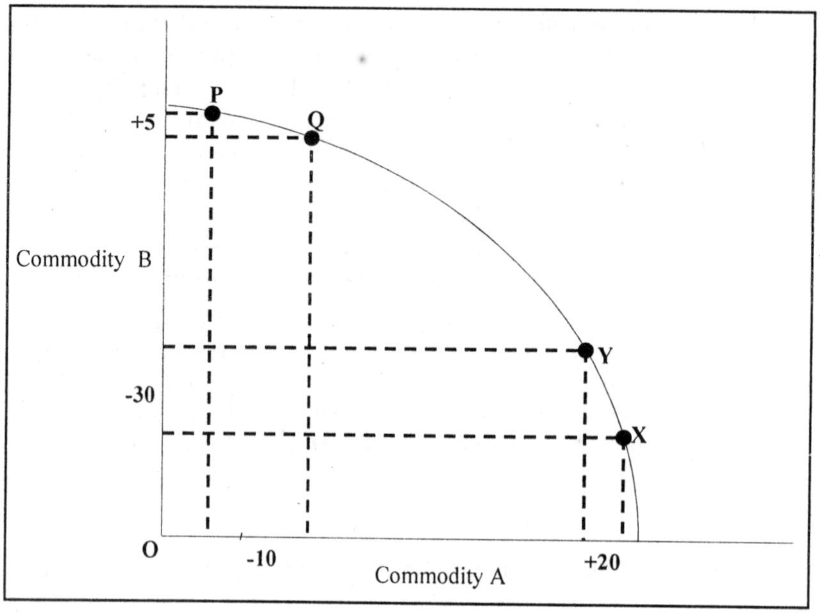

Figure 4.3: Concave shaped PPF – Increasing opportunity costs.

$$MRPS_{BA} = \frac{\text{Change in number of units of replaced commodity A}}{\text{Change in number of units of added commodity B}} = \frac{\Delta A}{\Delta B}$$

From the Figure 4.3, the $MRPS_{AB}$ along the movement from Y to X on PPF is 1.5. This implies that, for every one unit increase in the production of commodity A, 1.5 units of commodity B production should be sacrificed. This indicates, the marginal opportunity cost of commodity A production is 1.5 units of commodity B or the marginal opportunity cost of commodity A in terms commodity B is 1.5 or the marginal opportunity cost of commodity B in terms of commodity A is 0.67.

Similarly, the $MRPS_{BA}$ along the movement from Q to P on PPF is 2.0. This implies that, for every one unit increase in the production of commodity B, 2.0 units of commodity A production should be sacrificed. This indicates the marginal opportunity cost of commodity B production is 2.0 units of commodity A or the marginal opportunity cost of commodity B in terms of commodity A is 2.0 or the marginal opportunity cost of commodity A in terms of commodity B is 0.5.

Reasons for Concave Shape of PPF

The example considered above demonstrates increasing opportunity costs ($MRPS_{AB}$ along P to Q = 0.5 and $MRPS_{AB}$ along Y to X = 1.5) and it is this principle of increase in opportunity cost that makes PPF (bulging out) concave to origin. This is the most common form of PPF. This is because of the following reasons:

- This non-linear shape represents a disparity in the factor intensities and technologies used for the production of two commodities. That means, if an economy specializes more and more in the production of one commodity (say commodity A) *i.e.*, moving from P to X (Figure 4.3), the opportunity cost of producing that commodity An increases. This is because, we are devoting more and more resources towards commodity A that are less efficient in producing it or more efficient in producing commodity B. To increase the production of commodity A, first the resources allocated towards commodity B (say, labour) will move to commodity A. At first, less skilled or less qualified labour or workers will move from the industry producing commodity B to the industry producing commodity A. So, at the beginning, the opportunity cost of producing commodity A will be low or the loss in commodity B production is small. But, later, if commodity A production is still specialized, more and more skilled labour or workers will move towards the firms producing commodity A from the firms producing commodity B. But, the workers may not transfer their skills easily from one commodity to another. This may lead to diminishing returns to a factor on commodity A. So, the opportunity cost of producing commodity A will increase, as due to the movement of more skilled workers from the firms producing commodity B, its production is adversely affected. So, this increase in opportunity cost as we move the down the PPF gives it a concave shape.
- In reality, resources are not equally productive across the production of two commodities under consideration. This is because, some of the equipments (tools, machinery etc) may be designed specifically to produce specific goods only. If such resources say, specific to commodity B are diverted towards the production of

commodity A, the resource efficiency will get affected on the desired commodity A. So, it leads to low MRs and this increases the opportunity cost. So, specificity of resources usage also contributes to increase in marginal opportunity cost, as we move either extreme on the PPF.

2. Straight Line PPF

In some cases, the PPF is a straight line or linear in shape. This is because, the resources are not specialized here and they can be substituted for each other with no added cost. That means, it is quite easy to divert the resources from producing one commodity to another. Say, if we divert the resources towards commodity A from commodity B, they are equally efficient in producing commodity A as in commodity B. This means that, every time if we transfer the resources from commodity B to commodity A, there is the same increase in output of commodity A and the same decrease in output of commodity B. This implies, we experience constant MRs to a factor (whereas, in concave shaped PPF, we experience diminishing MRs). This further indicates, there is constant marginal opportunity cost. Products requiring similar resources (bread and pastry, two different brands of tea) will have an almost straight line PPF and hence, constant opportunity costs. The important aspect in case of linear PPF is that, all the resources can be treated equally efficient. As shown in Figure 4.4, in all cases, for every 10 units increase in the production of commodity A, 25 units of commodity B will be sacrificed, indicating that, $MRPS_{AB}$ is 2.5. This implies that, for every one unit increase in the production of commodity A, 2.5 units of commodity B production should be sacrificed. This indicates the marginal opportunity cost of commodity A production is 2.5 units of commodity B or the marginal opportunity cost of commodity

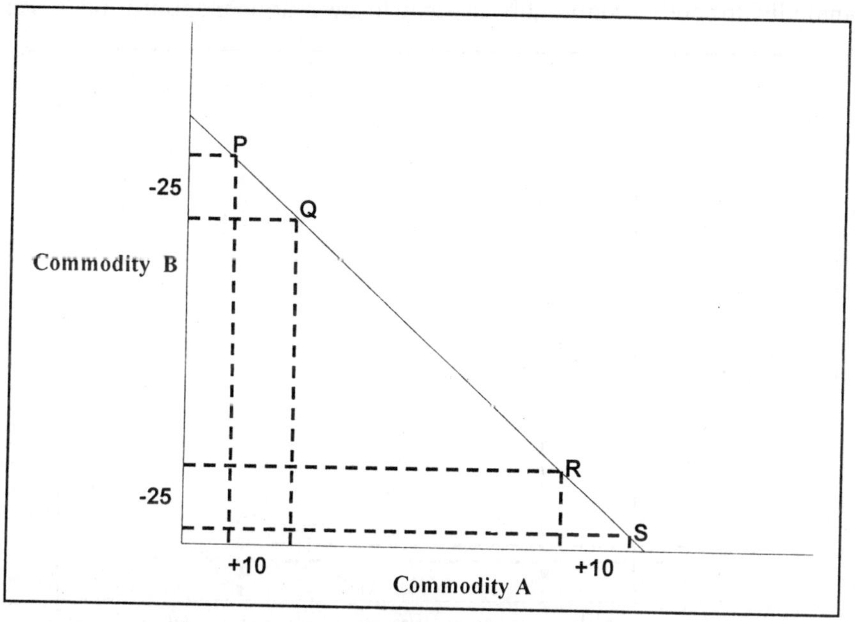

Figure 4.4: Straight line PPF – Constant opportunity costs.

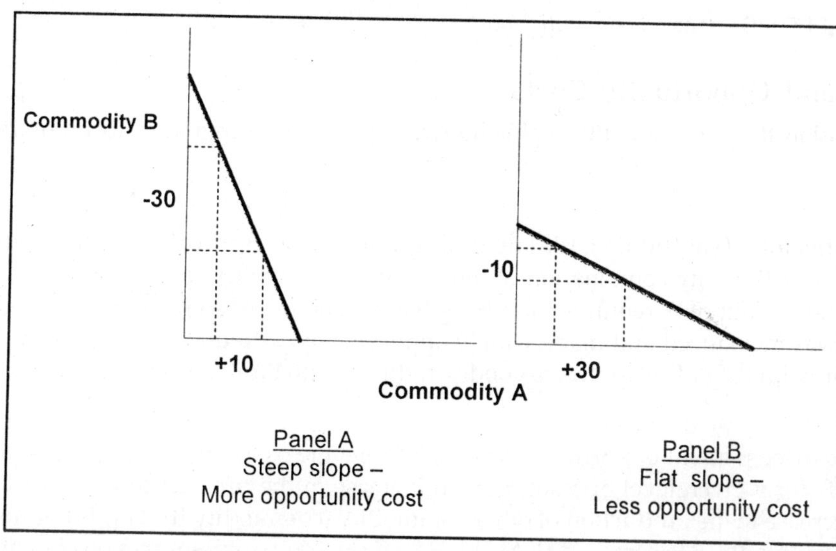

Figure 4.5: Straight line PPF with different slopes.

A in terms commodity B is 2.5 or the marginal opportunity cost of commodity B in terms of commodity A is 0.4. It is important to note that, if the PPF has steeper slope (Panel A of Figure 4.5), the marginal opportunity cost is higher and if the PPF has flat slope (Panel B of Figure 4.5), the marginal opportunity cost is lower, when the added product is increased in its magnitude.

3. Convex Shape PPF

This shape is also not common like linear PPF. But, this convex shaped PPF is possible, if a firm producing a commodity enjoys economies of large scale. Due to this, the firm produces the commodity more efficiently or at low cost. Hence, there will be less sacrifisation on the part of that commodity the firm produces. As shown in Figure 4.6, between P and Q points on PPF, initially, for 10 units increase in the production of commodity A, 50 units of commodity B will be sacrificed, indicating that, $MRPS_{AB}$ is 5.0. This implies that, for every one unit increase in the production of commodity A, 5.0 units of commodity B production should be sacrificed. This indicates the marginal opportunity cost of commodity A production is 5.0 units of commodity B or the marginal opportunity cost of commodity A in terms commodity B is 5.0 or the marginal opportunity cost of commodity B in terms of commodity A is 0.2. But, if we move along the PPF from PQ to RS points, it indicates for 10 units increase in the production of commodity A, only 5 units of commodity B will be sacrificed, indicating that, $MRPS_{AB}$ is only 0.5. This might be due to greater specialization and economies of large scale (say, mass production, specialization of labour, use of modern technology) are involved in the production of commodity B, and hence less sacrifisation of commodity B. This implies, decreasing opportunity costs in the production of commodity A and this gives convex shape to PPF (bulging in towards origin). Sometimes, due to quick adjustment of resources transferred from one commodity to another commodity also lead to low marginal opportunity costs.

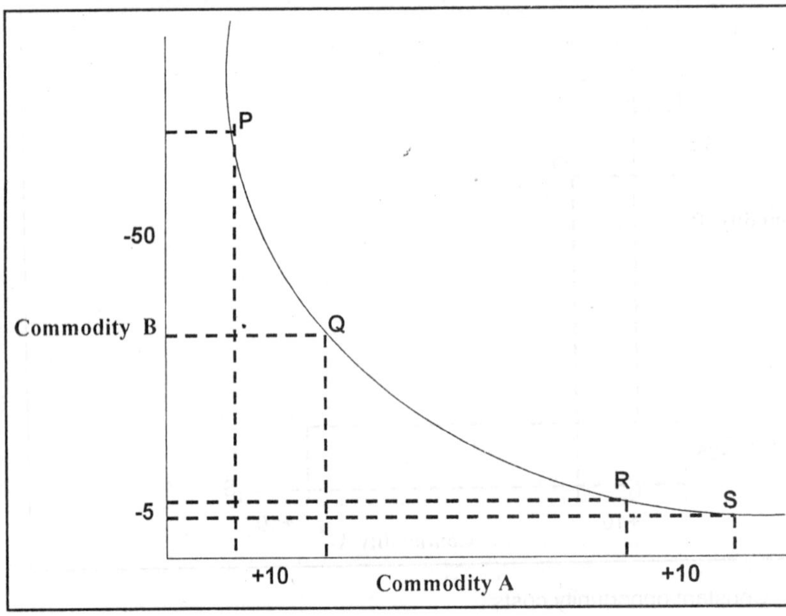

Figure 4.6: Convex shaped PPF – Decreasing opportunity costs.

4.4.4. Substitution and Opportunity Costs

From the above discussion, the principle of PPF highlights two important concepts *viz.*, Substitution and Opportunity costs.

Substitution

The PPF hints at the principle of substitution. We know, the given level of resources can be alternatively used either to produce commodities A or B or any combination of the two goods. If we try to produce a combination of two commodities, every additional unit of A requires sacrifising the production of some units of B and *vice versa*. This principle of substitution was introduced and made widely applicable by Alfred Marshall in a variety of economic activities. In the present principle of PPF, it applies to goods produced with limited resources.

Opportunity Cost

It is high time now to discuss about opportunity cost, as it forms the basis for offering explanation to different shapes of PPFs. On a PPF, for a given level of resources, an increase in the production of a one commodity (say, commodity A), entails decrease in the production of other commodity (commodity B). This is because, the resources must be transferred from commodity B to commodity A. This decline in the use of resources on commodity B affects its production and thereby, this sacrifice is the cost on the part of economy. This cost is called opportunity cost because increasing the production of commodity A entails losing the opportunity to produce some quantity of commodity B. So, opportunity cost is measured in terms of the number of units of the commodity B sacrificed or forgone due to increase in the production commodity A. It is defined as the cost equivalent to the amount of a product that would have been otherwise produced or the amount of resources which could have been otherwise used in the next best productive activities. This opportunity cost of a factor is also called as the Transfer earnings of the factor or the cost equivalent to the returns from the next best alternative foregone. In the words of Furguson, *'the alternative or opportunity cost of producing one unit of commodity 'A' is the amount of commodity 'B' that must be sacrificed in order to use that resources to produce 'A' rather than 'B'.'* So, opportunity cost means opportunity lost. In real life, a labour move from one occupation to another occupation in search of high income opportunities. The opportunity cost is the real cost of production and is a theoretical concept. It differs from the market or money cost of production. If labour takes up the new assignment, he loses the remuneration of the present job. The remuneration lost is termed as opportunity cost. For example, a labour is employed in a sugar factory and gets a salary of Rs. 5000 per month. He shifts himself to do a business, so he will be losing Rs. 5000 per month and it will be his opportunity cost. If he gets more than Rs. 5000 per month in the business, he will stick to it. If he gets lesser than Rs. 5000 in the business, he will close the business and returns back to the earlier job. This concept explains that, we will shift to alternative assignments when we are assured that, we shall not be getting lesser than the opportunity cost. This opportunity cost includes both explicit and implicit costs (discussed under section 15.1).

In the present example, the concept of opportunity cost is applied to two commodities produced with limited resources. But, it can be generalized in various ways. This is because, though resources are scarce, each unit of a resource is capable of being alternatively used in various production programmes. Thus, the alternative uses of resources highlights the concept of opportunity cost. When a resource unit is used in some productive activity, some other activity for which this resource unit could have been used will have to be forgone. That is the opportunity cost or the cost of the highest valued alternative of that particular choice. For example, an acre of land is capable of serving different purposes and the respective revenue that each purpose would yield is given as here under.

- ✰ Cultivation of paddy – Rs. 25000
- ✰ Cultivation of maize – Rs. 28000
- ✰ Cultivation of groundnut – Rs. 37000

Thus, when an acre of land is actually used for the cultivation of maize, it will yield Rs. 28000 returns and its opportunity cost is Rs. 25000 *i.e.*, the next best missed returns from paddy. Similarly, when the land is used for cultivating groundnut, it yields Rs. 37000 and its opportunity cost is Rs. 28000, *i.e.*, cultivation of maize. The difference, if any, between actual returns and opportunity cost is called economic rent. Thus, when land is used for cultivating maize or groundnut, the amount of economic rent is Rs 3000 and Rs. 9000 respectively. On the other hand, if the land is used for cultivating paddy, then both the returns from paddy and opportunity costs are identical and there is no economic rent. Thus, the concept of opportunity cost has wider applications besides offering notable explanation to PPF.

We can differentiate PPFs based on the behaviour of opportunity cost. In case of concave PPF, the opportunity cost will increase as we move along the PPF. In linear PPF, the opportunity cost remains same at any combination of production possibilities of two commodities. In case of convex shaped PPF, the opportunity cost will decrease as we move along the PPF.

4.4.5. Slope of PPF

The slope of the PPF at any given point indicates the MRPS or Marginal Rate of Transformation or Marginal Opportunity Cost. This indicates, by how much the quantity of one commodity is replaced for an increase in the quantity of the added commodity. So, it numerically describes the rate of transformation of one product to another product due to reallocation of resources. It is also called the 'Marginal Opportunity Cost' of a commodity, as the slope of PPF helps to analyze the opportunity cost of producing the commodity A in terms of commodity B, if the output of commodity A is increased and *vice versa*. For example, in Figure 4.3, between points Y and X, the marginal opportunity cost of commodity A in terms of commodity B is 1.5. So, in terms of marginal opportunity cost, it measures how much of commodity B is given up for an increase in quantity of commodity A and *vice versa*. This marginal rate of opportunity cost or transformation can be expressed in terms of either commodity. Since, the shape of a PPF is commonly drawn as concave from the origin, it represents increasing marginal opportunity cost and the reasons are discussed earlier.

4.4.6. Position and Shifts in PPF

The main determinants of the position of the PPF include the availability of resources and their efficiency, state of technology, discovery of new natural resources, increase in population (say, due to immigration), rendering quality training to employees, greater investment on capital goods such as machines and equipment etc. The points on PPF

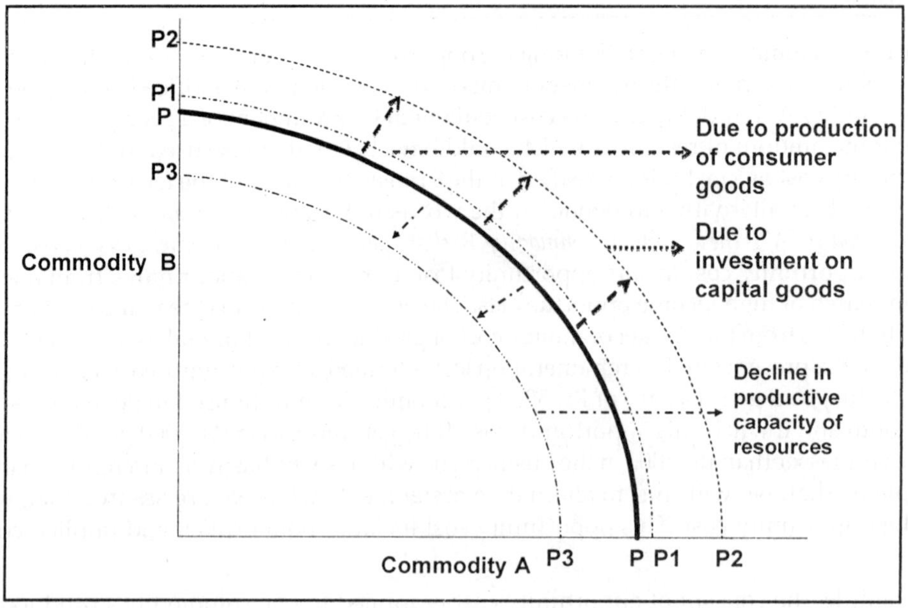

Figure 4.7: Shifts in PPF.

indicate the maximum limits of production combinations of two commodities that can be achieved in the short run (PP curve in Figure 4.7). However, in the long run, if the above mentioned factors show positive influence, the PPF shifts away from the origin indicating that, the capacity of the economy to produce two commodities increases, thereby, leading to economic growth (P_1P_1 and P_2P_2 curves). Conversely, if the factors showed negative influence, the total economy's productive capacity decreases and thereby, the PPF shifts towards left or towards origin (P_3P_3 curve).

The PPF is also helpful to illustrate the extent to which an economy is producing for,the present or the future (Figure 4.7). If the economy gives importance in the production of capital goods to consumer goods, it invests more on tools and machinery with a long term objective. This is shown by a large outward shift of the PPF (P_2P_2 curve), as the economy can produce more goods in the future by using tools and machinery. On the other hand, if the economy focuses more on consumer goods, it leads to a smaller outward shift in the PPF, because the investment on the machinery and technology is less (P_1P_1 curve).

It is very important to study the differences between movement of the economy from a point within the frontier to the point on the frontier *i.e.*, say from point M to P or Y or N or X or R in Figure 4.2 and movement of the economy from one PPF to another PPF (say, from PP to P_2P_2 in Figure 4.7). In both the cases output of goods will increase, but the former involves full employment of given resources, while the latter involves growth in recourses. The former case is dealt by short run macro economic theory, the latter is dealt by the theory of economic growth. To bring the economy from a point below the frontier to the point on to the frontier, aggregate demand should be increased to overcome depression in the economy. To increase the aggregate demand, there should be full employment of available labour and capital resources in the economy and this will remove unemployment and under utilization of the productive resources in the economy. However, to make the PPF curve shift outward, it is just not possible by mere increasing the aggregate demand. Rather, it needs accumulation of capital resources, investment and technological progress.

4.4.7. Reallocation of Resources Based on PPF

The PPF simply indicates various production possibilities of two commodities in an economy at a given level of resources usage. As explained earlier, any combination of the commodities produced at any point on the PPF is considered productively efficient. So, PPF indicates only what can be produced under different combinations. But, this again pose a problem to ascertain which combination is the best combination for the economy, *i.e.*, whether we have to produce more of commodity A and less of commodity B and *vice versa.* The decision regarding the production of desired combination of commodities depends upon the nature of the economic system.

In a free market economy, the decision regarding the production of commodities is influenced by the two market forces *viz.*, demand and supply. If there was a high level of demand for commodity A compared to commodity B, then the firms concentrate more on the production of commodity A by devoting more labour, materials etc., and even new firms also enter the industry to produce commodity A. So, the resources allocation towards commodity B will be decreased. So, market forces influences the allocation of resources across the commodities and the fluctuations in market forces calls for reallocation of resources between the selected commodities. For example, on the PPF shown in

Figure 4.2, say, due to increase in demand for commodity A, we move along PPF from Y to X, and thereby, we devote more resources towards the production of commodity A and hence, less of commodity B will be produced.

In a planned economy, the decision about what to produce is determined by Government instructions and directives. For example, the Government may place the order that, more factories and employees should produce commodity A rather than commodity B. In this case, the reallocation of resources from, say, Y to X is not determined by demand, but by the Government orders. This may even happen, if the Government does not trust market forces in directing the production mechanism in the economy. For example, a free market mechanism may direct the firms to produce more of a commodity, which is harmful to human health (say, cigarettes, alcoholic drinks etc.,) and in such case, the Government may intervene for proper allocation of resources in the economy. But, the major constraint here is, sometimes, the directions given by the Government may not match the actual requirements of the people in the economy. In such cases, the production of unwanted goods will be more than the goods actually demanded by the people.

4.4.8. Application of PPF

The following are the important areas of application of PPF:

- PPF helps to address the unemployment problem in the economy. As shown in the Figure 4.8, if the economy operates at point M, it implies less resources are utilized in the economy, and thereby P_1P_1 PPF depicts substantial unemployment of resources in the economy. This is because, the economy has the capacity to increase the output of the two commodities on to the PPF P_2P_2. So, if the idle resources are employed and used efficiently, the economy turns productive efficient by producing the two commodities on P_2P_2 frontier. If the economy operates at point A on P_2P_2 frontier, it implies more of commodity B is produced, if operates at point C more of commodity A is produced and if operates at point B, more of both the commodities are produced.

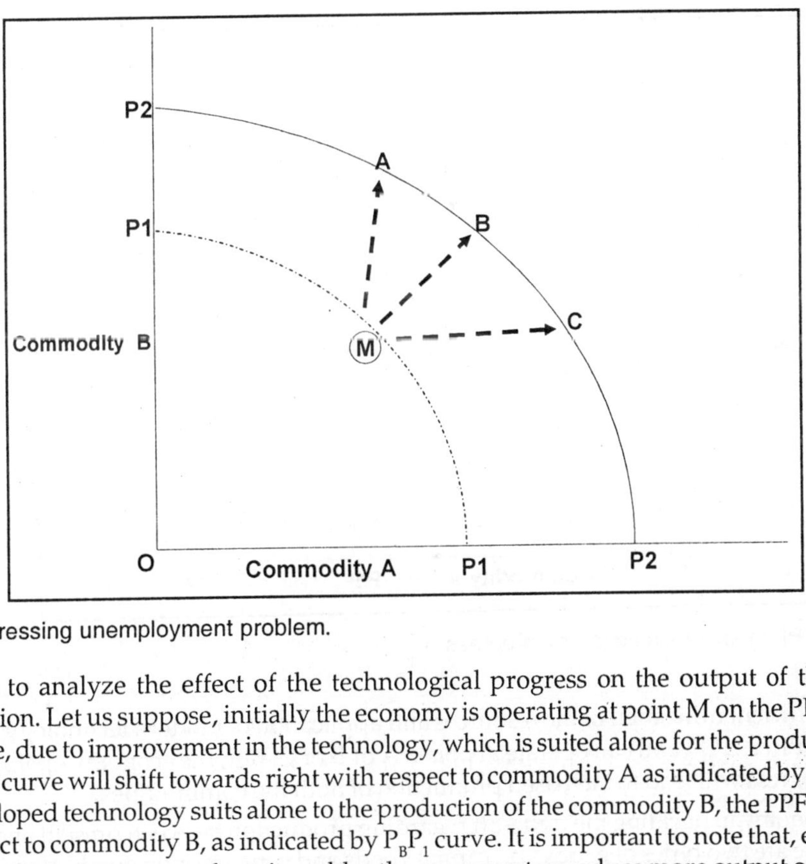

Figure 4.8: PPF- Addressing unemployment problem.

- PPF helps to analyze the effect of the technological progress on the output of the commodities under consideration. Let us suppose, initially the economy is operating at point M on the PPF, P_1P_1 (Figure 4.9). Let us suppose, due to improvement in the technology, which is suited alone for the production of the commodity A, the PPF curve will shift towards right with respect to commodity A as indicated by P_1P_A curve. Conversely, if the developed technology suits alone to the production of the commodity B, the PPF curve will shift upward with respect to commodity B, as indicated by P_BP_1 curve. It is important to note that, even if the technological progress is related to one product, it enables the economy to produce more output of both the commodities. Say, for example, if the technological progress suits to production of efficient capital goods (commodity A), it helps to produce more of consumer goods (commodity B) in the economy. If the technological progress suits to the production of both the commodities, it implies more economic progress and thereby, PPF will shift outwards from P_1P_1 to P_2P_2 and the economy operates at point C, thereby, producing increased output of both the commodities.
- The study of PPF helps to ascertain the extent of economic progress in increasing the output of the selected commodities under consideration. As shown in the Figure 4.10, if the economy is at S initially, through making economic progress, it will move to point M on the PPF, P_1P_1. Further improvement in the economic

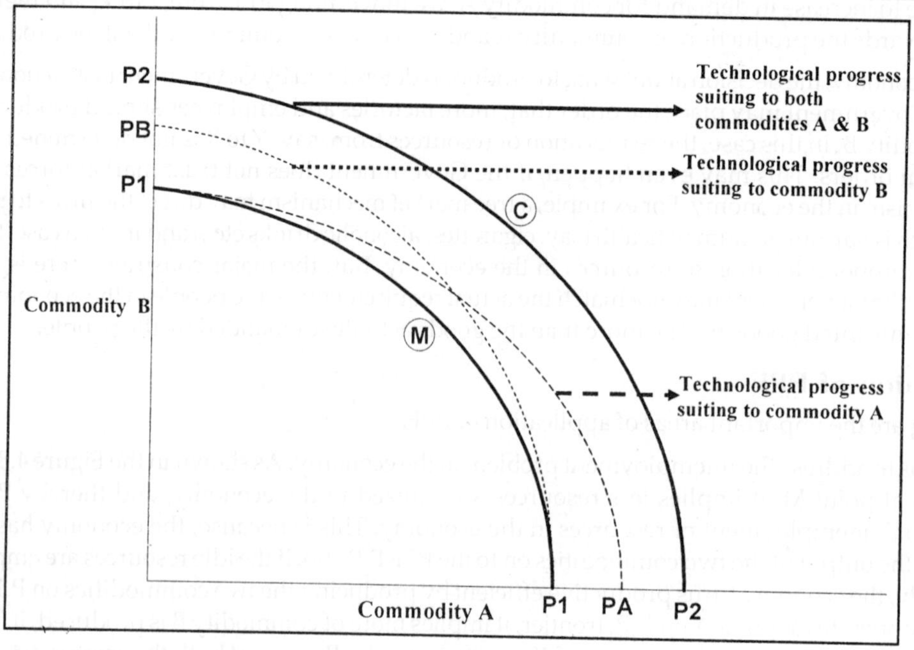

Figure 4.9: Shift in PPF due to improvement in production technology.

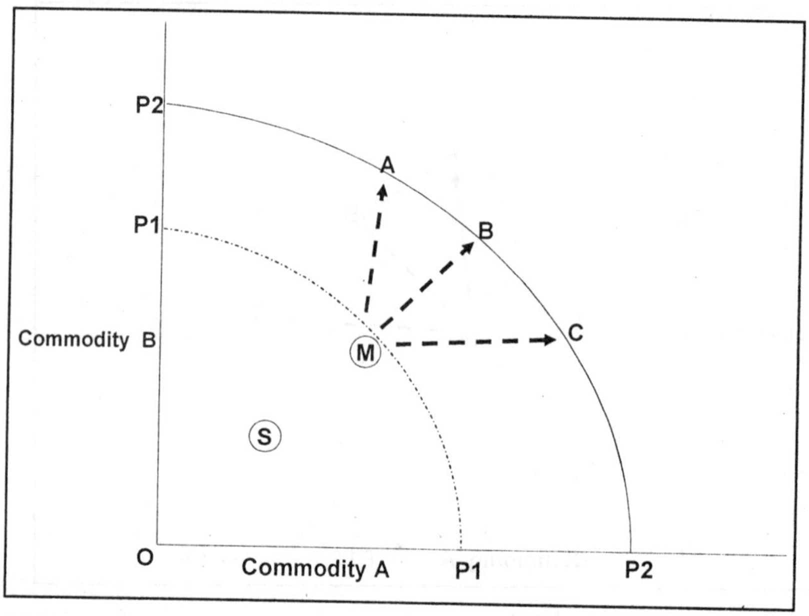

Figure 4.10: Application of PPF to ascertain economic progress.

progress due to efficient utilization of the resources and application of modern technology, the economy can move to higher PPF *i.e.*, P_2P_2 and can operate at points A or B or C. But, the economy operates at 'B' because, the economic progress will lead to increased production of both the commodities.

☆ PPF helps the economy in deciding the trade-off regarding production between present and future goods. In an economy, there are two types of goods *viz.*, capital goods and consumption goods. If the economy prefers to produce more of capital goods, the resources allocation towards production of consumer goods will decrease and *vice versa*. But, if the economy gives more preference to capital goods relative to the consumer goods with reference to resources allocation, the economy can produce more of both the goods in future (long run). This leads to greater shift in the PPF towards right from P_1P_1 to P_3P_3 (Figure 4.11). On the other hand, if the economy gives more preference to consumer goods relative to capital goods, there will be a smaller shift in the PPF towards right from P_1P_1 to P_2P_2.

☆ The PPF is an important concept of Modern Economics. The concepts of scarcity of resources and alternative use of resources of Robbins's definition can be explained with the help of PPF. The concept of PPF explains, for a given level of resources, the economy operates on a particular PPF. Since the resources are scarce, if we

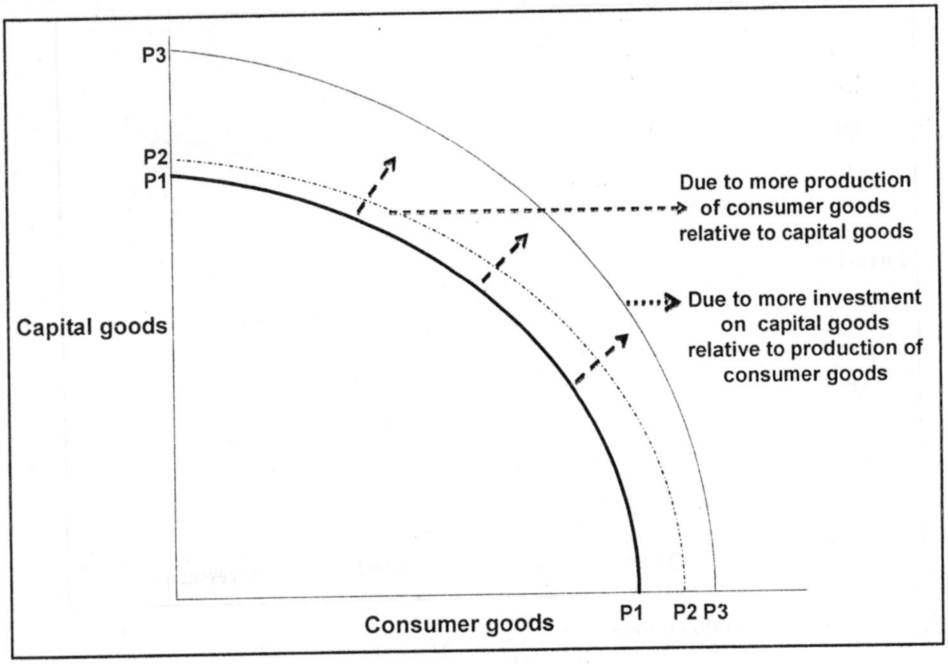

Figure 4.11: Application of PPF – Capital goods vs Consumer goods.

increase their allocation on one particular commodity, the output of other commodity gets affected. Thus, if we move along the PPF, it implies the trade-off between the selected commodities. If the given resources and technology are fully utilized, the economy operates on the PPF and thereby, produces both the commodities efficiently. But, the economy cannot produce the commodities outside the frontier, as the resources level is not sufficient. This indicates that, due to scarcity of the resources, the economy has to allocate them very efficiently among the selected commodities, so as to ensure maximum productive efficiency.

☆ PPF helps to analyze the central problems of the economy *viz.*, what, how and for whom to produce. Regarding 'what to produce', the PPF indicates what type of commodities should be produced in the economy and in what quantities. That means, a point on the PPF indicates the quantities of two commodities to be produced. It is clear from the Figure 4.2 that, if all the resources and technology of an economy are devoted to commodity A, then OQ_A output of commodity A will be produced and none of B would be produced. Similarly, if all the resources and technology of an economy are devoted to commodity B, then OQ_B output of commodity B will be produced and none of A would be produced. At point X on the PPF, the economy produces Q_{A1} quantity of commodity A and Q_{B1} quantity of commodity B. Similarly at point Y on the PPF, the economy produces Q_{A2} quantity of commodity A and Q_{B2} quantity of commodity B. All the points on the frontier, such as X, Y, P, N and R are the points of maximum productive efficiency, *i.e.*, no more output can be achieved from the given resources, as the commodities are fully utilizing the allocated resources. Thus, operating at different points on the PPF helps to draw different production possibility combinations of two commodities and this helps in guiding the allocation resources.

The other important central problem of the economy, 'how to produce' implies, what resources combination should be used in the economy to produce desired commodities in requisite quantities with the main objective of cost-minimization. If the economy operates below the frontier (say, at M in Figure 4.2), it implies, the resources are not fully utilized are not used efficiently in the production programme. This also indicates lower production of both the commodities in the economy, which will lower the welfare and standard of living of the people. So the concept of PPF guides the economy towards the efficient allocation of resources, so that it operates on the frontier and thereby, leads to productive efficiency of resources.

However, the other central problem of the economy, 'for whom to produce' is not directly revealed by the PPF, but we can obtain some knowledge about the distribution of goods from the frontier. It is important to note that, the pattern of distribution of goods in the economy depends upon the pattern of the production of goods, which in turn, depends upon the pattern of demand of the consumers. We can draw a PPF taking the combination of necessaries and luxurious goods in the economy. As shown in the Figure 4.12, at point A on the frontier, the economy produces more of luxurious goods, OQL_1 and less of necessaries, OQN_1. This implies, the economy is favouring rich sections of the society by producing more of luxurious goods and this widens the gap between rich and poor in the economy. On the other hand, if the economy operates at point B

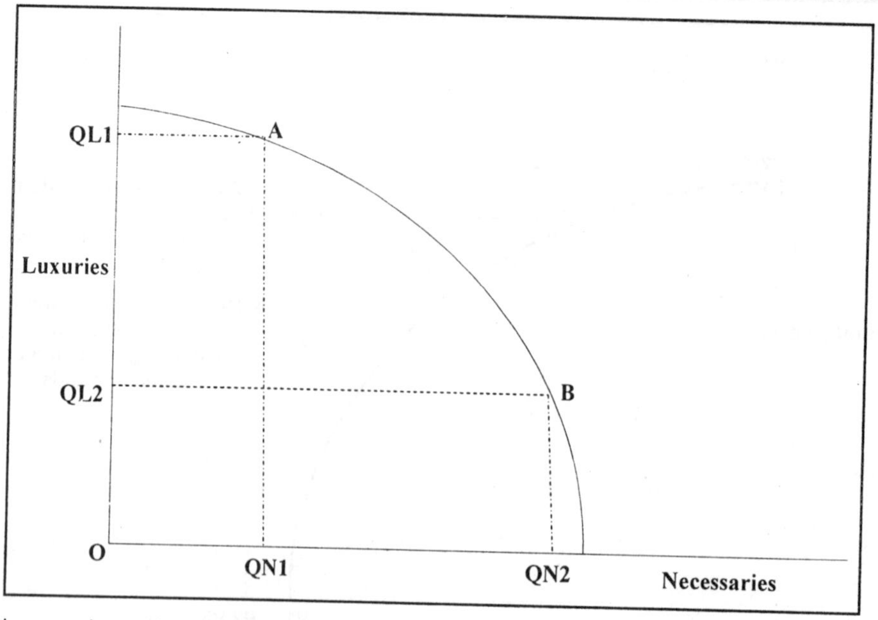

Figure 4.12: PPF – Necessaries vs luxurious goods.

on the frontier, it indicates more necessaries are produced compared to luxurious goods and this ensures equi-distribution of national income. This pattern of production of goods in the economy, *i.e.*, how much of luxurious goods and how much of necessaries would be produced depend upon the pattern of demand in the economy. This pattern of demand again depends upon the pattern of income distribution in the economy. That means, the more is the equi-distribution pattern of income in the economy, the more of necessaries and less of luxurious goods will be produced and *vice versa*.

☆ This concept helps to explain the consumption pattern outside the PPF. We know, PPF represents what an economy can produce two commodities efficiently for a given level of resources. However, it is possible to consume outside of the frontier through international trade. It is known that, some countries have comparative advantage in producing some commodities and other countries do enjoy comparative advantage for some other commodities. This facilitates to participate in international trade, so that, all the countries get mutually benefitted. This participation in international trade helps the country to import more quantity of desired commodities at low opportunity cost and the same is explained through Figure 4.13. Assume the country is operating at point 'N' of PPF, where the country is producing 50 units of rice and 70 units of maize. If the country gives up 10 units of maize, some of the resources gets diverted to rice production and its production will be increased to 55 units, *i.e.*, by five units only. This indicates more opportunity cost. On the other hand, if the country enters the international trade and identify a foreign country, which is having less comparative advantage in terms maize production, but enjoys more comparative advantage in terms of rice production, then the country can receive more than five units of rice in return. So, through participation in international trade, the country exchanges 10 units of maize and in return it imports 30 units of rice. So, the country now operates at V instead of at X. So, due to international trade, a country can operate outside the PPF.

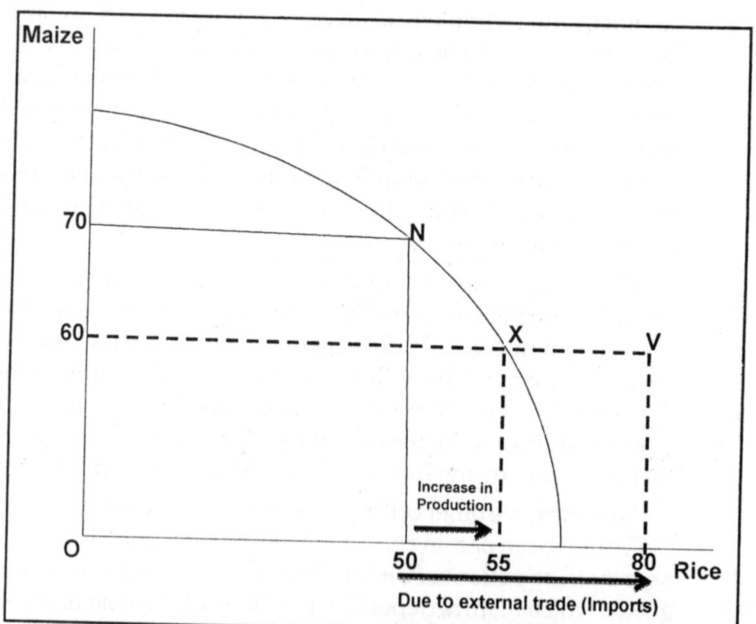

Figure 4.13: Consumption outside PPF – Participation in International trade.

Part II
Demand Theory

5

Theory of Consumer Behaviour

Economic science, is rather considered as an exact science compared to other social sciences like History, Ethics, Politics, Sociology, Psychology etc., as it uses the measuring rod in the form of 'money' for measuring the utility. However, this measuring rod *i.e.*, 'money' is not an exact unit of measurement of utility. Consumer means buyer of finished goods and services for satisfying his wants. We, generally, express the value of a commodity in terms of 'price' in monetary concept. But, this measurement of utility in terms of 'price' serves only as a rough measure. This is because, say, there are two consumers A and B purchasing rice in the market. If consumer A pays Rs. 30 for rice and if consumer B pays Rs. 35 for the same, it implies, the utility of good is more for consumer B than consumer A. This is because, if the utility in the commodity is higher, the consumer prefers to pay high price for it. Thus, utility of a commodity can be measured in terms of 'price' of the commodity. But, this approach may not be correct always. This is because, say another consumer C in the economy pays Rs. 45 for the same commodity. This does not mean that, the consumer C gets higher utility from rice compared to consumers A and B. This is because, consumer C might have paid higher 'price' to rice due to the following reasons:

- ✰ He might not be aware about the meaningful price of the rice in the market,
- ✰ He might be rich enough to pay high price for the commodity,
- ✰ He might fear about shortfall of that commodity in the market in the near future.

So, straight away we cannot adjudge that, high price implies high utility of the commodity. So, price of the commodity roughly indicate its utility and not in true sense. In spite of this defect, price (money) is used as measuring rod of utility. On this ground, it made Economics an exact science compared to other social sciences, as they have no measuring rod. So, Economists, while employing this measuring rod to measure utility derived by the consumer, are aware of the possible imperfections and limitations and hence, framed meaningful assumptions.

Techniques Employed to Study the Consumer Behaviour

In the earlier pages, it was mentioned that, Microeconomics is otherwise called as Price theory, as it mainly deals with the subject of pricing mechanism of both factors and products. We know price of the commodity is determined by two market forces *viz.*, demand and supply. The Classical Economists laid more emphasis on the 'supply' concept in determining the price of the commodity and neglected the approach from the 'demand' side. Because of this lop sided approach, they could not arrive at the price theory scientifically. The demand concept of price theory was later developed by Stanely Jevons, Alfred Marshall, Walras and Carl Menger. This demand concept developed through time can be divided into two approaches *viz.*, Utility analysis or Marshallian approach and IDC analysis or Hicksian approach. This Hicksian approach has been further refined by Paul A Samuelson by formulating 'Revealed Preference Theory'. These theories are discussed in-detail in the ensuing chapters.

6

Utility Analysis or Marshallian Approach of Consumer Behaviour

We know, human wants are unlimited and they vary in their intensity. This is because, the resources at the consumers are scarce and they have alternative uses. This makes the consumers not indifferent across the wants they wish to fulfill in the economy. Since, he is different across the wants, first he prepares the scale of preferences, prioritize the wants and allocates the resources to fulfill the urgent wants. This implies choice making among the wants. In the pursuit of fulfilling the wants, consumer may deal with one commodity or more than one commodity at a time. The important aspect here is, some economic forces will guide him to fulfill the wants. They include price of the commodity, utility level contributed by the commodity, MU of money etc. In this chapter, we analyze these economic forces, which contribute to consumer's equilibrium with respect to one commodity and more than one commodity. Jevon (1835-1882) was the first Economist, who introduces the concept of 'utility' in Economics. Later, Alfred Marshall and Pigou popularized this utility approach to explain the consumer behaviour or the theory of demand. It is also known as Neo-Classical utility analysis or Cardinal utility analysis or Marginal utility analysis or Marshallian utility analysis.

6.1. Basic Assumptions of Utility Analysis

Before going into the details about various concepts of Utility analysis in explaining the consumer behaviour, it is essential to describe the basic assumptions on which the whole Utility analysis rests. The following are the main assumptions or premises of Utility analysis:

(a) Cardinal Measurability of Utility

This cardinality approach implies utility is measurable or quantifiable. That means, if a consumer consumes a commodity, he can express utility in quantitative terms. For example, if consumer A consumes one mango he can express the utility or satisfaction derived by him in terms of numerical units *i.e.*, say, 10 units. This measurement of utility in numerical terms helps the consumer to quantify the utility derived by him in successive consumption of different units of the same commodity, compare the utility gained across the commodities and across the consumers etc. For example, if consumption of five mangoes yielded utility of 25 units to consumer A and 30 units to consumer B, it implies mango offered more utility to consumer B than consumer A. Likewise, utility gained can also be compared across the commodities. According to Marshall, money serves as the measuring rod of utility. But, according to other Neo-Classical Economists, utility is expressed as a quantity measured in hypothetical units called 'Utils'. That means, in the above example, the consumer A gets 25 utils of utility in consuming five mangoes. This unit of measurement of utility is purely imaginary.

(b) Independent Utilities

This assumption implies 'additive concept' of the utilities. That means, if a consumer consumes different units of the same commodity (say, mango), the Total Utility (TU) derived by him is simply the sum of all the utils (MU) of different units of the same commodity. That means, the TU gained by the consumer in the consumption of different units of same commodity basically depends upon the quantity or number of items of the same commodity consumed and not of the other commodities. Further, if the consumer consumes more number of commodities (say, five commodities), the TU derived by the consumer is the sum total of utilities gained by him in the consumption of five commodities. This clearly indicates that, the utilities derived from different units of the same commodity and from different commodities are additive and their addition gives TU derived by the consumer.

(c) MU of Money is Constant

MU refers to additional utility gained by the consumer. Here, the constancy of MU of money implies that, if the consumer goes on spending the money in purchasing different units of the same commodity or purchasing more than one commodity, the additional utility the consumer derives from additional unit of money spent on the commodities remains same, even though the quantity of money with the consumer diminishes by the successive purchases made by him. This assumption is essential because, this approach considered MU derived from a commodity is measured in terms of money. If the MU of money varies with successive purchases made by the consumer, it cannot yield correct measurement regarding the MU derived from the commodity in terms of money. It is important to note that, this assumption differs from MU derived from the consumption of additional unit of a commodity. This approach assumes that, MU of consumption of additional unit of a commodity diminishes with the successive consumptions of the commodity, but MU of money remains constant throughout, when the individual is spending money on a commodity and due to which the amount of money with him decreases in his successive purchases. This assumption of constancy of MU of money was first proposed by Daniel Bernoulli, but later, Marshall adopted this in his approach. But, this assumption was criticized on two aspects:

- When the price of commodity falls, the real income of the consumer rises or the money paid by the consumer to purchase the commodity also decreases and thereby, the MU of money will also fall.
- If the consumer goes on spending the money to purchase different units of the same commodity or different commodities, the amount of money with him diminishes and hence, the MU of money will increase rather than remaining constant. For example, if the consumer is having Rs. 500 in his pocket, initially he will spend the money rather lavishly on different items like shirt, go to hotel, watch a movie etc. But, when the amount of money left with him is only Rs. 50, he will spend the same very cautiously. This implies MU of money increases with successive purchases made by the consumer.

But, Marshall ignored these criticisms and assumed constancy of MU of money, as he opined that, the consumer's expenditure on any one thing or commodity constitute only a small share or part of his whole expenditure.

(d) Introspective Behaviour

The utility analysis assumes introspection, that is, from one's own experience, it is possible to draw inferences about other person. That means, this assumption shows that, people react in the same way (*i.e.*, derive same level of satisfaction) with respect to consumption of commodity under identical situations. However, this assumption is purely a guess work or intuition or the result of long lasting experience. So, this assumption advocates that, it is possible to make a correct guess work about the behaviour of other consumers' mind based on one's own mind. This assumption is important to analyze the behaviour of all the human beings in similar situations.

(e) Rational Behaviour

The consumer is rational in selecting the right commodity or best combination of commodities for deriving maximum satisfaction. This implies, the consumer is having perfect knowledge about the choices he make regarding the selection of a commodity or different commodity combinations that are open to him.

(f) Perfect Competition

The consumer has complete knowledge about the availability of commodities and price information. The prices of the commodities should be stable.

Based on the above basic assumptions or premises, the entire Marshallian analysis of consumer behaviour is explained through four basic concepts *viz.*, LDMU, LEMU, the Law of Demand and the Concept of Consumer's Surplus.

6.2. Concepts of Utility

Before discussing the concepts of Marshallian analysis in explaining the consumer behaviour to derive maximum satisfaction, it is essential to understand the basic terms involved in these concepts and they are discussed as under:

1.Total Utility (TU)

It refers to the aggregate sum of psychological satisfaction or benefit that an individual gains from consuming a given amount of goods or services at a given point of time. The amount of a person's TU corresponds to the person's level of consumption. Usually, the more the person consumes, the larger his or her TU will be. The higher a consumer's TU, the greater the level of satisfaction he derives in the consumption process. For example, if a consumer consumes ten mangoes, then the TU is the sum of satisfaction of consuming all the ten mangoes. This can also be explained through the concept of MU. Suppose, a consumer prefers to eat mangoes and from the consumption of first mango, he derives more satisfaction. If he eats the mangoes continuously, every successive unit of mango consumed will yield lesser and lesser satisfaction to him, because each successive unit satisfy partial hunger. The same explanation is shown through Table 7.1 of Chapter 7. So, the concept of TU is the sum of MUs derived from the consumption of successive units of the commodity. Mathematically, TU=ΣMUs. It is also clear from the Table 7.1 that, TU will be maximum for the consumer, when MU derived for that unit of commodity is equal to zero. So, by consuming nine mangoes the consumer can derive maximum satisfaction (TU=188 utils), where MU for the ninth unit of mango is equal to zero.

2. Marginal Utility (MU)

It refers to the additional or extra utility derived by the consumer in consuming additional unit of a commodity. It is, otherwise, defined as the change in the TU resulting from a one-unit change in the consumption of a commodity per unit of time. The basic assumption of Marshallian analysis is that, the MU derived by the consumer continuously decreases for successive units of consumption of a commodity. As evident from the Table 7.1, if the consumer goes on consuming the mangoes, the MU derived by him decreased continuously, as his hunger was partially satisfied on the consumption of successive units of commodity. So, MU goes on decreasing, becomes zero (indicating consumption of commodity yields no satisfaction to the consumer) and even turns negative (indicating consumption of commodity yields dissatisfaction to the consumer). Mathematically, MU is expressed as,

☆ $MU = \Delta TU/\Delta Q$

where, ΔTU = Change in TU, ΔQ = change in quantity consumed

☆ $MU_n = TU_n - TU_{n-1}$

where, MU_n = MU of 'nth' commodity, TU_n = TU of n units, TU_{n-1} = TU of n-1 units.

TU usually increases as more of a commodity is consumed by the consumer, while MU usually decreases with each additional increase in the consumption of a commodity. This is demonstrated by the concept of LDMU. Since, there is a certain threshold level for maximum satisfaction, the consumer will no longer receive the same pleasure from consumption once that threshold is crossed. In other words, TU will increase at a slower pace as an individual increases the quantity consumed. The relation between TU and MU is explained through Table 6.1.

Table 6.1: Relationship between TU and MU in consumption of a commodity.

MU Derived	*TU Derived*
Refers to utility derived from each additional unit of commodity	Refers to the utility gained from all the units of commodity consumed.
Declines and remain positive	Increases
Zero	Maximum
Negative	Decreases

It is important to note that, MU is regarded as an inverse function of supply of commodity. That means, if supply of the commodity is increased, MU decreases or *vice versa*. If the supply is unlimited, the MU is zero. The concept of MU is also influenced by the relationship between the commodities. Say, if the two commodities are good substitutes like tea and coffee, rice and wheat and if the quantity of tea is increased, its MU decreases and thereby, the MU of coffee will increase. On the other hand, if the two commodities are good complements, say, butter and bread, an increase in supply of butter will decrease its MU and thereby, MU of bread also decreases. Further, the price of the commodity is determined by the MU and not by TU. This is because, as the supply of the commodity to the consumer increases, the MU for each and every unit of the commodity decreases and thereby, the consumer wishes to pay less and less for the commodity. To be more exact, price is directly proportional to MU contributed by each unit of the commodity. So, the preferences of the consumer to pay for the commodity are in accordance with the MU contributed by each unit of the commodity. So, the consumer goes on purchasing the commodity, so long as the MU derived in terms of money is greater than price of the commodity and he reaches equilibrium, when equity is ensured between these two. This is well-explained through the Table 7.3 of Chapter 7 in discussing the LDMU concept.

Compared to TU, the concept of MU is widely useful to ascertain the consumer's equilibrium. Once analyzing the consumer's equilibrium in terms of MU derived in terms of money and price of the commodity, it is possible to prepare MU schedules of consumer in accordance with the price changes of the commodity. This, in turn, facilitates to derive the demand curve for the commodity.

3. Average Utility (AU)

AU is obtained by dividing TU by the total number of units of the commodity consumed. Mathematically, AU = TU/Q, where TU is total utility and Q is number of units of the commodity consumed. Generally, change in AU is smaller than (or at the most equal to) the change in MU, as change in AU is spread over all the units of the commodity, whereas change in MU is spread over one additional unit of the commodity consumed. AU curve is downward sloping and will not turn negative like MU, as the consumer stops consuming the commodity at the point of equilibrium.

4. Negative Utility

If the consumption of a unit of a commodity is carried to excess, then instead of giving any satisfaction, it may cause dissatisfaction to the consumer. The MU in such cases is negative. If the unit of a commodity yields negative (marginal) utility to the consumer, TU declines. In the Table 7.1, the consumption of 10th mango yields negative (marginal) utility (-10 utils), thereby, TU declines.

5. Initial Utility

It is the utility derived from the initial or the first unit of the commodity. In the Table 7.1, initial utility is 40 utils. So, initial utility is equal to TU in the consumption of first unit of the commodity.

6. Zero Utility

When the consumption of a unit of a commodity makes no addition to the TU, then it is the point of zero utility. In Table 7.1, the TU after the consumption of 8th mango is 188 utils and it remains same even after the consumption of 9th mango. Thus, the consumption of 9th mango results in no increase in TU. Thus, it is the point of zero utility. As explained earlier, the TU will be maximum, when the consumption of a commodity yields zero (marginal) utility.

6.3. Consumer's Equilibrium or Maximization of Consumer's Satisfaction

A consumer is said to be in equilibrium, if he derives maximum satisfaction in the consumption process. That means, it is the point of consumption of commodities, where the consumer is not willing to alter his consumption pattern or expenditure. Since the consumer derives maximum satisfaction at this point of consumption pattern of commodity, any change in the consumption will yield less satisfaction to the consumer.

A. Consumer's Equilibrium when the Commodity is Available Free of Cost to the Consumer

If the commodity is freely available to the consumer, he will not hesitate to consume the same commodity till he reaches the point of maximum satisfaction. That means, he will consume the same commodity till the point of satiety, which occurs when MU is equal to zero or TU is maximum. In case of certain commodities, the point of satiety is reached quickly, while in others it takes more time.

B. Consumer's Equilibrium when Commodity is Priced to the Consumer

Case-1: One-commodity Consumer's Equilibrium

On the other hand, if the commodity is not available at free of cost to the consumer, which is the common practical situation, the consumer tries to attain maximum satisfaction taking into consideration several aspects like MU derived from each unit of the commodity, price of the commodity, MU of money, MU derived from the commodity in terms of money (MU derived from the commodity/MU of money), MU of expenditure (MU derived from the commodity/Price of the commodity), prices of the substitutes etc. So, in this case, the consumer will not consume the commodity till the point of satiety. Rather, he will go on consuming the commodity units, as long as, MU derived from the commodity in terms of money is higher than price of the commodity or MU of expenditure is higher than MU of money. The consumer will reach the point of equilibrium or attain maximum satisfaction when MU derived in terms of money is equal to the price paid for each unit of the commodity or MU of expenditure is equal to MU of money. Since, the consumer is rational, he will not pay more for the commodity than the MU received in terms of money from each unit of the commodity. As the consumer knows, if he goes on consuming the commodity, the MU derived from each unit of the commodity diminishes, and hence, he will be very careful in allocating the resources on the commodity and he stops purchasing the commodity when the MU derived in terms of money is equal to its price. The same is explained through the Figure 6.1.

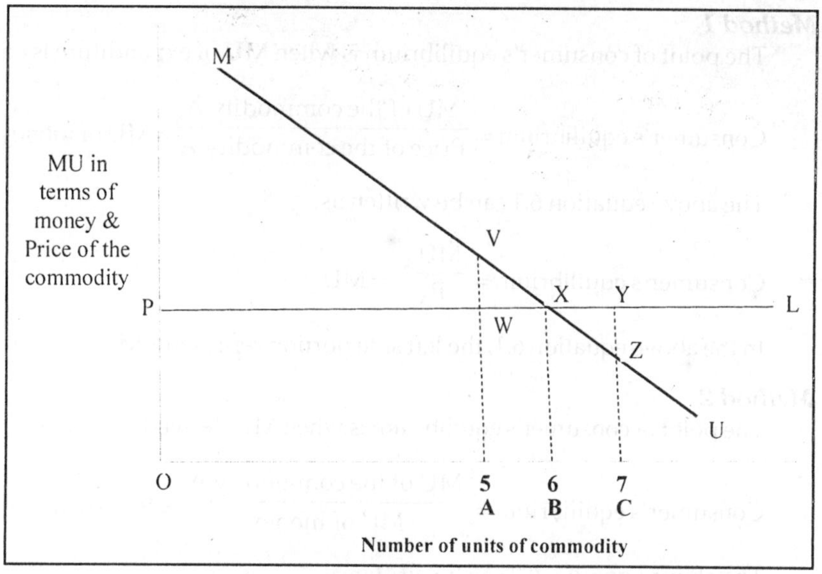

Figure 6.1: Point of consumer's equilibrium.

In the Figure 6.1, along X-axis, number of units of commodity and along Y-axis, price and MU derived from the commodity in terms of money are taken. As the MU diminishes continuously, due to continuous consumption of the same commodity, the MU curve (in terms of money) is a continuous downward sloping curve from left to right. Since the price of the commodity is constant at a given point of time, it is represented by a straight line 'PL' parallel to X-axis and the price is OP per unit. The consumer purchases the first five units of the commodity, since the MU derived in terms of money from each unit of the commodity is higher than the price paid for the commodity, OP. At fifth unit of the commodity, the MU derived in terms of money is VA and price paid is OP. So, he is left over with a surplus satisfaction of VW over the price paid for the commodity. At the sixth unit of the commodity, the MU derived in terms of money, BX is equal to price paid for the commodity, OP. So, this is the point of consumer's equilibrium or the point of maximum satisfaction to the consumer and at this point, there is no surplus satisfaction to the consumer. If the consumer consumes seventh unit of the commodity, he derives the MU in terms of money as CZ, but he pays the same price OP and thereby, he incurs loss of YZ, as the MU derived in terms of money from the seventh unit is less than price paid for the commodity. Since consumer is rational, he is not prepared to consume beyond sixth unit of the commodity, as the MU in terms of money still diminishes below the price of the commodity. So, the consumer reaches the point of equilibrium at sixth unit of the commodity and thereby, he stops consuming at OB units or six units of the commodity.

If the consumer consumes OA units of commodity, he is left over with surplus satisfaction of VWX. If he consumes OB unit of commodity, there is scope for realizing this left over surplus satisfaction, VWX. So, at sixth unit of commodity or consuming OB units of commodity, the consumer derives maximum satisfaction, as there is no scope for surplus satisfaction or left over satisfaction. On the other hand, if the consumer consumes seventh unit of the commodity, as mentioned earlier, the price paid for the commodity is more than MU derived from the commodity in terms of money and thereby, he incur loss in satisfaction to the tune of XYZ. So, moving in either way from OB unit or from sixth unit of the commodity, will lead to diminution of total satisfaction. So, consumer's equilibrium refers to the situation, in which a consumer gets maximum satisfaction for the given level of income and price of the commodity. In the words of Alfred Marshall, *'consumer's equilibrium is that state of consumer's demand, which he thinks to be the best and which he does not want to alter'*. The above explanation can be summarized in the following way:

At OA Units of Purchase of Commodity

Total satisfaction gained by the consumer = OMVA

Total cost incurred by the consumer in purchasing the commodity = OPWA

Surplus satisfaction enjoyed by the consumer = PMVW

Surplus satisfaction left over with the consumer = VWX

At OB Units of Purchase of Commodity

Total satisfaction gained by the consumer = OMXB

Total cost incurred by the consumer in purchasing the commodity = OPXB

Surplus satisfaction enjoyed by the consumer = PMX

Surplus satisfaction left over with the consumer = 0 (zero)

At OC Units of Purchase of Commodity

Total satisfaction gained by the consumer = OMZC

Total cost incurred by the consumer in purchasing the commodity = OPYC

Loss in satisfaction = XYZ

So, based on the above analysis, we can derive the point of consumer's equilibrium from the formulae given below.

Method 1

The point of consumer's equilibrium is when MU of expenditure is equal to MU of money

$$\text{Consumer's equilibrium} = \frac{\text{MU of the commodity A}}{\text{Price of the commodity A}} = \text{MU of money} \qquad \textit{Equation 6.1}$$

The above equation 6.1 can be written as,

$$\text{Consumer's equilibrium} = \frac{MU_A}{P_A} = MU_M$$

In the above equation 6.1, the left side portion represents MU of expenditure or Weighted MU.

Method 2

The point of consumer's equilibrium is when MU derived in terms of money is equal to price of the commodity. So,

$$\text{Consumer's equilibrium} = \frac{\text{MU of the commodity A}}{\text{MU of money}} = \text{Price of the commodity A} \qquad \textit{Equation 6.2}$$

The above equation 6.2 can be written as,

$$\text{Consumer's equilibrium} = \frac{MU_A}{MU_M} = P_A$$

In the above equation 6.2, the left side portion represents MU derived in terms of money.

Method 3

There is another way of expressing the point of consumer's equilibrium *i.e.*, through studying the net gain in the consumption of commodity. That means, the larger the net gain between TU in terms of money and total expenditure, the greater is the satisfaction (Explained in the Table 7.4 of Chapter 7).

Case 2: Two Commodities or More than One Commodity Consumer's Equilibrium

The same principle, as discussed above, also applies, if the consumer deals with more than one commodity to arrive at the point of consumer's equilibrium or the point of maximum satisfaction. For example, if the consumer deals with two commodities A and B, the prices of the two commodities are P_A and P_B, MUs derived from the two commodities is MU_A and MU_B and assumed constancy of MU of money is MU_M, then the point of consumer's equilibrium in case of two commodities is given by the following methods:

Method 1

If the consumer deals with more than one commodity, he will be in equilibrium when he equalizes the MU of expenditure or weighted MUs of all commodities and the consumer will go on purchasing commodities till the MU of expenditure of each commodity becomes equal to the MU of money to him.

$$\text{Consumer's equilibrium} = \frac{\text{MU of the commodity}}{\text{Price of the commodity}} = \text{MU of money} \qquad \textit{as per Equation 6.1}$$

The above equation 6.1 can be written as,

$$\text{Consumer's equilibrium} = \frac{MU_A}{P_A} = MU_M \qquad \textit{for commodity A}$$

$$\text{Consumer's equilibrium} = \frac{MU_B}{P_B} = MU_M \qquad \textit{for commodity B}$$

In the above equations, the left side portion represents MU of expenditure.

So, the point of consumer's equilibrium in dealing with both the commodities A and B is given by the formula:

$$\frac{MU_A}{P_A} = \frac{MU_B}{P_B} = MU_M \qquad \textit{Equation 6.3}$$

This is called Two commodity equilibrium or Principle of Proportionality. The above equation 6.3 implies that, in case of two commodities A and B, the consumer derives maximum satisfaction when MU of expenditure on both the commodities are equal to MU of money.

In the above equation, if $(MU_A/P_A) > (MU_B/P_B)$, then the consumer will go on substituting commodity A in place of commodity B. This results in fall in MU_A due to its excessive consumption and finally MU_A/P_A decreases and gets equal to MU_B/P_B and the consumer reaches the point of equilibrium. But, it is important that, the point of equality between MU_A/P_A and MU_B/P_B can be achieved not only once but also at different levels of expenditure. This is influenced by the size of money income with the consumer.

If the consumer deals with more than two commodities, then the point of equilibrium will be given by,

$$\frac{MU_A}{P_A} = \frac{MU_B}{P_B} = \frac{MU_C}{P_C} = \frac{MU_N}{P_N} = MU_M \quad \textit{Equation 6.4}$$

The above equation can be written as

$$\frac{MU_A}{MU_B} = \frac{P_A}{P_B} \text{ and } \frac{MU_B}{MU_C} = \frac{P_B}{P_C} \text{ and so forth} \quad \textit{Equation 6.5}$$

This indicates that, if the consumer deals with more than one commodity, he will be in equilibrium when he equalizes the ratios of MUs of commodities with the ratio of corresponding prices for each pair of commodities consumed.

Method 2

If the consumer deals with more than one commodity, he will be in equilibrium when he equalizes the MU derived from the commodities in terms of money with their respective prices.

$$\left(\frac{MU_A}{MU_M}\right) = P_A \quad \textit{Consumer's equilibrium for commodity A,}$$

where, $\left(\frac{MU_A}{MU_M}\right)$ is MU derived from commodity A in terms of money.

$$\left(\frac{MU_B}{MU_M}\right) = P_B \quad \textit{Consumer's equilibrium for commodity B,}$$

where, $\left(\frac{MU_B}{MU_M}\right)$ is nothing but MU derived from commodity B in terms of money.

Method 3

There is another way of expressing the point of consumer's equilibrium *i.e.* through studying the net gain across the consumption of commodities. That means, the larger the net gain between TU in terms of money and total expenditure across the commodities, the greater the satisfaction (Explained in Table 8.3).

7

Law of Diminishing Marginal Utility (One Commodity Equilibrium)

The LDMU concept is mainly based on two important facts. First, since the resources are limited with the consumer, it is not possible to satisfy all the wants. But, the available resources with the consumer may be sufficient enough to satisfy a particular want. However, the extra satisfaction the consumer derives from the successive units of consumption of a commodity will gradually decrease and finally a stage is reached, when the consumer prefers no further consumption of commodity. Secondly, different goods under consideration are not perfect substitutes for each other in the satisfaction of a particular want.

7.1 Law

Satisfaction of human wants follow some very important laws in Economics and one of them is the LDMU. This law describes a familiar and fundamental tendency of human behaviour. This law is very natural and hold good for every product and service. So, it is considered as the Fundamental psychological law. We know, one of the characteristics of human wants is its limited intensity. That means, if we have more of any thing, the intensity of our desire for that thing will diminish for successive units. So, when a person increases consumption of a commodity, while keeping consumption of other commodities constant, there is a decline in the MU that person derives from consuming each additional unit of that commodity. If the process of consumption is continued for some time, eventually a stage is reached, when the consumer fails to get any utility from the consumption of additional units of that commodity. If the consumer still continues with the consumption of the commodity, he will develop a dislike for the commodity. In this situation, he will derive a negative MU from subsequent units that he consumes. This tendency is seen in the consumption of almost every commodity and this law is based on the important characteristics of human wants *i.e.*, satiable. This generalization of satiable wants is known as the LDMU. This law operates because, if the people go to the market, they purchase the commodities without attaching equal importance across the commodities they prefer to purchase from the market. That means, they purchase more quantity of some commodities and less of other commodities. So, they are not indifferent across the commodities. This is influenced by two major factors *viz.*,

- ☆ The preferences of the consumers for the commodities is in accordance with their tastes, habits, fashions etc.
- ☆ The available quantity of the commodity with him before purchasing the same in the market. If the consumer has more of that commodity before making purchases, generally he prefers to purchase less quantity in the market. So, this explains, the MU of a commodity diminishes, as the consumer possess larger quantities of it.

This forms the background for the law, as the more of a commodity an individual possess or consumes, the smaller is the increase in TU, other things remaining constant.

H.H. Gossen contributed initially for the development of these ideas. The other Economists like Bentham, Jevons, Karl Menger etc., supported his views. Jevons called this law as Gossen's First Law. In the words of Gossen, *'the magnitude of one and the same satisfaction, when we continue to enjoy it without interruption, continually decreases until satiation is reached'*. But, Alfred Marshall perfected these ideas and made it as a law. In the words of Marshall, *'the additional benefit which a person derives from a given increase of his stock of a thing diminishes with every increase in the stock that he already has'*. Prof. Boulding stated the law as, *'a consumer increases the consumption of any one commodity, keeping constant the consumption of all other commodities, the marginal utility of the variable commodity must eventually decline'*. According to Chapman, *'the more we have of a thing the less we want additional increments of it or the more we want not to have additional increments of it'*. Richard G.Lipsey stated that, *'the utility that any household derives from successive units of a particular commodity will diminish as total consumption of the commodity increases, the consumption of all other commodities being held constant'*. To put in simple words, if a consumer goes on consuming successive units of the same commodity, the MU or additional utility derived by the consumer goes on diminishing. So, this law explains ordinary consumption behaviour of a consumer. But, the demerit of the law is that, it does not state the rate of decline of MU *i.e.*, whether the MU will decline at increasing rate or decreasing rate or at constant rate. However, it is difficult to state because, the operation of the law is purely subjective, varies from one commodity to another, person, time, place and other factors.

7.2 Assumptions

The LDMU expresses an important relation between utility and quantity consumed of a commodity. It reveals that, the MU which a consumer derives from the consumption of each additional unit of a commodity keeps decreasing with every increase in the stock of the commodity which he already has. This law has universal application, but subjected to certain assumed conditions and they are discussed here under.

- ☆ ***Rationality:*** Rationalisation is a term used especially in 1930's of schemes for reorganizing industries where there was excess capacity due to a fall in the demand for the product. The consumer should be rational in selecting the commodity that gives him maximum satisfaction, subject to the availability of his income. The consumer has perfect knowledge about the product and its MU. The choice for the commodity among various commodities by him is determined by his own evaluation or self-interest.
- ☆ ***Cardinal utility:*** The Cardinal measurability of utility implies that, utility is measurable and it can be quantified. The utility derived from the commodity can be added, subtracted, multiplied and divided. The unit of measurement of utility is 'utils'. However, Marshall argues that, *'the amount of money which a person is prepared to pay for a unit of good rather than go without it, is a measure of utility he derives from that good'*.
- ☆ ***Ordinary commodity:*** The commodity under consideration should be an ordinary one. The law may not be applicable, if the individual collects extra-ordinary goods like rare diamonds, rare stamps, rare coins, rare paintings of different types or varieties. However, if same type of the above mentioned extra-ordinary goods are collected, again the law operates.
- ☆ ***A particular want is satiable:*** This law assumes that, though human wants are unlimited, but each single want can be completely satisfied. But, if the consumer gets more and more units of the same commodity, his desire for additional units of that commodity goes on falling and finally, a point is reached, when the consumer no longer consumes additional units of that commodity.
- ☆ ***Deals with only one commodity:*** This law is based on a single commodity consumption mode. That means, a consumer consumes only one commodity at a time. For example, if the first unit of consumption is rice and the second is mango, this law will not operate, as the second consumption unit is more sweeter than the first.
- ☆ ***Suitable quantity of consumption:*** The commodity units should be of suitable size and they should not be very small. For example, milk should be consumed in glass and not in spoonfuls. This is because, every unit meant for continuous consumption should be of identical quantity, and if the quantity is too small, it is not possible to measure same quantity in each case.
- ☆ ***Constant MU of money:*** Marshall's MU analysis presupposes that, MU of money is constant. This assumption was first introduced by D. Bernoulli. It was later adopted by Marshall in his analysis. This assumption is very crucial in Marshallian analysis. The validity of assumption is that, the money is used as a measuring rod of utility. To be a unit of measurement, utility derived from money should remain constant throughout, whether the stock of money with the consumer is rising or falling. If the MU of money changes with the increase or decrease in income with the consumer, it cannot yield accurate measurement of the MU derived from the commodity in terms of money. In the words of Tapas Maunder, *'if money is supposed to provide the measuring rod of utility, then evidently as with all measuring rods its unit must be invariant, it must measure the same amount of utility in all circumstances'*. So, the law assumes that, even if the consumer goes on spending the money in purchasing the commodity, the MU of money used for purchasing commodity remains constant.

☆ *Diminishing MU from commodity:* If the consumer goes on consuming the commodity, in each successive stage of consumption, the MU derived from the commodity decreases in a given time period.

☆ *Utility is additive:* The utility derived from the commodity is measurable and additive. That means, in each successive stages of consumption, the commodity yields MU to the consumer and to get the TU, these MUs of different units of the commodity can be summated. So, we can say, TU = ΣMUs.

☆ *Consumption should be continuous:* There should not be any time lag in the process of consumption of successive units of commodity. That means, the consumption process should be continuous. If there is discontinuity in consumption, the intensity of want gets revived and hence, for the successive units, the MU will not decrease, but may rather increase. For example, if one mango is consumed in morning and another in evening, the second mango may provide higher MU. This is because, the MU derived from the commodity depends on the intensity of desire in having the commodity. So, this law is true only at a given point of time.

☆ *Identical units of the commodity:* The units of commodity consumed in successive stages should be identical and homogeneous in all respects without any change in quality, taste, flavour, colour, size etc. If the units are dissimilar in nature, for the successive units, the MU will not decrease, but may increase. If, for example, the first mango is fully ripened and the second is less ripened, then the first mango will give more utility than the first.

☆ *Standard unit of consumption:* The unit consumed should be of the standard unit. Say, for example, if the commodity under consideration is mangc, all the mangoes should be of same variety, same ripening stage, same size etc. Similarly, a cup of tea, a bottle of cool drink etc., represent standard units of consumption.

☆ *No changes in tastes, habits, customs and fashions:* These should be assumed constant for the effective operation of the law. If any of these changes, the consumer may suddenly switch over to any other commodity and this makes the law inoperative.

☆ *Price of the commodity:* The consumer should have unlimited income, so as to spend on the desired commodity to achieve maximum satisfaction. Further, the income level should remain constant. Say if income of the consumer increases, then there is ac chance on the part of the customer to switch over to superior commodity and this makes the law inoperative.

☆ *Income of the consumer:* The consumer should have unlimited income and it must remain constant. Say, if income of the consumer increases, then there is a chance on the part of the customer to switch over to superior commodity and this makes the law inoperative.

☆ *No substitutes:* Different commodities are not perfect substitutes for each other in the satisfaction of a particular want. As such, the MU derived from the commodity will decline as the consumer consumes additional units of the same commodity.

☆ *Normal person:* The consumer should be an economic man, who acts rationally. There should be no change in the character of the consumer. His mental condition should be stable and his behaviour should be consistent. For example, if the consumer is addicted to wine or drugs, then every successive unit of wine or drug will give higher MU, which is against the operation of the law.

☆ *No change in weather:* The weather should remain unchanged. If weather changes, the preferences and demand for a certain commodity will increase, and this makes the law inoperative.

☆ *Divisibility:* The commodity should be perfectly divisible. This facilitates to measure the number or quantity of commodity that gives maximum satisfaction to the consumer. However, in case of durable consumer goods, it is not possible to calculate the point of maximum satisfaction, as their utility spreads over a period of time. For example, a farmer will not purchase five ploughs, ten harvesters, seven tractors for his personal consumption.

☆ *No change in other peoples' stocks:* This law assumes that, there should not be any change in the other peoples' stocks. For example, if the area under paddy is increased in a village, the utility of establishing a second rice mill by the miller may increase. If the number of telephone connections in a village increases, the utility of a second phone in a house may increase.

☆ *No change in the consumers' other possessions:* For the effective operation of the law, there should not be any change in the consumers' other possessions. If a person is having ink, it is of less utility to him. But, if a pen is provided to him, the utility of ink will increase. So, the change in stock with the person, may add utility for a particular commodity.

7.3 Explanation of the Law

The law can be explained with the help of an hypothetical example (Table 7.1) Suppose, a person want to consume mangoes. In the first instance, he consumes one mango and this gives him great satisfaction of 40 utils. This is called as Initial utility. He then consumes second mango and this yields him 38 utils of satisfaction. So, the satisfaction (MU) yielded by second mango is less than that of first mango, as the extent of desire to consume second mango decreases with the satisfaction received from the first mango. So, the TU gained after consuming second mango is 78 utils, *i.e.*, 40+38 = 78 utils. If he consumes third mango, the satisfaction (MV) derived will be less than that of second mango and so on. The third mango yields MU of 36 utils and hence, the TU gained after consuming third mango is 114 utils *i.e.*, 40+38+36 = 114 utils. So, MU is the addition made to TU by consuming an additional unit of the commodity or it is the loss in TU, if one unit less is consumed. So, MU for third mango can be calculated as $MU_n = TU_n - TU_{n-1}$ *i.e.*, 114-78 = 36 utils and so on for other mangoes consumed. If the consumer goes on consuming the mangoes, the MU derived from each additional mango consumed will diminish and on consuming ninth mango, the MU derived is zero. This zero utility implies, the consumer is not getting any satisfaction in the consumption of ninth mango. It is the point of consumer's equilibrium or maximum satisfaction, as when MU is zero, TU is maximum (188 utils) for the consumer. This point of zero MU is called Satiety point. If the consumer is forced further to consume tenth mango, it yields dissatisfaction or disutility, as MU turns negative (-10 utils) and thereby, causing TU to decline. Here, it may be noted that, the MU of the successive mangoes consumed diminishes not because these mangoes are of inferior in quality than earlier units, but because, by the time the successive units are consumed, the desire of consumer to have extra units is slightly decreased in each case. Rather, we assumed that, all the mangoes meant for consumption are exactly alike. So, a rational consumer will not consume tenth mangoes, even if the commodity is offered at free of cost, as it yields negative MU. According to Professor Boulding, the following are the two important reasons, why the MU declines with increase in consumption of the commodity:

Table 7.1: MU and TU derived by the consumer in the consumption of mangoes.

Number of Mangoes Consumed	*MU (Utils)*	*TU (Utils)*
0	—	0
1	40	40
2	38	78
3	36	114
4	28	142
5	20	162
6	16	178
7	6	184
8	4	188
9	**0**	**188**
10	−10	178

- ☆ We know, a particular human want is satiable. Which-so-ever commodity a consumer continues to consume, sooner or later the point of satiety will be reached for a particular want, where MU will become zero. This is because, if a person goes on consuming the commodity, his hunger or desire to have more will be decreased, and thereby, the MU derived from each additional unit will decrease.
- ☆ In practical life, the commodities are not perfect substitutes, rather they are imperfect substitutes. So, a particular commodity will satisfy one want only. Hence, the MU diminishes, as the stock increases.

The above explanation holds good, when the mango is offered at free of cost to the consumer and when the money resources are abundant with the consumer. Then, the consumer shows no hesitation to consume nine mangoes because, he derives maximum satisfaction (TU=188 utils) after consuming nine mangoes. In short, when mango is available free of cost, a consumer increases consumption of the mangoes, so long as the additional units yield him positive MU and consumes upto the point when MU is equal to zero. The same explanation is shown through Figure 7.1. Along OX, number of mangoes consumed is taken and along OY, utility derived (utils) is taken. It is evident that, if the consumer goes on consuming mangoes, the MU diminishes continuously. Hence, TU increases and it reaches maximum, when MU becomes zero. This is seen after consuming ninth mango and this implies the consumer derives maximum satisfaction when he consumes nine mangoes in a continuous process. MU curve falls continuously from left to right, whereas TU curve rises upward from left to right and reaches maximum at Satiety point *i.e.*, where MU is equal to zero. It can also be seen that, the TU curve rises rather slowly because, MU gradually diminishes and when MU turns negative, it actually shrink with consumption of additional units. However, the above explanation holds good, when the commodity is available free of cost to the consumer. But, when the commodity is scarce and priced, the consumer may not consume nine mangoes and the consumer's equilibrium is achieved taking into consideration MU of money, MU of expenditure, MU derived from the commodity in terms of money etc. The same is explained through the Tables 7.2 to 7.4.

Consumer's Equilibrium when the Commodity is Priced

A rational consumer will consume less than ten mangoes. This is because, if he consumes ten or more than ten mangoes, then each and every unit of mango consumed will yield dissatisfaction to the consumer. If the mangoes are supplied to the consumer at free of cost, then the consumer consumes nine mangoes, as for ninth unit of mango, the MU is zero, and thereby, TU is maximum (188 utils). But, we know mangoes are not supplied to the consumer at free of cost in the market. So, the number of mangoes consumed by the consumer not only depends upon the MU and TU derived from it, but also by the income level of the consumer, price of the commodity and MU of money. Let us assume, MU of money is constant at eight utils per rupee and price of the mango is Rs.2/- in the market.

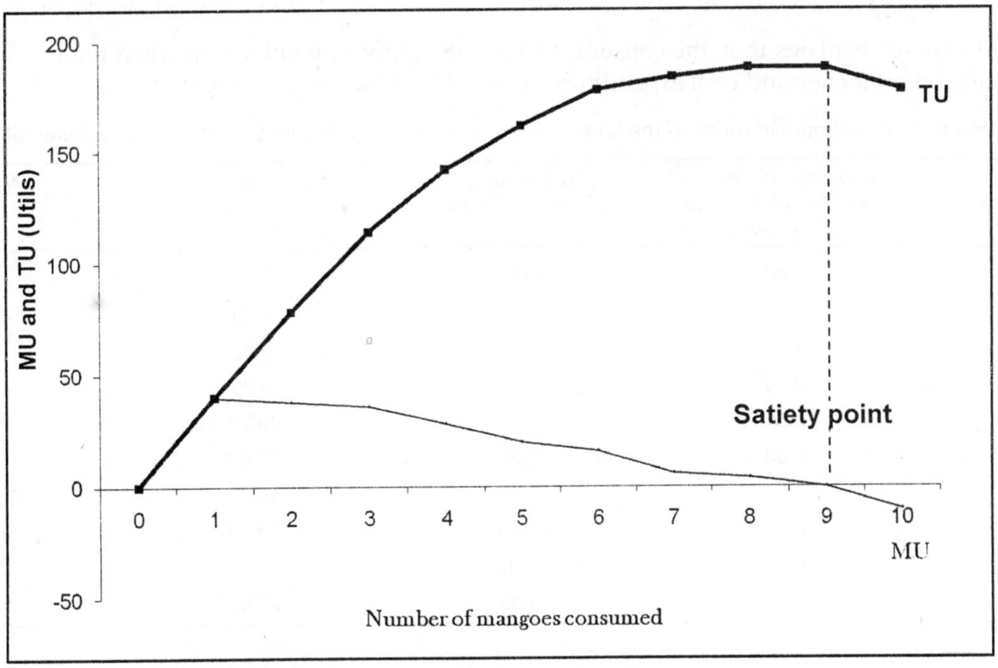

Figure 7.1: MU and TU derived by the consumer in the consumption of mangoes.

Method 1

This method employs the formula *viz.*, MU of expenditure = MU of money, which is the point of consumer's equilibrium. Let us re-write the earlier Table 7.1 in terms of MU of expenditure or weighted MU to arrive at the point of consumer's equilibrium and it is shown through Table 7.2. It is clear that, the point of consumer's equilibrium lies at sixth unit of mango, as at this point, MU of expenditure (8.0) is equal to MU of money (8.0). If the consumer consumes seventh mango, the MU of his expenditure (3.0) is lower than the MU of money (8.0) and this implies wastage of money resources. So, we can define consumer's equilibrium as the point, where MU of expenditure is equal to MU of money.

The above analysis clearly indicate that, the consumer's equilibrium depends upon consumer's behaviour in relation to his income, tastes and preferences, price of the commodity in the market, MU derived from the commodity, MU of money etc. If these factors changes, the consumer equilibrium also changes.

Method 2

This method employs the formula: MU derived in terms of money = Price of the commodity and this depicted in Table 7.3. It is evident that, a rational consumer consumes six mangoes, as at the sixth unit of mango, MU derived in terms of money (Rs. 2.00) is equal to price of the commodity (Rs. 2.00). If he consumes seventh mango, it is loss on the part of consumer, as the MU derived in terms of money (Rs. 0.75) is less than price paid for the mango (Rs. 2.00).

Table 7.2: Consumer's equilibrium in terms of MU of expenditure or weighted MUs and the MU of money in the consumption of mangoes.

Number of Mangoes Consumed	*MU of Expenditure or Weighted MU (MU of commodity/Price of commodity)*
1	20.0
2	19.0
3	18.0
4	14.0
5	10.0
6	**8.0**
7	3.0
8	2.0
9	0.0
10	–5.0

Table 7.3: Consumer's equilibrium in terms MU derived in terms of money and price of the mango

Number of Mangoes Consumed	*MU Derived in Terms of Money (MU of commodity/MU of money)*
1	5.00
2	4.75
3	4.50
4	3.50
5	2.50
6	**2.00**
7	0.75
8	0.50
9	0.00
10	–1.25

Method 3

This methodology explains that, the consumer attains the point of equilibrium when there is largest net gain between TU in terms of money and total expenditure depicted in Table 7.4. It is clear that,

Table 7.4: Consumer's equilibrium in terms of the largest net gain (in terms of money) derived in the consumption of mangoes.

Number of Mangoes Consumed	*MU Derived in Terms of Money (MU of commodity/ MU of money) (Rs/-)*	*TU Derived in Terms of Money (Rs/-)*	*Total Expenditure (Rs/-)*	*Net Gain (Rs/-)*
1	5.00	5.00	2.00	3.00
2	4.75	9.75	4.00	5.75
3	4.50	14.25	6.00	8.25
4	3.50	17.75	8.00	9.75
5	2.50	20.25	10.00	10.25
6	**2.00**	22.25	12.00	**10.25**
7	0.75	23.00	14.00	9.00
8	0.50	23.50	16.00	7.50
9	0.00	23.50	18.00	5.50
10	–1.25	22.25	20.00	2.25

- ☆ The rational consumer will consume six mangoes, where the net gain is maximum (Rs.10.25) in terms of difference between TU in terms of money and total expenditure incurred by the consumer.
- ☆ From the first unit of mango and upto sixth unit, this net gain showed increasing trend and reached the maximum at sixth unit of mangoes.
- ☆ At first unit of mango, the MU derived in terms of money is Rs.5 and for this the consumer is paying only Rs.2. So, the net gain to the consumer after consuming first unit of mango is 5-2 = 3. So, he will purchase the first unit of mangoes.
- ☆ With respect to second unit of mango, the consumer gets MU in terms of money worth of Rs. 4.75, but he pays only Rs.2/- for it. So, he will purchase second unit of mango, because he is the gainer. So, the total net gain derived by the consumer by consuming two mangoes is given by the formula:

Total net gain at the end of second unit of mango = (TU gained in terms of money after consuming second unit of mangoes) – (Total money expenditure incurred in purchasing two mangoes)

$= (5 + 4.75) - (2 \times 2)$

$= 9.75 - 4.00$

$= 5.75$

Based on the above data, we can generalize the formula for total net gain to the consumer after consuming 'n' units of commodity.

Total net gain after consuming 'n' units of commodity =

$$\text{TU gained in terms of money after consuming 'n' units of commodity} - \text{Total money expenditure incurred in purchasing 'n' units of commodity}$$

- ☆ In the same way, the consumer is the gainer after consuming third, fourth and fifth units of mango because, he is deriving more MU in terms of money when compared to price paid for each unit of the commodity.
- ☆ However, at sixth unit of mango, a different situation emerges. This is so because, MU derived in terms of money (Rs. 2.00) gets equal to price of the commodity/unit (Rs.2.00). Here, the consumer may or may not purchase the commodity. But definitely he will not purchase seventh unit of mango, because MU in terms of money (Rs.0.75) is lower than price of the commodity per unit (Rs.2.00). So, the consumer prefers to purchase five or six mangoes. But, he prefers to have six mangoes because, the MU in terms of money gets equal to price of the commodity and at this point the net gain received by the consumer is maximum at sixth unit of mangoes (Rs.10.25). But, a critical examination of the Table 7.4 shows that, total net gain is maximum even after consuming five mangoes, but the principle is not satisfied *i.e.*, MU in terms of money (Rs.2.50) is higher than price of the commodity (Rs. 2.00). This is also due to the fact that, the mangoes are not perfectly divisible, and hence, the consumer cannot purchase and consume 5.99 mangoes and so he will consume 6 mangoes.

On the other hand, if the commodity is perfectly divisible, there will be only one level of consumption at which the consumer's equilibrium will be achieved on the basis of above formula.

It is important that, proper care regarding the application of assumptions and concept of LDMU is essential to judge its validity. This can be explained with the help of an example. Suppose, a farmer 'A' is cultivating cotton crop in a village. He has purchased a tractor and he rent it out for 50 farmers in the village. In view of the profitability of cotton in that area, 25 more farmers shifted towards cotton cultivation. So, farmer 'A' gets more rent now, as he can rent the tractor for more number of farmers (75) cultivating cotton in that area. This implies, the 'utility' of tractor has increased to farmer 'A', as he earns more rent from the tractor now. Hence, we conclude, the concept of LDMU is not applicable in this case. But, if we critically go through the example, it indicates that, the assumptions and application of the law have not been properly maintained. In this example, if the farmer 'A' purchases second tractor, the operation of the law can be demonstrated. This is because, farmer 'A' finds lesser MU from additional tractor taking into consideration several aspects like initial cost of the tractor, less number of farmers shifted towards cotton cultivation, increases in repairs and maintenance costs, payment of extra tractor driver wages etc. So, in the demonstration of the law, care should be taken in properly applying the assumptions of the law. The assumptions and concepts should not be changed. If this is done properly, the law has universal application.

7.4 Exceptions or Limitations of the Law

There are some exceptions or limitations regarding the applicability of the concept of LDMU and they are discussed below.

- ☆ ***Abnormal persons***: For abnormal persons like chain smokers, drunkards, misers etc., the concept of LDMU is not applicable. This is because, say in case of drunkards, the more they drink, more is the satisfaction for them. Of course, a stage is also reached for them, where they drink less and less and eventually stop drinking after a certain period. If this is not so, the drunkard will continue to drink for an indefinite period. Further, the standard unit for drinking liquor may be different for different people, say one peg for an individual, one bottle for another etc. On the basis of the standard unit, every one will get diminishing utility of liquor.
- ☆ ***Rare collections:*** The more the number of collections that are special for the individuals say diamonds, old currency, stamps etc., the more is the satisfaction derived by the individuals. However, some opine that, even in case of rare collections also, the LDMU operates. This is because, in case of rare collections, if the individual collects same type of stamps or coins, this law will certainly operate. But, in general, in stamps or coins collection, the individual collects 'different varieties' of stamps or coins, which is against the assumption of application of LDMU concept. So, an individual does not spend an increasing amount of money in acquiring the same type of stamps or coins.
- ☆ ***Not applicable to money:*** It is generally opined that, the greater the amount of money an individual possesses, the greater will be the utility to him, as he can purchase more number of goods and services. Hence, the MU of money does not diminish, but actually rises. But, this opinion is not valid. This is because, when a person becomes rich, he spends his additional income on purchasing comforts and luxuries, which may not give him so much utility as basic necessaries of life. We generally observe in the economy that, rich people do not care much about money as the poor people do. Thus, as the money possessed by the individual diminishes, the MU of money will increase. Thus, the MU of money will not remain constant and this disproves the assumption of Marshallian analysis. However, the truth is that, the MU of money declines with richness, but never falls to zero because, the consumer allocate the same across a wide range of commodities.
- ☆ ***First time consumption of commodity:*** If the consumer consumes the commodity for the first time, it may yield more and more satisfaction, instead of diminishing MU. For example, if a person is exposed to use internet on the computer for the first time in his life, the MU will increase instead of diminishing.
- ☆ ***Books reading:*** Reading of more books gives more knowledge and in turn greater satisfaction.

7.5 Practical Importance

From the above analysis, we can conclude that, the LDMU, like other laws of Economics, is simply a statement of tendency. It holds good provided other factors remain constant. However, this law has great practical importance in the field of Economics and it is highlighted below.

- ☆ ***Basis for the law of demand:*** The LDMU and the Law of demand are very closely related to each other. In fact, the LDMU forms the basis for the Law of demand. According to the LDMU, the more we consume of a commodity, the less is the MU derived from each additional unit consumed. In other words, if a person consumes more and more of a particular commodity, the MU of the successive units of the commodity diminishes. So, every consumer, while buying a particular commodity, compare the MU of the commodity in terms of money with the price of the commodity, which he has to pay. If the MU of the commodity in terms of money is higher than that of price, he purchases that commodity. As he buys more and more, the MU of the commodity in terms of money diminishes, as MU derived from successive units begins to diminish. Then, he wishes to pay lesser amount for the successive units (as MU contributed by the successive units diminishes) and tries to equate the MU of the commodity in terms of money to the price of the commodity, where he will be in equilibrium. From this, it is also clear that, if a person (seller) wishes to increase the sale of a commodity, he must lower its price so that, it induces the consumers to purchase large quantities of commodity. This is because, when price is reduced, it makes the consumer to purchase the commodity in large quantities, as by purchasing large quantities only, the MU of the commodity in terms of money will greatly fall and it finally gets equated to the lower price quoted for the commodity. So, the consumer reaches the equilibrium point at large quantities of purchase of the commodity. This is what explained by the Law of demand *i.e.*, more quantity purchased (demanded) at lower price of the commodity. So, the LDMU highlights the inverse relationship between quantity demanded and price of the commodity. Thus, it is because of the diminishing MU, that the demand curve slopes downward from left to right. From this, we can conclude that, LDMU forms the base for the Law of demand and they both are closely inter-related.

☆ ***Basis for consumer's surplus concept:*** The theory of consumer's surplus is also based on the LDMU. As explained earlier, the LDMU states that, the more we consume of a commodity, the less is the MU derived from each additional unit consumed. So, every consumer, while buying a particular commodity, compares the MU of the commodity in terms of money with the price of the commodity, which he has to pay. The consumer goes on purchasing the commodity, as long as MU of the commodity in terms of money is higher than the price of the commodity. This difference between the MU of the commodity in terms of money and price paid for each unit of the commodity constitute surplus gained by the consumer for that unit of the commodity. In general, the consumer is prepared to pay for the commodity upto the MU in terms of money it contributes to him, as it is the point of consumer's equilibrium. So, for earlier units of the commodity, the consumer is willing to pay more than what he actually pays. This excess of the price which the consumer is willing to pay for the commodity at different successive units of consumption, but not actually paid for the commodity constitute the consumer's surplus or economic measure of surplus satisfaction. For the total number of units purchased by the consumer, the aggregate consumer's surplus can be expressed as TU in terms of money - Total amount spent on the commodity. So, the LDMU forms the basis for consumer's surplus concept.

☆ ***Importance to the producers:*** Producers in the economy are guided by the operation of LDMU. We know, producers in the economy constantly change the design, pattern and the package of the commodities, so that, they have more attractive appearance and appear as 'new goods' to the consumers. Or else, the consumers would think that, they are purchasing and using the same commodity, over and over. In such a situation, the LDMU operates in the minds of the consumers, and hence, they purchase or demand less for such commodities. So, the concept of LDMU guides the producers in producing goods and services with attractive features, so as to maximize their profits.

☆ ***Importance to the consumers:*** The concept of MU guides the consumer in allocating his scarce resources on the commodities he wishes to consume. We know, the consumer has scarce resources and he has to allocate them on the commodities, so as to derive maximum satisfaction. In case of one commodity, it guides the consumers to attain the equilibrium when the MU gained in terms of money is equal to the price paid for the commodity. If the consumer deals with more than one commodity, then he will try to allocate the resources on the commodities in such a way that, the MU derived from a commodity in terms of money is equal to the price of that commodity and the same condition should be satisfied across all the commodities or MU of expenditure across the commodities should be equal. So, this concept of MU guides the consumer in fixing his scale of preferences of commodities in the economy, in order to derive maximum satisfaction. This concept is nothing but LEMU, and thus LDMU, forms the basis for LEMU.

☆ ***Importance in taxation policy:*** Sometimes, it is pointed out that, the LDMU concept is not applicable to money. We generally presume that, if a person collects more and more money, the desire to accumulate further more money also increases. But, this view is only superficial. In reality, the concept of LDMU is also applicable to money. It is true that, if a person, gets more and more money, its MU gradually diminishes, though it does not reach to zero or become negative, as human wants are unlimited. For example, a person who earns Rs.25000 per month attaches less importance to Rs.10. But, for a man who gets Rs.1000 per month, the value of Rs.10 to him is very high. He will be very cautious in spending each and every unit of rupee on the goods and services. So, we can say, MU of money is high for poor people in the society compared to rich people. Hence, in planning taxation policies, high income people are taxed more compared to poor people. This system of taxation is called Progressive taxation and is therefore, based on the LDMU. So, the concept of LDMU ensures proper re-distribution of income in the economy from the rich sections of the society to the poor people through spending the tax proceeds on social services and this will promote the economic welfare of the community. This is explained through the Figure 7.2.

In the Figure 7.2, on X-axis, money income and on Y-axis, MU derived from income are taken. Taking these two parameters, MU curve for money is drawn and it is convex to the origin in accordance with the LDMU concept. Assume, OL is the income of the poor person and OH is the income of rich person in the economy. Let the Government resorted for progressive taxation, wherein, rich person taxed more compared to poor person. Here, assume that, the poor person is not taxed taking his base level of the income into consideration as per the tariff rates followed by the Government and more tax is imposed on the rich person. As shown in the Figure 7.2, due to the imposition of tax, the income of the rich people is declined to OH_D or declined by H_DH and this income received by the Government in the form of tax is diverted for the welfare of the poor person. So, now, the income of the poor person increases from OL to OL_I or by LL_I. Since, all the money received from rich person is diverted towards poor person, H_DH is equal to LL_I. So, the loss of total satisfaction by the rich person is $H_DH_2H_1H$ and the gain in total satisfaction by the poor person is $LL_1L_2L_I$ due to increase in income of the poor person. It is a known fact that, the values of MU at the higher level of MU curve are higher and at lower level of MU curve, they are lower, as the MU curve slope downward from left to right and the curve is more steep at the beginning and becomes flat at the down end. So, when the money is transferred from rich person to poor person, it is being utilized on the commodities that yields more MU to the poor person compared to the commodities used by rich person that yields less MU. So, the gain in TU is more for poor person and loss in TU is comparatively less for rich person. So, the net TU of both the persons taken together will increase. So, by applying this concept of LDMU, the Economists advocate the Government to implement progressive taxation policy, which leads to re-distribution of income among the poor and thereby, raise the economic welfare of the society. So, this concept forms the basis for socialism. However, this rise in economic welfare will be experienced only, if the investments are made by the State with respect to production of public goods meant for poor people. On the other hand, if the income is re-distributed among the poor, the economic welfare of the society cannot be on expected lines, as for each individual, the scale of preferences will vary.

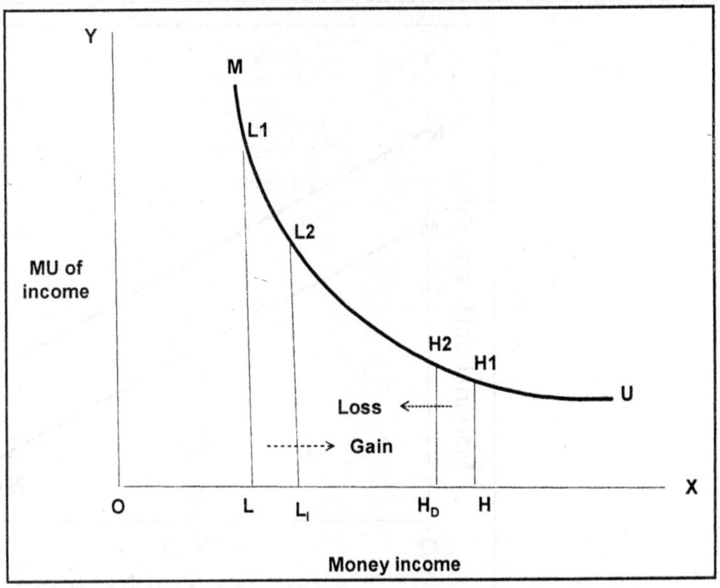

Figure 7.2: Applicability of LDMU concept in the field of taxation.

☆ ***Explain Diamond-Water paradox or Paradox of value:*** The famous 'diamond-water paradox' of Adam Smith can be explained with the help of this law. Adam Smith could not explain the diamond-water paradox, which means, why water which is so very essential and useful to life has such a low or no price, while the diamond which is comparatively less necessary, have such a high price. This concept of LDMU has offered satisfactory explanation to the above paradoxical situation, as MU is having crucial significance in explaining the determination of the prices of commodities. According to Modern Economists, it is not the TU, but rather MU that determines the price of the commodity. Since water is not a scarce commodity, it has only value-in-use and no value-in-exchange. So, being a free good, its MU is low and hence, offered at low price in the economy. On the other hand, the diamond is a scarce commodity and hence, it possesses both value-in-use and value-in-exchange. So, its relative MU is very high and hence, high price is offered to this commodity in the economy. In the words of Prof. Samuelson, '*the more there is of a commodity, the less the relative desirability of its last little unit becomes, even though its total usefulness grows as we get more of the commodity*'. So, the concept of LDMU explains, why water has low price as compared to a diamond, though it is more useful than the diamond.

In the Figure 7.3, MU_W and MU_D are the diminishing MU curves drawn for water and diamond respectively. These two curves are downward sloping from left to right, as explained by the concept of LDMU. The MU_W curve is above MU_D curve because, from the initial units of consumption of water, the consumer derives more satisfaction compared to initial units of diamond, as water being a necessary good for human life. As we

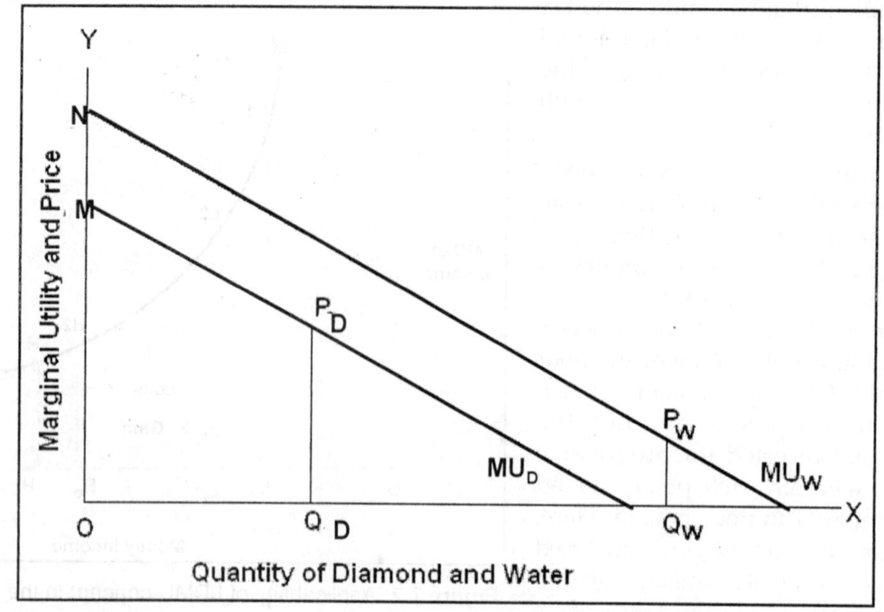

Figure 7.3: Diamond–Water Paradox.

know, the availability of water is more (being a free good), compared to diamond, let us suppose, OQ_W is the quantity of water available and OQ_D is the quantity of diamond available in the economy. At these quantities of OQ_W and OQ_D, the MU derived from water is less compared to diamond (as $OQ_W > OQ_D$). Accordingly, the price we pay for water is less compared to diamond, as the price of the commodity is influenced by its MU. So, it is clear that, water will be paid a low price of P_W, whereas, diamond will be paid a high price of P_D. So, the TU derived from water is given by ONP_WQ_W and it is comparatively higher than TU derived from diamond *i.e.*, OMP_DQ_D. So, this infers the following interesting points:

- For water, TU is more compared to diamond, but available at low price in the economy.
- For diamond, TU is less compared to water, but offered at high price in the economy.
- MU of water is negligible on account of its abundant supply in most of the places, as the later units pull down the market value of all units. (Say, if one bucket full of water is given, after we have used as much as we required, the remaining quantity would have almost no utility). So, less price or sometimes even no price is offered for the use of water.
- MU derived is notoriously high for diamond compared to water, and hence, high price is paid for diamond. If people had to do without diamond, there would be no suffering or inconvenience. But, the utility of a single diamond is appreciable, since its presence will make a significant difference on the part of human life.
- High TU infers more value-in-use to water (less value-in-exchange), whereas, high MU infers, high price to the commodity *i.e.*, diamond (more value-in-exchange).

Thus, the controversy of diamond-water paradox is resolved through MU and TU analysis. From this analysis, we can also infer the difference between value-in-use and value-in-exchange of a commodity besides this it is also explains, why other commodities like air, sunlight etc., have greater value-in-use (high TU), but little value-in- exchange (low MU).

☆ ***Basis for the LEMU:*** The concept of LDMU forms the basis for LEMU. As the MU derived from each successive unit of consumption of commodity decreases, the consumer will try to maximize the satisfaction at the point where, MU is equal to zero or TU is maximum. But, this point of consumer's equilibrium can be achieved only, when the commodity is available free of cost or when there is unlimited money with the consumer. But, these two are not common phenomena in the economy. So, the consumer must utilize his scarce resources efficiently and hence, he substitute one commodity for other commodity and consume the commodities and achieve the point of equilibrium, where MU derived in terms of money for a commodity is equal to the price paid for that commodity and the same condition should be satisfied across all the commodities or MU of expenditure is same across all the commodities, which is nothing but the concept of LEMU.

☆ ***Explain the law of supply:*** LDMU concept explains the phenomenon in value theory that, the price of a commodity falls, when its supply increases in the economy. This is because, with the increase in the stock of a commodity, its MU diminishes and thereby, price decreases.

- ***Allocating the commodity among different uses:*** This law guides the consumer to allocate the commodity among different uses, where he gains more MU. This is because, if the consumer possesses a little quantity of commodity (say, coal), he will put it to satisfy most urgent want. On the other hand, if he possess large quantity of the commodity, he begins to use the same for less and less urgent uses. So, as the consumer has a scale of preferences, he will try to allocate his limited resources first to fulfill most urgent want, from which he derives highest MU. On the other hand, when a person has more of a commodity, he will put to the same to satisfy less and less urgent wants and the MU derived from such wants is low.

7.6 Criticisms

The methodology relating to the concept of LDMU has been subjected to a severe criticism by the Modern Economists on the following grounds:

- Since utility is purely a psychological phenomenon, it is not measurable either in terms of 'utils' or in terms of money. It is purely imaginary.
- The basic assumption that, 'constancy of MU of money' is not valid, as MU of money increases with increase in money expenditure and MU of money declines with richness, but never falls to zero. So, this law equally holds good for money.
- In practical situation, no person will consume a single commodity upto the point of satiety, that too without any time lag in the process of consumption. In practical life, the resources are limited with the consumer and he will not allocate all the resources on the same commodity upto the satiety point.
- This law assumes that, the consumer consumes only one commodity at a time. Moreover, he has to take successive units of the same commodity. But, this is unrealistic, as in real life, a consumer consumes more than one commodity at a time.
- As utility itself is capable of varying from person to person, MU derived from the consumption of a good cannot be measured accurately.

8

Law of Equi-Marginal Utility (More than One Commodity Equilibrium)

We have already seen that, the focus of Economics is to utilize the limited resources efficiently among unlimited or prioritized wants. In this context, the concept of LEMU will guide the consumers in deriving maximum satisfaction from the available limited resources. The LEMU is an extension to the LDMU to two commodities or more than one commodity. This law explains the behaviour of a consumer in allocating his limited income among various goods and services, so as to derive maximum satisfaction. This law closely follows the Robbins's definition of Economics in various aspects *viz.*, the resources are scarce with the consumer, resources have alternative uses, resources can be allocated among different goods and services, the consumer has scale of preferences among unlimited wants, consumer aims to derive maximum satisfaction from limited resources etc. So, this law is more realistic compared to LDMU and it occupies an important place in the cardinal utility analysis. This law was propounded by H.H Gossen and hence, it is popularly called as Gossen's second law and later developed by Alfred Marshall. The LEMU is also known by various names *viz.*, the Law of Substitution, the Law of Maximum Satisfaction, the Law of Indifference, Principle of marginal comparisons, Law of consumer's demand, Principle of Equi-Marginal Returns and the Proportionality Rule. This law is known as the Law of Maximum Satisfaction because, the consumer tries to derive maximum satisfaction from his limited resources in such a way that, the MU of expenditure on one commodity is the same as the MU of expenditure on another use. It is also known as the Law of Substitution because, consumer substitutes one commodity for the other till he gets the maximum satisfaction from his limited resources. A consumer ever spends money at his disposal in accordance with respective intensities of his various wants the compares the amount that he has to spend or. a given article and the utility that he drives. Where money is this spent, the consumers constantly replace one article by another article of consumption and the other by the third. At the margin the last unit of money spent on various articles of consumption givens the same or more or less the same utility.

8.1. Law

Alfred Marshall stated that, *'if a person has a thing which can be put to several uses, he will distribute it among these uses in such a way that, it has the same marginal utility, for if it had a greater marginal utility in one use than in another, he would gain by taking away some of it from the second use and applying it to the first'.*

In the words of Gossen, *'if it is impossible to gratify all wants to the point of satiety, it is necessary in order to obtain maximum satisfaction, to discontinue the satisfaction of different wants at the point at which their intensity has become equal'*.

Boulding stated this law as *'in dividing a fixed quantity of anything among a number of different uses, just so much will be apportioned to each use, to cause the gain involved by transferring a unit of dividend into one use to be just equal to the loss involved in the uses from which the unit of dividend is withdrawn. It obviously applies no matter how many the numbers of uses to which the dividend can be put. It applies no matter what the dividend is'*.

Prof. Silverman stated the law as *'a consumer tends to regulate his purchases of different articles in such a way that, the marginal satisfaction derived from them are equal'*.

This law is stated by Lipsey as *'the household maximizing the utility will so allocate the expenditure between commodities that the utility of the last penny spent on each item is equal'*.

From the above definitions, we can infer that, the consumer will be in equilibrium, when he distributes his given money income among various commodities in such a way that, the MU of expenditure derived from the last rupee spent on each commodity is the same. To put it in simple words, satisfaction will be maximum by using the limited resources in such a way that, the MU of expenditure derived from the last unit of money resource spent on each commodity is more or less equal. Since this law is based on the LDMU, some of the assumptions of LEMU coincide with assumptions of LDMU.

8.2. Assumptions

The main assumptions of the LEMU are as under.

- ☆ ***Independent utilities:*** The MUs of different commodities are independent of each other and diminishes with more and more purchases.
- ☆ ***Constant MU of money:*** Even if the consumer goes on spending his limited resources on different commodities, the MU of money remains constant.
- ☆ ***Cardinal utility:*** Utility is cardinally measurable in terms of 'utils'. This utility is not only quantifiable, but also comparable.
- ☆ ***Rational:*** Every consumer will act rationally in selecting and purchasing the commodities with his limited resources.
- ☆ ***Limited money income:*** The consumer has limited amount of money income to spend on the commodities under consideration.
- ☆ ***No changes in tastes, habits, customs and fashions:*** These should be assumed constant for the effective operation of the law.
- ☆ ***Prices of the commodities:*** They should remain constant and known to the consumer.
- ☆ ***Income of the consumer:*** It should remain constant and he can spend his income in small amounts.
- ☆ ***No influence on price:*** The consumer is one among the many buyers in the market economy, so that, he is powerless to alter the market price.
- ☆ ***Divisibility:*** The commodities under consideration should be divisible. This facilitates to measure the number or quantities of commodities that gives maximum satisfaction to the consumer.
- ☆ ***More than one commodity:*** There should be more than one commodity, as the consumer tries to substitute one commodity for the other, so as to derive maximum satisfaction with the available limited resources.

8.3. Explanation of the Law

The doctrine of LEMU can be explained with the help of following hypothetical example (Table 8.1). Suppose, a consumer is having Rs. 24 in his pocket and he wishes to allocate this limited money among two commodities *viz.*, mangoes and apples. Assume the price of mango is Rs. 2/unit, price of apple is Rs. 4/unit, MU of money is 12 utils per rupee. The MUs in terms of expenditure derived from both these commodities are shown in Table 8.1. A rational consumer would like to get maximum satisfaction from his limited money of Rs. 24. He can spend this money in three different ways, as given below.

- ☆ Rs. 24 may be spent on mangoes only.
- ☆ Rs. 24 may be spent on apples only.
- ☆ Some amount on the purchase of mangoes and remaining amount on the purchase of apples.

If the consumer spends all his money on the purchase of 12 mangoes, he gets 198 utils of TU and if he consumes mangoes upto the point of satiety *i.e.*, 10 mangoes only, he derives 222 utils of TU. Similarly, if the consumer spends all his money on the purchase of six apples, he gets 204 utils of TU and if he consumes apples upto the point of satiety *i.e.*, five apples, he derives 212 utils of TU. However, in order to make the best use of the limited resources, he adjusts his expenditure based on the MU of expenditure with respect to the commodities under consideration.

Table 8.1: Consumer's equilibrium in terms of MU of expenditure or Weighted MUs and the MU of money in the consumption of mangoes and apples.

Units of Commodities	*MU of Mangoes (Utils)*	*MU of Expenditure with Respect to Mangoes (MU of mangoes/Price of mango)*	*MU of Apples (Utils)*	*MU of Expenditure with Respect to Apples (MU of apples/Price of apple)*
1	40	20	84	21
2	38	19	60	15
3	36	18	48	**12**
4	28	14	20	5
5	26	13	0	0
6	24	**12**	-8	-2
7	16	8		
8	10	5		
9	4	2		
10	0	0		
11	-8	-4		
12	-16	-8		
	TU = 198 utils		TU = 204 utils	

Consumer's Equilibrium

The following three methods can be employed to ascertain the point of consumer's equilibrium:

Method 1

This method employs the formula *viz.*, MU of expenditure = MU of money, which is the point of consumer's equilibrium. As shown in the Table 8.1, the MU of expenditure with respect to mangoes (12) is equal to MU of expenditure with respect to apples (12), when consumer purchases six mangoes and three apples and the consumer will go on purchasing these two commodities till the MU of expenditure on each commodity becomes equal to the assumed MU of money to him (12 utils/rupee). So, this combination of purchase of six mangoes and three apples amounts to Rs. 24 (*i.e.*, (6 x 2) + (3 x 4) = 24) and this combination yields highest possible TU equal to = 384 utils (192 utils from consumption of six mangoes and 192 utils from the consumption of three apples). Any other method of allocation of limited money resources among these two commodities will not yield the satisfaction more than 384 utils. So, the consumer reaches equilibrium by consuming six mangoes and three apples. In practical situation, it is very difficult to attain the equilibrium through getting the equality in terms of MU across the commodities and hence, computation of MU of expenditure with respect to the commodities under consideration and then studying their equality across the commodities is more logical and appropriate to analyze the point of maximum satisfaction, as prices across the commodities are not uniform in the economy.

Regarding the pattern of expenditure of limited money resources across the commodities, first the consumer prefers to purchase one apple, as the MU derived in terms of expenditure (21) is more than MU derived in terms of expenditure from the first mango (20). Later, he purchases first three mangoes in a sequence, as now, the MU derived in terms of expenditure in each case (20, 19 and 18 respectively) is more than MU derived in terms of expenditure from the second apple (15). Next, the consumer prefers to purchase second apple, as the MU derived in terms of expenditure (15) is more than MU derived in terms of expenditure from the fourth mango (14). Now, the consumer prefers to purchase two more mangoes *i.e.*, fourth and fifth mangoes in a sequence, as the MU derived in terms of expenditure in each case (14 and 13 respectively) is more than MU derived in terms of expenditure from the sixth mango (12) and third apple (12). Now, the consumer allocates the remaining money between sixth mango and third apple, where MU derived in terms of expenditure is equal in each case (12) and is equal to MU of money (12 utils/rupee). Then, the consumer will not purchase seventh mango or fourth apple, as his limited money was exhausted by spending Rs.12 on mangoes and Rs.12 on apples and moreover, the MU derived in terms of expenditure for the seventh mango (8) and fourth apple (5) are less than MU of money. So, it is clear that, the sensible consumer spends Rs. 12 each on mangoes and apples, thereby, purchases six mangoes and three apples to derive maximum satisfaction. It is also clear from the Table 8.1 that, MU of expenditure with respect to mangoes (MU_M/P_M) and MU of expenditure with respect to apples (MU_A/P_A) are equal at the purchases of six mangoes and three apples and they both are equal to MU of money (12 utils/rupee). So, this is the point of consumer's equilibrium. That means,

$$(MU_M/P_M) = (MU_A/P_A) = \text{MU of money}$$
$$12 = 12 = 12$$

Method 2

This method employs the formula: MU derived in terms of money = Price of the commodity and this is explained through Table 8.2. It is clear from the Table 8.2 that, a rational consumer consumes six mangoes and three apples to derive maximum satisfaction, as the MU derived in terms of money at the sixth unit of mango and at third unit of apple (Rs. 2.00 and Rs. 4.00 respectively) is equal to the respective prices of the two commodities (Rs. 2.00 and Rs. 4.00 respectively). If he consumes seventh mango and fourth apple, it is loss on the part of consumer, as the MU derived in terms of money (Rs. 1.33 and Rs. 1.67 respectively) goes below the prices of the two commodities (Rs. 2.00 and Rs. 4.00 respectively).

Table 8.2: Consumer's equilibrium in terms MU derived in terms of money and prices of the mango and apple.

Units of Commodities	*MU of Mangoes (Utils)*	*MU Derived in Terms of Money (MU of mangoes/ MU of money)*	*MU of Apples (Utils)*	*MU Derived in Terms of Money (MU of apples/MU of money)*
1	40	3.33	84	7.00
2	38	3.17	60	5.00
3	36	3.00	48	**4.00**
4	28	2.33	20	1.67
5	26	2.17	0	0.00
6	24	**2.00**	-8	-0.67
7	16	1.33		
8	10	0.83		
9	4	0.33		
10	0	0.00		
11	-8	-0.67		
12	-16	-1.33		

Method 3

The other method of explaining the consumer's equilibrium is through studying the total net gain in purchasing both the commodities. This analysis (Table 8.3) reveals that, the net gain in the purchase of mangoes is highest at Rs.4.00 *i.e.*, at sixth mango and similarly, the net gain in the purchase of apples is highest at Rs.4.00 *i.e.*, at third apple. So, the total net gain is Rs. 8.00, which is the highest than any other combination of these two commodities. So, consumption of six mangoes and three apples will yield maximum satisfaction to the consumer, as the total net gain with reference to the difference between TU gained in terms of money and total expenditure incurred on the two commodities is the highest. A close examination of the Table 8.3 reveals that, at fifth unit of mango and second unit of apple, the net gains are equal as sixth unit of mango and third unit of apple, but the consumer goes upto sixth unit of mango and third unit of apple because, in the former case, the Principle of Proportionality is not satisfied. Moreover, the commodities under consideration are not perfectly divisible.

The concept of LEMU can be well-explained through the Figures 8.1 and 8.2. MU_M/P_M and MU_A/P_A are the MU curves in terms of expenditure with respect to mangoes and apples respectively. Both the curves are downward sloping with increase in consumption of these two commodities. It is clear from the Figure 8.1 that, by consuming six mangoes and three apples, their MUs in terms of expenditure are equal to MU of money and this is the point of consumer's equilibrium. The Figure 8.1 is also presented in a different way (Figure 8.2) to denote the consumer's equilibrium. As shown in the Figure 8.2, the consumer will be in equilibrium by purchasing six mangoes and three apples, where MU of expenditure of mangoes and MU of expenditure of apples are equal and at these points, they are also equal to MU of money.

On the other hand, if any one of the above commodities are available free of cost to the consumer (say, mango), the consumer chooses that and consume it till the point of satiety *i.e.*, he consumes ten mangoes, but in case of apples, he consumes three apples in accordance with the methodology discussed earlier.

Table 8.3: Consumer's equilibrium in terms of the largest net gain (in terms of money) derived in the consumption of mangoes and apples.

Units of Commodities	*MU of Mangoes (Utils)*	*MU Derived from Mangoes in Terms of Money (MU of mangoes/ MU of money)*	*MU of Apples (Utils)*	*MU Derived from Apples in Terms of Money (MU of Apples/ MU of money)*	*Total Expen-diture on Mangoes (Rs/-)*	*Total Expen-diture on Apples (Rs/-)*	*TU Gained from Mangoes in Terms of Money (Rs/-)*	*TU Gained from Apples in Terms of Money (Rs/-)*	*Net Utility Gained from Mangoes in Terms of Money (Rs/-)*	*Net Utility Gained from Apples in Terms of Money (Rs/-)*
1	40	3.33	84	7.00	2	4	3.33	7.00	1.33	3.00
2	38	3.17	60	5.00	4	8	6.50	12.00	2.50	4.00
3	36	3.00	48	**4.00**	6	12	9.50	16.00	3.50	**4.00**
4	28	2.33	20	1.67	8	16	11.83	17.67	3.83	1.67
5	26	2.17	0	0.00	10	20	14.00	17.67	4.00	-2.33
6	24	**2.00**	-8	-0.67	12	24	16.00	17.00	**4.00**	-7.00
7	16	1.33			14		17.33		3.33	
8	10	0.83			16		18.17		2.17	
9	4	0.33			18		18.50		0.50	
10	0	0.00			20		18.50		-1.50	
11	-8	-0.67			22		17.83		-4.17	
12	-16	-1.33			24		16.50		-7.50	

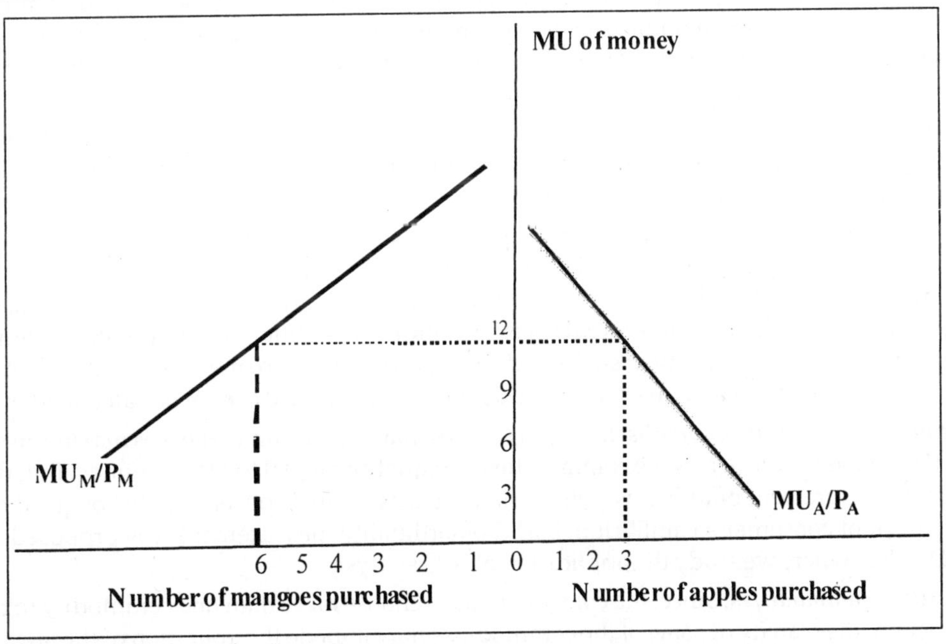

Figure 8.1: Consumer's equilibrium with respect to mangoes and apples (Method 1).

8.4. Limitations

Like other laws of Economics, this LEMU is also subjected to various limitations. It is absolutely not possible to follow this law accurately for making expenditure pattern by the consumer on different commodities and therefore, he may not obtain maximum satisfaction in all cases. This is because of the following limitations:

- ☆ ***Effect of fashions and customs:*** This law may become inoperative, if fashions and customs changes in the society. For example, due to change in fashions, the people are forced to switch over to other commodities, but they may yield less MU compared to earlier commodity. In such case, the consumer will be forced to spend on the commodity that yields low MU to him. So, the consumers cannot transfer the money resources from the less advantageous uses to the more advantageous uses because, they are forced by the changes in fashions

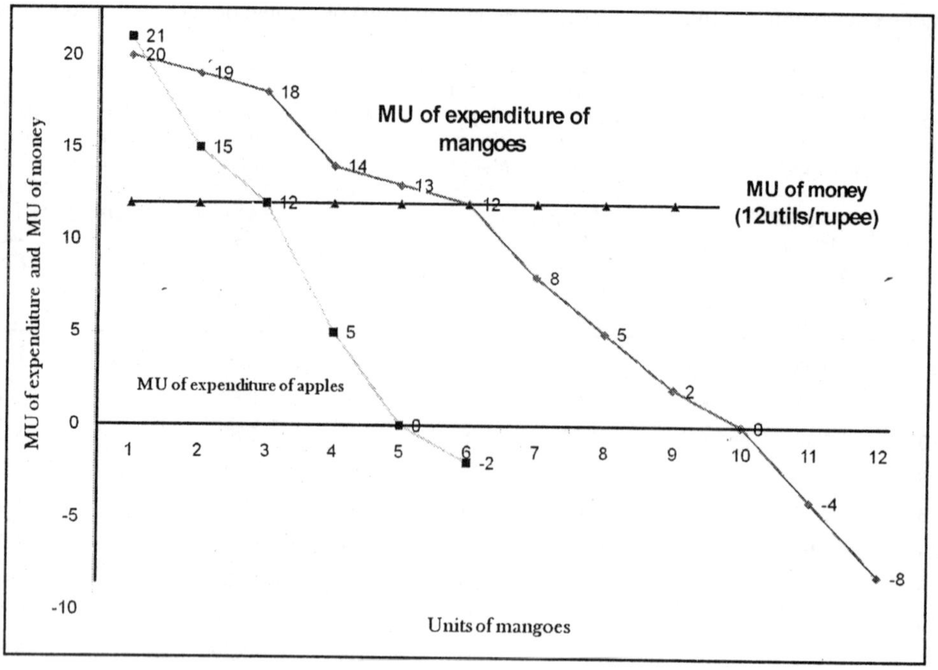

Figure 8.2: Consumer's equilibrium with respect to mangoes and apples (Method 1).

and customs in the society. Moreover, the consumers are not always calculative with respect to MUs of expenditure across the commodities and they simply allocate the money resources governed by fashions and customs in the society. Similarly, if there is no perfect freedom between various alternatives due to traditions in the society, the operation of law may be impeded.

- ☆ ***Ignorance or Carelessness:*** Sometimes, the consumers due to lack of awareness, allocate their money resources on the commodities that yield less MU of expenditure than the MU of money to him. So, they do not equate MU of expenditures across the commodities.
- ☆ ***Indivisible units:*** The units of the commodity should be perfectly divisible for the effective operation of this law. Otherwise, it is difficult to achieve equality of MU of expenditure across the commodities under consideration. So, indivisibility of certain durable goods like tractor, machinery tools, cars etc., is a great obstacle in the way of equalization of MU of expenditure across different commodities. Normally, a consumer will purchase only a single unit of such goods. So, it is ridiculous to prepare an individual MU schedule for such commodities. Similarly, marginal productivity of factors also cannot be calculated exactly.
- ☆ ***Cardinal measurement is unrealistic:*** The law states that, the consumer derives maximum satisfaction when the MU of expenditure across the commodities are equal and equal to MU of money. But, this again calls for cardinal measurement of utility, which is unrealistic, as utility is purely a psychological phenomenon. So, the concept of consumer's equilibrium based on ordinality measurement was proposed by J.R.Hicks and R.G.D.Allen, where we study the application of IDC analysis.
- ☆ ***Scarcity of commodities:*** Since, the consumer has to choose more than one commodity to derive maximum satisfaction, the scarcity of some of the commodities may impede the operation of the law.
- ☆ ***Consumer not perfectly 'rational':*** This law assumes perfect 'rational' behaviour on the part of the consumer. But, in practical situation, the consumer is not so rational in choosing the perfect combination of commodities, which gives him maximum satisfaction. This is because, every time he cannot calculate the MU of expenditure on different commodities and he will not try to equate the same across the commodities. Moreover, the multiplicity of wants and strong influence of the advertisement on the consumer may prevent him in choosing the right combination of commodities. So, under these conditions, we cannot expect the consumer to make rational choices based on the concept of LEMU. However, this law will operate in the purchases of basic necessaries of life compared to luxury commodities.
- ☆ ***MU of money is not constant:*** In practical situation, MU of money is not constant. Rather, it increases with increase in spending of money or as the quantity of money increases with the consumer, its MU will diminish.

- ☆ *Non-availability of commodities:* This law is not applicable, when the commodities of consumers' choice are not available in the market. So, it makes the consumer to adjust himself with the commodities that yields low MU in terms of money compared to price.
- ☆ *Calculation of weighted MU is impossible:* A consumer deals with a wide variety of goods. So, in every case, it is not possible for him to compute the weighted MUs. Moreover, utility itself is subjective and psychological feeling of satisfaction.
- ☆ *Free goods or unlimited supply of goods:* This law is inoperative, when the commodities are available free of cost or having unlimited supply in the economy. This is because, the consumer generally consumes such commodities upto the point of satiety and hence, no problem arises regarding the substitution of these commodities.
- ☆ *No freedom to choose:* If there is no perfect freedom between various alternatives, the operation of law may be impeded.

8.5. Practical Importance

In spite of several limitations, the law has more practical importance in almost every field of economic enquiry. This is because, every consumer consciously or unconsciously try to get the maximum satisfaction from his limited resources and this law guides the consumer in this direction. Not only in case of consumption process, this law has equal application in the fields of production, exchange and distribution.

- ☆ *In the field of consumption:* As discussed above, the concept of LEMU guides the consumer in planning his limited resources allocation among the commodities, in such a way that, MU of expenditure across the commodities are equal or equality of MU of expenditure of each commodity with MU of money, so as to derive maximum satisfaction. This law not only explains the number of units of various commodities to be purchased by the consumer, but it also helps the consumer in allocating income of the consumer into multiple uses of the same commodity. Based on the above principle, the consumer frame his family budget and tries to substitute one commodity for the other, till he derives maximum satisfaction. So, maximization of welfare of the household is the ultimate objective of the application of this law. This is mathematically expressed in Equation 6.4.
- ☆ *In the field of production:* The concept of LEMU is widely applicable in the field of production and guides the farmer in allocating his limited resources among the selected enterprises, such that, the Marginal Value Product of factor (MVPx) derived from each of the resource spent across different enterprises is same, so as to ensure maximum profits in the production programme. The farmer continues to substitute one resource for the other, till this objective is achieved. Through this, the farmer ensures the optimal combination of the resources and is said to be in equilibrium. Here, the MVPs of all the resources are proportional to their respective prices. If there are 'n' resources, the following condition will be satisfied in the equilibrium:

$$= \frac{\text{MVP of resource A}}{\text{Price of resource A}} = \frac{\text{MVP of resource B}}{\text{Price of resource B}} = \frac{\text{MVP of resource 'n'}}{\text{Price of resource 'n'}} \text{ and so on.}$$

  In the case of inequality in the above equation, the farmer will substitute the resource, which gives him more productivity for the resource, which gives him less productivity. This process of substitution will continue, till the ratios for the resources become equal. The farmer pays remuneration to the resources according to their MVPs. This is explained in-detail in Chapter 24.
- ☆ *In the field of exchange:* This concept is widely used in the field of exchange, so as reduce price-inequality of a commodity in all the markets. In markets, where there is scarcity of commodities, the prices will be very high and *vice versa*. So, the consumers start substituting less scarce goods in place of more scarce goods. As a result, the prices of more scarce goods will gradually decline in the market due to decrease in demand. Likewise, this concept of LEMU is also applicable in case of barter system of economy, where goods are exchanged for other goods and the process of exchange continues, till the utility derived from the extra unit of the commodity consumed, becomes equal to the utility lost from the extra unit of the commodity sacrificed by the consumer. This condition is true for both the parties taking part in exchange.
- ☆ *In the field of distribution:* We know that production is the result of joint efforts of different factors of production such as land, labour, capital and management. Their remuneration in the production programme in terms of rent, wages, interest and profits respectively is determined according to the principle of marginal productivity. This principle has a vital role to play in the 'Theory of distribution'. In distribution, we study the remuneration paid to different factors of production employed in the production programme. Generally, an entrepreneur will substitute one factor for another factor, so as to minimize the cost in the production programme. He will substitute one factor for another till the cost of employing each factor equals their MVPs resulting from its use.

- *Planning for the present and future:* The concept of LEMU guide the consumer in the allocation of resources towards fulfilling present wants and future wants. This implies, it helps the consumer in allocating his limited income between consumption expenditure and saving. In deciding the problem, he will compare the present utilities with the future utilities. The consumer saves more, if he thinks that, future utilities are more important than present utilities and *vice versa*. That means, he will substitute savings in place of present consumption, if future utilities are more important relative to present consumption. The distribution of income between present consumption and saving is the best in which the utility derived from the marginal unit of saving is equal to the utility derived from the marginal unit of consumption.
- *In public-finance:* The Government too is guided by this LEMU. We know, the modern state is a welfare state, which undertakes various developmental programmes for enhancing the social welfare. The Government can realize the objective of public finance to achieve maximum social advantage through the application of law of substitution. This principle guides the Government in spending the limited resources in potential areas, which yields maximum social advantage based on two important aspects *viz.*, the MR are equal from all lines of expenditure and the rates of taxation should be devised in such a way that, marginal sacrifice of all tax payers is equal.

On the similar lines, discussed above, the LEMU concept also guides the individual in the allocation of his time between work and leisure. So, in short, the LEMU is applicable to all the problems of allocation of scarce resources, so as to derive maximum satisfaction.

8.6. LEMU vis-à-vis LDMU

The concept of LEMU is almost similar to LDMU, but differs from LDMU in the following aspects (Table 8.4):

Table 8.4: Differences between LDMU and LEMU.

LDMU	*LEMU*
LDMU deals with a single commodity	LEMU deals with more than one commodity.
Resources level with the consumer is unlimited.	Resources level with the consumer is limited.
The consumer can go upto the point of satiety, as the resources level is unlimited with him (provided the commodity is supplied free of cost).	The consumer cannot go upto the point of satiety, as the resources level is limited with him.
The consumer derives maximum satisfaction, when MU is equal to zero or at the point of satiety.	The consumer derives maximum satisfaction, when the weighted MUs are same across the commodities or when weighted MU of each commodity is equal to MU of money or when MU derived in terms of money is equal to the respective prices of the commodities under consideration.
The concept of substitution of one commodity by another will not arise in LDMU.	The concept of substitution of one commodity by another will arise in LEMU, so as to maximize the satisfaction from limited resources.
The satisfaction derived through the application of LDMU concept is lower compared LEMU concept, as he will not substitute one commodity for another commodity.	The consumer derives higher satisfaction through the application of LEMU concept, as he substitutes one commodity with higher MU in place of another commodity with lower MU.
LDMU concept aims at maximization of consumer's satisfaction through consumption of a single commodity.	LEMU concept ensures twin objectives *viz.*, efficient utilization of resources and maximization of consumer's satisfaction.

8.7. Derivation of Demand Curve from Utility or Marshallian Analysis

The assumption that, all commodities exhibit diminishing MU has a simple implication for demand curves *i.e.*, they are all negatively sloped. This is because, at the point of consumer's equilibrium, if price of the commodity A decreases, in order to restore the equilibrium, as per the equation 6.2, the MU derived from the commodity A should be decreased. For this, the consumer has to purchase more of the commodity A. This clearly shows that, there exists inverse relationship between quantity of commodity purchased or demanded and price of the commodity. The same principle holds good even in case of more than one commodity equilibrium. So, both the concepts *i.e.*, LDMU and LEMU can be employed to derive the demand curve from the MU analysis. Before going into the details, it is once again essential to recollect the basic assumptions of utility analysis *viz.* utility function of different commodities are independent of each other and constancy of MU of money. Based on these fundamental assumptions, it is possible to derive demand curve in two cases *viz.*, in case of single commodity based on the concept of LDMU and in case of multi-commodity based on the concept of LEMU. Let us discuss one by one here under.

8.7.1. Derivation of Demand Curve in Case of Single Commodity Based on the Concept of LDMU

Method 1

The point of consumer's equilibrium is, when the MU of the expenditure is equal to MU of money. This is given by, (MU derived from commodity A / Price of the commodity A)=MU of money → As per Equation 6.1.

The left side portion of the above equation is called MU of expenditure. So when the price of the commodity A falls in the market, the left side component of the above equation will be greater than MU of money and hence, the consumer's equilibrium gets disturbed. So, to ensure equality between the two components in the above equation, MU derived from the commodity A should be decreased, as MU of money is assumed constant in utility analysis. So, to decrease MU derived from the commodity, the consumer has to consume more *i.e.*, he has to purchase or demand more for the commodity A. So, this infers that, when price of the commodity is decreased in the market, quantity demanded for that commodity extends, indicating an inverse relationship between these two. The same explanation is depicted in the Figure 8.3 in Panels A and B.

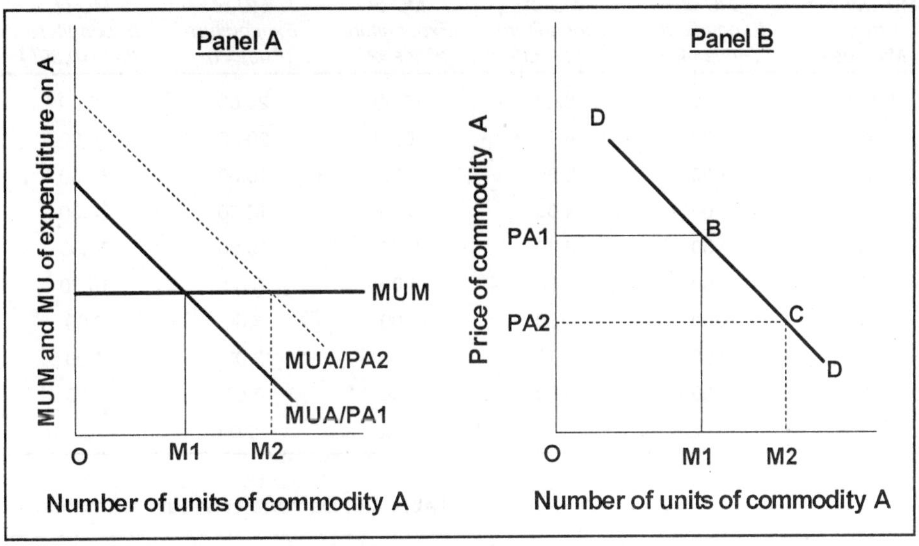

Figure 8.3: Method 1- Diminishing MU analysis and Demand curve.

In the Panel A, on X-axis, number of units of the commodity A and on Y-axis, MU of money and MU of expenditure are taken. As MU of money is assumed to be constant, MU_M is a straight line parallel to X-axis and MU of expenditure curve for the commodity is downward sloping from left to right as MU_A/P_{A1}, indicating that, as the consumer goes on consuming the commodity, the MU derived by the consumer will be diminishing for each and every successive unit of consumption of the commodity. This MU_A/P_{A1} curve will cross the MU_M curve at OM_1 quantity of commodity A, indicating that, for a given level of price of the commodity P_{A1}, the consumer consumes OM_1 quantity, at which the MU of expenditure is equal to the MU of money, thereby, the consumer is at equilibrium.

Let us suppose, if the price of commodity A decline in the market from P_{A1} to P_{A2}, then the MU of expenditure of commodity A will rise (left side component of above equation 6.1) and hence, the MU_A/P_{A1} curve will shifts upward or towards right as MU_A/P_{A2}. This curve will again cross the MU_M curve at OM_2 quantity of commodity consumed, indicating that, more quantity of commodity is consumed by the consumer due to fall in price of the commodity and again the equality of MU_A/P_{A2} with MU_M is ensured. This indicates that, with the fall in price of the commodity, more quantity is consumed or demanded by the consumer. To show this, we consider Panel B in the Figure 8.3, taking number of units of commodity or quantity demanded or consumed on X-axis and price of the commodity on Y-axis. As explained in the Panel A, when the price of the commodity is OP_{A1}, the quantity demanded is OM_1. This is indicated by point B on the demand curve DD in Panel B. If price of the commodity is decreased to OP_{A2}, the quantity demanded is increased to OM_2, and the same is indicated by point C on the DD curve in Panel B. Likewise, we can plot different points for different prices and different quantities commodity A demanded. If we connect the points say B and C, we get a downward sloping curve measuring the relationship between price of the commodity A and quantity demanded of commodity A, which is nothing but a demand curve DD for commodity A. By this, we proved the law of demand is based on the concept of LDMU.

Mathematical Illustration

The following example is considered to prove the derivation of demand curve from the concept of LDMU through assuming the hypothetical data regarding the consumption of mangoes by the consumer (Table 8.5). If the consumer goes on consuming the mangoes, he reaches the point of satiety at ninth mango. But, since the commodity is priced in the market and money income is limited with the consumer (under practical situations), he will not go upto the point of satiety and spends his limited money on the consumption of mangoes in framing the relationship MU of expenditure (MU_E) and MU of money (MU_M). Assume, MU_M is 4.00 utils/rupee. For a given level of price of the commodity *i.e.*, Rs.4.00, the MU_E is computed, and now, the point of consumer's equilibrium is at third mango, where MU_E is equal to MU_M. In the Table 8.5, the MU_E is computed at different prices, so as to ascertain the point of consumer's equilibrium. It is evident that, when price of mango is decreased to Rs. 3.00, Rs. 2.00, Rs. 1.00 and Rs. 0.50, the consumer attained the points of equilibrium at the consumption of fourth, sixth, seventh and eighth mangoes respectively. The same is shown through Figure 8.4. So, this implies that, at different prices of mangoes, the consumer reached equilibrium by consuming different units of mangoes and the same is shown through demand schedule (Table 8.6). By plotting the graph between these two variables *viz.*, prices of the mango and quantities of mangoes demanded at various prices, we can derive the downward sloping demand curve (Figure 8.5).

Table 8.5: Consumer's equilibrium based on LDMU approach to derive demand curve.

Number of Mangoes Consumed	*MU Derived from Mangoes*	*MU of Expenditure (Price=4/-)*	*MU of Expenditure (Price=3/-)*	*MU of Expenditure (Price=2/-)*	*MU of Expenditure (Price=1/-)*	*MU of Expenditure (Price=0.50/-)*	*MU of Money*
1	24	6.00	8.00	12.00	24.00	48.00	4.00
2	20	5.00	6.70	10.00	20.00	40.00	4.00
3	16	**4.00**	5.30	8.00	16.00	32.00	4.00
4	12	3.00	**4.00**	6.00	12.00	24.00	4.00
5	10	2.50	3.30	5.00	10.00	20.00	4.00
6	8	2.00	2.70	**4.00**	8.00	16.00	4.00
7	4	1.00	1.30	2.00	**4.00**	8.00	4.00
8	2	0.50	0.70	1.00	2.00	**4.00**	4.00
9	0	0.00	0.00	0.00	0.00	0.00	4.00
10	–5	–1.30	–1.70	–2.50	–5.00	–10.00	4.00

Table 8.6: Demand schedule for mango.

Prices of Mango	*Number of Mangoes Demanded*
4.00	3
3.00	4
2.00	6
1.00	7
0.50	8

Method 2

As discussed earlier, in LDMU concept, the consumer derives maximum satisfaction at the point of satiety, which occurs when MU is equal to zero or TU is maximum. But this is not a common phenomenon in the practical situation, as the commodity is priced in the market and moreover, the income with the consumer is limited. So, the consumer goes on the consuming the commodity as long MU derived in terms of money is greater than price of the commodity and finally he reaches the equilibrium when the equality is ensured between these two. So, the point of consumer's equilibrium is given by:

(MU derived from the commodity A/MU of money) = Price of the commodity A- → As per the Equation 6.2

From the above equation 6.2, it is clear that, when price of the commodity falls in the market, the left hand side portion of the equation is greater than price of the commodity. So, to ensure equality between the two components of the equation, the left side component should be reduced, as price of the commodity in the market cannot be varied, as it influenced by demand and supply forces and not by the individual. So, invariably left side component has to be reduced to ensure consumer's equilibrium. As the MU of money is assumed to be constant in Utility analysis, the numerator in the left side component *i.e.* MU derived from the commodity has to be reduced. This is possible by purchasing or consuming more of the commodity. So, the discussion finally infers that, if the price of the commodity A falls in the market, quantity purchased or demanded of the commodity will extend, so as to reduce MU derived from the commodity, implying an inverse relationship between price and quantity demanded of the commodity by the consumer. The same explanation is presented through the Figure 8.6 in Panels A and B.

In the Panel A, on X-axis, number of units of the commodity A and on Y-axis, MU derived in terms of money are taken. As per the concept of LDMU, the MU derived in terms of money is a downward sloping curve from left to right. This negative slope implies, as the consumer consumes larger quantities of commodity A, its MU derived in terms of money diminishes, as MU of money is assumed constant in Utility analysis. At OM_1 quantity, the MU derived in terms of money is MU_1. This should be equal to price OP_1 by definition (as per the equation 6.2). That means, the consumer

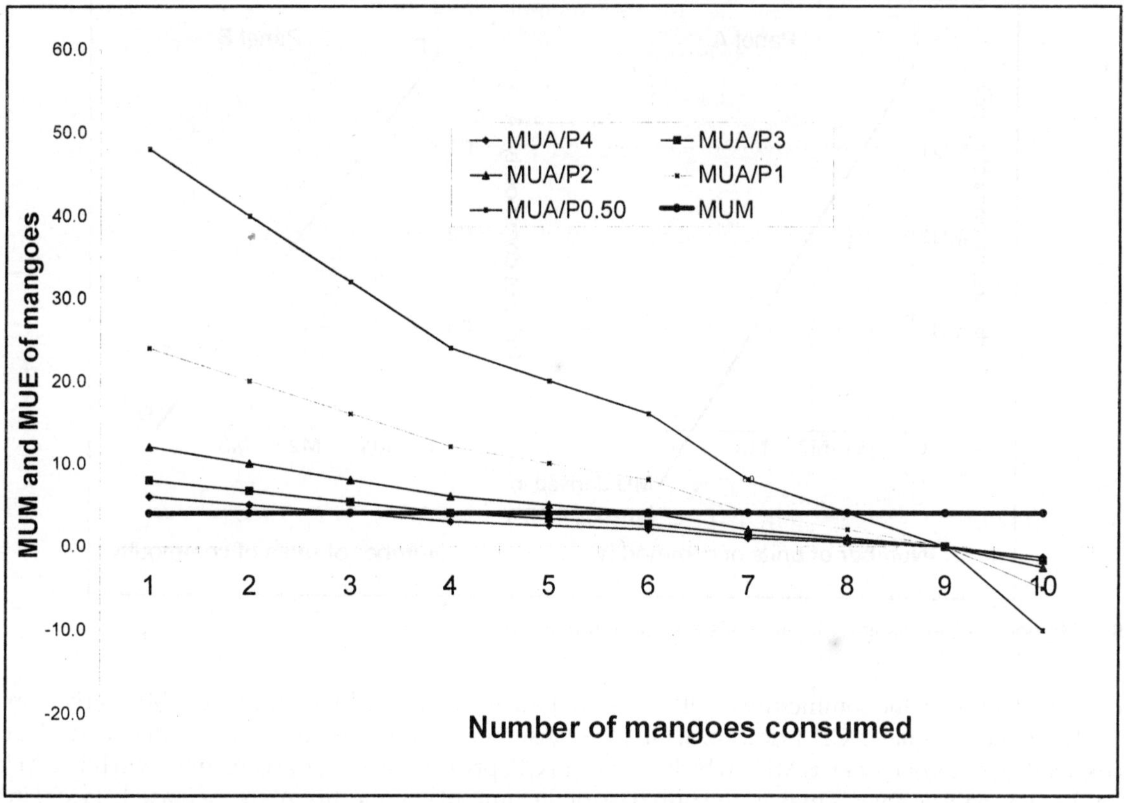

Figure 8.4: Consumer's equilibrium based on LDMU approach to derive demand curve.

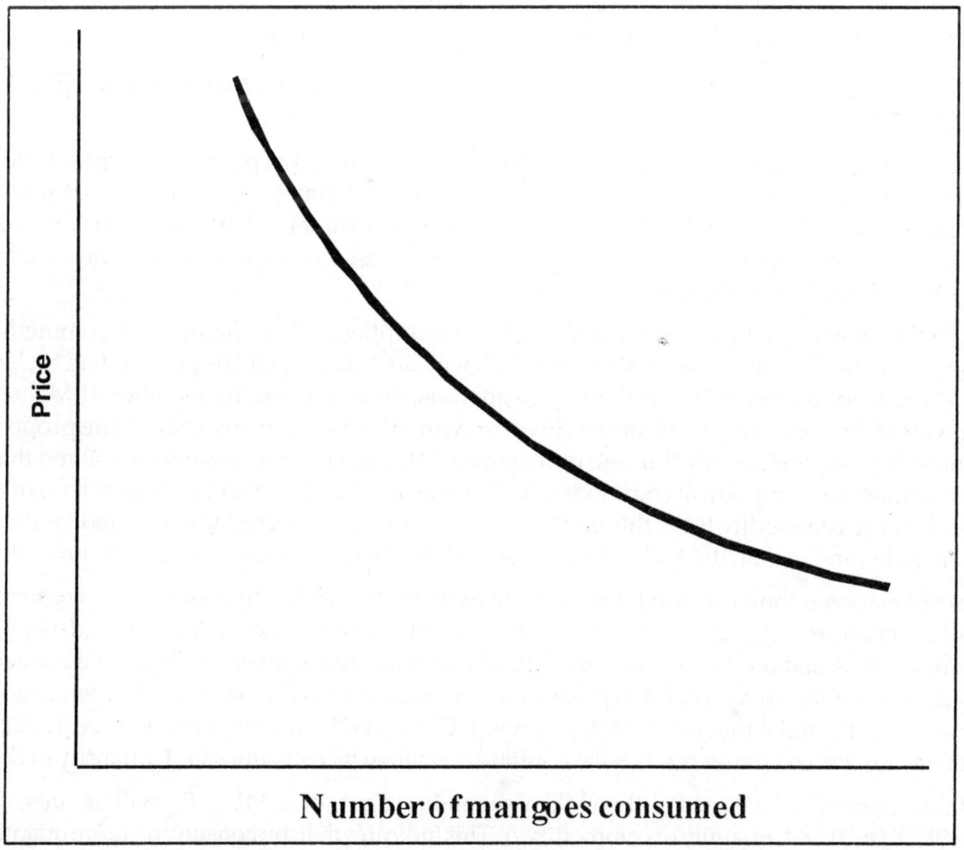

Figure 8.5: Derivation of demand curve for mangoes - LDMU approach.

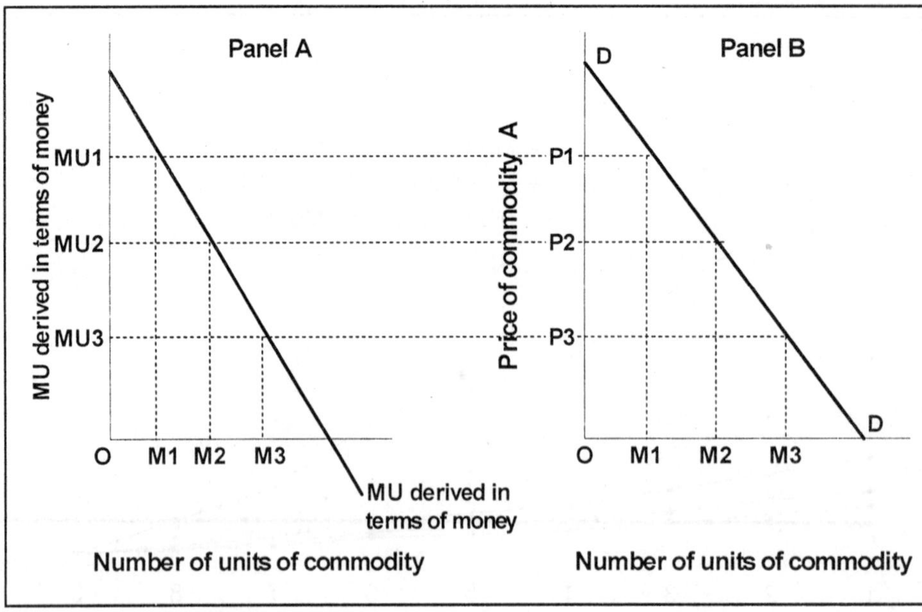

Figure 8.6: Method 2 - Diminishing MU analysis and Demand curve.

demands OM_1 quantity of the commodity at OP_1 price. In the same way, at OM_2 quantity, the MU derived in terms of money is MU_2, which is equal to OP_2 price. So, at OP_2, the consumer will buy OM_2 quantity. Similarly, at OM_3 quantity, the MU derived in terms of money is MU_3, which is equal to OP_3 price. So, at OP_3, the consumer will buy OM_3 quantity and so on. So, the analysis reveals that, as the purchase of the units of commodity A are increased, its MU in terms of money will decrease. So, at diminishing price, the quantity demanded of commodity A increases and the same is shown through Panel B. This analysis indicates that, demand curve will slope downward from left to right and if the price of the commodity falls, other things remaining the same, the quantity demanded of a commodity will extend and *vice versa*. The negative section of the MU curve in terms of money (Panel A) does not form part of the demand curve (Panel B), since negative quantities demanded does not make sense in Economics.

8.7.2. Derivation of Demand Curve in Case of more than One Commodity Based on the Concept of LEMU

Let us assume, the consumer deals with two commodities A and B and as per the Principle of Proportionality, the consumer will be in equilibrium when MU of expenditure across the commodities are equal for the last rupee spent on the commodities or when $MU_A/P_A=MU_B/P_B=MU_M$ *i.e.*, a rational consumer will be in equilibrium, when he equalizes the MU_M with MU of expenditure of each of the commodity under consideration or when MU derived in terms of money for each commodity is equal to its respective price.

Suppose, in the above equation, given *ceteris paribus* assumption, when the price of commodity A falls in the market, the equality in the above equation get disturbed. That mean, MU_A/P_A will be greater than MU_B/P_B and MU_M. Of course, MU_M is assumed to be constant in the Utility analysis. So, to ensure the equality of MU of expenditure of commodity A with MU of expenditure of commodity B or with MU of money, invariably, the proportion of MU_A/P_A should be reduced. This is possible only through reducing the MU derived from commodity A and this can be ensured through more and more consumption of commodity A. That means, the consumer has to purchase or demand more of commodity A relative to commodity B. So, this implies, with the decline in price of the commodity, the consumer has to purchase or demand more, so that, the MU will decrease and finally the consumer's equilibrium will be restored.

The same explanation is shown through Figure 8.7 in two Panels A and B. In Panel A, we measure the relationship between MU of expenditure, MU_M and number of units or quantity of commodity A consumed. On X-axis, number of units of the commodity A and on Y-axis, MU_M and MU of expenditure are taken. As MU_M is assumed to be constant, MU_M is a straight line parallel to X-axis. If the price of the commodity A is P_{A1}, the MU of expenditure is a downward sloping curve as MU_A/P_{A1} and it touches the MU_M curve at 'C' *i.e.*, at OM_1 quantity of commodity A consumed by the consumer. That means, the consumer reaches the equilibrium, when he consumes OM_1 quantity of the commodity A.

If price of the commodity A falls to P_{A2}, then MU of expenditure curve *i.e.*, MU_A/P_{A2} will rise upward and it touches the MU_M curve at 'D' *i.e.*, at OM_2 quantity of commodity A. This indicates that, by consuming more quantity of commodity A, *i.e.*, OM_2, again the consumer's equilibrium is restored at point 'D'. If still the price of commodity A falls to P_{A3}, the MU of expenditure curve, MU_A/P_{A3} will shift further upward and while sloping downward, it crosses the MU_M curve at 'E',

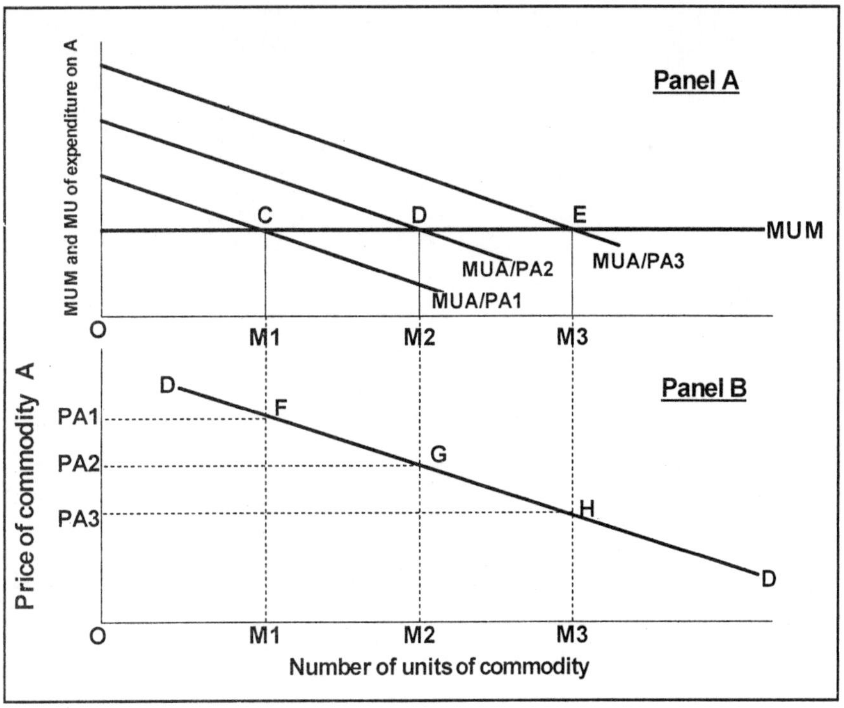

Figure 8.7: LEMU analysis and derivation of Demand curve.

i.e., at OM_3 quantity of commodity A. This again implies that, by purchasing more quantity of commodity A *i.e.*, OM_3 due to deckube in price of commodity A, again the point of consumers equilibrium is restored at point 'E'. The above explanation clearly indicates that, with the fall in price of the commodity A, the consumer will purchase more of the commodity A, so that the MU derived by the consumer will fall and again the consumer's equilibrium will be restored.

The same explanation is shown in the Panel B, but on Y axis, we take price of the commodity A and on X-axis again, the number of units of commodity or quantity consumed or purchased or demanded by the consumer are taken. The corresponding equilibrium points of Panel A will be extended downward to the Panel B. So, the corresponding point of C,D,E of the Panel A will now become F,G,H in Panel B and if these points are joined, it gives a downward sloping demand curve DD. It is important that, in the upper Panel A, the point C indicates that, the consumer buys OM_1 quantity of commodity A when the MU_A/P_{A1} is equal to MU_M. But, in the lower Panel B of the curve DD, the corresponding point F indicates that, OM_1 quantity of commodity demanded at OP_{A1} price directly. Likewise, the points F and G also indicate OM_2 and OM_3 quantities purchased by the consumer at OP_{A2} and OP_{A3} prices respectively. Again in Panel A, when the price the commodity is decreased, MU of expenditure curve will shift upward from MU_A/P_{A1} to MU_A/P_{A2} and this shift shows that, quantity purchased is increased from OM_1 to OM_2. But, this is shown directly in Panel B *i.e.*, when price of the commodity is decreased from OP_{A1} to OP_{A2}, the quantity demanded increases from OM_1 to OM_2. So, by drawing Panel B, it is convenient to study the variations in quantity demanded due to changes in price of the commodity directly compared to Panel A. So, the discussion finally reveals that, due to fall in price of the commodity A, the quantity demanded will extend and this leads to fall in MU derived from the commodity and again the consumer's equilibrium will be restored, such that, MU_M becomes equal to MU of expenditure of commodity A and MU of expenditure of commodity B and so on. It further concludes that, the demand curve slope downward from left to right and the working force behind this downward sloping of demand curve is the diminishing MU.

8.8. Shortcomings of Utility Analysis or Marshallian Approach of Consumer Behaviour

The explanation given earlier emphasizes the importance of Utility analysis in arriving at consumer's equilibrium both in the cases of one commodity and more than one commodity dealt by the consumer. Both LDMU and LEMU concepts have more practical applications and they are popularly regarded as useful concepts in explaining the human tendency with respect to their consumption behaviour. Further, the analysis is very useful in deriving the demand curve from both the concepts of LDMU and LEMU. However, this approach suffers from serious limitations, as it is based on several unrealistic assumptions. It is felt appropriate by the author to discuss these limitations or shortcomings, while comparing this Utility analysis with IDC analysis in the ensuing pages.

9

Indifference Curve Analysis–Consumer Optimization without Measurable Utility

In IDC approach, the basic objective of studying the consumer behaviour remains same as in the last section *i.e.*, achieving consumer's equilibrium, but here, the critical difference is that, the utility is not measurable or quantified. The consumer is assumed to maximize his satisfaction by allocating the given budget among the selected commodities. The consumer is aware of different levels of satisfaction that he can derive by consuming different combinations of commodities at different levels of income and prices of the commodities. But, as mentioned earlier, it is not possible to quantify the level of satisfaction, rather he can rank the derived satisfaction from low to high. So, the consumer simply allocates his limited budget among the related commodities and makes the choice regarding the best combination of commodities that yields him maximum satisfaction. This IDC approach was originally due to the Italian Economist, Vilfredo Pareto (1848–1923), who severely criticized the concept of cardinal utility. He stated that, utility is neither quantifiable nor addible. It can, however, be compared. He, therefore, suggested that, the concept of utility should be based on Ordinality approach. J.R.Hicks and R.G.D.Allen, following the footsteps of Pareto, popularized the technique of IDCs and thereby, replaced the cardinal utility approach by using ordinal utility function. This IDC approach is also termed as Modern approach of consumer behaviour and it is a more advanced form of geometric device to explain the consumer's equilibrium. This concept was discussed in-detail by Prof. Hicks in his work, *'Value and Capital'* in 1939.

9.1. Concepts in IDC Analysis

In dealing with consumer's behaviour, this IDC technique mainly comprises of two basic components *viz.*, IDC and Price line. These two components are essential to analyze the point of maximization of consumer's satisfaction. Before understanding these two components, it is essential to discuss the important assumptions of IDC technique.

Assumptions

The following are the assumptions formulated under IDC analysis:

- *Rational behaviour of the consumer:* The consumer acts rationally in making decisions regarding the purchase of two commodities under consideration and he substitute one commodity for another, so as to derive maximum satisfaction.
- *Ordinal measurement of utility:* Unlike in cardinality approach, utility cannot be quantified. Rather, it is expressed ordinally. That means, the utility can be ranked towards lower to higher.

☆ ***Two commodities:*** IDC analysis deals with two commodities to explain the consumer's behaviour in deriving maximum satisfaction. Other things being equal, the consumer always prefers more of any one commodity to less of that same commodity.

From the Figure 9.1, it is clear that, with reference to combination A, OQ_R quantity of rice and OQ_W quantity of wheat are preferred by the consumer for his consumption, for the given level of income and prices of the two commodities. As per the above assumption, the choice made by the consumer *i.e.*, combination A with respect to OQ_R quantity of rice and OQ_W quantity of wheat is superior to the combinations lie below 'A', but inferior to the combinations lie above 'A'. This infers that, combination 'A' is superior to combinations that have less of both the commodities and inferior to combinations that have more of both the commodities (except on the boundaries, where they have more or less of one commodity and the same amount of the other commodity).

☆ ***Diminishing Marginal Rate of Substitution (MRS):*** The principle of diminishing MRS is assumed, while dealing with the consumption behaviour of the consumer with respect to the two commodities under consideration. That means, if the consumer goes on consuming one commodity, it substitutes less and less of other commodity.

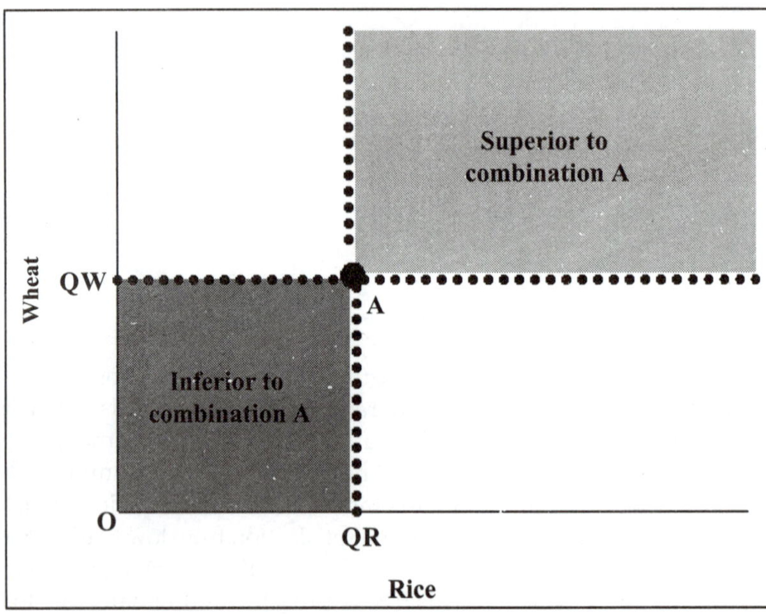

Figure 9.1: Consumer's preferences to different combinations of rice and wheat.

☆ ***Commodities are highly divisible:*** This assumption implies easy substitution of two commodities under consideration, which further ensures smooth and continuous IDC that facilitates easy analysis for maximizing consumer's satisfaction.

☆ ***Consistency in choice:*** The consumer is assumed to be 'consistent' in his behaviour with respect to consumption of commodities. For example, if the consumer prefers more of commodity rice (Q_R) compared to commodity wheat (Q_W) of a particular combination at a given instance, his behaviour remains consistent with respect to the choices he makes. That means, during another period of time, his preferences for the combination of these two commodities does not change. Symbolically, it can be expressed as, $Q_R > Q_W$ or $Q_W < Q_R$.

☆ ***Consumer's preferences not self contradictory:*** This assumption is otherwise referred as assumption of Transitivity. This is an important assumption for making consistent choices among a large number of combinations of commodities. For example, if a consumer prefers combination A of two commodities rice and wheat over combination B and B to C, it implies, the consumer is also preferring combination A over combination C. Symbolically, it can be expressed, if A > B and B > C, then A > C. Similarly, if the consumer is indifferent between combinations A and B and B and C, then he is altogether different between A and C. Symbolically, if A = B and B = C, thereby, A □ C. This feature implies IDCs does not cross each other.

☆ ***Commodities are good substitutes:*** The commodities under consideration should be good substitutes. As the consumer possesses limited income, he has to allocate the same between the two commodities. If he allocates more resources to one commodity say rice, he purchases less of wheat and *vice versa*. He goes on substituting one commodity for the other till he derives maximum satisfaction. The satisfaction can be maintained at the same level by consuming more of one commodity and less of the other commodity. So, many combinations of the two commodities are possible and he is indifferent among these possible combinations, which yield same level of satisfaction on a particular IDC. So, consumer based on his tastes and preferences, income and prices of the commodities, tries out to find the best combination that gives him maximum satisfaction and to achieve this best combination, the two commodities should be good substitutes.

- ☆ *Perfect competition:* The consumer should have complete price information about the commodities and possible combinations of two commodities. The prices of two commodities should be stable. This is because, say, if prices of the two commodities are increased in the market, he may switch over to other cheap commodity and thereby, the analysis is not feasible.
- ☆ *Scale of preferences:* The consumer has a definite scale of preferences and he is indifferent among the different combinations of commodities. He can rank his preferences either in ascending or descending order. It is important that, the scales of preferences of consumer are not influenced by the market prices of the commodities. That means, he will not consider high priced commodities as superior commodity and low priced commodity as an inferior one.
- ☆ *Tastes, habits and preferences of the consumer:* They are assumed to be stable throughout the analysis.
- ☆ *Non-satiety:* The consumer is not over supplied with the commodities under consideration. But, the consumer prefers more to less of the commodities (Monotonicity). So, the consumer's preferences exhibit non-satiation. That means, if A and B combinations have identical quantity of commodity rice, but combination A has a little more of commodity wheat than combination B then, A is preferred to B. So, the preferences are said to be 'symmetric', as if A is preferred to B, the consumer cannot prefer B to A. This property of non-satiation will give negative slope to the IDC curve.
- ☆ *Weak ordering:* This hypothesis was formulated by J.R.Hicks. According to this assumption, the consumer is not so particular with reference to any particular combination, but rather he is indifferent among the possible combinations of the commodities. However, the concept of Strong ordering of Revealed preference theory replaced this Weak ordering hypothesis.
- ☆ *Continuity:* This assumption implies, the consumer can choose to consume any amount of the commodity.
- ☆ *Completeness of preferences:* In situations, where a consumer is confronted with any two commodities, he is able to determine his order of preference or indifference between the two. So, the consumer can rank various combinations of two commodities. That means, either A preferred to B, or B preferred to A, or indifferent between the two.

By going through the above assumptions, we can infer that, IDC approach has retained some of the assumptions of Utility analysis like rationality of consumer behaviour, perfect competition and continuity (in the IDC analysis, it refers to the ranking of possible combinations of commodities).

9.2. IDC

An IDC indicates the various combinations of two commodities (among which the consumer is indifferent), which yield same level of satisfaction to the consumer. A.L. Meyers defined IDC as *'a schedule of various combinations of goods, which will be equally satisfactory to the consumer concerned'*. In the words of Prof. Leftwitch, *'a single IDC shows the different combinations of 'X' and 'Y' that yield equal satisfaction to the consumer'*. So, an IDC curve is also called as Iso-utility curve. Hicks and Allen who popularized this concept criticized the earlier version of Utility analysis on various aspects *viz.*, the consumer will not deal with a single commodity in the consumption process, the utility cannot be quantified and MU of money will not remain constant. They opined that, the consumer deals with a combination of related commodities, 'utility' concept is purely subjective and immeasurable and MU of money increases as the consumer goes on spending the money and will not remain constant. So, they introduced the concept of ordinalism, wherein, utility is not quantified, but can be ranked on the basis of scale of preferences. So, the aim of IDC analysis is to analyze how a rational consumer chooses between the various combinations of related commodities, so as to maximize his satisfaction.

We know, the consumer chooses his best combination of commodities based on his income, prices and tastes and preferences over the possible combinations of the commodities. It is generally assumed that, the consumer has a well-defined preference to the set of all possible combinations of commodities. This facilitates him to compare his preferred combination over other possible combinations of commodities. It is further assumed that, the consumer can compare any two commodities and between these two commodities, he can have more of one to less of the other and he remains indifferent with respect to any combination of commodities on the same IDC. That means, an IDC shows all possible combinations of two commodities that yield same level of satisfaction to the consumer. This makes the consumer indifferent between any combinations of two commodities on the same IDC regarding the level of satisfaction. In other words, utility or satisfaction is held constant along an IDC. Furthermore, it is also assumed that, the consumer can rank the combinations of two commodities in the order of preferences over them. Suppose, a consumer is having Rs. 60 and he has to allocate this money between two commodities *viz.*, rice and wheat. Assume, the price of rice is Rs. 5/kg and price of wheat is Rs. 6/kg. The possible combinations of rice and wheat the consumer can purchase with the given amount of money include, (10.8kg, 1.0kg), (9.6kg, 2.0kg), (8.4kg, 3.0kg), (7.2kg, 4.0kg), (6.0kg, 5.0kg), (4.8kg, 6.0kg), (3.6kg, 7.0kg), (2.4kg, 8.0kg), (1.2kg, 9.0kg), (1.0kg, 9.2kg), (2.0kg, 8.3kg), (3.0kg, 7.5kg), (4.0kg, 6.7kg), (5.0kg, 5.8kg), (7.0kg, 4.2kg), (8.0kg, 3.3kg), (9.0kg, 2.5kg), (10.0kg, 1.7kg) and (11.0kg, 0.8kg). Besides these combinations, the consumer can prefer other combinations like (0kg rice, 10kg wheat) and (12kg rice, 0kg wheat), but these combinations violate the basic assumption of IDC technique that, 'the consumer should be rational in dealing with both the commodities'. In these two cases *viz.*, (0kg, 10kg) and (12kg, 0kg), the consumer prefers only one commodity, rather than their combinations. Further, he can purchase other combinations like (2kg, 3kg), (3kg, 2kg), (4kg, 5kg), (5kg, 4kg) etc., but in these cases, the consumer cannot spend all the money on these combinations, as it costs him less than Rs.60. However, the consumer cannot afford to buy combinations like (5kg, 8kg), (2kg, 9kg), (7kg, 5kg) etc., as such combinations costs him more than Rs. 60 at the given prices. So, among the possible scale of preferences, the consumer can rank his combinations as per his tastes and preferences, income and prices of the two commodities. The Table 9.1, which shows different combinations of rice and wheat commodities that yield same level of satisfaction to the consumer at a given level of income and given prices of two commodities is called Indifference schedule. A.L.Meyers defined Indifference schedule as *'a schedule of various combinations of goods that will be equally satisfactory to the individuals concerned'*.

Table 9.1: Indifference schedule of consumer with reference to rice and wheat commodities.

Ranking of Combinations	*Combinations of Rice and Wheat*
A	10.8kg, 1.0kg
B	9.6kg, 2.0kg
C	8.4kg, 3.0kg
D	7.2kg, 4.0kg
E	6.0kg, 5.0kg
F	4.8kg, 6.0kg
G	3.6kg, 7.0kg
H	2.4kg, 8.0kg
I	1.2kg, 9.0kg
J	1.0kg, 9.2kg
K	2.0kg, 8.3kg
L	3.0kg, 7.5kg
M	4.0kg, 6.7kg
N	5.0kg, 5.8kg
O	7.0kg, 4.2kg
P	8.0kg, 3.3kg
Q	9.0kg, 2.5kg
R	10.0kg, 1.7kg
S	11.0kg, 0.8kg

It is evident from the Table 9.1 that, the prioritized preference of the consumer is to purchase or consume 10.8 kilograms of rice and one kilogram of wheat. A close examination of the Table 9.1 reveals that, among these two commodities, the consumer is weighing more importance to rice compared to wheat in the order of assigning preferences among the possible combinations of rice and wheat in accordance with his tastes and preferences, income and prices of the two possible commodities. This implies that, among possible combinations of two commodities, the consumer can prefer more of one commodity to the less of the other through the process of substitution. That means, one commodity is considered as an Added commodity and the other as a Replaced commodity in the substitution process. But, as per the assumption of IDC technique, the consumer is indifferent between these various combinations on the same IDC, as any combination of these two commodities yield same level of satisfaction to the consumer. Further, the combinations also indicate that, the commodities are perfectly divisible.

9.2.1. Monotonic Preferences

These are the preferences of the consumer for different combinations of two commodities, such that, the consumer will not remain indifferent among different combinations. That means, the assumption of Monotonic preferences is opposite to IDC analysis because, in Monotonic preferences, the consumer generally prefers a combination of commodities, in which, there is more quantity of atleast one of the commodities and no less of the other commodity, as compared to any other possible combinations. For example, in the Table 9.1, if the consumer has monotonic preference, he prefers combination (6.0kg, 5.0kg) than any other combination because, this combination has more of both the commodities, rice and wheat. This preference (6.0kg, 5.0kg) is followed by (7.2kg, 4.0kg), (8.4kg, 3.0kg), (9.6kg, 2.0kg), (10.8kg, 1.0kg), (4.8kg, 6.0kg), (3.6kg, 7.0kg) and so on. So, in Monotonic preferences, the consumer prefers more quantity to less quantity and thereby, he remains different among different combinations of commodities.

To compare the IDC technique and Monotonic preferences of the consumer, go through the following example. Say, if there are different combinations of rice and wheat commodities like (10.8kg, 1.0kg), (9.6kg, 2.0kg), (8.4kg, 3.0kg), (7.2kg, 4.0kg), (6.0kg, 5.0kg), (4.8kg, 6.0kg) and (3.6kg, 7.0kg), in case of IDC analysis, any of these combinations yield same level of satisfaction, thereby, the consumer is indifferent among all the given possible combinations. However, if the consumer has Monotonic preference, the first preferred combination is (6.0kg, 5.0kg) followed by (7.2kg, 4.0kg),

(8.4kg, 3.0kg), (9.6kg, 2.0kg), (10.8kg, 1.0kg), (4.8kg, 6.0kg) and (3.6kg, 7.0kg). So, in Monotonic preferences, the consumer is not indifferent among the possible combinations of commodities.

9.2.2. Properties of IDC

An IDC shows different combinations of two commodities, (among which the consumer is indifferent) that yield same level of satisfaction. The main attributes or properties or characteristics of IDC are discussed here under.

1. IDCs are Negatively Sloped

The IDC slope down from left to right. That means, the slope of IDC is convex to origin or negatively sloped. It slopes downward because, as the consumer increases the consumption of rice commodity (added commodity), he has to give up certain units of wheat commodity (replaced commodity), so as to derive the same level of satisfaction (Table 9.2 and Figure 9.2). This also implies that, the two commodities under consideration are imperfect substitutes. To explain the concept clearly, a hypothetical example presented through Table 9.2 and Figure 9.2 reveals that, all the five combinations of rice and wheat will yield same level of satisfaction to the consumer. If the consumer prefers the first combination, he consumes one kilogram of rice and 13 kilograms of wheat. In the second combination, he consumes one more kilogram of rice and he is prepared to give up four kilograms of wheat, in order to derive or maintain the same level of satisfaction. That means, for increasing the consumption of one kilogram of rice, he sacrificed four kilograms of wheat. Likewise, when the consumer reaches the fifth combination, he sacrifices one kilogram of wheat for increasing one kilogram of rice consumption. So, the rate of substitution of rice for wheat is in diminishing fashion, and this gives the convex shape to IDC.

The Figure 9.2 indicates that, the consumer is indifferent among different combinations of rice and wheat, as any combination of these two commodities on the same IDC represents equal level of satisfaction. For example, at combination B on the IDC, the consumer is satisfied with two kilograms of rice and nine kilograms of wheat. The consumer is equally satisfied at combination C, where he consumes three kilograms of rice and six kilograms of wheat on the same IDC. So, this indicates that, any combination of rice and wheat on the same IDC will yield same level of satisfaction to the consumer. This Figure 9.2 also reveals that, it is only on the negatively sloped IDC, the different combinations of commodities of rice and wheat will yield the same level of satisfaction and it makes the consumer indifferent. It is

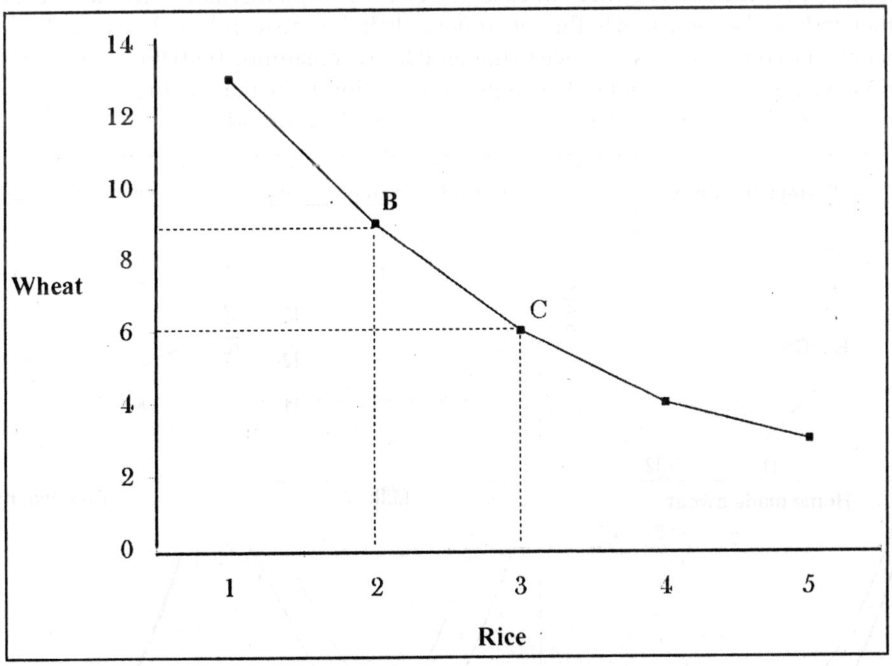

Figure 9.2: Negatively sloped IDC.

important that, every point on IDC represent a different combination of both rice and wheat, but the consumer is indifferent between any two points on the same IDC in terms of satisfaction. So, all the combinations are equally desirable to the consumer in terms of level of satisfaction. But, the consumer is different among the combinations of two commodities taking into consideration his tastes and preferences, income level, prices of the commodities etc. Say, for example, if the consumer prefers to have more of rice and if price of rice is lower than wheat, he consumes more of rice compared to wheat and thus, he prefers the combinations comprising more units of rice and less units of wheat, though the satisfaction he derives at any point on the IDC remains same. So, we can infer that, IDC is a locus of different combinations of two commodities, which yield same level of satisfaction to the consumer.

Table 9.2: Indifference schedule of a consumer with reference to rice and wheat commodities indicating diminishing MRS.

Combinations	*Rice (Kg) (Added commodity)*	*ΔRice (ΔR)*	*Wheat (kg) (Replaced commodity)*	*ΔWheat (ΔW)*	*MRS Rice for Wheat* MRS_{RW}= (ΔW/ΔR)
A	1		13		–
B	2	1	9	–4	4
C	3	1	6	–3	3
D	4	1	4	–2	2
E	5	1	3	–1	1

Different Possible Shapes of IDCs

The shape of IDC discussed above (convex) illustrate the substitution relationship between the commodities under consideration. Likewise, different shapes of IDCs represent different types of relationships between the commodities. They are discussed below.

(a) Perfect Substitutes

In case of good substitutes like rice and wheat (discussed above), tea and coffee etc., the IDC is convex in shape. However, in case of perfect substitutes like owned seed and purchased seed, home made sweet and purchased sweet etc., the IDC is a straight line connecting the two axes (Panel 9.3A). This implies, MRS is constant or will not vary regarding how much of each of the commodity the consumer consumes. That means, the parallel straight lines of IDCs for different levels of satisfaction represent that, the consumer would be willing to increase the satisfaction level at a fixed ratio.

(b) Perfect Complements

These are the commodities, which combine in a fixed proportion for deriving a given level of satisfaction. In such a case, the IDC is right angled or 'L' shaped (Panel 9.3B). For example, to have a cup of milk, we have to combine milk and sugar in a definite proportion. The MRS in case of perfect complements is zero, as the commodities will not substitute each other and yield satisfaction to the consumer only if they are combined in a fixed proportion. To increase the level of satisfaction, the consumer has to move to higher IDC by consuming both the commodities in large quantities in a fixed proportion. No movement to higher IDC is possible by simply increasing the quantity of consumption of one commodity without increasing the quantity of consumption of other commodity.

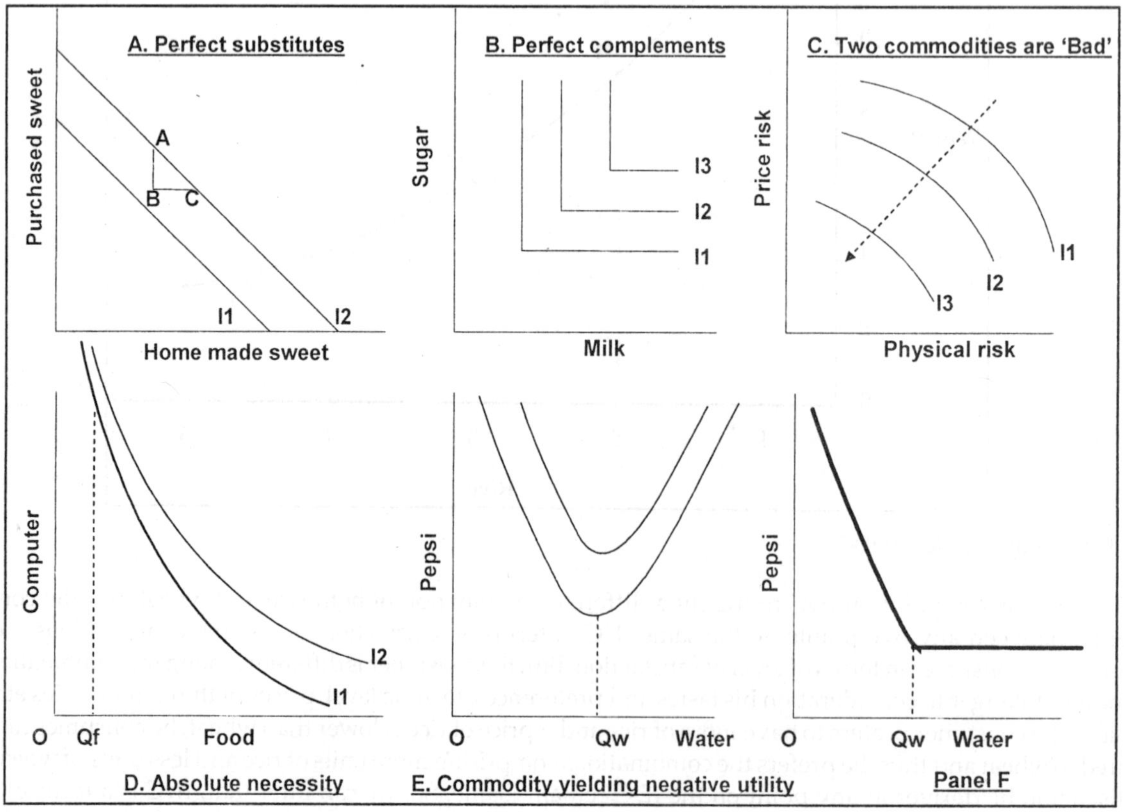

Figure 9.3: Different shapes of IDCs.

(c) When Two Commodities are Not Desirable or 'Bad'

When the two commodities under consideration are 'Bad' it implies, the less the consumer prefers them, the greater is the satisfaction. Say, for example, the lesser the storage risk and price risk of the agricultural commodities, the greater is the benefit to the farmers. So, we can say, both storage risk and price risk are 'Bads' on the part of the farmers. In such cases, the IDCs will have a concave shape to the origin (Panel 9.3C). So, as the farmer moves towards the origin or in the south-west direction, his benefits or satisfaction will increase *i.e.*, if he moves to lower IDCs. So, the farmer realizes more satisfaction at the point of origin, where the quantities of both 'Bads' are zero.

(d) An Absolute Necessity

If a commodity is considered essential upto a certain level, then it demands more and more sacrifice on the part of other commodity. So, upto the essential level of that commodity, the MRS of other commodity will be more, but with the increased availability of that essential commodity, the MRS on the part of other commodity will gradually decrease. This is explained through the Panel 9.3D. We know, food is the most essential commodity for human life. So, upto OQ_F quantity of food, the substitution on the part of other commodity (computer) will be more, being food is an essential commodity for human survival. Beyond OQ_F level of food, the MRS on the part of computer will gradually decrease, as with increased availability of food, less and less sacrifisation will be there with respect to computer. In such cases, the IDCs will have steep slope upto OQ_F quantity of food and later, the curve will fall gradually.

(e) Commodity Yielding Negative Satisfaction after Certain Level of Consumption

If the consumer goes on consuming a commodity, the marginal satisfaction derived from each and every unit of a commodity diminishes and even yields dissatisfaction after certain level of consumption. For some commodities like water, the consumer attains dissatisfaction rather quickly compared to others, as it is comparatively less tasteful. Say, for example, if the consumer is faced with two commodities water and Pepsi, and if he goes on consuming water, the MRS of water for Pepsi will gradually decrease. But, after certain level of consumption of water, it yields dissatisfaction to the consumer and it will not substitute Pepsi and hence, the IDC instead of falling downwards will now take upward turn, thus giving a U shape to the IDC. As shown in the Panel 9.3E, beyond OQ_W of volume of water consumed, the IDC will have a positive slope, indicating that, the consumer gets negative satisfaction from consuming extra water. This upward sloping of IDC beyond OQ_W implies that, the consumer derives dissatisfaction from consumption of water, which is compensated by the increase in consumption of Pepsi, so that, total satisfaction on IDC remains same. Similarly, if Pepsi is consumed more by the consumer, he reaches the point of satiation at some point, beyond which, now the consumer consumes more of water, so as to compensate the total satisfaction on the IDC. On the other hand, if the consumer can dispose off the extra unwanted units of water at no cost, then beyond OQ_W, the IDC becomes horizontal (Panel 9.3F).

(f) Neutral Good

A neutral good is a good, which will not influence the consumer's satisfaction. When a commodity gives no satisfaction to the consumer at all, then he will be unwilling to sacrifice even the smallest amount of other commodity to obtain any quantity of the commodity in question. This is a case, when a vegetarian consumer dealing with a vegetarian food and buffalo meat. In such a case, the consumer will not sacrifice even a smallest amount of vegetarian food for the sake of buffalo meat. The IDCs are horizontal straight lines or parallel to the X-axis (Panel A of Figure 9.4) on which neutral good is considered (in this case, buffalo meat for vegetarian consumer). By having more and more quantity of neutral good on X-axis, it makes no difference to the consumer, but only increasing the quantity of consumption of vegetarian food on Y-axis, the consumer moves to higher IDC. So, the direction of preference will be upward towards the north indicating that, the consumer will move to higher IDC, only if he increases the consumption of vegetarian food, irrespective of the quantity of neutral good (buffalo meat) offered to him. In case, if the neutral good is taken on Y-axis, the IDCs will be parallel to Y-axis and the direction of movement to higher IDCs is towards the east (Panel B of Figure 9.4).

However, such parallel IDCs are not possible because, it violates the basic assumption of IDC technique, as, in this case, the consumer is not dealing with two commodities and for the vegetarian consumer, the two commodities under consideration *viz.*, vegetarian food and buffalo meat are not related commodities. But, if we consider a non-vegetarian consumer, the IDC will have a convex shaped curve.

(g) One Commodity Desirable Over the Other

If the consumer faces with the situation such that, one commodity is better or desirable (Computer) over the other (Pollution), the IDCs will have the shape as shown in the Panel A of the Figure 9.5. As IDC I_2 yields more satisfaction compared to IDC I_1, it implies that, an increase in number of computers causes an increase in satisfaction, since it makes the consumer to move onto the higher IDC I_2, but an increase in the amount of pollution results in a decrease in satisfaction.

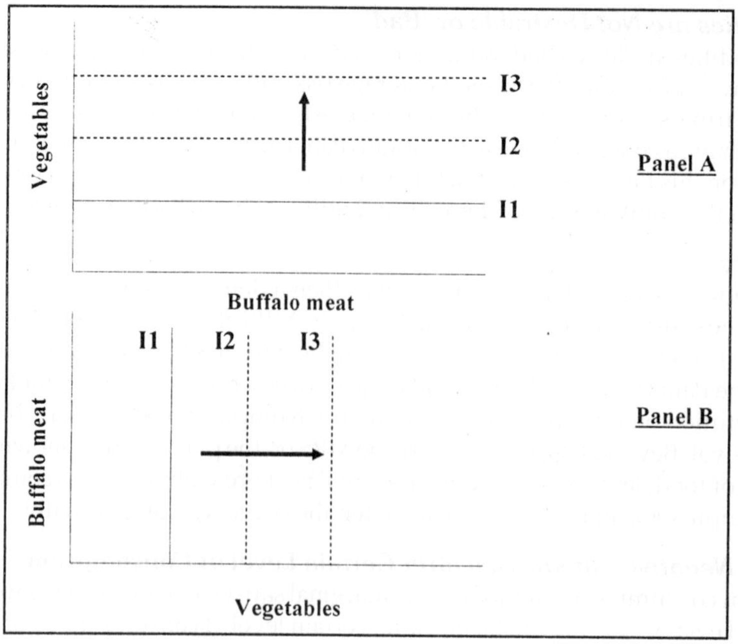

Figure 9.4: Shapes of IDCs – Case of a Neutral good.

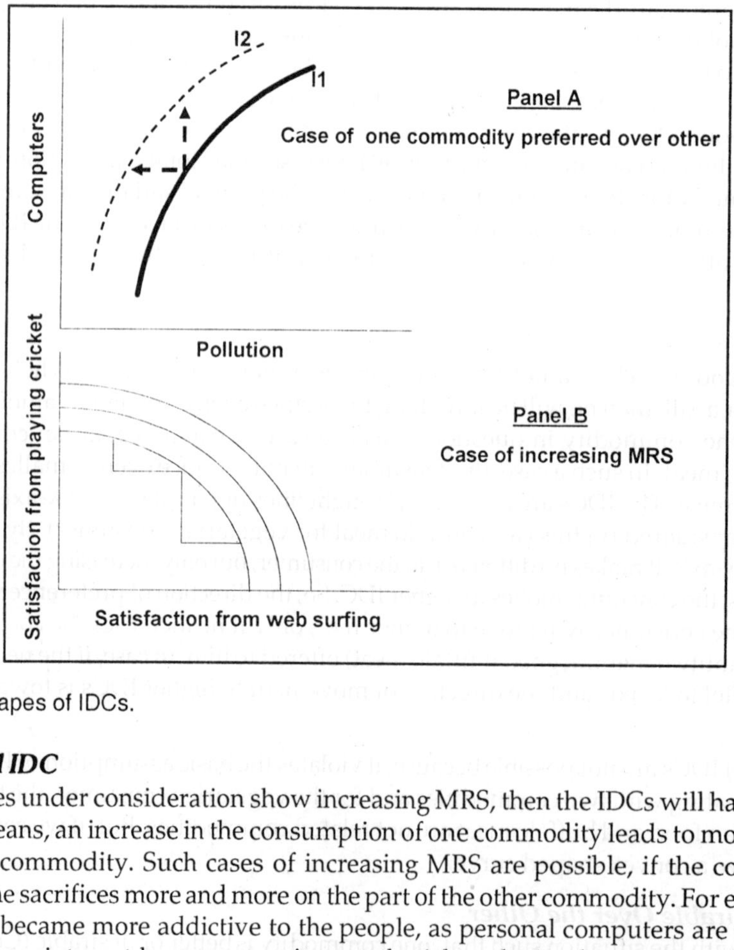

Figure 9.5: Different shapes of IDCs.

(h) Concave Shaped IDC

If the commodities under consideration show increasing MRS, then the IDCs will have a concave shape (Panel B of Figure 9.5). That means, an increase in the consumption of one commodity leads to more and more sacrifisation on the part of the other commodity. Such cases of increasing MRS are possible, if the consumer is addictive to one commodity, thereby, he sacrifices more and more on the part of the other commodity. For example, in the recent period, in India, web surfing became more addictive to the people, as personal computers are connected to internet. If we consider, web surfing or internet hours as one item and cricket playing hours as another item, the more time an individual searches the web, it yields more and more marginal satisfaction to him and thereby, it leads to increasing MRS on the part other item *i.e.*, cricket hours. That means, the more time an individual searches the web, the more additional or marginal satisfaction he derives and thereby, the more additional satisfaction of playing cricket will be foregone by the consumer. This is because, a person can play cricket continuously for about three to four hours, but after that, his satisfaction level will fall gradually. But, this is not the case with web surfing because, the net work of web links continues to grow and the more number of web sites the person explores, the greater the satisfaction he derives. So, the loss in marginal satisfaction with reference to playing cricket for a longer period of time will be much more compared to marginal gain of satisfaction from internet surfing. So, this increasing MRS will give concave shape to the IDCs.

The extent of slope of the IDC depends upon the extent of relationship between the two commodities. The greater the substitution relationship between the two commodities, the less is the convexity of IDC and *vice versa*. That means, when the commodities are perfect substitutes, the IDC will be a straight line and when the commodities are good substitutes, the IDC will have a convex shape. On other hand, when the substitution between the commodities is completely absent, rather, they are perfect complements, the IDC will have a right angle or 'L' shape. So, we can say that, the greater the complementarity between the commodities, the greater the convexity of the IDC. It is important to note that, within the same IDC, there is a possibility of having both complementarity and substitution relationship between the two commodities. That means, the commodities may be good substitutes and good complements up to a certain range on the IDCs. For example, we know computer monitor and Central Processing Unit (CPU) are perfect complements. But, we can use old CPU for new monitor and also new CPU for old monitor. That means, to a certain extent or level of satisfaction, we can increase the consumption (purchase) of new monitors without increasing the consumption of new CPUs and similarly, we can increase the consumption of new CPUs without increasing the consumption (purchase) of new monitors. The same is explained through the Figure 9.6, wherein, between AB portion of the IDC I_1, the commodities are substitutable, but beyond AB portion of the IDC I_1, the commodities are perfect complements. So, this implies that, throughout the same IDC I_1, we cannot say that, both the commodities are perfect complements or perfect substitutes, as sometimes both the relationships may exist.

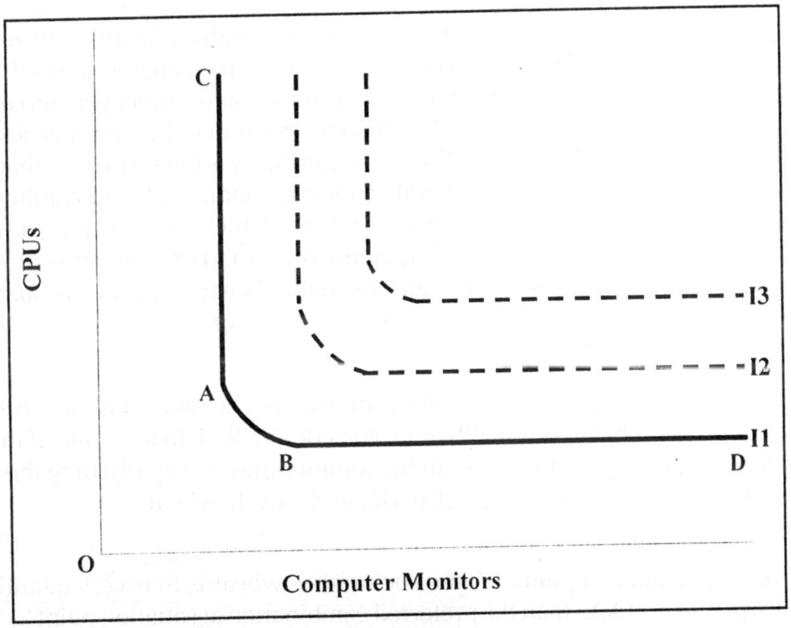

Figure 9.6: IDCs with ranges of both complementarity and substitution.

Different Zones of 'Goodness', 'Badness' and Point of Bliss

The discussion, so far, revealed that, the consumer prefers more quantities of commodities to less quantities and as a result of this, the IDC slopes downward from left to right. As shown in the Panel A of the Figure 9.7, the preferable combinations of commodities for different levels of satisfaction can be conveniently divided into four zones. Assume that, the consumer preferred the 'A' combination of both rice and wheat commodities. Keeping the assumption in mind, we can explore the following aspects:

- If the consumer is at point 'A' on the indifference map, then all possible combinations of the two commodities in Zone I and the boundary are more preferred by him because, these combinations have more of both the commodities or more of one commodity and the same amount of the other commodity compared to point 'A' chosen.
- The combinations of commodities in Zone III and the boundary are less preferred to the point 'A' by the consumer because, at point 'A', the consumer consumes more quantities of both the commodities.
- The combinations of commodities in Zone II and Zone IV are less preferable or equally preferable to the point 'A' considered. Since, the zones II and IV are the only zones that could have a point of indifference to the point 'A' chosen, the IDCs must have a negative slope.

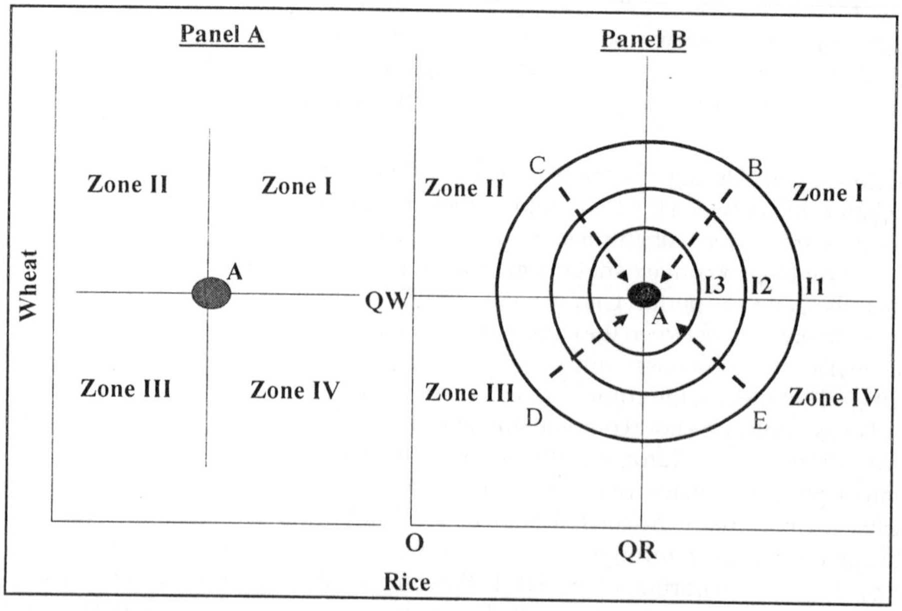

Figure 9.7: Different zones of Goodness and Badness and Point of Bliss (Satiation).

From the above explanation, we can study about the Goodness or Badness of different zones and the Point of Bliss and to have clear understanding, please see Panel B of the Figure 9.7. Like Panel A, the entire area of the preferential combinations of two commodities is divided into four zones. We know, if a consumer goes on consuming the commodity, it yields less and less satisfaction to him and thereby, at some point or other, he reaches the point of maximum satisfaction. This is called Satiation point or Point of Bliss. If the consumer is forced to have that commodity beyond that point, it will yield dissatisfaction to him. So, this implies that, the commodity is 'Good' for the consumer upto the point of satiation and beyond that, it becomes 'Bad' to the consumer. In the Panel B, we see that, the point of satiation is at 'A' and the consumer reaches that point by consuming OQ_R quantity of rice and OQ_W quantity of wheat and this is the most preferred combination by the consumer. Let us examine the preferential behaviour of the consumer for different quantities of commodities in different zones and his movement towards the point of satiation.

Zone I

The IDCs have a negative slope. At point 'B', both the commodities, rice and wheat are in excess quantity than their combination at satiety point. So, both the commodities are considered 'Bad' in this zone. If the consumer moves from IDC I_1 to I_2 and I_3, he will decrease the quantities of both the commodities and by reaching the Point of Bliss 'A', he will have the desired combination of commodities *i.e.*, OQ_R of rice and OQ_W of wheat.

Zone II

The IDCs have a positive slope. At point 'C', the commodity wheat is in excess quantity (*i.e.*, > OQ_W), but the commodity rice is less in quantity (<OQ_R) than the preferred combination at satiation point 'A'. So, we consider, wheat is a 'Bad' commodity and rice a 'Good' commodity in this zone. So, in order to maximize his satisfaction, *i.e.*, to reach the point of satiation 'A', the consumer has to move from IDC I_1 to I_2 and I_3 and while reaching the point of satiation, he will reduce the quantity of consumption of wheat upto OQ_W and he will increase the quantity of consumption of rice upto OQ_R, and thereby, the point of satiation is reached.

Zone III

The IDCs have a negative slope. At point 'D', both the commodities viz, rice and wheat are in less in quantity than the preferred combination 'A'. So, both the commodities are considered 'Good' in this zone. So, to maximize the satisfaction, the consumer will increase the consumption of both the commodities till OQ_R of rice and OQ_W of wheat and hence, he moves from IDC I_1 to I_2 and I_3, thereby, the point of satiation is reached.

Zone IV

The IDCs have a positive slope. At point 'E', the commodity rice is in excess quantity (>OQ_R) and commodity wheat is in less quantity (<OQ_W), than the preferred combination at satiation point 'A'. So, we consider rice is a 'Bad' commodity and wheat is a 'Good' commodity in this zone. So, to maximize the satisfaction *i.e.*, to reach the point of satiation 'A', consumer has to consume more of wheat (upto OQw) and less of rice (upto OQ_R) and thereby, he moves from IDC I_1 to I_2 and I_3 and derives maximum satisfaction at the Point of Bliss.

So, the above analysis infer that, in selecting the combinations of commodities, the consumer tries to move towards the point of satiation, so as to maximize his satisfaction. Taking the 'Goodness' and 'Badness' of the commodities under consideration, the consumer tries to increase the 'Goodness' further or decrease the 'Badness' further, so that, he moves towards the point of satiation or Bliss point.

2. Slope of IDC indicates Diminishing MRS of Commodities

As studied through Table 9.2, if the consumer goes on consuming rice commodity, he is sacrificing less and less of wheat commodity in the consumption process. This is in accordance with the concept of LDMU. The concept of LDMU explains that, when the consumer goes on consuming a commodity (rice), the MU derived from that commodity gradually diminishes and thereby, the additional units of rice will yield less additional satisfaction than the previous units. So, in order to maintain same level of satisfaction, it replaces less and less of other commodity (wheat). In the present case, in the first combination, the consumer consumes one kilogram of rice and 13 kilograms of wheat to derive certain level of satisfaction. In the second combination, he consumes one more kilogram of rice and he is prepared to give up four kilograms of wheat in order to maintain the same level of satisfaction. In case of third combination, he gives up only three kilograms of wheat (less than combination B) for one more kilogram of rice consumption. In the fourth combination, the sacrifice of wheat consumption is still lowered to two kilograms only for an increase of one more kilogram of rice consumption. Likewise, in the fifth combination, the sacrifisation of wheat is still lowered to one kilogram only, for an additional consumption of one more kilogram of rice. So, as the consumer moves from combination A to E, the rate of substitution of commodity rice (Added commodity) for commodity wheat (Replaced commodity) goes down. In other words, as the consumer consumes more and more units of rice, he is prepared to forego less and less of wheat in each successive case. This behaviour of compulsory fall in the consumption of one commodity (wheat) due to increase in consumption of other commodity (rice) is called Substitution. In the consumption process, for each and every additional unit of consumption of rice, it leads to compulsory sacrifisation of some additional units of wheat consumption and this rate of substitution is called MRS. Since, the MRS is at diminishing fashion, it is called as diminishing MRS. It is otherwise defined as 'the rate at which the consumer must gives up one commodity (wheat) in order to obtain or consume an additional unit other commodity (rice). According to Bilas, '*the MRS of X for Y is defined as the amount of Y the consumer is just willing to give up to get one more unit of X and to maintain the same level of satisfaction*'. According to Hibdon, '*the law of diminishing MRS states that, the consumer will be willing to forgo smaller and smaller units of Y in order to have successive additional units of X*'. So, the MRS is calculated between two commodities (rice and wheat) placed on an IDC, which displays a frontier of equi-satisfaction for each combination of these two commodities. This concept of MRS was introduced by Prof. J.R. Hicks and Prof. R.G.D. Allen to replace the concept of diminishing MU. They are of the opinion that, consumer being rational, tend to substitute one commodity for the other, so as to derive the same level of satisfaction. The MRS is also called as Marginal Rate of Transformation. According to Prof. Hicks, the following are the two reasons for diminishing MRS:

- ☆ Since each particular want is satiable, the consumer goes on consuming one commodity and in this process, the marginal satisfaction derived from each and every unit of the commodity will diminish continuously. So, the consumer will be prepared to sacrifice less and less amount of the other commodity in order to obtain more and more amount of this commodity, in order to maintain same level of satisfaction.
- ☆ The commodities are imperfect substitutes for one another. Since, they are not perfectly substitutable, every time an increase in consumption units of one commodity will not lead to substitution of other commodity in equal amounts, but substitute in decreasing amount due to the operation of LDMU concept and thereby, leading to diminishing MRS of commodities.

The concept of diminishing MRS is illustrated in Figure 9.8. As the consumer moves from A to E, it indicates the willingness to substitute commodity rice for commodity wheat diminishes. This means that, as the amount of commodity rice is increased by equal amounts, the consumption of commodity wheat diminishes by smaller amounts. Thus, the MRS of rice for wheat is the quantity of wheat commodity that the consumer is willing to give up to gain a marginal increase of rice commodity. This diminishing MRS contributes negative (convex) slope to the IDC. The greater the fall in MRS, the greater is the convexity of the IDC.

Initially, at point A, the consumer is prepared to have more of wheat commodity compared to rice commodity. Since, rice commodity is scarce here, he will give more value to rice commodity than wheat commodity and hence, he will come down the IDC, thereby, substituting rice commodity for wheat commodity. That means, when the consumer is at higher or top most position of IDC, the economic significance is more for rice commodity. On the other hand, when the consumer is down the IDC, he will consume more of rice and less of wheat. So, the economic significance of wheat commodity will increase now and hence, he will not substitute rice for wheat. That means, the rate of substitution of wheat by consuming more of rice will gradually decrease, as the consumer moves down the IDC because, while coming down the IDC, the economic significance of wheat increases gradually. This phenomenon of changing economic significance between the commodities at the top and lower positions of IDC leads to diminishing MRS.

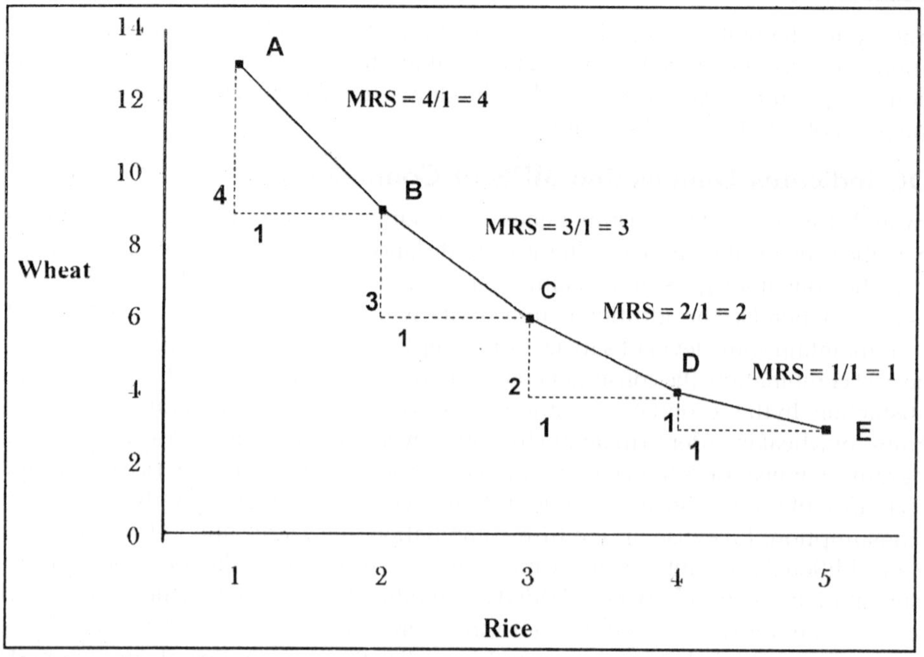

Figure 9.8: IDC showing diminishing MRS of commodities.

The concept of MRS is explained by the following formula:

$$MRS = \frac{\text{Change in number of units consumed of replaced commodity (wheat)}}{\text{Change in number of units consumed of added commodity (rice)}}$$

So, MRS of rice for wheat is given by, $MRS_{RW} = \frac{\Delta W}{\Delta R}$ *Equation 9.1*

The same concept is explained through Table 9.2 and Figure 9.8. Please note, the convention is to ignore 'minus' sign.

MRS of Commodities and Shapes of the IDCs

Having studied different shapes of IDCs in the earlier pages, it is essential to know that, the MRS of commodities also influences the shape of IDCs.

- *MRS is diminishing:* If the MRS is diminishing, it implies, both the commodities are good substitutes and the shape of IDC is convex to origin (Figure 9.2).
- *MRS is increasing:* If the MRS between commodities is increasing, the shape of IDC is concave to origin. (Panel B of Figure 9.5)
- *MRS is constant:* If the MRS between commodities is constant or infinite, it implies the commodities are perfect substitutes and the shape of IDC is a straight line connecting the axes. (Panel A of Figure 9.3)
- *MRS is zero:* If the MRS between two commodities is zero, then the two commodities are said to be perfect complements and the IDC is right angled or 'L' shaped. (Panel B of Figure 9.3)

However, among the possible shapes of IDCs discussed above, convex shaped IDC is more common, as diminishing MRS is the general principle of IDC technique and other shapes such as concavity, straightness and L-shaped IDCs are exceptional or rare cases.

Importance of MRS

The diminishing MRS may seem intuitively akin to diminishing MU, however, for the latter, we hold consumption of all but one commodity constant, while in diminishing MRS, we have more of one commodity compensating for fewer of the other. This concept of diminishing MRS is superior to diminishing MU concept of Marshallian analysis. This is because, Marshallian approach is mainly based on the introspective cardinalism, wherein, utility is measured or quantified and moreover, it is a single commodity analysis. However, the Hicksian approach is an improvisation over the Marshallian analysis, in the sense that, it is based on ordinalism in measuring the utility and further studies the consumer behaviour with reference to combination of commodities without assuming constancy of MU of money. The following are the important merits of this MRS concept of Hicksian approach:

- ☆ It guides the consumer at which rate the commodities under consideration are substituting each other. The greater the substitution, the lesser the saturation of consumer with reference to the added commodity.
- ☆ It is relative, as it measures the utility of a commodity in relation to another commodity and not in isolation. Hence, it guides the consumer in selecting the right combination of related commodities.
- ☆ Like the LDMU, this diminishing MRS used in IDC analysis provides useful insight into market demand and the law of demand. If a commodity generates less and less marginal satisfaction, it implies, the consumer is having more of that commodity by that time and the less marginal satisfaction implies, the consumer is willing to pay a relatively lower price, which forms the basis for the law of demand.

Though the concept of diminishing MRS is considered as parallel to the concept of diminishing MU, there lies some important differences between these two and they are as under.

- ☆ The concept of diminishing MRS has no subjective element in it, whereas, the concept of diminishing MU involves the subjective element. So, the concept of MRS is an objective one. It simply tells us, how much quantity of commodity a consumer should be sacrificed, if he goes on increasing the consumption of other commodity. So, the concept of MRS does not involve cardinal approach of measurement of utility.
- ☆ The concept of diminishing MRS assumes, there is substitution relationship between the commodities, whereas in case of diminishing MU concept, the utilities of the commodities are assumed to be independent.
- ☆ The concept of diminishing MRS will not make invalid assumption regarding the constancy of MU of money, unlike the diminishing MU concept.
- ☆ The concept of MRS deals with related commodities in combinations and any combination of the two commodities on the same IDC yield same level of satisfaction. However, in case of diminishing MU concept, we consider that, a particular want is completely satiable and hence, it laid more emphasis on individual commodity analysis that too considering independent utilities.
- ☆ The concept of MRS is based on ordinality approach, unlike the concept of diminishing MU.

3. Position of IDC

The position of IDC with reference to origin depends upon the level of satisfaction derived by the consumer. Higher the level of satisfaction, the farther away the position of IDC from the origin and *vice versa*. So, it implies that, a consumer will prefer to have the combination of commodities that lie on a higher IDC to the combination that lies on a lower IDC, as higher IDC indicates larger quantities of the two commodities. So, larger quantities of the two commodities provide greater satisfaction than the smaller quantities of the same commodities. Therefore, greater is the distance of an IDC from the point of origin, higher it will be in the consumer's preferential order. In other words, an IDC that lies above and to the right of another IDC denotes preferred combination of commodities and thus yields higher satisfaction. In the Figure 9.9, the consumer would prefer to lie on IDC I_5 rather than IDC I_1, as higher IDC denotes larger combinations

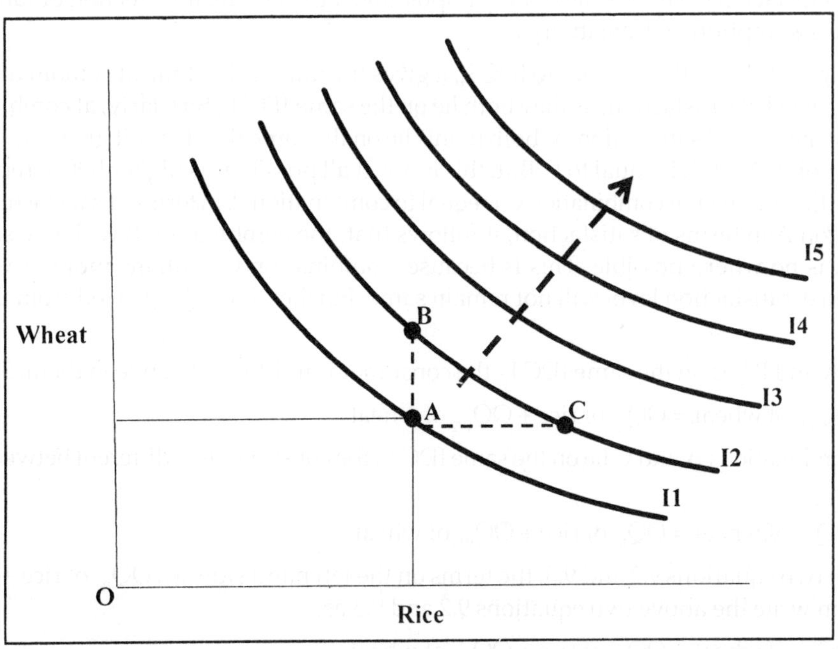

Figure 9.9: IDC map and Position of IDC indicating level of satisfaction.

of two commodities. So, the order of satisfaction levels is denoted as $I_5>I_4>I_3>I_2>I_1$. For example, points 'B' or 'C' or any point between them on IDC I_2 have more of at least one commodity without having less of the other commodities compared to point 'A' on IDC I_1. So, these points on IDC I_2 represent higher satisfaction levels compared to any point (combination of commodities) on IDC I_1. Since, any point on the same IDC represent same level of satisfaction, so all points on IDC I_2 imply higher satisfaction as compared to all points on IDC I_1. Thus, farther the distance of IDC away from the origin, suggests higher level of satisfaction.

Indifference Map

We can draw any number of IDCs by considering more quantities of both the commodities. But, the consumer remains indifferent among different combinations of the two commodities on each IDC. So, we can say, an indifference map is a geometrical expression of a number of indifference schedules on the same graph. So, an indifference map comprises of a set of IDCs of different levels of satisfaction. The indifference map is also called as Preference map. In the Figure 9.9, if the consumer moves along the arrow or in north-east direction, he is said to be climbing a 'utility mountain' *i.e.*, moving from lower utility levels to higher utility levels, as indicated by IDCs I_1 to I_5. Each point on the map represents the same elevation. But, the important aspect is that, an IDC simply indicate all possible combinations of rice and wheat that provide same level of satisfaction to the consumer, but it does not indicate exactly how much satisfaction is derived by the consumer from these combinations. This is because, the IDC approach assumed the concept of ordinality in dealing with consumer's satisfaction and no quantitative measurement of utility is possible.

In IDC analysis, we need not make any assumptions regarding the extent of differences in satisfaction between the successive IDC curves. But, we simply assume that, the satisfaction yielded by IDC I_5 is more than IDC I_4, which in turn exceeds the satisfaction level attached to IDC I_3 and so on. The concept of Indifference map further signifies that,

- In between two IDCs, there can be any number of IDCs indicating different levels of satisfaction to the consumer.
- In the Figure 9.9, we assigned the numbers to different IDCs such as I_1, I_2, I_3, I_4 and I_5. But, these numbers have no special relevance in the IDC analysis, as they are purely arbitrary. We can assign the numbers to IDCs, either in ascending order or descending order, but the important aspect to remember is, the farther the distance of IDC from the origin, the greater the level of satisfaction it yields to the consumer.

4. The IDCs Do Not Intersect or Tangent to Each Other

The IDCs do not intersect each other, as different IDCs will indicate different levels of satisfaction. As shown in the Figure 9.10, at the point of the intersection 'A' of IDC I_1 and IDC I_2, it implies that, the level of satisfaction yielded by the two IDCs are equal, which is absurd and impossible. As discussed earlier, higher IDC I_2 indicates combinations of larger quantities of two commodities and lower IDC I_1 indicates combinations of smaller quantities of the same two commodities. So, different combinations of two commodities in smaller quantities (I_1) and different combinations of two commodities in larger quantities (I_2) will not yield same level of satisfaction. But, at point A, it implies, both the IDCs are yielding same level of satisfaction, which is absolutely impossible. Further, the intersection or tangency of two IDCs implies violation of the assumption of 'transitivity'.

As shown in the Figure 9.10, with reference to IDC I_2, it gives the impression that, at combinations 'A' and 'C', the consumer derives same level of satisfaction, as they both lie on the same IDC I_2. Similarly, at combinations 'A' and 'B', the consumer derives same level of satisfaction, as both points lie on the same IDC I_1. So, it gives the impression that, the satisfaction derived at point 'A' on I_2 is equal to I_1. But, this is not at all possible, as higher IDC I_2 represent higher level of satisfaction than IDC I_1. Moreover, if combination C is equal to combination A in terms of satisfaction and combination B is equal to combination A in terms of satisfaction, it follows that, the combination B will give same satisfaction as combination C, which is no where possible. This is because, combination C contains more of wheat compared to combination B and hence, satisfaction level will not remain same. Further, it can be proved from the equations given below.

As combinations A and B lies on the same IDC I_1, the consumer is indifferent between them. So, we can write,

$$OQ_{R1} \text{ of rice} + OQ_{W1} \text{ of wheat} = OQ_{R2} \text{ of rice} + OQ_{W2} \text{ of wheat} \quad \textit{Equation 9.2}$$

Similarly, as the combinations A and C lie on the same IDC I_2, the consumer is indifferent between them. So, we can write,

$$OQ_{R1} \text{ of rice} + OQ_{W1} \text{ of wheat} = OQ_{R2} \text{ of rice} + OQ_{W3} \text{ of wheat} \quad \textit{Equation 9.3}$$

Since, in the above two equations 9.2 and 9.3, the terms on the left hand side *i.e.*, OQ_{R1} of rice + OQ_{W1} of wheat are same and hence, we can write the above two equations 9.2 and 9.3 as,

$$OQ_{R2} \text{ of rice} + OQ_{W2} \text{ of wheat} = OQ_{R2} \text{ of rice} + OQ_{W3} \text{ of wheat} \quad \textit{Equation 9.4}$$

From the above equation 9.4, we can conclude that, OQ_{W2} of wheat is equal to OQ_{W3} of wheat, which is impossible. So, no two IDCs will intersect or tangent to each other.

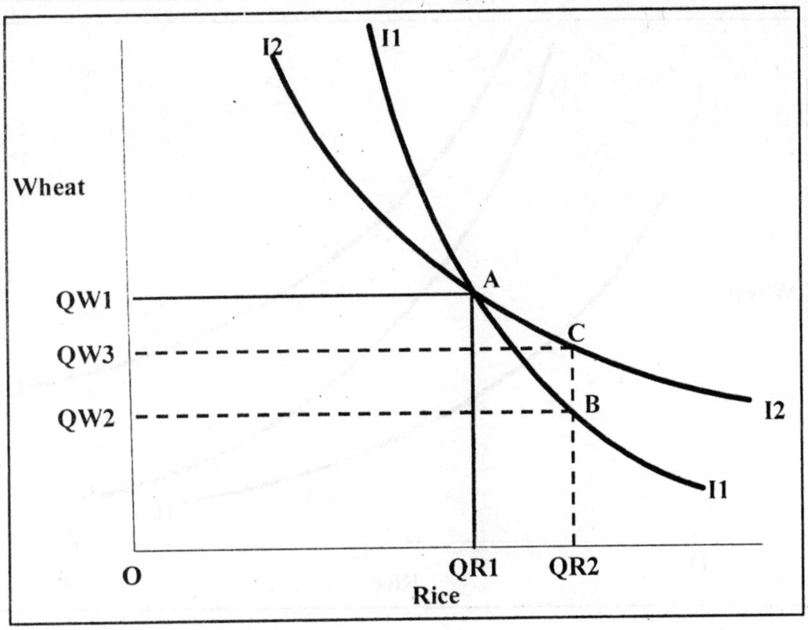

Figure 9.10: IDCs do not intersect to each other.

5. IDC is Asymptotic to Both the Axes

It implies, the IDC will not touch either of the two axes. This is because, the basic assumption of IDC analysis is that, the consumer is rational and he purchases different combinations of two commodities. If IDC touches one of the axes, it means, the consumer prefers only one of the two commodities. Since, he is not supposed to purchase only one commodity, the IDCs will not touch either of the two axes. In the Figure 9.11, it is shown that, the IDC (broken line) touches both the axes *i.e.*, at point Q_A on X-axis and at point Q_B on Y-axis. So, at point Q_A it gives the impression that, the consumer purchase only OQ_A quantity of rice and no amount of wheat. Similarly, at point Q_B, the consumer purchases only OQ_B quantity of wheat and no amount of rice. So, such IDCs are against the basic assumption that, 'consumer is rational in purchasing both the commodities in different combinations'. However, this behaviour of purchasing only one commodity by the consumer refers to Monomania, which indicates that, the consumer is altogether different with reference to one commodity and he is interested only in other commodities, which is against the basic assumption of IDC analysis.

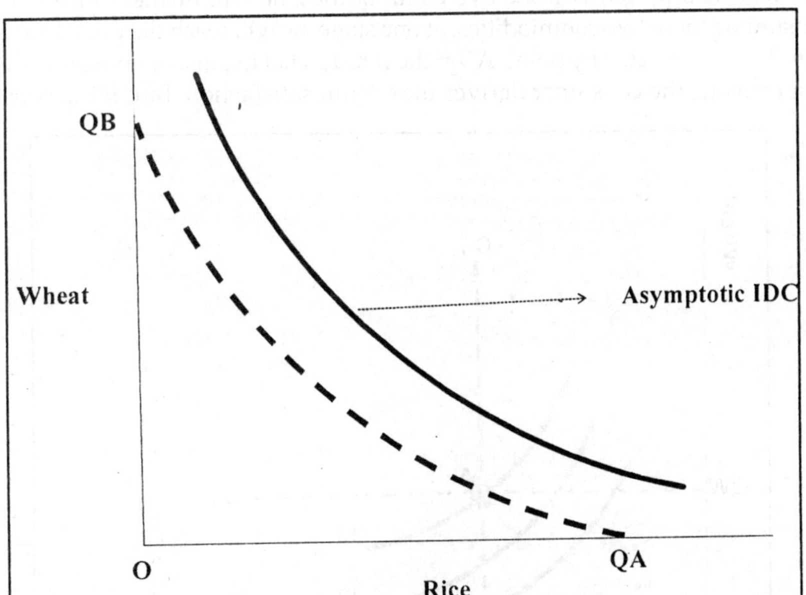

Figure 9.11: IDC is asymptotic to both the axes.

6. IDCs not Necessarily Parallel to Each Other

Since the commodities under consideration are related commodities, IDCs will not be parallel to each other. On the other hand, if the commodities under considerations are perfect substitutes and perfect complements, the IDCs remain

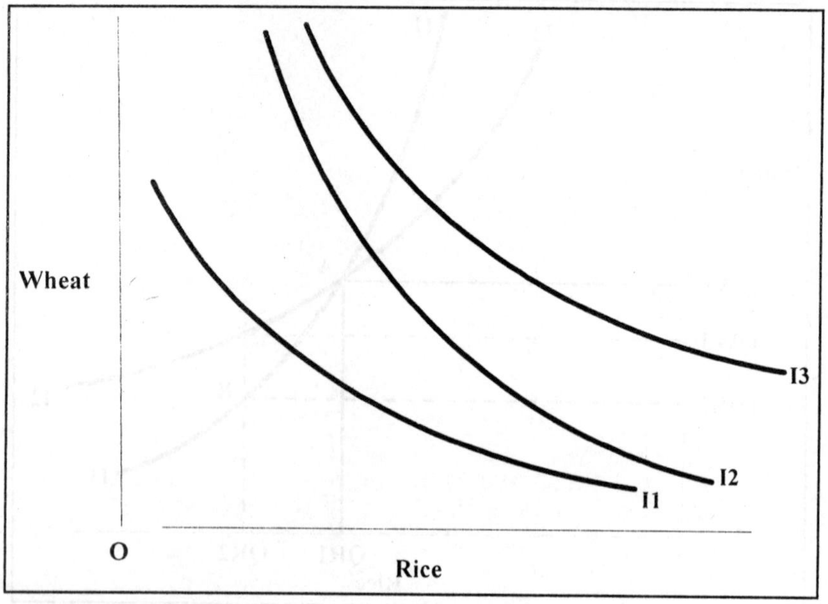

Figure 9.12: IDCs are not parallel to each other.

parallel to each other in an indifference map. If the assumptions of Marshallian analysis *viz.*, independent utilities of the commodities, absence of income effect and constancy of MU of money are introduced in this IDC analysis, the IDC curves in a map will be placed at equi-distance and they remain parallel to each other. But, since the commodities are related (imperfect substitutes) and the MRS differs with reference to quantities of commodities consumed, the IDCs will slope downward from left to right, but may not remain parallel to each other. The same is explained through Figure 9.12 and the three IDCs I_1, I_2 and I_3 are not parallel to each other. The steeper the slope of the IDC curve, the greater the MRS of two commodities and hence, the IDCs may not be parallel to each other.

7. IDCs are in the Form of Circles

In reality, the IDCs are in the form of circles as shown in the Figure 9.13. That means, while increasing the consumption of two commodities, the consumer moves from one circle (IDC) to another circle (IDC). If the consumer increases the consumption of both rice and wheat, he gains more satisfaction and thereby, he moves from IDC I_1 to IDC I_2. If still he increases the consumption of these two commodities, he will further jump to higher IDC I_3. But, if the consumer goes on consuming these two commodities, at one stage, he will reach the point of satiation, where he derives maximum satisfaction. This is indicated by point 'A' on the IDC I_3. That means, by consuming OQ_R and OQ_W quantities of rice and wheat respectively, the consumer derives maximum satisfaction. But, if the consumer still increases the

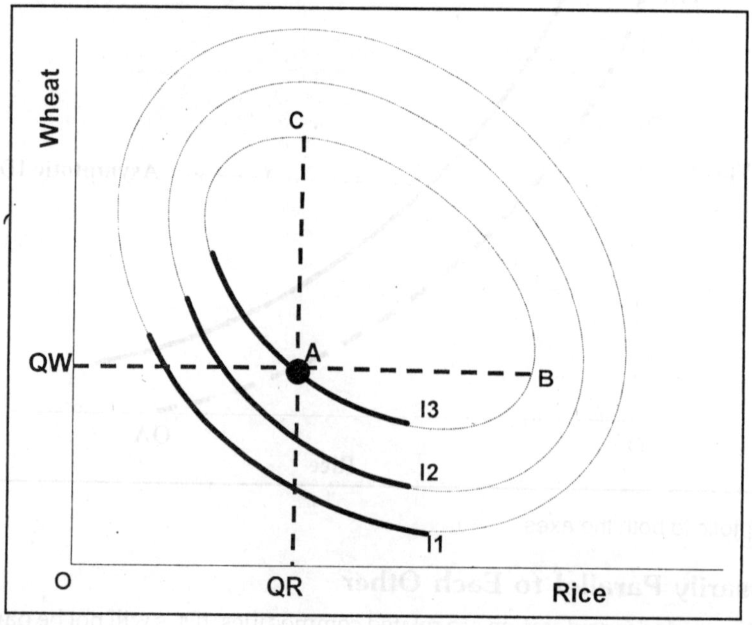

Figure 9.13: IDCs are circular in shape.

consumption of both rice and wheat, *i.e.*, beyond OQ_R and OQ_W he will derive negative satisfaction. Say, if the consumer increases the consumption of rice beyond $OQ_{R'}$, he will move to the dotted portion of the IDC I_3 say to point 'B', where he gets dissatisfaction with respect to rice commodity. So, to compensate this, he will try to increase the consumption of wheat, but by this also he derives dissatisfaction again, as he moves to dotted portion of the IDC I_3 at point 'C'. So, either by increasing the consumption of rice or wheat or both beyond point 'A', the consumer derives dissatisfaction, as he will be on the dotted portion of IDC I_3. So, for maximizing the consumer's satisfaction, only the convex portion of the IDC (I_3) curve is the effective region. The movement to the dotted portion of the IDC (I_3) implies negative satisfaction.

9.3. Price Line

From the indifference map alone, it is not possible to compute the consumer's equilibrium, as it simply indicate the various combinations of two commodities under consideration by the consumer for deriving different levels of satisfaction. On the same IDC, different combinations of two commodities will yield same level of satisfaction. But, the best choice of combination of commodities the consumer will make, however, depend on his income, relative prices of the two commodities, tastes and preferences of the consumer etc. So, to ascertain the point of consumer's equilibrium, we should have information about the income of the consumer and prices of the two commodities, which helps to derive the price line of the consumer. So, this concept of studying the price line of the consumer is essential for knowing the consumer's equilibrium. A price line represents the various combinations of two commodities, which can be purchased by the consumer with a given money income and at given prices of commodities. This price line is drawn as a continuous line and it identifies the options from which the consumer can choose the combination of commodities. It also indicates the limitations of consumer's choices for different combinations of commodities. The price line is also called as Budget line, Price Opportunity line, Expenditure line, Price-Income line etc. It is also called Consumption possibility line, as it indicates the possible combinations of two commodities that are preferred for consumption by the consumer.

For example, a consumer is having an income of Rs.60 and he prefers to have two commodities *viz.*, rice and wheat. The price of rice is Rs. 5/kg and the price of wheat is Rs. 6/kg. Given the income and the prices of the two commodities, the consumer prefers to have 13 possible combinations of two commodities as shown through Table 9.3 and Figure 9.14.

The Table 9.3 shows that, if the consumer spends all his income on wheat (Combination A), he purchases 10 kilograms of wheat at Rs. 6/kg thereby, nothing is left to purchase rice. Similarly, if the consumer allocates all his money on rice (Combination M), he purchases 12 kilograms of rice at Rs. 5/kg and nothing is left for the purchase of wheat. However, the intermediate purchases of the combinations of rice and wheat from B to L shows the mixes of both rice and wheat commodities purchased worth of his income Rs. 60/-. For example, at combination B, the consumer purchases one kilogram of rice and 9.2 kilograms of wheat by allocating all his income (Rs. 60) on both the commodities. The same is explained through Figure 9.14.

Table 9.3: Attainable combinations of rice and wheat by the consumer for a given income and prices of the commodities.

Combinations	*Rice (kg)*	*Wheat (kg)*
A	0.0	10.0
B	1.0	9.2
C	2.0	8.3
D	3.0	7.5
E	4.0	6.7
F	5.0	5.8
G	6.0	5.0
H	7.0	4.2
I	8.0	3.3
J	9.0	2.5
K	10.0	1.7
L	11.0	0.8
M	12.0	0.0

In the Figure 9.14, the line PL indicates the price line of the consumer, indicating various combinations of rice and wheat commodities purchased by the consumer at a given level of income and at given prices of two commodities. It shows 13 possible combinations of rice and wheat purchased by the consumer from A to M. Combination A indicates that, the consumer has purchased 10 kilograms of wheat by allocating all his money *i.e.*, Rs. 60 on wheat only and nothing of rice is purchased. Similarly, point M shows the consumer has purchased 12 kilograms of rice by allocating all his money *i.e.*, Rs. 60 on rice only and nothing of wheat is purchased. The remaining points on the price line *i.e.*, from B to L indicates various combinations of both rice and wheat the consumer can purchase at given income and prices of the commodities.

Another interesting point we can ascertain from the price line is that, the consumer can purchase any combination of commodities that lies inside the price line. In the Figure 9.14, combination R of rice and wheat commodities is feasible to purchase by the consumer, but some money or income will be saved on the part of the consumer. This is because, purchase of rice and wheat commodities at combination R involves outlay of money less than Rs.60. On the contrary, to purchase any combination of commodities of rice and wheat outside the price line, such as, N requires an outlay

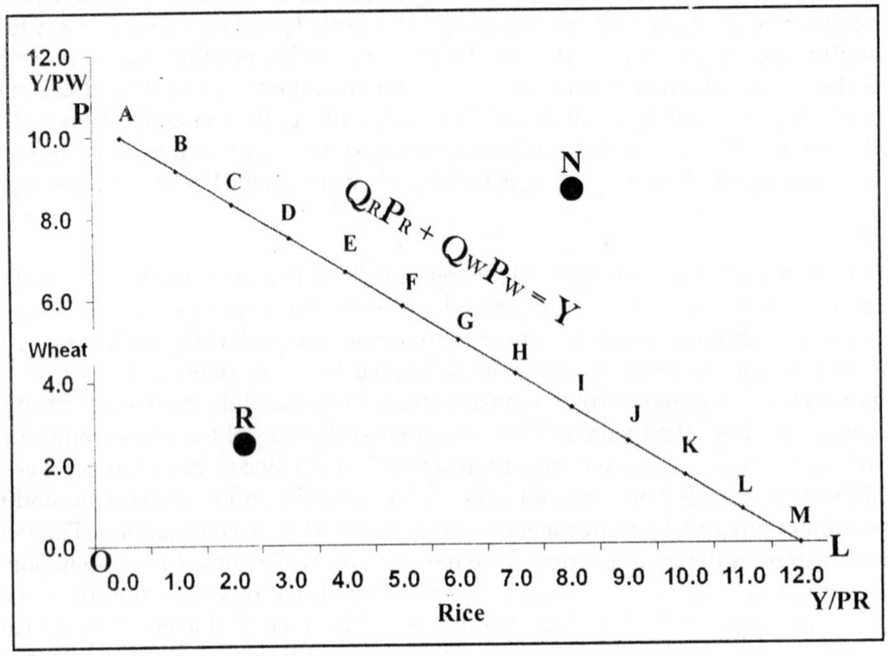

Figure 9.14: Price line and Budget set of the consumer for rice and wheat combinations.

larger than the consumer's income of Rs.60, thereby, such combination is not attainable. So, total spending for the commodities can fall short of the income of the consumer, but should not exceed it. This is called Budget constraint and the set of combinations of commodities that the consumer can purchase with his given income and at given prices of two commodities is called Budget set.

To arrive at this budget constraint and budget set of the consumer, let us assume that, the income of the consumer is Y and the prices of two commodities *viz.*, rice and wheat are P_R and P_W respectively. If the consumer wishes to purchase Q_R and Q_W quantities of rice and wheat respectively as the preferred combination, he has to spend Q_RP_R amount of money on rice and Q_WP_W money on wheat respectively. So, the consumer to purchase Q_R and Q_W quantities of both rice and wheat, he will have to spend $Q_RP_R + Q_WP_W$ amount of money. That means, the consumer can purchase this combination of commodities, if he possess at least $Q_RP_R + Q_WP_W$ amount of money. Of course, the consumer can prefer any combination of rice and wheat commodities at given prices and given income of the consumer, provided the costs incurred on them is less than or equal to the income he possess. So, we can infer that, the consumer can select any combination of commodities such that, $Q_RP_R + Q_WP_W = Y$. This inequality is called the consumer's budget constraint. So, a budget constraint means, what a consumer can purchase is constrained by his income level. The slope of the budget constraint measures the rate at which the consumer can trade off one commodity for another at the relative prices of the two commodities. So, budget constraint is determined by both the income of the consumer and the relative prices of the commodities.

The set of combinations of commodities available to consumer at his given budget is called the budget set and the space, which indicates the total budget set of the consumer below the price line, refers to budget space. So, in the Figure 9.14, the area POL represents the budget space. So, budget set represents the various combinations of rice and wheat commodities that the consumer can buy with his income at the given prices of commodities and it is shown through Table 9.3. Since, both the commodities are perfectly divisible, the consumer's budget set would consists of all combinations of rice and wheat, such that, Q_R and Q_W are any numbers greater than or equal to zero and $Q_RP_R + Q_WP_W = Y$. All the combinations of rice and wheat that fall below the price line in the positive quadrant are included in the budget set. But, on the price line, any combination of the rice and wheat commodities amounts to total money income or budget available with the consumer given by, $Q_RP_R + Q_WP_W = Y$, hence, the name Budget line. So, points below the budget line represent the combinations of rice and wheat commodities, which costs strictly less than income Y. The budget line, $Q_RP_R + Q_WP_W = Y$ can also be expressed with respect to Q_R as $(Y/P_R) - (Q_WP_W/P_R)$. Similarly, the budget line, $Q_RP_R + Q_WP_W = Y$ can also be expressed with respect to Q_W as $(Y/P_W) - (Q_RP_R/P_W)$. So, the budget line is a straight line with horizontal intercept Y/P_R and with vertical intercept Y/P_W. The horizontal intercept (Y/P_R) represents the combination of commodities that the consumer purchases, if he spends all his income on commodity rice *i.e.*, he purchases only rice and no amount of wheat. Similarly, the vertical intercept (Y/P_W) represents the combination of commodities that the consumer purchases, if he spends all his income on commodity wheat, *i.e.*, he purchases only wheat and no amount of rice (Figure 9.14).

9.3.1. Slope of the Price Line

The slope of the price line indicates the price ratio of the two commodities. The slope of price line can be studied through the following three methods:

Method 1

The slope of price line can be studied through the Figure 9.14. In the Figure 9.14, OL units of rice can be exchanged with OP units of wheat, as the money value with respect to OL units of rice and OP units of wheat is same. So, one unit of rice is equal to (OP/OL) units of wheat. This is the same thing as the tangent of angle PLO. So, the rate of exchange of the rice and wheat is thus given by tangent PL. This PL is the price line. The slope of the price line can be derived mathematically, as given below:

- ✰ Along X-axis, the total quantity of rice purchased by the consumer is OL at a given price of P_R, if the whole income of the consumer is allocated on the rice commodity. This is given by $Y = OL \times P_R$. So, $OL = Y/P_R$
- ✰ Similarly, along Y-axis, the total quantity of wheat purchased by the consumer is OP at a given price of P_W, if the whole income of the consumer is allocated on the wheat commodity. This is given by $Y = OP \times P_W$. So, $OP = Y/P_W$
- ✰ As discussed earlier and from the Figure 9.14, slope of the price line is given by OP/OL. So, from the equation 9.5 given below, the slope of price line indicates the price ratio of two commodities.

  $OP/OL = (Y/P_W)/(Y/P_R) = (P_R/P_W)$ *Equation 9.5*

In the given example, as the price per unit of rice is Rs.5 and price per unit of wheat is Rs.6, the slope of price line is 0.83. Since, the prices are assumed constant at a given time, the price line is a straight line.

Method 2

We know, the slope of the price line measures the amount of change in one commodity (say, wheat) due to per unit change in other commodity (say, rice) along the price line. In the Figure 9.15, let us consider two points A and B on the price line, where point A represents Q_R and Q_W combinations of two commodities and point B represents the movement down the price line given by $Q_R+\Delta Q_R$, $Q_W-\Delta Q_W$. So, the price line with reference to these two points A and B is given as follows:

At point A: $(P_R \times Q_R) + (P_W \times Q_W) = Y$ *Equation 9.6*

At point B: $(P_R \times (Q_R + \Delta Q_R)) + (P_W \times (Q_W - \Delta Q_W)) = Y$ *Equation 9.7*

Subtracting Equation 9.6 from Equation 9.7, we get,

$P_R\, \Delta Q_R - P_W\, \Delta Q_W = 0$ *Equation 9.8*

By rearranging terms in Equation 9.8, we get the slope of the price line as

$(\Delta Q_W/\Delta Q_R) = (P_R/P_W)$.

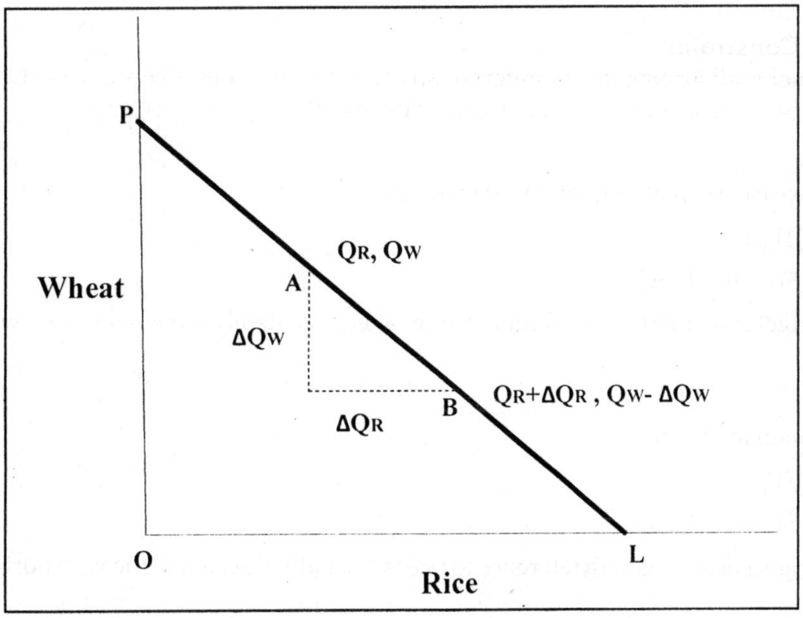

Figure 9.15: Method 2 – Estimation of slope of the price line.

Method 3

The slope of the price line can be simply estimated through computing the ratio between vertical intercept and horizontal intercept of the price line. As discussed earlier, we know, price line can be denoted as $(P_R \times Q_R) + (P_W \times Q_W) = Y$.

- ☆ ***Horizontal intercept of price line:*** The above price line can be expressed with respect to Q_R as
  $(P_R \times Q_R) = Y - (P_W \times Q_W)$.
  So, $Q_R = (Y/P_R) - (Q_W P_W/P_R)$ *Equation 9.9*
  In the above equation 9.9, (Y/P_R) is the horizontal intercept of the price line with respect to rice commodity.
- ☆ ***Vertical intercept of price line:*** Similarly, we can compute the vertical intercept of price line. The price line can be expressed with respect to Q_W as
  $(P_W \times Q_W) = Y - (P_R \times Q_R)$.
  So, $Q_W = (Y/P_W) - (Q_R P_R/P_W)$ *Equation 9.10*
  In the above equation 9.10, (Y/P_W) is the vertical intercept of the price line with respect to wheat commodity.
- ☆ So, slope of price line = Vertical intercept/Horizontal intercept = $(Y/P_W)/(Y/P_R) = (P_R/P_W)$

Notice that, the slope of the price line depends only on the ratio of the prices of two commodities and not on their absolute values. In this example, the price of rice is Rs. 5/kg and price of wheat is Rs. 6/kg. So, the ratio of the prices of rice and wheat is 1.0:1.2 and thereby, if the consumer wants to increase the consumption of rice by one more unit, he has to sacrifice 1.2 units of wheat. On the other hand, if the consumer increases the wheat consumption by one more unit, he has to sacrifice 0.83 units of rice. Similarly, if we consider prices of rice and wheat as Rs. 10/kg and Rs. 12/kg respectively, the consumer must still forego 1.2 units of wheat for increasing the consumption of one more unit of rice and 0.83 units of rice for increasing the consumption of one more unit of wheat. So, as long as the price of wheat is 1.2 times the price of rice, the consumer must forgo the same amounts, if he increases the consumptions of either rice or wheat by one more unit. This guides that, the slope of the price line is guided by the price ratio of the commodities rather than their absolute values. So, this concept also illustrates the opportunity cost of rice consumption vis-à-vis wheat consumption and it can be written as:

(P_R/P_W) = Opportunity cost of rice in terms of wheat or Opportunity cost of wheat in terms of rice.

So, the above explanation infers that, any change in income and changes in the prices of both rice and wheat in the same proportion will not affect the price ratio, thereby, the slope of price line will remain unaffected. This discussion, further helps to clarify the distinction between money prices and relative prices. Here, both P_R and P_W are the money prices of rice and wheat respectively, while the ratio P_R/P_W is a relative price.

It is important to note that, both IDC and price line have negative slopes, but the MRS indicated by the slope of IDC diminishes as we move down the curve, whereas, the price ratio indicated by slope of price line remains same throughout the price line.

Computation of Budget Constraint

For example, a consumer with income Rs. 60 prefers to buy two commodities, rice and wheat, at price Rs.5/kg of rice and Rs.6/kg of wheat. So, we can write the budget constraint as, $(P_R \times Q_R) + (P_W \times Q_W) = Y$.

Vertical Intercept

For ease of graphing, we can solve the equation for wheat as:

$$Q_W = (Y/P_W) - (Q_R P_R/P_W)$$

$$Q_W = (60/6) - (5Q_R/6) = 10 - 0.83Q_R$$

So, the consumer's budget constraint with reference to wheat is a negatively-sloped line with vertical intercept 10 and slope –0.83.

Horizontal Intercept

We can solve the equation for rice as:

$$Q_R = (Y/P_R) - (Q_W P_W/P_R)$$

$$Q_R = (60/5) - (6Q_W/5) = 12 - 1.20Q_W$$

So, the consumer's budget constraint with reference to rice is a negatively-sloped line with horizontal intercept 12 and slope –1.20.

9.3.2. Position and Shifts in Price Line

The position of price line from the origin indicates the money income with the consumer. The closer the price line to the origin, the lesser the money income with the consumer and *vice versa*. However, the position of price line with respect to origin is inversely influenced by the prices of the two commodities. That means, if the prices of both the commodities are increased, the price line moves towards origin and *vice versa*. So, both income of the consumer and prices of the commodities are the influential factors in determining the position of price line with respect to origin and their influences are discussed here under.

(a) Changes in Income and No Change in Prices of the Commodities

In this given situation, the price line shifts from the original position. With the rise in income, the price line shifts away from the origin *i.e.*, towards right side or upward in a parallel position. Conversely, with the fall in income, the price line shifts towards the origin *i.e.*, towards left side or downward in a parallel position.

A rise in the level of income of the consumer and no change in the prices of the commodities, will shift the price line towards right side or away from the origin from the original position. This is because, with increase in income, the consumer can purchase more of both the commodities than before, as the cost of one commodity in terms of the other remains the same. Conversely, a fall in the level of income and no change in the prices of the commodities, will shift the price line towards left side or towards origin from the original position. This is because, with decrease in income, the consumer can purchase less of both the commodities than before. Figure 9.16 reveals that, due to rise in money income with the consumer, the price line PL shifts away from the origin or parallel upwards as P_IL_I. So, the consumer is able to purchase more of both the commodities rice and wheat. Conversely, due to fall in income, the price line PL shifts towards the origin or parallel downwards as P_DL_D. So, the consumer will purchase less of both the commodities.

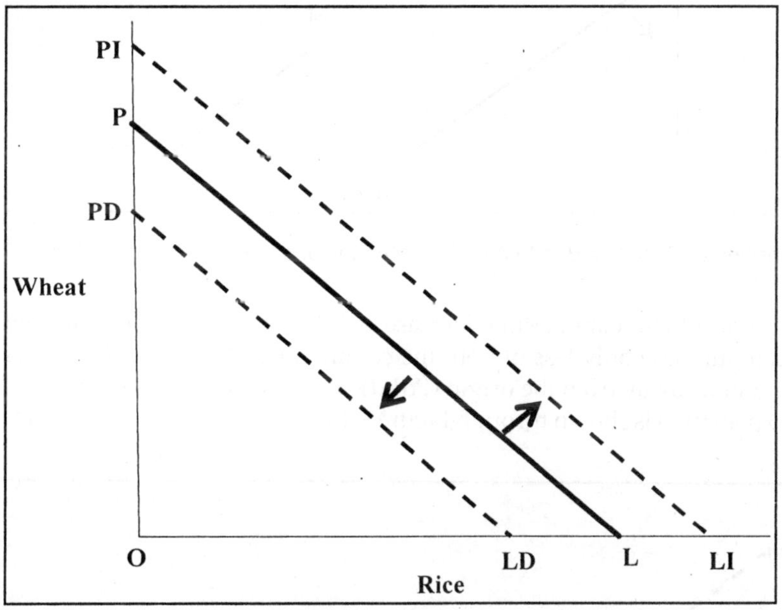

Figure 9.16: Shifts in price line due to changes in money income with the consumer.

(b) Changes in Prices and No Change in Income of the Consumer

Now, let us assume, the prices of the commodities under consideration *viz.*, rice and wheat fluctuate in the market, whereas, the income of the consumer remains same.

Changes in Price of Rice Commodity

In the first case, assume that, price of rice is increased in the market. Then the consumer purchases less of rice than before. So, the new price line (broken line in Panel A of Figure 9.17) becomes more steeper than before and touches the axis closer towards origin on which the commodity rice is considered (X-axis), indicating that, the relative price of rice is costlier.

Now assume that, the price of rice commodity is decreased in the market. Then the consumer purchases more of rice than before. So, the new price line (broken line in Panel A of Figure 9.17) becomes more flatter than before and touches the axis away from the origin on which the commodity rice is considered (X-axis), indicating that, the relative price of rice is cheaper.

Changes in Price of Wheat Commodity

In the first case, assume that, price of wheat is increased in the market. Then the consumer purchases less of wheat than before. So, the new price line (broken line in Panel B of Figure 9.17) touches the axis closer towards origin on which the commodity wheat is considered (Y-axis), indicating that, the relative price of wheat is costlier.

Now assume that, the price of wheat is decreased in the market. Then the consumer purchases more of wheat than before. So, the price line (broken line in Panel B of Figure 9.17) touches the axis away from the origin on which the commodity wheat is considered (Y-axis), indicating that, the relative price of wheat is cheaper.

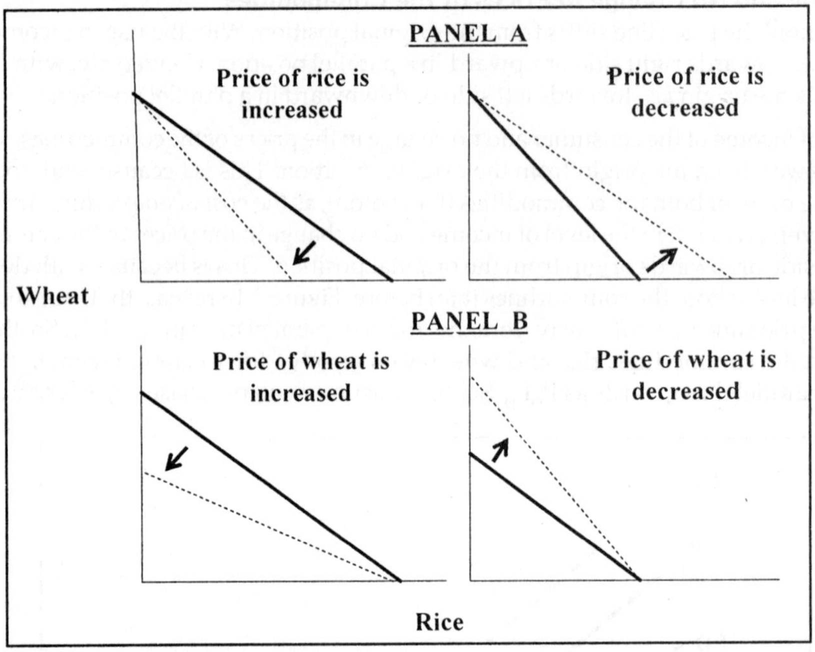

Figure 9.17: Shifts in the position of Price line due to changes in prices of one of the commodities.

In some cases, the prices of both the commodities increase in the market. So, the price line moves towards the origin (PiLi), as the consumer can purchase only less of both the commodities. However, in case of price fall of both the commodities, the price line shifts away from the origin (PdLd), as the consumer can now purchase more of both the commodities. The same explanation is shown through Figure 9.18.

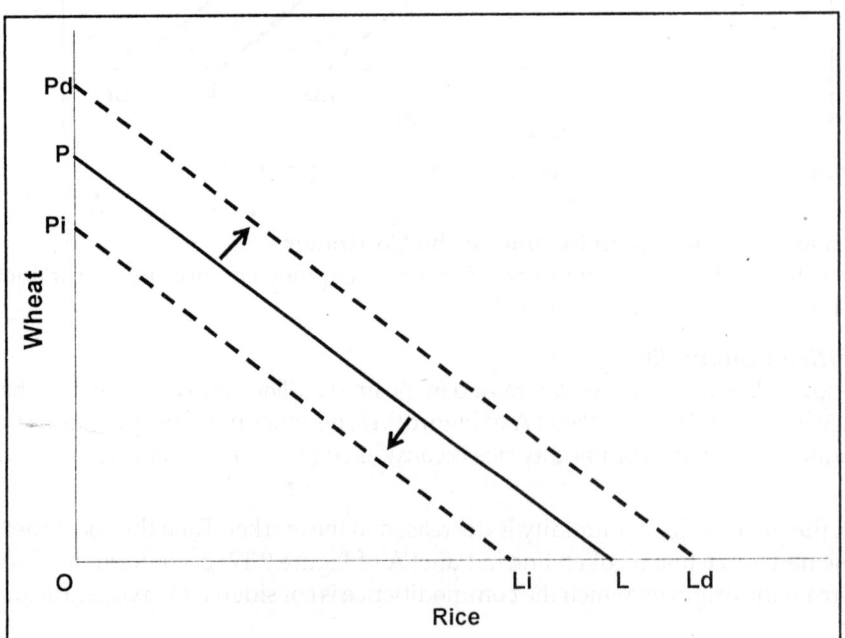

Figure 9.18: Shifts in price line due to fluctuations in prices of both the commodities.

The above analysis indicates that, both income of the consumer and prices of the two commodities are the two major determinants of the position of the price line from the origin. Further, the analysis advocates the importance of price line in determining the consumer's equilibrium.

9.4. Consumer's Equilibrium

For consumer's equilibrium, we should first know about the scale of preferences of the consumer for rice and wheat commodities. It guides the consumer to draw an indifference map and this forms the basis for analyzing the consumer's equilibrium. The above discussion on IDC and price line reveals two important aspects *viz.*, the IDC infers 'what the consumer want to do' or 'what the consumer is willing to buy' *i.e.*, it guides the consumer in drawing the scale of preferences of different combinations of commodities and the price line guides the consumer, 'what he can do' or 'what the consumer is able to buy' *i.e.*, it helps him to select any combination of commodities that lie below or on the price line, so as to maximize his satisfaction in the consumption process. So, when the IDC and the price line are combined, we find the quantities of each commodity the consumer is both willing and able to buy. This infers that, consumer face trade-offs in his purchase decisions and he tries out to make the best choice in selecting the right combination of commodities suiting to his income level, tastes and preferences and prices of the related commodities. So, in order to make the best choice of combination of two commodities (rice and wheat), the consumer must combine the IDCs of different levels of satisfaction (what combination he would like to consume) and price line (what he can afford). So, the indifference map and price line of the consumer helps to find out the combination of commodities that will lead to consumer's equilibrium. The aim of consumer is to attain maximum satisfaction and this is possible by reaching the highest IDC possible. So, the consumer's equilibrium refers to the point of maximum satisfaction derived by the consumer at a particular combination of commodities, given his income and prices of the commodities in the market. The consumer is said to be in equilibrium at a point, where the price line touches the highest attainable IDC in an indifference map from below.

9.4.1. Assumptions

The following assumptions are made to arrive at the point of the consumer's equilibrium:

- ✰ ***Consistency of choice:*** This is the basic assumption, as the consumer should be consistent in making choices with respect to two commodities under consideration.
- ✰ ***Rationality:*** The consumer should be rational in selecting the best combination of commodities, so as to derive maximum satisfaction, at the given income and prices of commodities.
- ✰ ***Utility is ordinal:*** The measurement of utility should be in ordinal terms *i.e.*, the utility cannot be quantified rather, it can be ranked from low to high according to the level of satisfaction derived from the consumption of different quantities of commodities.
- ✰ ***Perfect competition:*** There is perfect competition in the product market, such that, the commodities are homogenous and without price fluctuations.
- ✰ ***Tastes, habits and preferences of the consumer:*** They are assumed to be stable throughout the analysis, so that, the commodities and their combinations preferred by the consumer will not alter.
- ✰ ***No change in income:*** The income of the consumer should remain same, such that, the consumer's preferential combinations of commodities will not alter. The consumer can spend his income in small amounts also.
- ✰ ***Given indifference map:*** The consumer's indifference map is given and there should not be any change in the scale of preferences of the combinations of commodities.
- ✰ ***Good substitutes:*** Commodities under consideration should be good substitutes, so that, the consumer can increase the consumption of one commodity at the expense of other commodity and this also facilitates to select the best combination of commodities.

9.4.2. Conditions to be Fulfilled to Arrive at Consumer's Equilibrium

The following three conditions should be fulfilled to arrive at consumer's equilibrium and they are discussed in-detail here under:

1. The price line should be tangent to the IDC from below
2. Slope of the price line should be equal to the slope of IDC
3. IDC should be convex to the origin

1. The Price Line Should be Tangent to the IDC from Below

The consumer's equilibrium position is only at a point, where the price line is tangent to the highest attainable IDC from below. This is explained through the Figure 9.19, where an indifference map comprises of three IDCs, I_1, I_2 and I_3

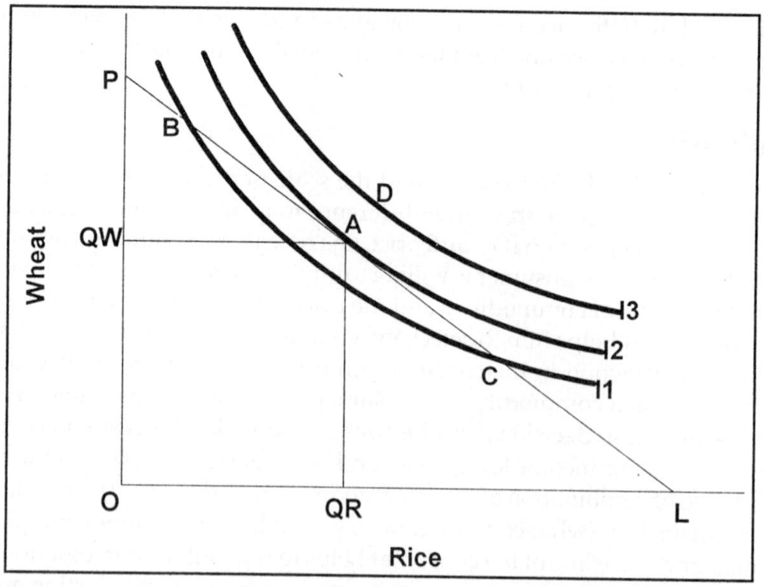

Figure 9.19: Consumer's equilibrium – Price line tangent to highest attainable IDC from below.

and PL is the price line of the consumer for his given income and given prices of rice and wheat commodities. This price line PL is tangent to the IDC I_2 at point A, where the consumer derives maximum satisfaction by consuming OQ_R quantity of rice and OQ_W quantity of wheat. Even though, the price line touches the IDC I_1 at points B and C, they are not the points of maximum satisfaction, as IDC I_1 yields less level of satisfaction compared to the IDC I_2. Moreover, price line is also touching higher IDC *i.e.*, I_2, thereby, definitely the consumer moves to higher IDC I_2 to derive much more satisfaction. As regards to point D on IDC I_3, no doubt, the consumer derives higher satisfaction compared to IDC I_2, but that combination of rice and wheat commodities is not attainable by the consumer with his price line. To put it simply, the income of the consumer is not sufficient enough to purchase the combination of rice and wheat commodities at point D on IDC I_3. That means, any combination of commodities lie above the price line is unattainable. So, we can infer that, point A on IDC I_2, where price line is tangent to the highest attainable IDC I_2 from below is the point of consumer's equilibrium *i.e.*, by consuming OQ_R quantity of rice and OQ_W quantity of wheat commodities, the consumer can derive maximum satisfaction.

For example, a consumer is having an income of Rs. 60. He prefers to have two commodities *viz.*, rice and wheat. The price of rice is Rs. 5/kg and the price of wheat is Rs. 6/kg. Given the income and the prices of the two commodities, the consumer prefers to have 13 combinations of two commodities *i.e.*, A to M and they are shown through Table 9.3. In the given Figure 9.20, the price line touches the highest possible IDC I_2 at A, where the consumer consumes six kilograms of rice and five kilograms of wheat and at this combination of commodities, he is satisfying his budget equation *i.e.*,

$$(P_R \times Q_R) + (P_W \times Q_W) = Y$$
$$= (5 \times 6) + (6 \times 5) = 60$$
$$= 60 = 60.$$

2. Slope of the Price Line Should be Equal to the Slope of IDC

The second condition essential for arriving at the point of consumer's equilibrium is that, the slope of the price line should be equal to the slope of highest attainable IDC from below. As explained through Figure 9.20, the price line PL of the consumer is tangent to the highest attainable IDC I_2 from below at point A. So, at point A on IDC I_2, the slope of the price line is equal to the slope of the IDC I_2, where the consumer consumes OQ_R and OQ_W quantities of rice and wheat respectively. We know, the slope of IDC indicates MRS of rice for wheat and slope of price line indicates price ratio of two commodities and the point of consumer's equilibrium is explained below.

$$\text{Slope of the IDC} = MRS_{RW} = (\Delta W)/(\Delta R)$$

$$\text{Slope of the Price line} = (P_R/P_W)$$

So, the point of consumer's equilibrium is when the slope of the IDC is equal to the slope of the price line and this is given by,

$$(\Delta W)/(\Delta R) = (P_R/P_W) \text{ or } MRS_{RW} = (P_R/P_W) \qquad \textit{Equation 9.11}$$

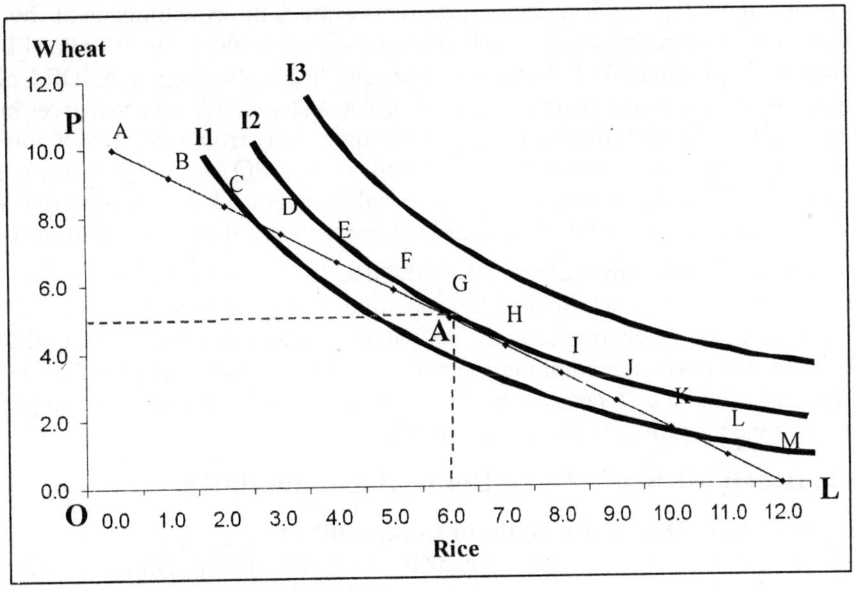

Figure 9.20: Consumer's equilibrium for rice and wheat combinations.

The above equation 9.11 infers that, at point A, what the consumer is willing to pay *i.e.*, his personal exchange rate between rice and wheat commodities (MRS_{RW}) is equal to what he actually pays *i.e.*, the market exchange rate P_R/P_W. So, at the point of consumer's equilibrium, the slope of IDC is equal to the slope of price line or the MRS_{RW} is inversely equal to the price ratio of the rice and wheat commodities. The point of consumer's equilibrium can also be stated that, the rate at which the consumer is willing to substitute rice commodity for wheat commodity is inversely equal to the price ratio between rice and wheat commodities.

3. IDC Should be Convex to the Origin

This is an important condition to be fulfilled for arriving at the point of consumer's equilibrium. When the price line is tangent to the highest attainable IDC, at the point of tangency, the IDC should be convex to origin and this is a necessary condition for consumer's equilibrium. In other words, the MRS of rice for wheat must be diminishing at the point of equilibrium. The same is depicted through the Figure 9.21, where at the point of consumer's equilibrium *i.e.*, at A, the shape of the IDC I_2 is convex to origin.

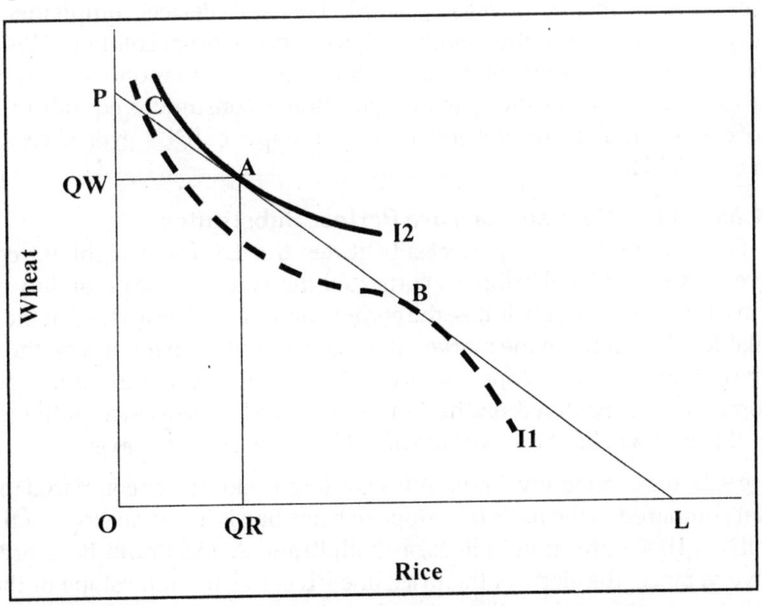

Figure 9.21: Consumer's equilibrium – Price line tangent to highest attainable convex IDC.

In the Figure 9.21, the price line PL touches IDC I_1 at points B and C, and IDC I_2 at point A. Since the point A lies on higher IDC I_2, the consumer definitely moves to higher IDC I_2. Moreover, a close examination of the Figure 9.21 shows that, at the point of tangency of price line and IDC I_1 at point B, the IDC I_1 is concave to the origin, where it implies MRS_{RW}

is increasing rather than diminishing. So, it gives the impression that, the consumer is at the minimum point of satisfaction at B, and if the consumer moves away from B towards either axis along the price line PL, he can increase his satisfaction and he can go even to higher IDC I_2. More over, the price line is also tangent to IDC I_1 at point C, where the IDC I_1 is convex to origin, but the consumer will move to higher IDC I_2 at point A, where it gives higher satisfaction to the consumer compared to IDC I_1. So, definitely the consumer moves away from point B and point C and reaches the higher IDC I_2 and comes to equilibrium at point A, where he purchases OQ_R and OQ_W quantities of rice and wheat respectively. So, this gives the information that, the consumer will be in equilibrium when price line is tangent to the highest possible IDC and that too, when that IDC is convex to origin, where it shows diminishing MRS of commodities.

The above explanation clearly infers that, at point A on the IDC I_2 (Figure 9.21), all the above three conditions are fulfilled or satisfied and hence, we can conclude that, the consumer is in equilibrium at point A, through consuming OQ_R and OQ_W quantities of rice and wheat respectively. So, we can define the consumer's equilibrium as a situation, where the consumer purchases a particular combination of two commodities with a given level of his income and at given prices of the commodities, such that, he is not willing to move away from the attained equilibrium situation or show no tendency to re-arrange his purchases for the commodities.

9.4.3. Inequality between MRS and Price Ratio of Commodities

Case 1: When MRS_{RW} is Greater than Price Ratio of Commodities

This is given by, $(\Delta W)/(\Delta R) > (P_R/P_W)$. So, to attain consumer's equilibrium or point of maximum satisfaction, the consumer should consume more of rice or less of wheat, such that, the equality between MRS and Price ratio of commodities is ensured and thereby, the consumer is at the point of equilibrium.

Case 2: When MRS_{RW} is Less than Price Ratio of Commodities

This is given by, $(\Delta W)/(\Delta R) < (P_R/P_W)$. So, to attain consumer's equilibrium or point of maximum satisfaction, the consumer should consume more of wheat or less of rice, such that, the equality between MRS and Price ratio of commodities is ensured and thereby, the consumer is at the point of equilibrium.

9.4.4. Interior and Corner Solutions of Consumer's Equilibrium

The explanation given earlier regarding the consumer's equilibrium with reference to three important conditions refers to Interior solution of the consumer's equilibrium. This is because, a close examination of the Figures 9.19 to 9.21 reveals, point A is the consumer's equilibrium and this point 'A' lies within or interior of the commodity space. The economic implication of this Interior solution is that, the consumption pattern of the consumer is diversified, as he is prefers the related commodities in combinations. This is the practical phenomenon we commonly see regarding the consumer behaviour in the economy. However, this interior solution is possible, when the IDC is an ordinary convex shaped curve. But, depending upon the relationships between the commodities, there are different shapes of IDCs such as straight line (Perfect substitutes), concave (increasing MRS), 'L' shape (Perfect complements) etc. The consumer's equilibrium achieved in such cases of commodities relationships refers to Corner solution. This is so called because, the consumer comes to equilibrium at one extreme of the IDC, where he consumes only one commodity instead of two commodities that are open to him. As a result, the essential condition of consumer's equilibrium *i.e.*, $MRS_{RW} = P_R/P_W$ is not fulfilled in such cases of corner equilibrium. Not only in special shapes of IDCs, even in case of convex shaped IDCs also, the corner's solution is possible.

1. Corner Solution, in Case, if the Commodities are Perfect Substitutes

As discussed earlier, if the commodities are perfect substitutes, the IDC is a straight line connecting the two axes. In such case, we can achieve consumer's equilibrium by corner solution only, as the possibility of tangency of price line to the IDC is not possible. In case of perfect substitutes, the corner solution will be arrived at the extreme of IDC, for the commodity, which is available at low price in the market. If the commodity considered on the X-axis is offered at low price in the market, then the corner solution of consumer's equilibrium will be at the extreme of IDC with respect to X-axis. Conversely, if the commodity considered on the Y-axis is offered at low price in the market, then the corner solution of consumer's equilibrium will be at the extreme of IDC with respect to Y-axis.

In Panel A of the Figure 9.22, the commodity, Brand A tea is offered at low price compared to Brand B tea and hence, the price line PL is more flat compared to the IDCs (*i.e.*, slope of price line is less than slope of IDCs). This price line PL will touch the IDC I_1 and IDC I_4 (IDCs are straight lines, as both Brand A and Brand B tea are perfect substitutes) at points 'P' and 'L' respectively. Since, the slope of the price line PL is less than the slope of the IDCs, then price line touches the highest possible IDC I_4 at 'L' and it is the point of consumer's equilibrium, *i.e.*, at one extreme of the IDC I_4 inferring corner solution with reference to Brand A tea on X-axis. Even though, the price line touches the IDC I_1 at 'P' on the Y-axis with reference to Brand B tea, but the consumer definitely moves to higher IDC I_4 *i.e.*, at point 'L', where he derives maximum satisfaction through corner solution. So, by arriving at corner solution of consumer's equilibrium,

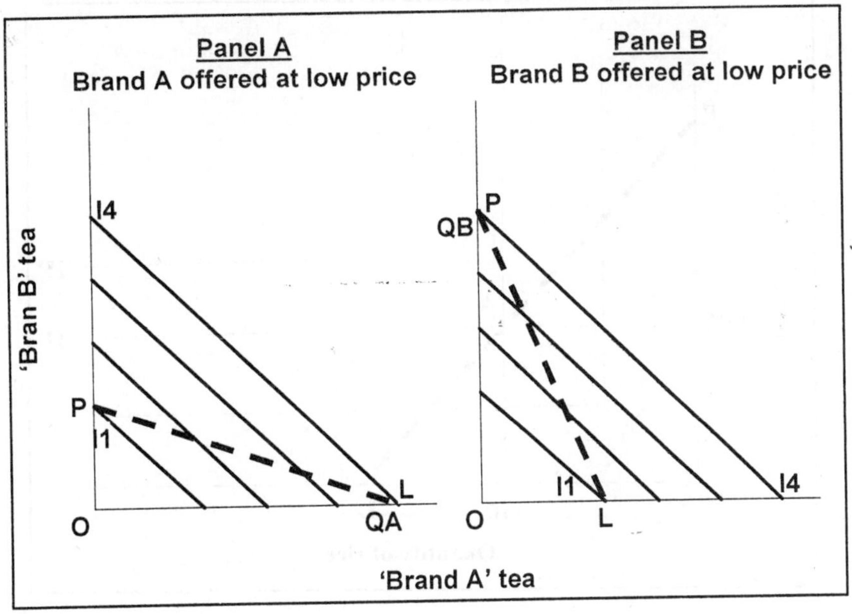

Figure 9.22: Corner solution of consumer's equilibrium – Case of Prefect substitutes.

the consumer purchases only Brand A tea on X-axis and none of Brand B tea on Y-axis. That means, consumer derives maximum satisfaction by corner solution by purchasing OQ_A quantity of Brand A tea and nothing of Brand B tea.

In Panel B of the Figure 9.22, the commodity, Brand B tea is offered at low price compared to Brand A tea. The price line PL is more steep or less flat compared to the IDCs (*i.e.*, slope of price line is more than slope of IDCs). This price line PL will touch the IDC I_1 and IDC I_4 at points 'L' and 'P' respectively. Since the slope of the price line PL is more than the slope of the IDCs, then price line touches the highest possible IDC I_4 at 'P' and it is the point of consumer's equilibrium, *i.e.*, at one extreme of the IDC I_4 inferring corner solution with reference to Brand B tea on Y-axis. Even though, the price line touches the IDC I_1 at 'L' on the X-axis with reference to Brand A tea, but the consumer definitely moves to higher IDC I_4 *i.e.*, at point 'P', where he derives maximum satisfaction through corner solution. So, by arriving at corner solution of consumer's equilibrium, the consumer purchases only Brand B tea on Y-axis and none of Brand A tea on X-axis. That means, consumer derives maximum satisfaction by corner solution by purchasing OQ_B quantity of Brand B tea and nothing of Brand A tea.

It is important to note in the above two cases, the consumer will not be in equilibrium between P and L on the price line. This is because, in case of Panel A, the extremity of price line PL *i.e.*, 'L' lie on the highest possible IDC I_4, where as, in case of Panel B, the extremity of price line PL *i.e.*, 'P' lie on the highest possible IDC I_4. So, as a result, in case of perfect substitutes, there will be a corner solution to the consumer's equilibrium, where the consumer will succumb to Monomania, *i.e.*, purchases only one commodity.

2. Corner Solution, in Case, if the Commodities are Perfect Complements

In case of perfect complements (say, rice and water in cooking rice) too, interior solution is not possible, as the price line cannot be tangent to a point of the right angled or L shaped IDCs. So, the price line will cut the right angled IDC at its corner point, such that, at no other point of IDC in the map, this consumer's equilibrium is possible. As shown in the Figure 9.23, the price line PL cuts the IDC I_2 at its corner point A and the consumer will be in equilibrium by purchasing OQ_R quantity of rice and OQ_W volume of water in cooking rice.

3. Corner Solution in Case of Concave Shaped IDCs

If the commodities substitute at increasing rate, then the shape of IDC is concave to origin. However, such increasing MRS of commodities is less practically seen. For example, web surfing hours and cricket playing hours substitute at increasing rate, as discussed under Panel B of Figure 9.5. When, the commodities substitute at increasing rate, the consumer will attain the point of equilibrium at one corner or extreme of the IDC and this is depicted through Figure 9.24.

As shown in the Figure 9.24, the IDCs I_1 to I_4 are concave in shape indicating increasing MRS and different levels of satisfaction. PL is the price line of the consumer and it is tangent to the IDC I_1 at point A. But, here the consumer is not in equilibrium because, at point A, the shape of the IDC is concave to the origin. This implies that, at point A, the consumer experiences minimum satisfaction and if he moves along the price line PL, he can increase and maximize his

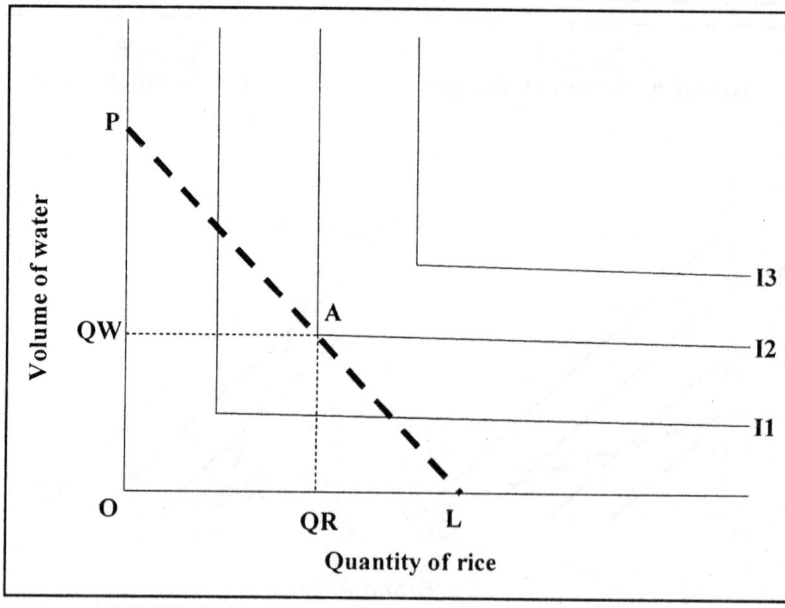

Figure 9.23: Corner solution of consumer's equilibrium – Case of Prefect complements.

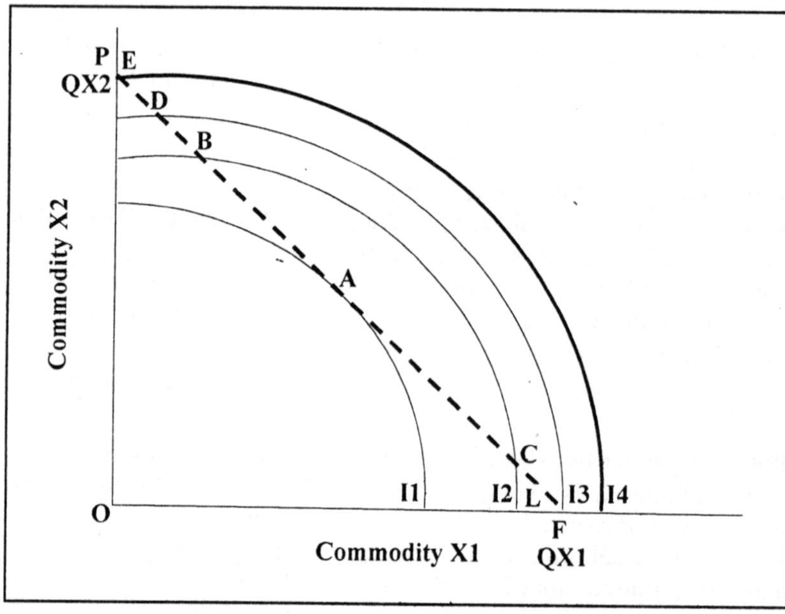

Figure 9.24: Corner solution of consumer's equilibrium – Case of Increasing MRS.

satisfaction, as the consumer can move to higher IDCs. This point 'A' is called as Point of minimum satisfaction by Professor J.R. Hicks. Accordingly, the consumer moves to higher IDC I_2, where the PL is tangent to IDC I_2 at points B and C. But, definitely he will move to point D on IDC I_3, where he can further increase his satisfaction. However, at point D, the situation is same as in situation A of IDC I_1, where the IDC I_3 is concave to origin, but the consumer is on the higher IDC I_3. So, if the consumer still moves along the price line PL, he can reach the higher IDC I_4 and the PL touches the IDC I_4 at one corner (Y-axis) at point E. So, the consumer attains the point of equilibrium at one corner of the IDC I_4, where he purchase OQ_{X2} quantity of commodity X_2 and none of commodity X_1. On the other hand, if the consumer wishes to purchase X_1 instead of X_2, his budget line will touch the IDC I_3 at F and thereby, he attains equilibrium by purchasing OQ_{X1} of commodity X_1 and none of commodity X_2. So, in these two cases, the consumer prefers only one commodity and none of other commodity. If the consumer prefers to have commodity X_2, he consumes OQ_{X2} quantity of commodity X_2, on the extremity of price line PL, *i.e.*, 'P', which touches the highest possible IDC I_4 at point E. But, if the consumer prefers to have commodity X_1, he consumes OQ_{X1} quantity of commodity X_1, as the extremity of the price line PL, *i.e.*, 'L' which touches the highest possible IDC I_3 at point F.

4. Corner Solution in Case of Convex Shaped IDCs

Even in case of convex shaped IDCs, there is possibility of getting corner solution of consumer's equilibrium. This is especially true in case of commodities like television, fridge, tractor, processing machinery etc., where such commodities are not so easy to purchase, as they are offered at high prices in the market economy. Even in such cases also, the IDC analysis helps us to draw the consumer's equilibrium.

For example, see Panel A of the Figure 9.25, where interior solution is not possible, as the price line is not tangent to any IDC, even when the curvature of IDC is convex to origin. Rather, the price line touches the IDC I_2 at the corner or extreme end at point A, indicating that, OQ_Y of commodity Y is purchased and none of commodity X is purchased. This is indicated by the fact that, since the price of commodity X is so high, the price line is much steeper than the IDCs of the two commodities. In economic terms, we can infer that, the MRS of X for Y is less than price ratio of X and Y commodities. This is given by. $MRS_{XY} < P_X/P_Y$ or $\Delta Y/\Delta X < P_X/P_Y$. That means, price of X is so high compared to MRS_{XY}. So, to ensure equilibrium, the consumer has to purchase more of Y or less of X and so we have corner solution of consumer's equilibrium at point A on IDC I_2, indicating that, the consumer reaches corner equilibrium without purchasing even one unit of commodity X.

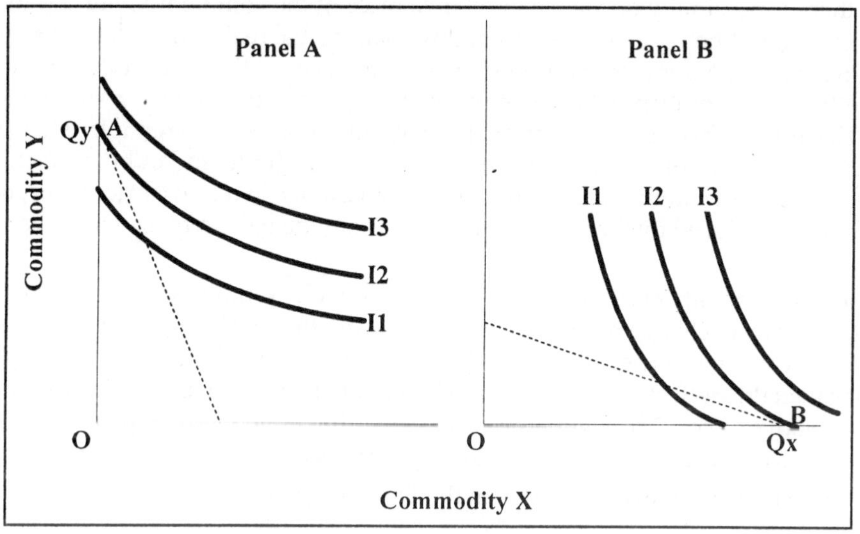

Figure 9.25: Corner solution of consumer's equilibrium – Case of Diminishing MRS.

In Panel B of the Figure 9.25, again the interior solution is not possible and thereby, we have corner solution for consumer's equilibrium at point B of the IDC I_2 on X-axis, indicating only commodity X is purchased and none of commodity Y is purchased. Here, the price of the commodity Y is very high and hence, the price line is more flat or less steeper compared to IDCs of the two commodities and it touches the IDC I_2 at point B on X-axis indicating the consumer's equilibrium. In economic terms, we can infer that, the MRS of X for Y is greater than price ratio of X and Y commodities. This is given by, $MRS_{XY} > P_X/P_Y$ or $\Delta Y/\Delta X > P_X/P_Y$. That means, price of X is so small compared to MRS_{XY} or price of Y is so high compared to MRS_{XY}. So, to ensure consumer's equilibrium, the consumer has to purchase more of X or less of Y and thereby, we have corner solution of consumer's equilibrium at point B on IDC I_2, indicating that, the consumer reaches corner equilibrium without purchasing even one unit of Y.

The analysis of corner solution of consumer's equilibrium with respect to both concave and convex shaped IDCs reveals the following interesting points.

- ☆ In case of concave shaped IDCs, the corner solution of consumer's equilibrium is inevitable. That means, only one commodity has to be purchased at the sacrifice of the other commodity.
- ☆ However, in case of convex shaped IDCs, the corner solution of consumer's equilibrium is not an inevitable or a compulsory situation. The corner solution is possible when the MRS of the commodities is too high or too low compared to the price ratio of the commodities. If MRS of two commodities is greater than price ratio of the commodities, the consumer will prefer to purchase the commodity considered on X-axis alone and none of the commodity considered on Y–axis. On the other hand, if the MRS of two commodities is less than price ratio of the commodities, the consumer will prefer to have the commodity considered on Y-axis and none of the commodity considered on X-axis.
- ☆ The concavity of the IDCs will make the consumer to purchase a single commodity in the consumption process. But, this is not the practical situation, as we generally observe the consumer will prefer the commodities in combinations. So, this gives the impression again that, IDCs of the consumers are mostly convex in shape.

9.5. Economic Effects in Consumer's Equilibrium

In arriving at the consumer's equilibrium, we generally make the following three important assumptions:

- ☆ Income of the consumer remains constant.
- ☆ Prices of the two commodities remains constant.
- ☆ Tastes, habits and preferences of the consumer remains constant.

But in practical situation, the above assumptions will not hold good, as they vary very frequently. Hence, it is essential to study the variations in the above assumptions, so as to ascertain their influence on consumer's equilibrium or point of maximum satisfaction. Thus, there are three main ways in which the conditions underlying the consumer's equilibrium in the indifference map gets disturbed.

- ☆ First, there is possibility of change in consumer's income, but prices of the commodities remain constant. Due to change in the money income, the consumer may become better-off (increase in income) or worse off (decrease in income). The result of this type of change is called as Income effect.
- ☆ Second, there is possibility of change in relative price of the commodity, but the consumer's real income remains constant. When price of the commodity changes, the real income of the consumer also changes. To keep the real income of the consumer constant, so as to trace out the effect of change in relative price alone on the quantity of commodity purchased, the price change is compensated by a simultaneous change is money income. The result of this type of change is called Substitution effect. For example, when price of the commodity X falls, the real income of the consumer will increase. In order to find out the substitution effect alone *i.e.*, change in quantity of commodity X purchased due to change in relative price alone, the consumer's money income must be reduced by the amount that cancels out the gain in real income that resulted due to fall in price.
- ☆ Third, there is possibility of change in price of the commodity, but money income of the consumer remains constant. But, due to change in price of the commodity, the real income of the consumer changes. In such situation, the consumer experiences both Income effect and Substitution effect. The result of this type of change is called Price effect. So, price effect includes both income effect and substitution effect. Thus, in price effect, the relative price of the commodity will change, but there is no compensating variation in income. So, the consumer's real income changes due to change in price of the commodity.

  We can summarize the above discussion in the Table 9.4.

Table 9.4: Economic effects of consumer's equilibrium.

Type of Effect	*Money Income*	*Price*	*Real Income*
Income effect	Changes	Constant	Changes
Substitution effect	Changes	Changes	Constant
Price effect	Constant	Changes	Changes

9.5.1. Income Effect-Income Consumption Curve (ICC)

We know that, the income of an individual vary from time to time and whenever there is an increase in income, the consumer feels much better off and purchase more quantities of commodities, thereby, derives more satisfaction and *vice versa*. This effect of change in income on the changes in quantities of commodities purchased or demanded, but assuming prices of the commodities and tastes and preferences of the consumer constant refers to Income effect. It is otherwise, defined as, the effect on the various quantities of the commodities purchased by the consumer due to changes in his income, but keeping the prices of the commodities and his tastes and preferences constant. So, when there is increase in income, the consumer's equilibrium point shifts from lower IDC to higher IDC and *vice versa*. This relationship between different levels of income and different quantities of commodities consumed can be traced out through ICC. This is explained through the Figure 9.26.

Suppose, there are two commodities rice and wheat and for the given level of income of the consumer and given prices of these two commodities, P_1L_1 is the price line. This price line P_1L_1 touches the IDC I_1 at point Q_1 (consumer's equilibrium), where the consumer purchases OM_1 quantity of rice commodity and ON_1 quantity of wheat commodity.

Let us suppose, the income of the consumer is increased (prices of the commodities and tastes and preferences of the consumer remains same), as a result, the consumer feels much better-off and hence, he purchases large quantities of both rice and wheat. So, P_2L_2 is the new price line of the consumer, which lies parallel to P_1L_1. This new price line touches the higher IDC I_2 at point Q_2 (consumer's equilibrium), where the consumer purchases OM_2 quantity of rice and

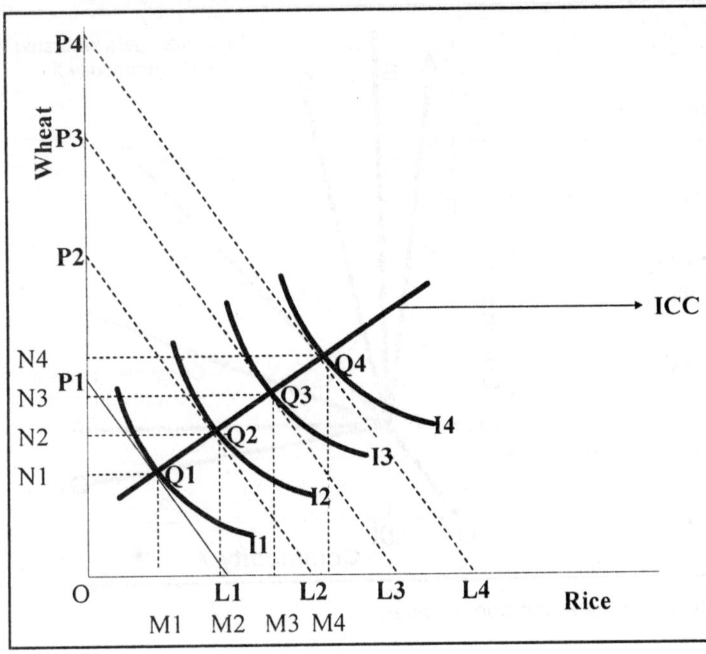

Figure 9.26: Upward sloping ICC – Case of a normal good.

ON_2 quantity of wheat. That means, if the income of the consumer increases, he purchases more of both rice and wheat commodities ($OM_2 > OM_1$ of rice commodity and $ON_2 > ON_1$ of wheat commodity).

If the income of the consumer is increased, P_3L_3 is the new price line and it touches still higher IDC I_3 at Q_3 (consumer's equilibrium), where the consumer purchases OM_3 quantity of rice and ON_3 quantity of wheat commodity. If the income of the consumer is further increased, P_4L_4 is the new price line and it touches still higher IDC I_4 at Q_4 (consumer's equilibrium), where the consumer purchases OM_4 quantity of rice and ON_4 quantity of wheat commodity. That means, if income goes on increasing with the consumer, he moves from lower IDC to higher IDC and hence, he will attain consumer's equilibrium at the highest possible IDC.

Now, if we connect all the consumer's equilibrium points *i.e.*, Q_1, Q_2, Q_3 and Q_4 by a line or curve, it indicates ICC. So, this ICC will trace out the income effect on the different quantities of commodities consumed by the consumer at different levels of income. It is otherwise defined as the locus of consumer's equilibrium points at various levels of income. The ICC curve slopes upward from left to right in the north-easterly direction, if both the commodities are 'normal'.

9.5.1.1. Types of Income Effect

Income effect can work to either increase or decrease in the quantity of a commodity demanded, depending on whether the commodity is normal or inferior. Accordingly, the income effect can be categorized into Positive income effect and Negative income effect.

- ☆ ***Positive income effect:*** Income effect is said to be positive, when more quantities of a commodity are purchased by the consumer due to rise in his income. In Figure 9.26, the income effect is positive with reference to both the commodities, as more quantities of both the commodities are purchased due to increase in income of the consumer. This positive income effect is seen in case of normal goods for which, a change in income causes a positive change in demand. That is, an increase in income causes an increase in demand and a decrease in income causes a decrease in demand for normal goods. So, the E_Y for a normal good is positive (greater than zero). So, if there is a positive income effect with reference to both the commodities (normal goods), the ICC will slope upwards, as shown by ICC curves of Figure 9.26 and C, D and E curves of Figure 9.27.
- ☆ ***Negative income effect:*** Income effect is said to be negative, when there is decrease in quantities of a commodity purchased, even though, there is rise in income on the part of the consumer. This negative income effect is seen in case inferior goods. For example, agricultural labour generally consumes sorghum or bajra or ragi as their staple food. But, when their income increases, they switch over to superior commodities like rice or wheat and spend less on sorghum or bajra or ragi, considering them as inferior goods. These goods are considered 'inferior' in the sense relating to affordability rather than a statement about the quality of the goods. So, when the income increases, the consumer prefers 'superior' goods to these inferior goods. But,

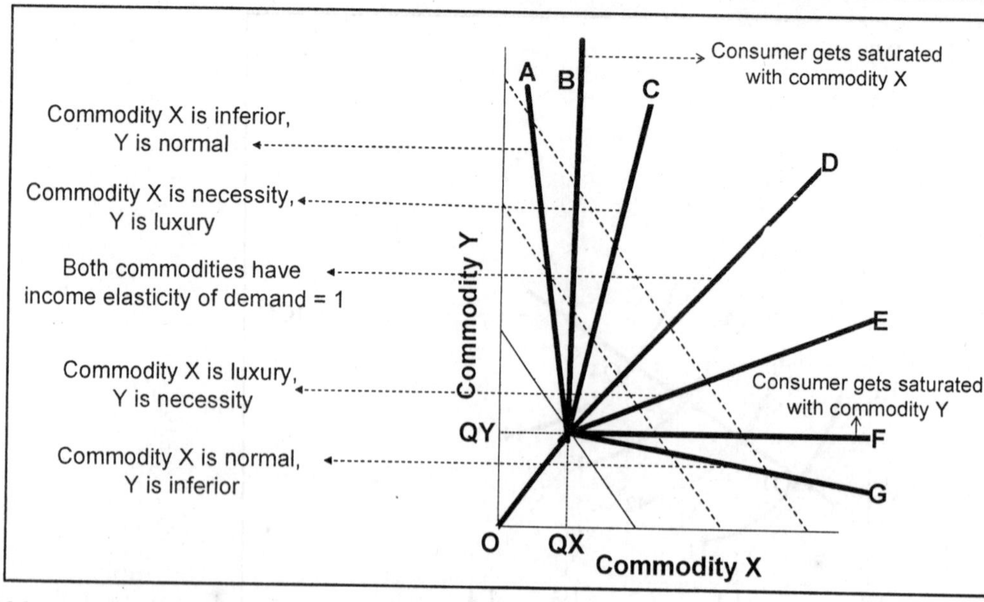

Figure 9.27: ICCs for different relationships between the commodities.

when the income of the consumer is low, they could not afford to purchase the superior goods and hence, they prefer inferior goods. The other examples of inferior goods were already discussed in Chapter 2). This negative income effect is seen in case of inferior goods for which, a change in income causes an opposite change in demand. That is, an increase in income causes a decrease in demand and a decrease in income causes an increase in demand for the inferior goods. So, the E_Y for an inferior good is negative. So, if there is a negative income effect with reference to commodity X (inferior good), the ICC will bend towards Y-axis, on which the other commodity Y (normal good) is taken (ICC 'A' in Figure 9.27), indicating that, more of commodity Y and less of commodity X are consumed with increase in income of the consumer. On the other hand, if the income effect is negative with reference to commodity Y (inferior good), the ICC will bend towards X-axis (ICC 'G' in Figure 9.27), on which the commodity X (normal good) is taken, indicating that, the consumer consumes more of commodity X and less of commodity Y with increase in income.

So, the concept of ICC helps to study the consumption behaviour of a consumer with reference to both normal goods and inferior goods, but the demerit is that, it cannot explain why the income effect is negative for an inferior good.

9.5.1.2. Different Shapes of ICCs

Different shapes of ICCs (Figure 9.27) are possible depending upon the nature of commodities under consideration and they are discussed here under.

Case 1: When Commodity X is Inferior and Commodity Y is Normal

As shown through the Figure 9.27, the ICC 'A' bends towards Y-axis on which normal good is considered, indicating that, with increase in income of the consumer, less of commodity X is purchased being inferior and more of commodity Y is purchased. That means, the E_Y is negative for commodity X (Figure 9.28), being an inferior good and it is positive for commodity Y, being a normal good.

Case 2: When Commodity X is Normal and Commodity Y is Inferior

In this case, the ICC 'G' bends towards X-axis on which normal good is considered, indicating that, with increase in income of the consumer, more of commodity X is purchased and less of Y is purchased, being inferior (Figure 9.27). That means, the E_Y is negative for commodity Y, being an inferior good and it is positive for commodity X (Figure 9.28), being a normal good.

It is important to note that, up to a certain level of consumption, the inferior good also behave like a normal good and after reaching that consumption level, the good become inferior to the consumer and this is shown through Figure 9.29. If commodity X is considered as an inferior good, upto OQ_X level of its consumption, it behaves like a normal good, but beyond that OQ_X level, the commodity X becomes inferior to the consumer and thereby, he purchases less and less, even as the income of the consumer increases. This is shown by the bending of ICC 'A' towards Y-axis indicating that, as the income of the consumer increases, the consumer upto OQ_X quantity of X consumes more of the commodity X, but later the quantity consumed of commodity X decreases even with the increase in the level of income of the consumer. But, regarding commodity Y, with increase in the level of income of the consumer, the quantity purchased of commodity Y increases continuously.

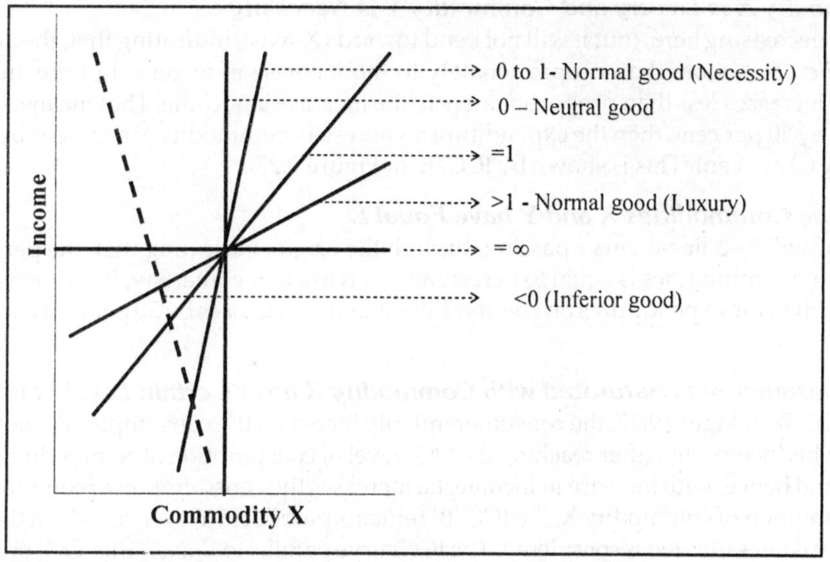

Figure 9.28: Income elasticity of demand for commodity X.

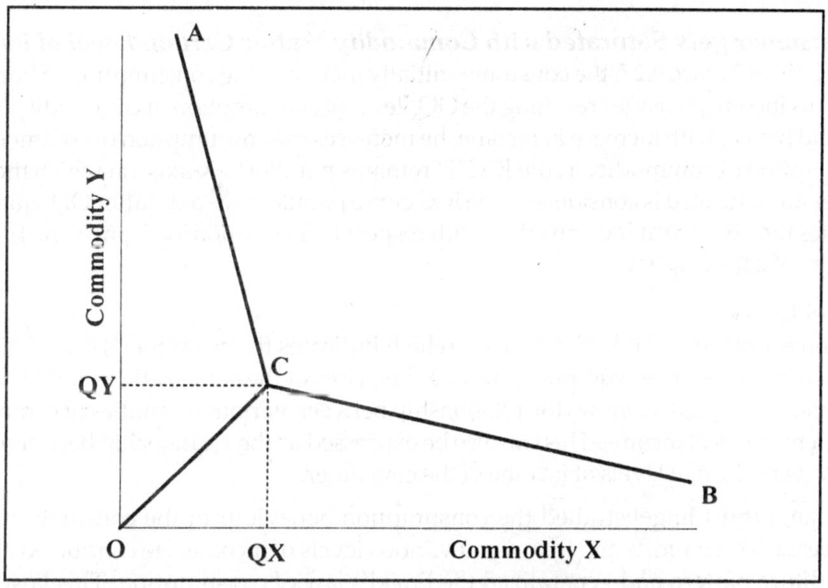

Figure 9.29: ICCs in case of normal and inferior goods.

On the other hand, if the commodity Y is considered as inferior and commodity X is a normal good, now the ICC 'B' will bend downward towards right *i.e*, towards the X-axis and commodity Y acts like a normal good upto OQ_Y level of consumption, beyond which it behaves like an inferior good. The other commodity X being a normal good, its consumption increases through out, as the income of the consumer increases.

Further, in the category of normal goods, there may be necessity and luxury goods. Necessity goods includes basic food commodities like rice, wheat, edible oils, sugar etc., and luxury goods includes jams, burger, pizza, butter, ice cream etc. For such necessary and luxury goods, the ICCs assume different shapes as discussed under the following cases:

Case 3: When Commodity X is Necessity and Commodity Y is Luxury

The ICC will rise upward, (but it will not bend towards Y-axis), indicating that, the amount spent on the luxury commodity Y increases more than proportionately to the increase in income, but the amount spent on the necessity commodity X also increases, but less than proportionately to the increase in income, thereby, ICC shows increasing slope upward. That means, say, if the income of the consumer increases by 50 per cent, then the expenditure on necessity commodity X increases by 20 per cent and on luxury commodity Y, by 60 per cent. This is shown by ICC 'C' of Figure 9.27.

Case 4: When Commodity X is Luxury and Commodity Y is Necessity

The slope of ICC is decreasing here, (but it will not bend towards X-axis), indicating that, the amount spent on the luxury commodity X increases more than proportionately to the increase in income, but the amount spent on the necessity commodity Y increases less than proportionately to the increase in income. That means, say, if the income of the consumer increases by 50 per cent, then the expenditure on necessity commodity Y increases by 20 per cent and on luxury commodity X by 60 per cent. This is shown by ICC 'E' of Figure 9.27.

Case 5: When Both the Commodities X and Y have Equal E_Y

In this case, the ICC will be a linear curve passing through the origin, indicating that, the percentage increase in expenditure on each of the commodities is equal to percentage increase in income. Say, if the income of the consumer increases by 50 per cent, then the expenditures on commodities X and Y increases by 50 per cent each. This is shown by ICC 'D' of Figure 9.27.

Case 6: When the Consumer gets Saturated with Commodity X after Certain Level of its Consumption

As shown by the ICC 'B' of Figure 9.27, the consumer initially increases the consumption of both the commodities X and Y with increase in his income, but after reaching the OQ_X level of consumption of commodity X, he gets saturated with the commodity X and hence, with increase in income, he increases the consumption of commodity Y only. Hence, after OQ_X level of consumption of commodity X, the ICC 'B' remains parallel to Y-axis, on which the other commodity Y for which the consumer is not saturated is considered. The ICC curve parallel to Y-axis (after OQ_X quantity consumption of commodity X) indicates the absence of income effect with respect to X commodity. Hence, the E_Y of commodity X is zero after its OQ_X quantity of consumption.

Case 7: When the Consumer gets Saturated with Commodity Y after Certain Level of its Consumption

As shown by the ICC 'F' of Figure 9.27, the consumer initially increases the consumption of both the commodities X and Y with increase in his income, but after reaching the OQ_Y level of consumption of commodity Y, he gets saturated with the commodity Y and hence, with increase in income, he increases the consumption of commodity X only. Hence, after OQ_Y level of consumption of commodity Y, the ICC 'F' remains parallel to X-axis, on which the other commodity X for which the consumer not saturated is considered. The ICC curve parallel to X- axis (after OQ_Y quantity consumption of commodity Y) indicates the absence of income effect with respect to Y commodity. Hence, the E_Y of commodity Y is zero after its OQ_Y quantity of consumption.

9.5.1.3. ICC and Engel Curve

The above discussion reveals that, an ICC is a curve, which indicates the consumer's equilibrium with reference two commodities at different levels of income, but assuming the prices of the commodities and tastes and preferences of the consumer held constant. So, an ICC shows the relationship between various quantities of commodities consumed by the consumer at different levels of income. This can also be expressed as the relationship between various quantities of commodities purchased at different levels of income of the consumer.

A German Statistician, Ernest Engel studied the consumption behaviour of the consumer on various types of commodities such as necessaries, comforts and luxuries at various levels of income. He conducted a number of studies regarding family expenditure pattern and generalized the Engel's law of consumption. This law was enunciated in Europe, but it has got universal application. This Engel's law comprises of three important propositions as given below

- *As income increases, the percentage or proportionate expenditure on the basic necessaries of life will decrease.*
- *As income increases, the percentage or proportionate expenditure on rent, fuel, light etc., will remain more or less same.*
- *As income increases, the percentage expenditure on comforts and luxuries will increase.*

Based on the above three propositions, we can generalize that, the Engel's law studies the relationship between various quantities of different commodities purchased at various levels of income of the consumer. So, there lies same concept behind the ICC and Engel's law and this helps to derive Engel's curve from ICC. So, an Engel's curve is the curve derived from ICC, which proves Engel's law indicating various quantities of commodities purchased by the consumer at different levels of income. Modern Economists also study the relationship between the expenditures incurred on the commodities at various levels of income and the curve so drawn is called Engel's expenditure curve. But, in Engel's curve, we study the relationship between various quantities of commodities purchased by the consumer at different levels of income. This analysis is discussed through Figure 9.30 in two Panels A and B. In Panel A of the Figure 9.30, the ICC is derived as discussed under Figure 9.26 for two commodities rice and wheat.

In Panel B the Figure 9.30, to derive Engel's curve, income of the consumer is taken on Y-axis and commodity rice is taken on X-axis. As the price lines drawn in Panel A represents the income levels of the consumer, the same income levels are taken along Y-axis in Panel B. As we assumed the prices of the commodities remains constant, the ratio of an income level of the consumer and the price of the commodity rice directly reveals the quantity of rice purchased by the consumer for a given level of income. So, the quantities of rice purchased by the consumer at different levels of income

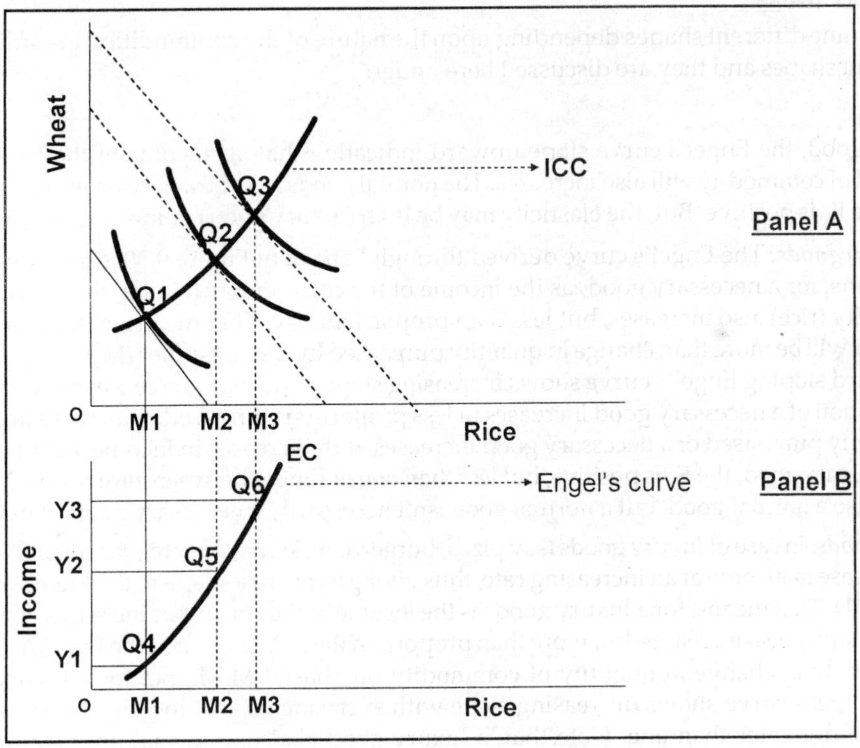

Figure 9.30: Derivation of Engle's curve from ICC.

will be computed and they are shown on X-axis. The equilibrium positions achieved in Panel A *i.e.*, Q_1, Q_2 and Q_3 with reference to ICC were extended downward to the lower Panel B and the corresponding equilibrium positions in Panel B are Q_4, Q_5 and Q_6 respectively, which represent different quantities of rice commodity purchased *viz.*, OM_1, OM_2 and OM_3 respectively at different levels of income OY_1, OY_2 and OY_3. If the points Q_4 Q_5 and Q_6 are connected, it represents Engel's curve 'EC' indicating various quantities of commodity rice purchased at different levels of income of the consumer.

Likewise, we can also derive Engel's curve for wheat commodity, taking income on the Y-axis and quantities of wheat commodity purchased on X-axis. It appears from both the Panels 'A' and 'B' that, both ICC and Engel's curve are identical, but they are different. This is because, in deriving ICC, we consider two commodities rice and wheat along X-axis and Y-axis respectively, whereas, in Engel's curve we consider income levels of consumer on Y-axis and quantities purchased of only one commodity (either rice or wheat) on X-axis.

Properties of ICC

The following are the properties or features of the ICC, as evident from the above Figures 9.26 to 9.30.

- The ICC gives the information about the various quantities of commodity consumed by the consumer at various levels of income, but keeping the prices of the commodities, tastes and preferences of the consumer constant.
- ICC can also be expressed as the relationship between various quantities of a commodity purchased at various levels of income of the consumer and this guides to derive E_Y for the commodity. Thus, an ICC helps to measure the E_Y indirectly. The shape of the ICC drawn between the two commodities infers the degree of E_Y and also indicates the nature of the commodities under consideration (Figure 9.28).

  Necessary goods: For necessary goods, the E_Y ranges from zero to one.

  Luxury goods: For luxury goods, the E_Y is more than one.

  Neutral goods: For neutral goods, the E_Y is equal to zero.

  Inferior goods: For inferior goods, the E_Y is negative or less than zero.
- It helps to derive an Engel's curve for the commodity.
- The ICC connects the points of consumer's equilibrium at various income levels of the consumer. These equilibrium points infer that, at different income levels, the consumer is deriving maximum satisfaction in consuming different quantities of related commodities. So, the pattern of equilibrium points on the ICC indicates the optimum pattern of consumption by the consumer at different income levels of the consumer.

Shapes of Engel's Curves

Since IC· ·ssume different shapes depending upon the nature of the commodities, accordingly, Engel's curves too assume different shapes and they are discussed here under.

1. Normal Goods

For a normal good, the Engel's curve slope upward indicating that, as income of the consumer increases, the quantity purchased of commodity will also increases. The normal goods may be a necessary good or a luxury good. For a normal goods, the E_Y is positive. But, the elasticity may be high (luxury good) or low (necessary good).

(a) *Necessary goods:* The Engel's curve derived through Panel B of Figure 9.30 is the case of a necessary good. That means, for a necessary good, as the income of the consumer increases, the quantity purchased of the commodity (rice) also increases, but less than proportionately. That means, change in income (Y_2Y_3) of the consumer will be more than change in quantity purchased by the consumer (M_2M_3). So, for a necessary good, the upward sloping Engel's curve shows increasing slope with every increase in income, as the quantity of consumption of a necessary good increases in less proportion compared to increase in income. That means, the quantity purchased of a necessary good increases with increases in income, but at a decreasing rate. So, for a necessary good, the E_Y is positive, but less than one or ranges between zero to one. Note that, a necessary good is also a normal good, but a normal good isn't necessarily a necessary good (it may be a luxury good).

(b) *Luxury goods:* In case of luxury goods (say pizza, burger, jam, ice creams etc), the quantity purchased increases with increase in income at an increasing rate, thus giving a concave shape to the Engel's curve (Panel A of the Figure 9.31). That means, for a luxury good, as the income of the consumer increases, the quantity purchased of a commodity also increases, but more than proportionately. As shown in the Panel A, the change in income Y_2Y_3 is less than change in quantity of commodity purchased M_2M_3. So, for a luxury good, the upward sloping Engel's curve shows decreasing slope with every increase in income. For a luxury good, the E_Y is positive and greater than one. Note that, a luxury good is also a normal good, but a normal good isn't necessarily a luxury good (it may be a necessary good).

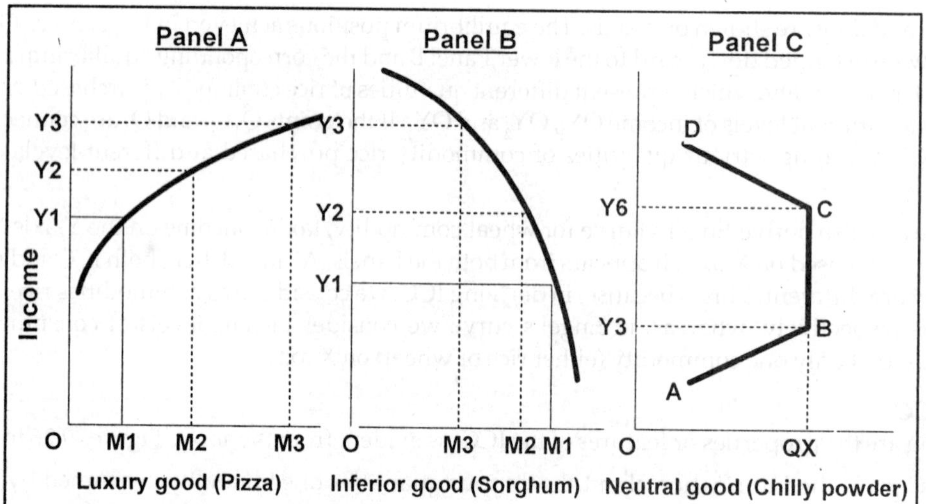

Figure 9.31: Engel's curves for different commodities.

2. Inferior Goods

In case of inferior goods such as sorghum, bajra, ragi etc., with increase in income of the consumer, the quantity consumed will decrease and hence, the Engel's curve is a backward bending curve, as shown in the Panel B of the Figure 9.31. As the income of the consumer increases from OY_2 to OY_3, the quantity of commodity purchased decreases from OM_2 to OM_3, indicating inverse relationship between income of the consumer and quantity of commodity purchased. So, for an inferior good, the E_Y is negative or less than zero.

3. Neutral Goods

A neutral good is a good for which the demand is income-inelastic *i.e.*, the E_Y is equal to zero. That means, the quantity of commodity purchased by the consumer remains same irrespective of the level of income of the consumer. So, the Engel's curve for a neutral good remains parallel to Y-axis. For example, the commodities like salt, chilli powder, edible oil, hair oil, tooth paste etc., are considered as neutral goods, whose quantity of purchase remains same irrespective of the level of income of the consumer. As shown in the Panel C of the Figure 9.31, the portion of the curve between B and

C segments represents the Engel's curve for a neutral good. This indicates that, even if the income of the consumer increases from OY_3 to OY_6, the quantity purchased by the consumer remains same *i.e.*, OQ_X. The Panel C also shows that, the portion AB of the Engel's curve represents the commodity is a normal good and between C and D, the commodity is an inferior good.

9.5.1.4. Derivation of Income Demand Curve from ICC

An ICC indicates various quantities of commodities consumed by the consumer at various levels of his income, but keeping the prices of the commodities and tastes and preferences of the consumer remains constant. The income demand also explains the same as explained by the ICC. Hence, we can derive an income demand curve from ICC.

As discussed in the Figure 9.26, we derived the ICC for the two commodities rice and wheat. To study the income demand curve with respect to rice commodity, from the ICC drawn, we extend the equilibrium points Q_1, Q_2, Q_3 and Q_4 to the lower Panel B as Q_5, Q_6, Q_7 and Q_8 equilibrium points, thereby, we arrive at the different quantities of rice commodity purchased at different income levels of the consumer, but at the same price of the rice commodity. As shown in the Panel B of the Figure 9.32, the income demand curve for rice shifts towards right, indicating that, due to increase in income of the consumer, more quantity of rice is demanded at the same price. A close examination of the Panel A and Panel B of the Figure 9.32 reveals that, in Panel A, the quantities of both the commodities rice and wheat are taken on X-axis and Y-axis respectively and in Panel B, on X-axis, the quantity of commodity under consideration (rice) is taken and on Y-axis, the price of the commodity rice is taken. Like wise, we can also derive income demand curve for wheat, by taking quantity of wheat purchased on X-axis and price of wheat on Y-axis.

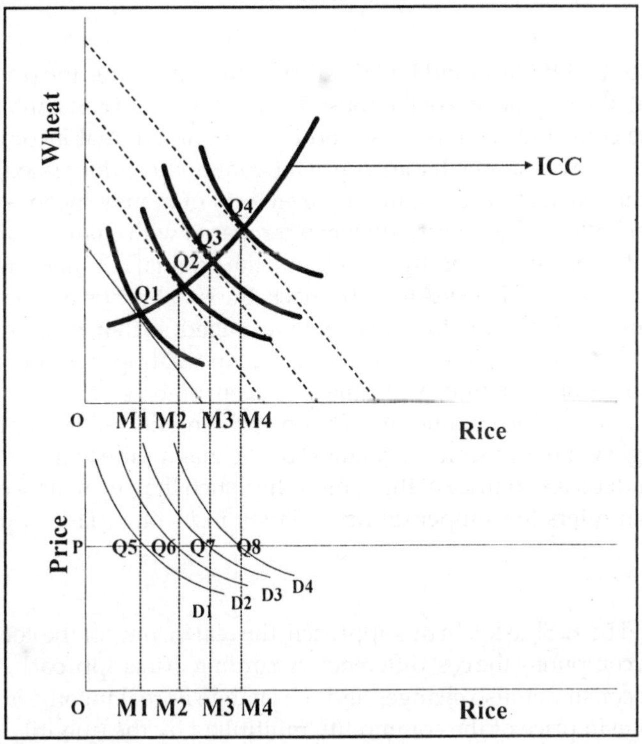

Figure 9.32: Derivation of income demand curve from ICC.

9.5.2. Substitution Effect

Whenever there is a change in price of the commodity, the tendency of the consumer is to substitute a cheaper commodity for a costlier or dearer commodity. For example, when there is an increase in price of a commodity say rice, the consumer substitute wheat for rice and thereby, he purchases more of commodity wheat. Similarly, when there is decrease in price of the commodity rice, now the consumer psychologically feels wheat is costlier in the market and thereby, he purchases more of rice and he substitutes rice for wheat. So, when price decreases (increases), substitution effect works to increase (decrease) quantity demanded of that commodity. This process of substitution of one commodity for another due to change in price of one of the commodities refers to substitution effect. It also refers to the change in the quantity purchased of a commodity due to change in the relative price alone, but keeping the real income of the consumer constant. To be more accurate, the substitution effect refers to the change in quantity purchased by the consumer due to change in price of one of the commodities and which involves substitution of a relatively cheaper commodity for a costlier commodity, keeping the price of the other commodity, real income and tastes and preferences of the consumer as constant. In the above definition, 'assuming real income constant' is an important aspect. This is because, say, if the price of the commodity rice is decreased in the market, the consumer purchases more of rice and substitutes rice for wheat for two reasons:

- ☆ The fall in price of rice makes the commodity more cheaper than wheat and thereby, the demand for the commodity rice extends and so, he substitutes rice for wheat (Substitution effect).
- ☆ Due to fall in the price of rice, now the consumer psychologically feels his real income was increased. So, the rise in real income also makes the consumer to purchase more quantity of rice (Income effect).

So, in order to compute the true substitution effect, we have to keep the real income of the consumer constant. In case of fall in price of the commodity, the real income of the consumer will increase and to trace out the substitution effect alone, the money income of the consumer should be reduced by an amount, so as to cancel out the indirect increase in real income and thereby, the real income of the consumer will remain constant. Conversely, in case of rise in price of the commodity, the real income of the consumer will decrease and to trace out the substitution effect alone, the money income of the consumer should be increased by an amount, so as to cancel the indirect decrease in real income and the real income of the consumer will remain constant.

There are two approaches to isolate the substitution effect from price effect *viz.*, Hicks-Allen approach and Slutsky approach. These two approaches also guide to decompose or split the price effect into income effect and substitution effect. They differ in their concepts with regard to nullifying the change in real income of the consumer back to the original level, so that, the substitution effect is isolated from the price effect and this facilitates to compare substitution effect and income effect due to changes in price of the commodity.

1. Hicks-Allen Approach

This approach was proposed by J.R.Hicks and R.G.D.Allen. In this approach, the price change is accompanied by change in money income, so that, the real income of the consumer is maintained constant. If the price of the commodity is decreased, real income of the consumer will increase and to maintain the real income constant, some amount of money income is taken away from the consumer. If the price of the commodity is increased, real income of the consumer will decrease and to maintain the real income constant, some amount of money income is given to the consumer. By doing so, the consumer is neither better-off nor worse-off when compared with changed priced situation. Since, the real income is maintained constant, the consumer is brought back onto the original IDC and the movement on the same IDC refers to substitution effect. Thus, in case of Hicks-Allen approach, the substitution effect will take place on the original IDC. The amount by which the money income of the consumer is changed, so that, he is neither better-off nor worse-off when compared with changed priced situation is referred as Compensating variation in income. It is, otherwise, explained as the change made to the original money income of the consumer, so that, it is just sufficient to compensate the consumer for the price variation in the commodity. That means, in Hicks-Allen approach, to trace out the true substitution effect alone, the original money income is changed by the magnitude to the extent of indirect change in real income either due to increase or decrease in price of the commodity, such that, it ensures or maintains the constancy of real income and this adjustment refers to Compensating variation in income. This approach is discussed in-detail through the Figure 9.35.

2. Slutsky Approach

This approach was proposed by E. Slutsky. In this approach, the real income of the consumer is brought back to the normal or original level, through computing the cost difference. According to this approach, when price of the commodity changes, the real income of the consumer also changes and the change in real income is equal to the cost difference, which is computed as the change in price of the commodity multiplied by the quantity of commodity purchased at initial or old price. In case, if the price of the commodity is decreased, the real income of the consumer will increase and it is adjusted back to the normal level, by deducting the cost difference from the increased real income of the consumer in the changed (decreased) price situation. Conversely, if the price of the commodity is increased, the real income of the consumer will decrease and it is adjusted back to the normal level, by adding the cost difference to the decreased real income of the consumer in the changed (increased) price situation. But, through adjusting the changed real income of the consumer in terms of cost difference (say, in case of fall in price of the commodity), the consumer will move to higher IDC, where he purchases more of cheaper commodity and less of dearer commodity. Thus, in case of Slutsky approach, the substitution effect will not take place on the original IDC, but on higher IDC in case of fall in price of the commodity and on lower IDC in case of rise in price of the commodity. This approach is discussed in-detail through the Figure 9.36.

The above discussion reveals that, a decline in price of the commodity always leads to increase in quantity demanded of the cheaper commodity and hence, the substitution effect is always negative. That means, the substitution effect will act in the opposite direction of price and thereby, it leads to increase in quantity purchased of the cheaper commodity.

9.5.3. Price Effect-Price Consumption Curve (PCC)

We know the prices of the commodities fluctuate frequently in the market and the existence of stable prices is only an abnormal or assumed situation. Let us suppose, price of rice commodity falls in the market, thereby, the consumer purchases more of rice commodity and less of wheat commodity. This effect of change in price of the commodity on the quantity of its purchase refers to price effect. It is otherwise defined as, the effect on the purchases of various quantities of a commodity due to change in price of that commodity, but keeping the price of other commodity, income level and tastes, habits and preferences of the consumer remains same.

When there is a change in price of commodity, the consumer feels either better-off (when price falls) or worse-off (when price increases) and accordingly, he shifts to a higher IDC or to a lower IDC respectively. Thus, price effect is shown as an extension and contraction of demand for a commodity when its price changes, but keeping the price of other commodity constant. The graphical illustration (Figure 9.33) shows that, at the given level of income of the

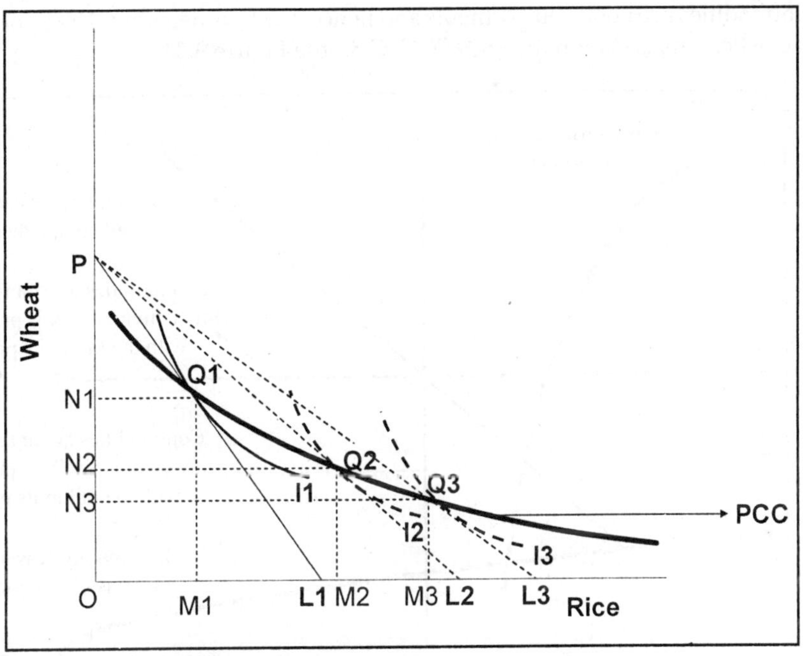

Figure 9.33: Downward sloping PCC – Case of a normal good.

consumer and at the given prices of the commodities (rice and wheat) and tastes and preferences of the consumer, PL_1 is the price line and it touches the IDC I_1 at Q_1 (consumer's equilibrium), where the consumer purchases OM_1 quantity of rice and ON_1 quantity of wheat commodities respectively.

Let us suppose, the price of rice commodity is deceased in the market and hence, the consumer prefers more of rice and less of wheat. So, the consumer wishes to substitute rice for wheat, due to decline in price of rice. So, PL_2 is the new price line and it touches the higher IDC I_2 at Q_2 (consumer's equilibrium), where the consumer purchases OM_2 quantity of rice and ON_2 quantity of wheat respectively. So, with the fall in price of rice commodity, the consumer prefers more of rice commodity (OM_2>OM_1) and he prefers less of wheat commodity (ON_2<ON_1). Assume, price of rice is still decreased in the market and hence PL_3 is the new price line and it touches the higher IDC I_3 at Q_3 (consumer's equilibrium), where the consumer purchases OM_3 quantity of rice and ON_3 quantity of wheat respectively. This increase in quantity of commodity rice purchased *i.e.*, M_1M_2, M_2M_3 due to fall in price of rice commodity refers to Price effect. This infers that, with the fall in price of rice commodity, the consumer prefers a cheaper commodity to a dearer commodity. The line connecting the equilibrium points Q_1, Q_2 and Q_3 constitute the PCC. It refers to the line connecting the consumer's equilibrium points at various levels of prices of rice commodity, but assuming the price of other commodity (wheat), income level and tastes, habits and preferences of the consumer remains same. Thus, a PCC traces out the price effect of the commodity.

Properties of PCC

The following are the properties or features of the PCC, as evident from the Figure 9.33:

- ☆ The PCC gives the information about the various quantities of a commodity purchased by the consumer at various levels of prices of that commodity, but keeping the price of other commodity, income of the consumer, tastes and preferences of the consumer constant. So, a PCC indicates the demand curve for a commodity indirectly.

☆ The relationship between various quantities of a commodity purchased at various levels of its price infers the E_P for the commodity. So, PCC helps to measure the E_P indirectly. It also helps to ascertain the E_C between the commodities. The shape of the PCC drawn between the two commodities infers the degree of price and cross elasticities of demand and also indicates the nature of the commodities under consideration. Depending upon the relationship between the two commodities, the PCC assumes different shapes and they are discussed below.

PCC slopes downward from left to right: If the PCC slopes downward smoothly from left to right (Figure 9.33), it implies both the commodities are good substitutes and also indicates the E_P is high with reference to the commodity for which the price is decreased. The commodities under consideration are normal goods and for a normal good, the E_P is more than one. Since, both the commodities are good substitutes, the E_C is positive.

PCC rises upward towards right: It implies, with the decrease in price of a commodity X or commodity Y, more of the quantities of both the commodities *i.e.*, both X and Y are purchased by the consumer. This indicates, both the commodities are good complements and hence, the E_C is negative. The E_P of commodity X is less than one or less elastic. This is shown through PCC 'C' in the Figure 9.34.

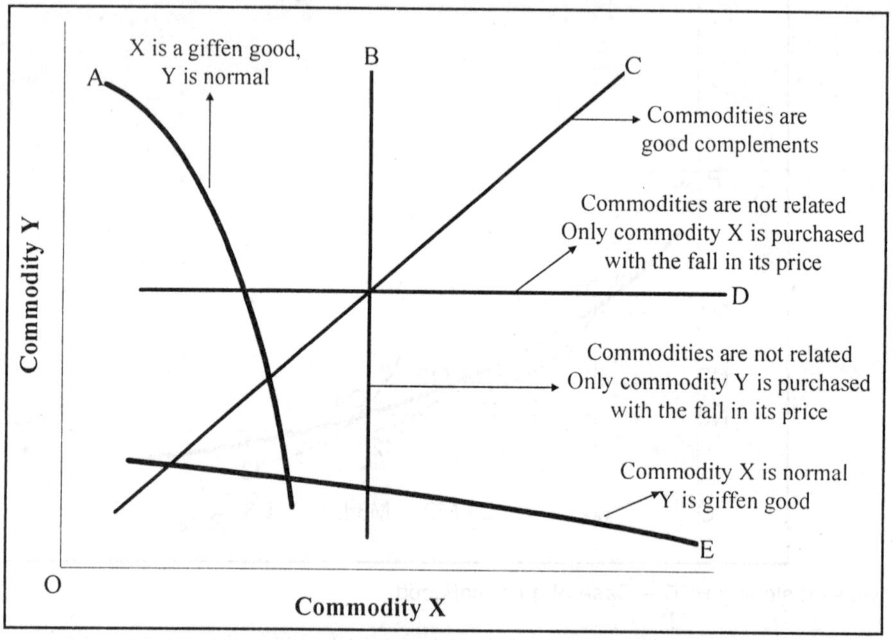

Figure 9.34: PCC for different relationships between the commodities.

PCC parallel to X-axis: It implies, with the fall in price of the X commodity, only commodity X is purchased by the consumer and none of commodity Y is purchased, indicating that, the two commodities are not related, thereby E_C is zero. The E_P with respect to commodity X is unity. This is shown through PCC 'D' in the Figure 9.34.

PCC parallel to Y-axis: It implies, with the fall in price of the Y commodity, only commodity Y is purchased by the consumer and none of commodity X is purchased, indicating that, the two commodities are not related, thereby E_C is zero. The E_P with respect to commodity Y is unity. This is shown through PCC 'B' in the Figure 9.34.

PCC bends towards X-axis: It implies that, with the fall in price of the Y commodity, less and less of it is purchased and more and more of commodity X is purchased. This indicates that, commodity Y is a giffen good. The E_C between X and Y is negative. This is shown through PCC 'E' in the Figure 9.34.

PCC bends towards Y-axis: It implies that, with the fall in price of the X commodity, less and less of it is purchased and more and more of commodity Y are purchased. This indicates that, commodity X is giffen good. The E_C between X and Y is negative. This is shown through PCC 'A' in the Figure 9.34.

☆ The PCC connects the points of consumer's equilibrium at various prices of a commodity. These equilibrium points infer that, at different prices of a commodity, the consumer is deriving maximum satisfaction in consuming different quantities of related commodities. So, the pattern of equilibrium points on the PCC indicates the optimum pattern of consumption by the consumer at different prices of the commodity over a period of time.

The price effect is considered as one of the important achievements of IDC analysis. This is because, it is an improvisation over the earlier Marshallian analysis, in the sense that, in IDC analysis, we study price effect of a commodity in terms of both income effect and substitution effect. In case of Marshallian analysis, only price effect was studied due to change in price of the commodity and income effect was not separated from the substitution effect. With regard to decomposition of price effect into income effect and substitution effect as separate components, the IDC analysis gained popularity in studying the consumer's behaviour. This concept of price effect was first studied by J.R. Hicks. So, price effect is the result of two other effects *viz.*, Income effect and Substitution effect and hence, we can infer that, Price effect=Income effect + Substitution effect.

Income Effect

It refers to the change in quantity purchased by the consumer due to change in his real income, as a result of change in price of the commodity. When the price of the commodity falls, the consumer feels he is better-off now, as his purchasing power increases in the market. So, he can purchase more of the commodity for which the price is decreased and this indicates, increase in his real income due to fall in price of the commodity. This increase in quantity of commodity purchased due to indirect increase in real income of the consumer, as a result of price change of the commodity refers to Income effect. So, income effect refers to increase in purchasing power of the consumer due to fall in price of the commodity. On the other hand, with the increase in price of the commodity, the real income of the consumer decreases, as he purchases only less quantity of the commodity. So, here, income effect refers to decrease in purchasing power of the consumer due to rise in price of the commodity.

Substitution Effect

It refers to change in quantity purchased of the commodity due to the commodity becoming cheaper or dearer in relation to other commodities. For example, if price of the commodity rice is decreased in the market, the consumer prefers more of rice and less of wheat and hence, he substitutes rice for wheat. On the other hand, if the price of the commodity rice is increased in the market, now the consumer substitutes wheat for rice. So, the substitution of one commodity for other commodity as a result of price change with reference to one of the commodities constitutes Substitution effect. It is, otherwise, defined as, the change in quantity demanded of the commodity due to the commodity becoming cheaper in relation to other commodity in the consumer's demand. To analyze the substitution effect, it is essential to maintain constancy of real income of the consumer due to changes in price of one of the commodities.

The splitting of price effect into income and substitution effects has the following merits:

- ☆ It helps in explaining the concept of Law of demand
- ☆ It has offered valuable explanation for Giffen's paradox
- ☆ It helps to understand the concepts of real income and money income of the consumer.
- ☆ It helps to know, whether income effect is stronger than substitution effect or substitution effect is stronger than income effect and thereby, helps to ascertain the nature of the commodity under consideration and its obeyance of Marshallian law of demand.

9.5.3.1. Decomposition of Price Effect into Income Effect and Substitution Effect

In the economic literature, there are two slightly different approaches for explaining the impact of a price change on the quantity demanded of the two commodities by the consumer. They include, Hicks-Allen approach and Slutsky approach. They are discussed in-detail in the ensuing pages.

1. Hicks-Allen Approach

This approach was proposed by J.R.Hicks and R.G.D. Allen. This approach aims at separating the substitution effect from the income effect through compensating the variation in income due to change in price of the commodity. In the Figure 9.35, at a given level of income of the consumer, given prices of rice and wheat commodities and at the given tastes and preferences of the consumer, PL_1 is the price line and it touches the IDC I_1 at Q_1, where the consumer purchases OM_1 quantity of rice and ON_1 quantity of wheat commodities.

Suppose, price of rice is decreased in the market and there is no change in the price of wheat. So, with the fall in price of rice, now the consumer prefers to have more of rice and less of wheat because of the following two reasons:

- ☆ Commodity rice becomes cheaper compared to wheat and thereby, the consumer substitutes rice for wheat (Substitution effect).
- ☆ When the price of rice is decreased in the market, the consumer psychologically feels his real income is increased and thereby, his purchasing power will increase and he purchases more of rice (Income effect).

So, due to fall in price of rice commodity, PL_2 is the new price line and it touches the IDC I_2 at Q_2, where the consumer purchases OM_2 quantity of rice and ON_2 quantity of wheat. So, due to fall in price of the rice commodity, the consumer purchases more of rice ($OM_2>OM_1$) and less of wheat ($ON_2<ON_1$). This movement of consumer from IDC I_1 to

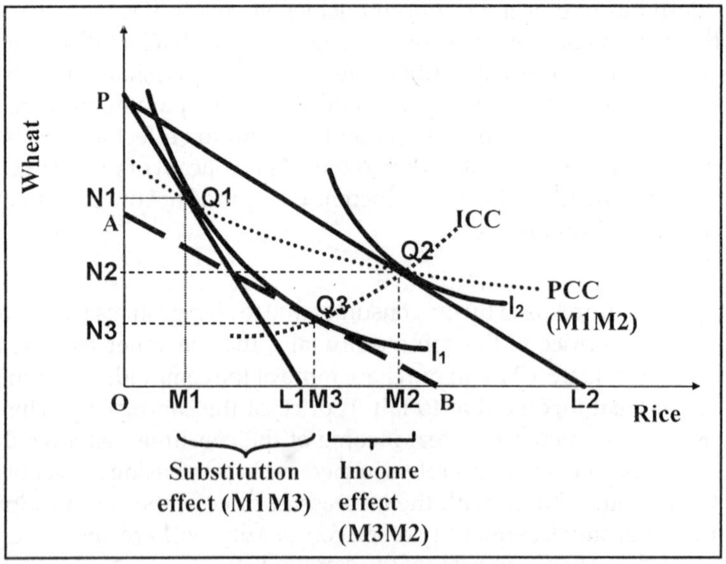

Figure 9.35: Hicks-Allen approach : Price effect = Substitution effect + Income effect (Normal good).

IDC I_2 *i.e.*, from Q_1 to Q_2 has been resulted due to fall in price of rice commodity, as indicated by the PCC and it refers to Price effect. But, this price effect is due to two reasons as explained above and so, the price effect can be split into two efforts *i.e.*, income effect and substitution effect.

So, to isolate the substitution effect from price effect, we have to withdraw or nullify the indirect increase in the real income (due to fall in price of the commodity) on the part of the consumer by taking away some money income, so as to cancel his gain in indirect increase in real income *i.e.*, to avoid income effect. The reduction in the money income of the consumer is to be made by such amount, which is sufficient enough to nullify the effect of indirect increase in real income (income effect) due to fall in price of the rice commodity and this enable the consumer to bring back onto the original IDC I_1. So, PA is the amount of money income with reference to wheat commodity and BL_2 is the amount of money income with reference to rice commodity is to be reduced or withdrawn, so that, the consumer will be brought back on to the original IDC I_1 and the movement on the same IDC I_1 will indicate only substitution effect. This amount by which the money income is reduced (PA is the amount of money income with reference to wheat commodity and BL_2 is the amount of money income with reference to rice commodity) is called Compensating variation in income. So, when some amount of money is taken away from the consumer, the new price line is AB and it lies below and parallel to PL_2. That means, by withdrawing some amount of money income from the consumer (at declined price), he is neither better-off nor worse-off compared to the original income situation (at initial price). So, AB is the hypothetical price line drawn parallel to PL_2 and it touches the original IDC I_1 at point Q_3, where the consumer purchases OM_3 quantity of rice and ON_3 quantity of wheat commodity ($OM_3>OM_1$ and $ON_3<ON_1$).

Thus, with the fall in price of rice commodity, the consumer has substituted rice for wheat and in order to purchase more of rice, the consumer has moved from equilibrium point Q_1 to equilibrium point Q_3 on the same IDC I_1 and this movement, which is due to substitution of rice for wheat on the same IDC I_1 refers to substitution effect. Substitution effect, thus, relates to the change in its relative price alone, but real income of the consumer is kept constant. Since the total price effect corresponds to the movement of the consumer from equilibrium point Q_1 on IDC I_1 to equilibrium point Q_2 on IDC I_2, the isolation of substitution effect *i.e.*, movement from equilibrium point Q_1 on IDC I_1 to equilibrium point Q_3 on IDC I_1, the remaining component of price effect *i.e.*, movement from equilibrium point Q_3 on IDC I_1 to equilibrium point Q_2 on IDC I_2 infers income effect. So, from the Figure 9.35, we can have the following findings:

- ☆ Movement of consumer from Q_1 on IDC I_1 to Q_2 on IDC I_2 or movement from OM_1 to OM_2 (*i.e.*, M_1M_2) refers to price effect.
- ☆ Movement of consumer from Q_1 to Q_3 on IDC I_1 or M_1M_3 (or N_1N_3) refers to substitution effect.
- ☆ Movement of consumer from Q_3 on IDC I_1 to Q_2 on IDC I_2 or M_3M_2 (or N_2N_3) refers to income effect.
- ☆ So, $M_1M_2 = M_1M_3 + M_3M_2$. This infers that, Price effect = Substitution effect+Income effect. So, the total price effect is positive inferring that, due to fall in price of the rice commodity, more quantity of rice (M_1M_2) is purchased.

So, this concept proposed by Hicks and Allen to study the decomposition of price effect into income effect and substitution effect mainly aims at compensating the indirect increase in real income of the consumer back to the original level and thereby, he was brought back onto the same or original IDC. The same approach is employed to study the decomposition of price effect into income effect and substitution effect for different goods *viz.*, inferior goods and Giffen goods in the ensuing pages through Figures 9.37 and 9.38 respectively.

2. Slutsky Approach

This approach was proposed by E. Slutsky. In this method, the increase in real income due to fall in price of the commodity is not nullified by compensating the price variation (as in Hicks-Allen method) and hence, this method does not attempt the consumer to stay on the original IDC to trace out the substitution effect. This is because, with the fall in price of the commodity, the consumer will purchase more of the cheaper commodity and tries to attain maximum satisfaction, thereby, he moves to higher IDC. The increase in purchasing power of the consumer due to fall in price of the commodity can be worked out from the following methodology:

Assume, price of rice is Rs. 15/kg and the consumer wishes to purchase 10 kgs of rice. So, total purchasing power or total cost incurred by consumer is 15x10=Rs.150. Suppose, price of rice is decreased to Rs 10/kg and if the consumer purchases same quantity *i.e.*, 10kgs units, the total purchasing power or total cost is 10x10=Rs.100. So, the increase in purchasing power or cost difference is 150-100=50. This implies, the consumer had an extra purchasing power of Rs.50 and he can purchase some more units of rice commodity, as its price was decreased in the market. The other way of expressing the same is given by:

Increase in purchasing power = Differences in prices of rice x Number of units of rice purchased at old price

So, increase in purchasing power = (15-10) x 10 = Rs.50

The increase in purchasing power of the consumer due to fall in price of the commodity can also be expressed as follows:

Cost difference = (Quantity purchased at old price x old price) - (Quantity purchased at old price x new price)

So, cost difference = (10 x 15) – (10 x 10) = Rs. 50

Thus, the change in the purchasing power of the consumer or cost difference can be worked out by,

- ☆ Multiplying the differences in prices of commodity and quantity of commodity purchased at old price
- ☆ Studying the difference between total cost of purchasing the commodity at old price and total cost of purchasing the same quantity of commodity at new price.

So, this increase in purchasing power or cost difference due to fall in price of the commodity should be removed from the money income, so as to trace out the substitution effect alone.

Thus, in Slutsky method, the cost difference method has adopted to study the increase in purchasing power on the part of the consumer due to fall in price of the commodity and it should be removed to isolate substitution effect from price effect. As shown in the Figure 9.36, for the given level of income of the consumer and prices of two commodities *viz.*, rice and wheat, PL_1 is the price line and it touches the IDC I_1 at Q_1, where the consumer purchases OM_1 quantity of rice and ON_1 quantity of wheat commodities.

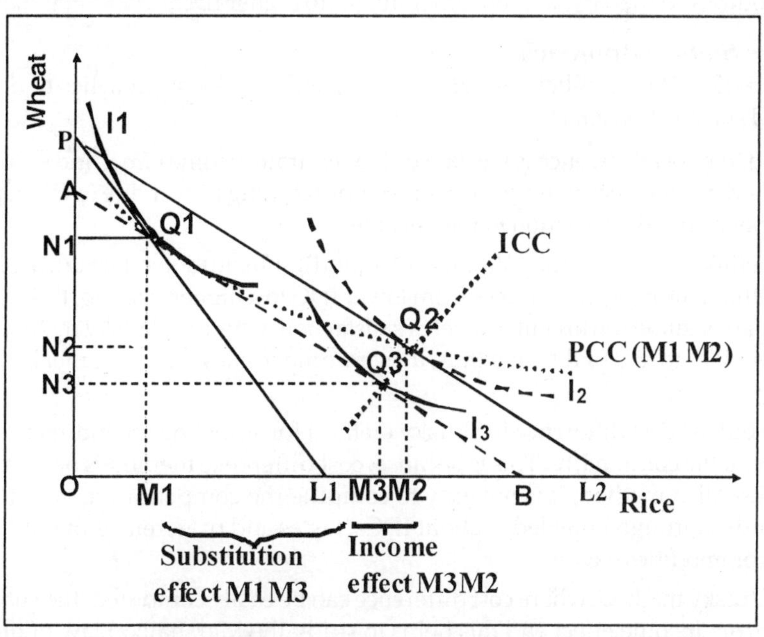

Figure 9.36: Slutsky approach : Price effect = Substitution effect + Income effect (Normal good).

Let us suppose, price of rice commodity is decreased in the market and price of wheat commodity, income of the consumer and tastes and preferences of the consumer remains same. Due to fall in price of the rice commodity, the new price line is PL_2 and it touches the higher IDC I_2 at Q_2, where the consumer purchases OM_2 quantity of rice and ON_2 quantity of wheat commodities. This movement from Q_1 to Q_2 represents price effect indicated by PCC. As a result, the consumer purchases extra M_1M_2 quantity of rice due to fall in its price. Now, in order to find out the true substitution effect, his increased real income (due to fall in price of rice) should be reduced back to the original level, such that, he can buy, if desires, the original combination at Q_1. Thus, the cost difference is worked out (as explained earlier) and it is equal to PA with reference to wheat commodity and BL_2 with reference to rice commodity and this cost difference should be removed from the PL_2 price line, so as to trace out only substitution effect. So, the new price line, AB is called as Slutsky price compensated line and it is drawn parallel to PL_2 and it passes through original combination or initial equilibrium, Q_1. That means, the indirect increase in real income due to fall in price of the commodity was compensated to original level and this is indicated by passing of AB compensated price line through original combination or initial equilibrium, Q_1 and this infers two important aspects *viz.*, if the consumer desires or wishes, he can purchase the original combination of commodities *i.e.*, OM_1 quantity of rice and ON_1 quantity of wheat. This implies that, income effect is removed now, by considering new price line, AB. But, the consumer will not purchase the combinations of rice and wheat at original combination Q_1 because, since the price of rice commodity is decreased, he will purchase more of rice and less of wheat and hence, he moves to a higher IDC I_3 and comes to equilibrium at Q_3, where the consumer purchases OM_3 quantity of rice and ON_3 quantity of wheat commodity ($OM_3 > OM_1$ and $ON_3 < ON_1$). This movement from Q_1 on IDC I_1 to Q_3 on IDC I_3 refers to substitution effect. This is because, with the fall in price of rice commodity, the consumer has substituted rice for wheat. So, in order to purchase more of rice, the consumer has moved from equilibrium point Q_1 on IDC I_1 to equilibrium point Q_3 on IDC I_3. So, this movement, which is due to substitution of rice for wheat *i.e.*, Q_1 to Q_3 or M_1M_3 refers to substitution effect.

Since, the total price effect corresponds to the movement of the consumer from equilibrium Q_1 on IDC I_1 to equilibrium Q_2 on IDC I_2 and the substitution effect refers to the movement from equilibrium Q_1 on IDC I_1 to equilibrium Q_3 on IDC I_3, then remaining component of price effect *i.e.*, movement from equilibrium Q_3 on IDC I_3 to equilibrium Q_2 on IDC I_2 infers income effect.

From the Figure 9.36, we can have the following findings:

- ☆ Movement of consumer from Q_1 on IDC I_1 to Q_2 on IDC_2 or movement from OM_1 to OM_2 (*i.e.*, M_1M_2) refers to price effect.
- ☆ Movement of consumer from Q_1 on IDC I_1 to equilibrium point Q_3 on IDC I_3 (M_1M_3 or N_1N_3) refers to substitution effect.
- ☆ Movement of consumer from Q_3 on IDC_3 to Q_2 on IDC_2 or M_3M_2 (or N_2N_3) refers to income effect.
- ☆ So, $M_1M_2 = M_1M_3 + M_3M_2$. This infers, Price effect = Substitution effect + Income effect.

So, the concept proposed by Slutsky to study the decomposition of price effect into income effect and substitution effect mainly aims at studying the cost difference or increase in the purchasing power on the part of the consumer due to fall in price of the commodity and thereby, the consumer moves to higher IDC.

Hicks-Allen Approach Vs Slutsky Approach

The following are the basic differences between Hicks-Allen and Slutsky approaches of decomposition of price effect into income effect and substitution effect:

- ☆ In Slutsky method, the cost difference is worked out to separate income effect and substitution effect. But, in Hicks-Allen method, the indirect increase in income or purchasing power due to fall in price of commodity is nullified by compensating the variation in real income.
- ☆ In Hicks-Allen method (Compensating Variation Method), substitution effect takes place on the same IDC, but in Slutsky method, the consumer moves from lower IDC to a higher IDC. So, this infers that, Hicks-Allen method restores the consumer on his initial level of satisfaction, whereas, Slutsky method makes the consumer to move from lower IDC to higher IDC and thus, it 'over compensates' the consumer by putting him on higher IDC.
- ☆ It is easier to find out the cost difference in Slutsky method for adjusting the indirect increase in real income due to fall in price of the commodity. This is because, cost difference method is based on observable market data. But, in Hicks-Allen method, it is not easy to compute the compensating variation in income. This is because, it demands thorough knowledge about IDCs, tastes and preferences of the consumer for different combinations of commodities etc.
- ☆ With the help of Slutsky method, where cost difference can be easily computed, the substitution effect can be easily separated from income effect and this helps to study the Marshallian law of demand. But, since the Hicks-Allen method (compensating variation method) is little more complicated, as it is not based on observable market data, its analysis is quite useful in the concepts of consumer's surplus and Welfare Economics.

9.5.3.2. Income and Substitution Effects and Nature of Goods

A close examination of the Figures 9.35 and 9.36 reveals that, the quantities of commodities purchased by the consumer are influenced by both substitution effect and income effect. It is important that, the substitution effect is always negative, as the fall in price of the commodity always leads to increase in quantity demanded for the commodity. That means, if the price of the commodity decreases, the consumer purchases more of that commodity and less of the other commodity. So, the consumer tries to substitute a cheaper commodity in place of a costlier commodity. This is called substitution effect, which is always negative. On the other hand, the income effect is uncertain. That means, if the income of the consumer increases and if the commodity under consideration is a normal good, then the consumer purchases more of that good. But, if the commodity under consideration is an inferior good, then the consumer purchases less of that good, even if his income increases. So, it implies that, income effect is positive for normal goods and negative for inferior goods. So, on comparison with substitution effect, we can infer that, substitution effect is always negative, but the income effect may be positive or negative depending upon the nature of the goods under consideration.

(a) Normal Goods

For normal goods, as explained earlier, the substitution effect is negative and income effect is positive. So, when the price of the normal good decreases in the market, the substitution effect acts in the direction that induces the purchase of more of that (cheaper) good in place of costlier good. Similarly, the income effect (due to increase in real income) also acts in the direction that induces the purchase of more of cheaper good. That means, substitution effect is negative and income effect is positive and they both act in the same direction, if the price of the normal good is decreased in the market. Thus, both the effects induces to purchase large quantities of a cheaper good, whose price is decreased in the market. So, in case of normal goods, for a price decrease,

- ☆ Both substitution effect (negative) and income effect (positive) act in the opposite direction of price.
- ☆ Both substitution effect (negative) and income effect (positive) act in the same direction of purchasing large quantities of a cheaper commodity.
- ☆ When price of the commodity decreases, the substitution effect contributes to an increase in the quantity demanded because, consumers substitute more of the cheaper commodity for other commodities.
- ☆ The reduction in price of the commodity, increases the consumer's ability to purchase the commodity due to increase in consumer's real income. Since, the commodity is normal, this increase in purchasing power, further increases the quantity of the commodity demanded through the income effect.
- ☆ This positive income effect complements the substitution effect and this increases the quantity demanded due to fall in price of the commodity.
- ☆ Quantity demanded always moves in the opposite direction of price
- ☆ Obey the Marshallian law of demand

An increase in the price of the normal good works in an equivalent fashion. If the price of the normal good increases, the consumer substitute more of other goods, whose prices are now relatively lower. So, the substitution effect reduces the quantity demanded for the dearer good. The increase in price also reduces the real income and thereby, causing consumer to reduce consumption of the costlier good *via* the income effect.

(b) Inferior Good

In case of an inferior good, the substitution effect is negative. The income effect is also negative because, even if the price of the inferior good is decreased, the consumer will not purchase more quantity of the inferior good (even though his real income is increased). But, for inferior goods, the negative income effect is weak compared to negative substitution effect, because only a small proportion of income is spent on the inferior good (negative E_Y) and thereby, it cannot outweigh strong negative substitution effect. The substitution effect will work to purchase more of an inferior good (as substitution effect is always negative), but income effect will work to purchase less of an inferior good, even if the price of the inferior good is decreased, as income effect is negative. So, in case of an inferior good, both substitution effect and income effect are negative and these two effects act in the opposite direction regarding the quantity of purchase of an inferior good, if its price was decreased in the market. That is, substitution effect leads to purchase of more quantity, but income effect leads to purchase of less quantity of an inferior good due to fall in its price. However, the quantity of purchase of an inferior good, if its price was decreased, depends upon the relative strengths of substitution effect and income effect. Since, the negative substitution effect is stronger than negative income effect, the consumer purchases more of an inferior good (cheaper good) and less of the other good (costlier good), if the price of inferior good is decreased in the market. So, in case of inferior goods, for a price decrease,

- ☆ Substitution effect is negative and acts in the opposite direction of price.
- ☆ When price of an inferior good decreases, the consumer will substitute more of the inferior good for other goods because, its price has fallen relative to those goods. The quantity demanded increases as a result of the substitution effect.

- ☆ The reduction in price of the inferior good increases the consumer's ability to purchase the good due to increase in consumer's real income. But, because, the good is inferior, this reduces quantity demanded. So, income effect is negative and acts in the direction of price.
- ☆ Thus, substitution effect acts in the opposite direction of price, but income effect acts in the direction of price.
- ☆ Strong negative substitution effect outweighs weak negative income effect and hence, quantity demanded always moves in the opposite direction of price.
- ☆ The higher real income (due to fall in price) leads to decrease in quantity demanded, but this cannot outweigh the increase in quantity demanded due to the strong substitution effect and thereby, net quantity demanded increases. So, income effect works against the substitution effect, but it cannot outweigh the substitution effect.
- ☆ Obey the Marshallian law of demand

So, the case of an inferior good is thus, different from that of normal good. The income effect of a price change works in a direction opposite to that of the substitution effect in the case of an inferior good, whereas it reinforces the substitution effect in the case of a normal good.

(c) Giffen Good

On the other hand, if the negative income effect is stronger than negative substitution effect, then the consumer purchases less of an inferior good (cheaper good) and more of other good (costlier good), even if the price of inferior good is decreased. This is the case of a Giffen good, named after Sir Robert Giffen. In case of giffen goods, for a price decrease, the negative income effect is strong compared to negative substitution effect because, a large proportion of income is spent on the giffen good (negative E_Y) and thereby, it outweighs negative substitution effect. We can infer that, all giffen goods are inferior goods (because, income effect is negative), but all inferior goods are not giffen goods (because, for giffen goods, income effect is stronger than substitution effect). So, in case of giffen goods, for a price decrease,

- ☆ Substitution effect is negative and acts in the opposite direction of price.
- ☆ Income effect is negative and acts in the direction of price.
- ☆ Strong negative income effect outweighs weak negative substitution effect and hence, quantity demanded always moves in the direction of price.
- ☆ The higher real income (due to fall in price) leads to decrease in quantity demanded and this outweighs the increase in quantity demanded due to the substitution effect and thereby, net quantity demanded decreases. So, income effect works against the substitution effect and it outweighs the substitution effect.
- ☆ Disobey the Marshallian law of demand

The following Tables 9.5 to 9.8 briefs out the above discussion regarding the relative strengths of substitution effect and income effect due to changes in price of different goods:

Table 9.5: Relative strengths of substitution effect and income effect due to fall in price of the different commodities.

Goods	*Substitution Effect*		*Income Effect*		*Movement of Substitution Effect and Income Effect with Reference to Direction of Price*	*Relative Strengths of Substitution Effect and Income Effect*	*Example*	*Price Effect*
	Nature	*Direction (Effect) with Reference to Price of the Good*	*Nature*	*Direction (Effect) with Reference to Price of the Good*				
Normal goods	Negative	Opposite direction	Positive	Opposite direction	Both substitution effect and income effect act in the opposite direction of price	Both effects complement each other and they act in the opposite direction of price	Rice	Quantity demanded increases hence, obeys the Marshallian law of demand
Inferior goods	Negative	Opposite direction	Negative	Same direction	Substitution effect act in the opposite direction of price, but income effect act in the direction of price	Strong negative substitution effect outweighs weak negative income effect	Jowar	Quantity demanded increases hence, obeys the Marshallian law of demand
Giffen goods	Negative	Opposite direction	Negative	Same direction	Substitution effect act in the opposite direction of price, but income effect act in the direction of price	Strong negative income effect outweighs weak negative substitution effect	Potato, Bread	Quantity demanded decreases hence, disobey the Marshallian law of dernand

Table 9.6: Relative strengths of substitution effect and income effect due to rise in price of the different commodities.

Goods	*Substitution Effect*		*Income Effect*		*Movement of Substitution Effect and Income Effect with Reference to Direction of Price*	*Relative Strengths of Substitution Effect and Income Effect*	*Example*	*Price Effect*
	Nature	*Direction (Effect) with Reference to Price of the Good*	*Nature*	*Direction (Effect) with Reference to Price of the Good*				
Normal goods	Negative	Opposite direction	Negative	Opposite direction	Both substitution effect and income effect act in the opposite direction of price	Both effects complement each other and they act in the opposite direction of price	Rice	Quantity demanded decreases hence, obeys the Marshallian law of demand
Inferior goods	Negative	Opposite direction	Positive	Same direction	Substitution effect act in the opposite direction of price, but income effect act in the direction of price	Strong negative substitution effect outweighs weak positive income effect	Jowar	Quantity demanded decreases hence, obeys the Marshallian law of demand
Giffen goods	Negative	Opposite direction	Positive	Same direction	Substitution effect act in the opposite direction of price, but income effect act in the direction of price	Strong positive income effect outweighs weak negative substitution effect	Potato, Bread	Quantity demanded increases hence, disobey the Marshallian law of demand

Table 9.7: Changes in quantity demanded based on relative strengths of substitution effect and income effect due to fall in price of the different commodities.

Goods	*Substitution Effect*	*Income Effect*	*Relative Strengths of Substitution Effect and Income Effect*	*Overall (Price Effect)*
Normal goods	Negative *i.e.*, increases quantity demanded	Positive *i.e.*, increases quantity demanded	Substitution effect and income effect complement each other	Increases quantity demanded and thereby, demand curve slopes downward
Inferior goods	Negative *i.e.*, increases quantity demanded	Negative *i.e.*, decreases quantity demanded	Strong negative substitution effect outweighs weak negative income effect	Increases quantity demanded and thereby, demand curve slopes downward
Giffen goods	Negative *i.e.*, increases quantity demanded	Negative *i.e.*, decreases quantity demanded	Strong negative income effect outweighs weak negative substitution effect	Decreases quantity demanded and thereby, demand curve slopes upward

Table 9.8: Changes in quantity demanded based on relative strengths of substitution effect and income effect due to rise in price of the different commodities.

Goods	*Substitution Effect*	*Income Effect*	*Relative Strengths of Substitution Effect and Income Effect*	*Overall (Price Effect)*
Normal goods	Negative *i.e.*, decreases quantity demanded	Negative *i.e.*, decreases quantity demanded	Substitution effect and income effect complement each other	Decreases quantity demanded and thereby, demand curve slopes downward
Inferior goods	Negative *i.e.*, decreases quantity demanded	Positive *i.e.*, increases quantity demanded	Strong negative substitution effect outweighs weak positive income effect	Decreases quantity demanded and thereby, demand curve slopes downward
Giffen goods	Negative *i.e.*, decreases quantity demanded	Positive *i.e.*, increases quantity demanded	Strong positive income effect outweighs weak negative substitution effect	Increases quantity demanded and thereby, demand curve slopes upward

The income effect derived through ICC (*i.e.*, changes in the income itself and not due to change in price of the commodity) across different types of goods *viz.*, normal, inferior and giffen goods is often confused with income effect realized (either increase or decrease in real income) due to change in price of one of the commodities. The following Table 9.9, illustrates the income effect for different goods based on the ICCs.

So, to conclude, the price effect involves two effects *viz.*, substitution effect and income effect and the relative strengths of the two effects will guide the consumer to purchase desired quantities of the commodities due to changes in price of one of the commodities.

Table 9.9: Income effects realized across different commodities due to change in income and not due to change in price of other commodity.

Goods	*Income Effect Due to Changes in Income itself (ICC)*
Normal goods	Positive
Inferior goods	Negative
Giffen goods	Negative

9.5.3.3. Relative Strengths of Substitution Effect and Income Effect and Nature of Goods

The information shown through Table 9.5 with reference to relative strengths of substitution effect and income effect for different types of goods is studied through graphical illustrations in the ensuing pages.

1. *Relative Strengths of Substitution Effect and Income Effect in Case of Normal Goods*

The graphical illustration shown through Figure 9.35 explain the relative strengths of substitution effect and income effect with reference to normal good. As shown in the Figure 9.35, when the price of the rice commodity is decreased in the market, the consumer purchases more of rice and less of wheat. As the substitution effect is always negative, the consumer always prefers to purchase cheaper good in place of costlier good. So, the quantity of rice purchased due to substitution effect was increased from OM_1 to OM_3. It is interesting to note from the Figure 9.35 that, the income effect is positive, as indicated by the upward sloping ICC, indicating that, more quantities of both the goods are purchased *i.e.*, $OM_2>OM_3$ with reference to rice and $ON_2>ON_3$ with reference to wheat, due to increase in real income of the consumer. That means, substitution effect is negative and income effect is positive and they both act in the same direction towards the purchase of more quantities of rice (cheaper good). This is further illustrated by the downward sloping PCC showing the combined effect of both substitution effect and income effect, indicating purchase of large quantities of rice due to fall in its price. So, the total price effect leads to purchase of large quantities of rice at lower price, thereby, obeying the Marshallian law of demand.

2. *Relative Strengths of Substitution Effect and Income Effect in Case of Inferior Goods*

As explained earlier, in case of inferior goods, the substitution effect is negative and the income effect is also negative. That means, both the effects act in the opposite direction because, substitution effect leads to purchase more of cheaper goods (inferior goods, when its price was decreased in the market), whereas, negative income effect leads to purchase of costlier goods and less of cheaper goods (inferior goods, if the price of the inferior goods is decreased in the market).

In the Figure 9.37, for the given level of income of the consumer and given prices of the two commodities *viz.*, jowar (inferior good on X-axis) and rice (normal good on Y-axis), PL_1 is the price line and it touches the IDC I_1 at Q_1, where he purchases OM_1 quantity of jowar commodity and ON_1 quantity of rice commodity. When the price of jowar (inferior good) is decreased in the market, PL_2 is the new price line drawn and it touches the IDC I_2 at Q_2, where the consumer purchases OM_2 quantity of jowar and ON_2 quantity of rice. By keeping the real income constant (after the fall in price of jowar), the consumer is brought back on to the original IDC I_1 at Q_3 with reference to compensated price line AB.

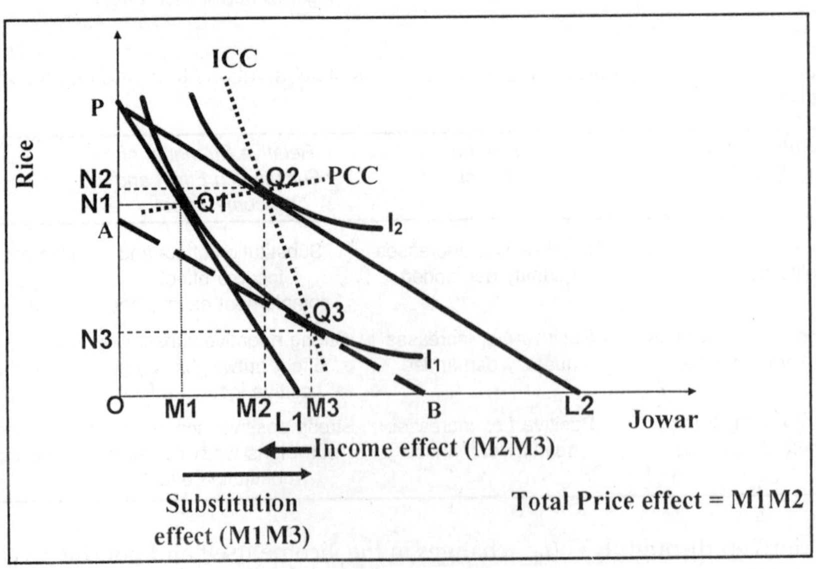

Figure 9.37: Relative strengths of Substitution effect and Income effect – Case of an Inferior good.

However, it is interesting to note that, since the commodity jowar is an inferior good, for which the income effect is negative, the ICC will bend towards Y-axis on which the commodity rice is taken. This indicates that, with increase in real income of the consumer due to fall in price of jowar, he consumes less of jowar (being an inferior good) and consumes more of rice. But, a close examination of the Figure 9.37 reveals that, substitution effect is always negative and it induces the consumer to purchase more of inferior good (jowar) due to fall in its price. In case of inferior goods, the strong negative substitution effect will outweigh the weak negative income effect and thereby, it induces the consumer to purchase more of jowar (inferior good) due to fall in its price. This can also be stated that, the negative income effect is small and is not sufficient enough to outweigh the strong negative substitution effect. This is because, the consumer spends a very small proportion of his income on the inferior goods. So, as a result, with the fall in price of an inferior good, the consumer purchases more of inferior good (jowar) and less of normal good. This infers that, the overall price effect leads to purchase of large quantities of jowar (inferior good) due to fall in its price. This is indicated by the upward sloping PCC and the movement from Q_1 to Q_2 *i.e.*, M_1M_2 indicates price effect. The income effect is negative and hence, the quantity purchased of jowar is decreased by M_3M_2. Had the income effect is absent, it makes the consumer to purchase OM_3 quantity of jowar. But, since the income effect is present and it is negative, it makes the consumer to purchase less quantity of jowar by M_2M_3. But, since the negative substitution effect is stronger than negative income effect, it makes the consumer to purchase more quantity of jowar, when its price was decreased and thereby, the net quantity purchased by the consumer is M_1M_2. So, the total price effect is M_1M_2 quantity of jowar purchased. The total price effect is positive *i.e.*, quantity purchased of jowar was increased by M_1M_2 due to fall in price of jowar.

The above discussion indicates that, with the fall in price of an inferior good, the consumer purchases more quantity of an inferior good, but the increase in quantity purchased by the consumer is limited to some extent due to negative income effect. But, still the consumer purchases more quantity of jowar (inferior good) due to fall in its price, as the strong negative substitution effect outweighs the weak negative income effect. The discussion reveals that, even inferior goods also obey the Marshallian law of demand like normal goods, as decrease in its price leads to more quantity of inferior good purchased, but the increase in quantity demanded is less compared to normal good due to the influence of negative income effect.

3. Relative Strengths of Substitution Effect and Income Effect in Case of Giffen Goods

The concept of Giffen goods was introduced by Sir Robert Giffen. According to Giffen, low paid workers in Great Britain consume both bread and meat. But, when the price of bread was increased in the market, the consumers still purchased more quantity of bread and they economized their expenditure on meat, as for them bread is the staple food. This is because, an increase in the price of the bread will reduce the purchasing power of the poor consumers and this will force them to reduce their expenditure on other expensive good, meat. So, even when the price of bread is increased, it was still their cheapest food in the market, thereby, they still purchase more quantity of bread compared to other still expensive food items. Conversely, when the price of the bread falls, the consumers purchase less of bread. This is because, when the price of the bread falls, the purchasing power of the consumer increases and since the consumer spends a large proportion of income on the bread, this increase in purchasing power leads to strong negative income effect (because, giffen good is an inferior good), which outweighs the weak negative substitution effect and this leads to decline in quantity purchased of bread even due to fall in price of the bread. Such goods like bread, which disobey the Marshallian law of demand are called Giffen goods. A close look at the above example reveals an important aspect that, bread is an inferior good compared to meat and hence, giffen goods also fall under the category of inferior goods. But, the major distinction between the inferior goods and giffen goods is that, inferior goods obey the Marshallian law of demand, whereas giffen goods disobey the Marshallian law of demand. This is because, in case of inferior goods due to fall in price, strong negative substitution effect outweighs the weak negative income effect, but in case of giffen goods, due to fall in price, strong negative income effect outweighs the weak negative substitution effect. That means, for giffen goods, a decrease in price leads to decrease in quantity demanded and *vice versa*. So, like in case of inferior goods, for giffen goods also, both substitution effect and income effect will act in opposite directions, but since strong negative income effect outweighs weak negative substitution effect (in case of fall in price), the giffen goods do not obey the Marshallian law of demand and the same is explained through Figure 9.38.

In the Figure 9.38, for the given level of income of the consumer and given prices of the two commodities *viz.*, bread (giffen good on X-axis) and meat (normal good on Y-axis), PL_1 is the price line and it touches the IDC I_1 at Q_1, where he purchases OM_1 quantity of bread commodity and ON_1 quantity of meat. When the price of bread (giffen good) is decreased in the market, PL_2 is the new price line drawn and it touches the IDC I_2 at Q_2, where the consumer purchases OM_2 quantity of bread and ON_2 quantity of meat. By keeping the real income constant (after the fall in price of bread), the consumer is brought back on to the original IDC I_1 at Q_3 with reference to compensated price line AB.

However, it is interesting to note that, since the commodity bread is a giffen good, for which the income effect is strongly negative, the ICC will bend towards Y-axis on which the normal (superior) goods, meat is taken. This indicates

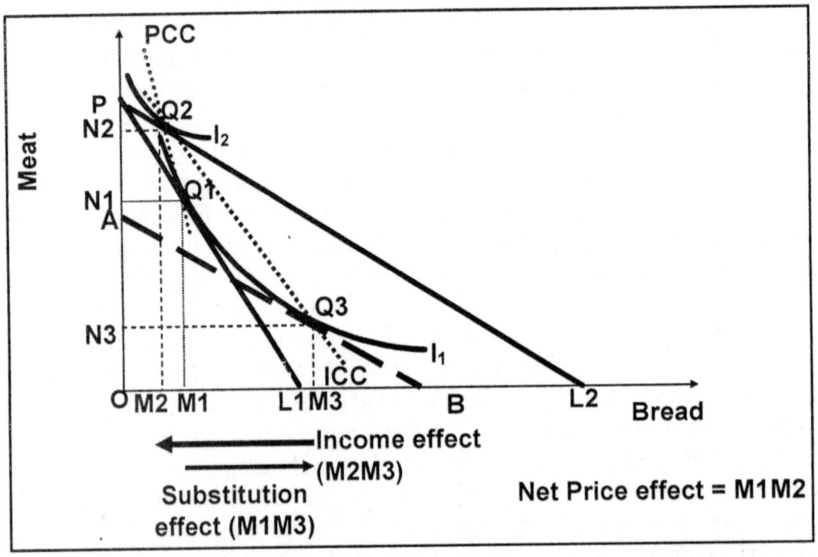

Figure 9.38: Relative strengths of Substitution effect and Income effect – Case of a Giffen good.

that, with increase in real income of the consumer due to fall in price of bread, he consumes less of bread (being a giffen good) and consumes more of meat. A close examination of Figure 9.38 reveals that, substitution effect is negative and it induces the consumer to purchase more of bread (giffen good) due to fall in its price. This is represented as M_1M_3 with reference to bread commodity. But, since, income effect is strongly negative (due to fall in price of bread), it makes the consumer to purchase less quantity of bread (giffen good) and this is shown as M_3M_2. Thus, strong negative income effect outweighs weak negative substitution effect, thereby, the consumer purchase less quantity of bread. This is indicated by M_1M_2 due to fall in the price of the bread. This movement from Q_1 on IDC I_1 to Q_2 on IDC I_2 or M_1M_2 is called price effect. This is further indicated by the upward sloping of PCC towards Y-axis and hence, the movement from Q_1 to Q_2 *i.e.*, M_1M_2 indicates price effect.

Thus, in case of giffen goods, due to fall in price, the substitution effect is always negative and it induces the consumer to purchase more of giffen good (bread). But, in case of giffen goods, the strong negative income effect will outweigh the weak negative substitution effect and thereby, it induces the consumer to purchase less of bread (giffen good). This can also be stated that, the negative income effect is very large and it outweighs the weak negative substitution effect. This is because, the consumer spends a large proportion of his income on the giffen goods. So, as a result, with the fall in price of a giffen good, the consumer purchases less of giffen good (bread) and more of superior good (meat). This infers that, the overall price effect leads to purchase of less quantities of bread (giffen good) due to fall in its price. This explanation clearly infers that, giffen goods disobey the Marshallian the law of demand and thereby, with the decrease in price of the giffen goods, quantity purchased of giffen good also decreases.

From the above analysis, we can infer the following criteria to consider the good as a 'giffen good':

- The good must be considered as an inferior good with a weak negative substitution effect
- The good must have large negative income effect and this will outweigh the weak negative substitution effect
- The proportion of income spent on the good should be large, so as to induce large negative income effect.

9.5.3.4. Income Effect and Substitution Effect and Law of Demand

- In case of normal goods, where the substitution effect is negative and income effect is positive and since, both the effects will work in the same direction, it leads to purchase of more quantities of a commodity when its price is decreased in the market. This infers that, the Marshallian law of demand is obeyed by the normal goods.
- In case of inferior goods, both the substitution effect and income effect are negative, but the strong negative substitution effect outweighs the weak negative income effect and thereby, the consumer purchases more of the inferior good, if its price decreases. That means, the Marshallian law of demand is obeyed by the inferior goods, but the increase in quantity purchased of the inferior good due to fall in its price is less compared to the normal good. This infers that, when the income effect is negative and relatively weaker than substitution effect, the law of demand is obeyed by the commodity (inferior goods).
- In case of giffen goods, like the inferior goods, both the substitution effect and income effect are negative, but the strong negative income effect outweighs the weak negative substitution effect and thereby, the consumer

purchases less of the giffen good, even if its price decreases. That means, the Marshallian law of demand is disobeyed by the giffen goods. This infers that, when the income effect is negative and relatively stronger than substitution effect, the law of demand is not obeyed by the commodity (giffen goods). So, in case of giffen goods, the quantity demanded varies directly with the price of the commodity.

The above discussion reveals that, in case of normal goods and inferior goods, we get a downward (negative) sloping demand curve, whereas, in case of giffen goods, we get an upward (positive) sloping demand curve. In this context, the IDC analysis is considered superior to Marshallian analysis, as in Marshallian analysis, the giffen goods are considered as an exception to the law of demand, but IDC analysis helps to explain the reasons why giffen goods disobey the law of demand. Since, Marshallian analysis has considered total price effect in explaining the law of demand without splitting the price effect into income effect and substitution effect, it could not explain the demand theory with reference to giffen goods. But, this is clearly analyzed through IDC approach, where the strong negative income effect (say, in case of price fall of giffen good) outweighs the weak negative substitution effect and this is responsible for the disobeyance of law of demand by the giffen goods. But, it is important to note that, the occurrence of giffen goods in the practical world is almost negligible. This is because, no consumer will prefer the same good even if the price of that good is increased in the market because, there are several substitutes available for that good in the market economy. That means, if the price of a particular good is increased in the market, the consumer will try to diversify his expenditure on other goods, which are relatively cheaper. For example, if a good X is a necessary to the consumer and if its price was increased in the market, the consumer will allocate less money on good X due to increase in its price and purchase less quantity of good X (being a necessary good) and allocate the remaining money on other goods say, Y, Z etc., as they are the good substitutes of X. In such case, the negative income effect on good X is smaller and it cannot outweigh the negative substitution effect. So, now the good X will obey the law of demand. Hence, in the practical world, due to the availability of substitutes for a particular good, the concept of giffen goods will not arise. So, the concept of giffen goods is possible only on theoretical note, but in practical sense, their existence is highly questionable.

Impact of a Price Increase

A price increase can be analyzed in exactly the same way as a price decrease. But, in case of price increase, the price line PL_1 pivots towards the origin as PL_2 and the remaining methodology pertaining to price effect and its components viz., income effect and substitution effect will be studied as in case of a price decrease situation. To maintain the real income constant, here some money is given back to the consumer to the extent of decrease in real income (Hicks-Allen approach) and cost-difference is worked out and the same is added to the PL_2 price line or real income (Slutsky approach). The above discussion in summarized below.

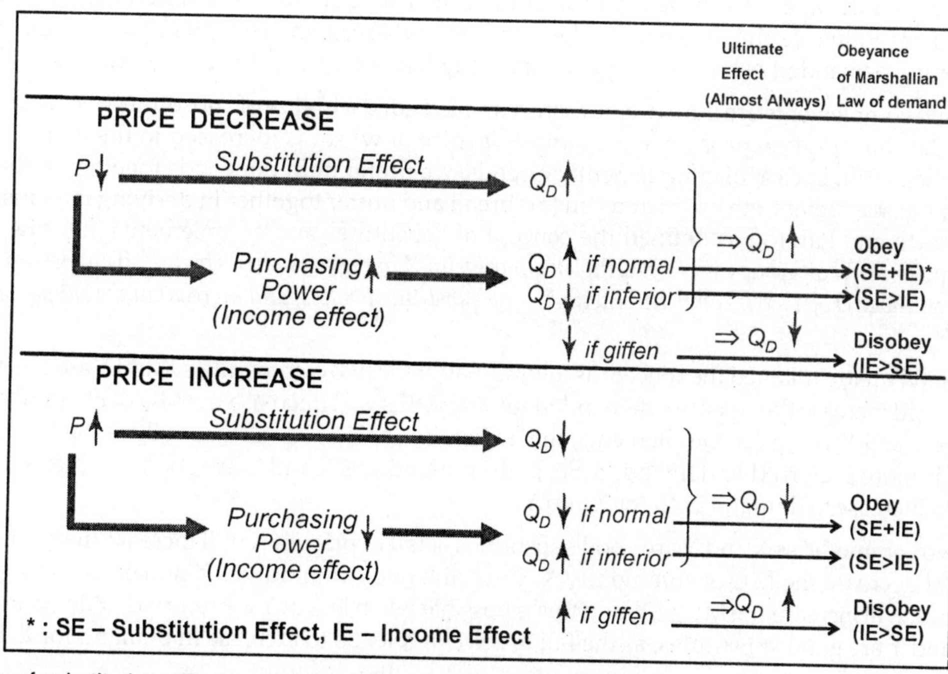

Relative strengths of substitution effect and income effect across the commodities

9.5.3.5. Substitutes and Complementary Relationships in IDC Analysis

IDC analysis will also allow us to see whether two commodities are substitutes or complements. In Panel A of the Figure 9.39, a fall in the price of rice has led to a change in the price line PL_1 away from the origin as PL_2. As a result, more of rice and less of wheat are purchased by the consumer. So, the commodities under consideration are substitutes.

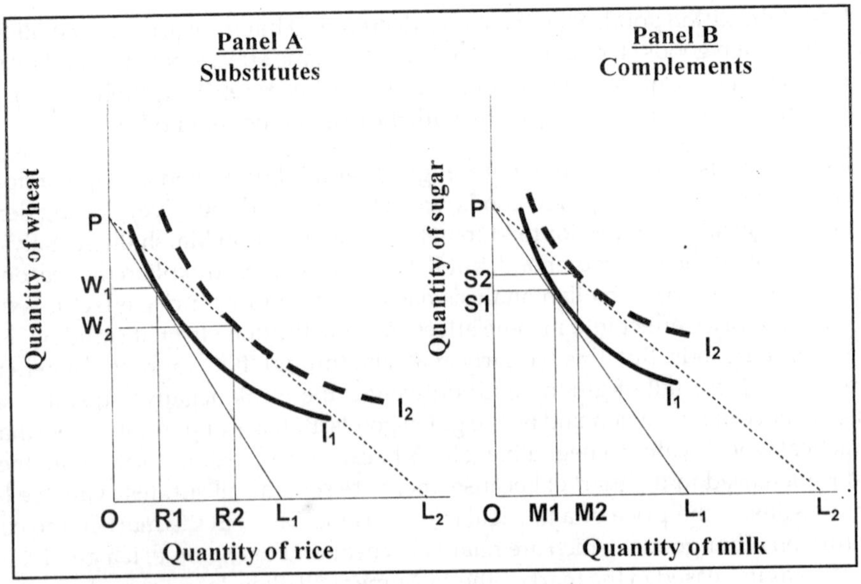

Figure 9.39: Substitutes and Complementary relationships between the goods.

In the Panel B of the Figure 9.39, due to the fall in the price of milk, the price line PL_1 pivots as PL_2 and this has led to an increase in the quantity demanded of both milk and sugar, thereby, both the commodities are complements.

Marshallian analysis could not offer satisfactory explanation regarding the substitution and complementarity relationships among the commodities, as it assumed utilities derived from the commodities are independent. But, it offered the explanation for substitutes and complements through studying the E_C. However, such invalid assumption was not framed in IDC analysis and hence, it offered satisfactory explanation for substitutes and complementary goods. We can say, the study of substitutes and complementarity relationships between the goods represent the case of related goods and the consumer altogether faces a different pattern of consumption behaviour compared to unrelated goods. Say, for example, the influence of income and prices of the commodities on the consumption in case of unrelated goods such as harvester and fertilizer, pesticide and computer etc., is altogether different compared to their influence of consumption of rice and wheat, bread and jam, jam and butter etc. This is because, say, the change in price of a pesticide in the market will not influence the demand for the computer, whereas, the change in price of jam, will definitely influence the quantity demanded of bread (complementary good) and butter (substitute good).

Substitutes and complements are the extreme cases of related goods. In case of substitutes, we replace one good for another good in the consumption process. For example, if price of wheat is increased in the market, the consumer prefers rice for wheat. But, in case of complementary goods, we prefer both the goods together in the consumption process. For example, we prefer both bread and jam (or) bread and butter together in deriving the satisfaction. Before Marshall, Edgeworth and Pareto has defined the concept of substitutes and complements in terms of MU and by assuming constancy of MU of money as, '*Y is complementary with X in the consumer's budget, if an increase in the supply of X (Y constant) raises the MU of Y; Y is competitive with X (or a substitute for X) if and increase in the supply of X (Y constant) lowers the MU of Y*'.

Edgeworth and Pareto explained the concept as follows. Say, if the two commodities X and Y are good complements, a fall in price of X will increase the quantity demanded for X and this will decrease the MU of commodity X. Since, the two commodities X and Y are good complements, the decrease in MU of X will also reduce the MU of Y and this is possible when Y is more preferred or demanded. So, both commodities X and Y are good complements, as the fall in price of X leads to increase in demand both for X and Y.

Say, if the two commodities X and Y are good substitutes, a fall in price of X will increase the quantity demanded for X and this will decrease the MU of commodity X. Since, the two commodities X and Y are good substitutes, the decrease in MU of X will increase the MU of Y, and this is possible when less of Y is preferred or demanded. So, both the commodities X and Y are good substitutes, as the fall in price of X leads to increase in demand for X, but decrease in demand for Y.

However, Prof.J.R.Hicks criticized the above version given by Edgeworth and Pareto on two grounds *viz.*, MU of money is assumed constant and total price effect is considered in explaining the consumption behaviour of related goods and it is not decomposed into income effect and substitution effect.

Hicks Version of Substitutes and Complements

According to Professor J.R. Hicks, the explanation offered for substitutes and complements based on the E_C is inadequate, as the explanation is based on total price effect. He opined that, the shapes of the IDCs often indicate the relationship between the commodities under consideration by the consumer.

1. ***Linear IDCs:*** The IDC is a straight line connecting the two axes with a constant negative slope (MRS) and this indicates the two commodities under consideration are perfect substitutes. In the Panel A of the Figure 9.3, AB=BC and this slope remains same throughout the IDC. Examples include, Brand A tea and Brand B tea, Brand A coffee and Brand B coffee, Shimla apple and Kashmir apple etc. In such cases, the consumer will purchase only one of the two goods, as MRS is equal to one and the consumer is said to have *Monomania* for a particular good, as he cannot distinguish one good over the other.
2. ***Convex shaped IDCs:*** The IDC will have a convex shape indicating the two commodities under consideration are good substitutes. In such cases, the MRS diminishes (Figure 9.8). Examples include rice and wheat, tea and coffee, jam and butter, sugar and jaggery etc.
3. ***Right angled or 'L' shaped:*** The IDC will have a right angled or 'L' shaped, if the commodities under consideration are perfect complements. As shown in the Figure 9.40, the movement from A to B with reference to commodity milk, will not induce any reduction with reference to commodity sugar *i.e.*, same quantity of OS_1 of sugar is consumed thereby, MRS_{MS}=0. Similarly, the movement from A to C with reference to commodity sugar, will not induce any reduction with reference to commodity milk *i.e.*, same quantity of OM_1 milk is consumed, thereby, MRS_{SM}=0. So, any movement along X-axis or Y-axis with reference to a particular commodity will not reduce or substitute the consumption of other commodity.
   If the consumer wants to derive higher level of satisfaction, he has to move from IDC I_1 to IDC I_2 and for this, he has to increase the consumption of both the commodities. That means, if the consumer wants to derive higher satisfaction at point D on IDC I_2 he has to consume more of both milk (OM_2>OM_1) and sugar (OS_2>OS_1). An increase in the consumption of milk *i.e.* from OM_1 to OM_2, without increasing the consumption of sugar (*i.e.*, the consumer will consume only OS_1), will not enable the consumer to reach IDC I_2 or will not enable to derive higher level of satisfaction.
4. ***Right angled or 'L' shaped IDC, but with a curvature at the bend:*** This shape of IDC will imply both the commodities under consideration are good complements, but not perfect complements. That means, within the same IDC, both complementarity and competition (substitution) relationship co-exists. The explanation for such type of IDC was already discussed with reference to Figure 9.6. It is evident from the Figure 9.6 that, between B and D with reference to computer monitors on X-axis and between A and C with reference to CPUs on Y-axis, the commodities are complements with MRS equal to zero, but between A and B portion of IDC, the commodities are good substitutes with diminishing MRS.

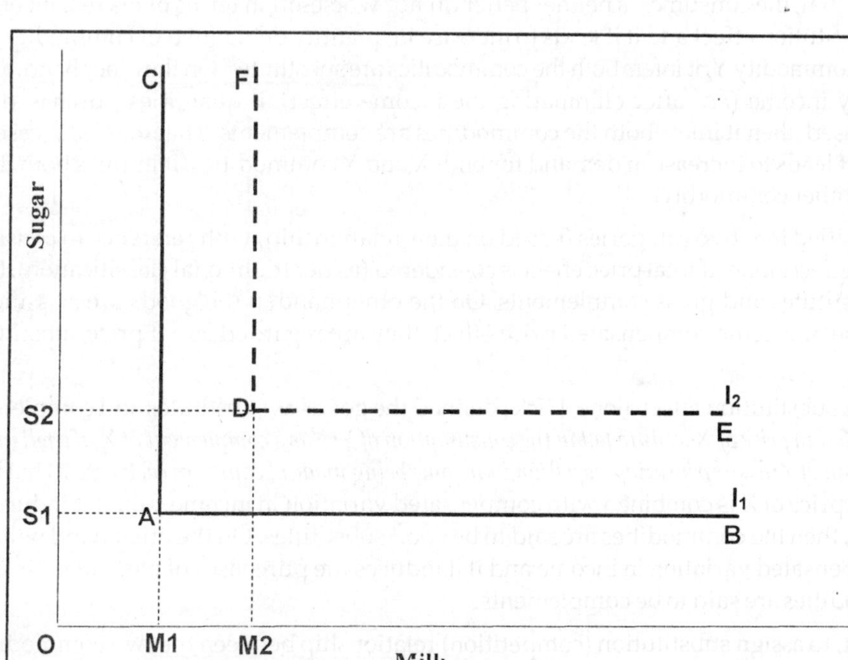

Figure 9.40: Right angle or 'L' shaped IDCs – Complementary goods.

Of course, the above explanation with reference to nature of the commodities as perfect substitutes, good substitutes, perfect complements and good complements was given based on the different shapes of IDCs. But, Hicks rightly pointed out that, it is impossible to find out at what degree of curvature of IDC, there is distinction between complementarity and substitution relationship between the commodities. Hence, Hicks proposed the classification of goods into complements and substitutes based on the splitting of price effect into income effect and substitution effect and only if income effect is eliminated by means of compensating variation in income, the substitution effect can be isolated and based on the strength of the substitution effect, we can ascertain the relationship between the goods.

In this context, Hicks criticized the earlier version of classifying the commodities into substitutes and complements based on the E_C. According to this traditional classification, two commodities milk and sugar are said to be complements, if the E_C between them is negative. This is because, if the price of commodity milk is decreased in the market, then its demand will increase. Then, commodity sugar, which is preferred along with milk, its demand also increases. So, both milk and sugar are complements and E_C is negative, as price of milk and quantity demanded of sugar move in opposite direction. On the other hand, two commodities rice and wheat are said to be substitutes, if the E_C between them is positive. This is because, if the price of the rice is decreased in the market, its demand will increase. Then, wheat is less preferred and its demand will decrease in the market. So, both rice and wheat are good substitutes and E_C is positive, as both price of rice and quantity demanded of wheat move in the same direction.

But, this classification is inadequate. This is because, if the two commodities X and Y are substitutes, and if the price of X commodity decreases, it may lead to increase in demand for both the commodities X and Y. This may happen, if total price effect is taken into consideration. That means, if the price of the commodity X decreases, it induces both income effect and substitution effect. Sometimes, the income effect is much stronger and it outweighs the negative substitution effect. So, due to strong income effect, the consumer can purchase more of commodity Y (even though price of X decreases) and he also purchases more of commodity X (due to negative substitution effect), thereby, the consumer purchase more of both the commodities X and Y due to fall in price of commodity X. That means, on the basis of total price effect, the commodities can be described as complements, even though they are substitutes. In this context, the traditional classification of commodities into substitutes and complements was severely criticized, as the classification was based on E_C considering total price effect. So, according to Hicks, this income effect should be removed and only substitution effect should be considered for classifying the commodities into substitutes and complements. He further suggested the elimination of income effect due to change in price of the commodity by means of compensating variation in income. So, the concept of IDC analysis proposed by Hicks guides to split the price effect into income effect and substitution effect and by eliminating the income effect by means of compensating variation in income, only substitution effect is left over and this alone guides to ascertain whether the commodities are substitutes or complements.

Let us suppose, there are two commodities X and Y and if price of X falls, (price of commodity Y remains same), the consumer purchases more of commodity X due to two reasons *viz.*, income effect and substitution effect. To ascertain the true relation between the two commodities X and Y, we have to eliminate this income effect by means of compensating variation in income, so that, the consumer is neither better off nor worse-off in terms of his real income. Now the price effect include only substitution effect and if it leads to increase in quantity demanded of commodity X and decrease in quantity demanded of commodity Y, it infers both the commodities are substitutes. On the other hand, after compensating the variation in money income (*i.e.*, after eliminating the income effect), if quantities purchased of both X and Y commodities are increased, then it infers both the commodities are complements. That means, in case of complements, if the substitution effect leads to increase in demand for both X and Y commodities, it implies both the commodities X and Y are substituting other commodity.

So, goods are classified into two categories (based on their relationship) with reference to considering total price effect and substitution effect alone. If total price effect is considered (as per traditional classification), then the goods are classified as gross substitutes and gross complements. On the other hand, if the goods are classified on the basis of substitution effect alone or income compensated price effect, they are regarded as net price substitutes and net price complements.

So, considering the substitution effect alone, Hicks defined the goods as substitutes and complements as *'I shall say Y is a substitute of X, if a fall in price of X leads to fall in the consumption of Y; Y is a complement of X, if a fall in the price of X leads to a rise in the consumption of Y; a compensating variation in income being made of course in each case'*. That means, according to Hicks, when a fall in price of X is combined with compensated variation in income, and if it induces the purchase of more of X and less of Y, then the commodities are said to be good substitutes. On the other hand, when a fall in price X is combined with compensated variation in income and if it induces the purchases of more of both the commodities X and Y, then the commodities are said to be complements.

It is important that, to assign substitution (competition) relationship between the two commodities, we need two commodities, but to analyze the complementarity relationship between the commodities, we should need more than two commodities. That means, to say two commodities are complements, when atleast one commodity should lie outside the group of complements, at whose expense, the substitution in favour of the group of complements can occur. In the words of Prof. Hicks, '*if the consumer is dividing his income between purchases of two goods only and cannot buy any other goods than these two, then there cannot be anything else but a substitution relation between the two goods. For if he is to get more of one of them and still be no better off than before, he must have less of the other. But, when he is dividing his income between more than two goods other kinds of relation becomes possible*'.

The above explanation clearly infers that, the relationship between the commodities is studied based on substitution effect alone. J.R. Hicks also analyzed the concepts of substitutes and complements using two commodity case and three commodity case. In his famous work, '*Value and Capital*', J.R.Hicks mentioned three commodities *viz.*, X,Y and money (M). Here, the third commodity, 'money' represents all other goods and it is, otherwise, termed as 'composite good'.

Analysis of Substitutes in Case of Two Commodities

For easy understanding, let us consider two commodities rice and wheat, as they are good substitutes. The analysis shown through Figure (9.35) reveals that, after the elimination of income effect, the substitution effect alone contributes to the purchase of more quantity of rice by M_1M_3. This clearly infers that,

- ✰ The two commodities are good substitutes, as due to fall in the price of the rice commodity, the quantity purchased of rice commodity is increased (M_1M_3) and quantity purchased of wheat commodity is decreased (N_1N_3).
- ✰ A relatively small change in price of rice commodity brings about large change in the quantity of rice demanded.

However, in case of perfect substitutes like Kashmir apples and Shimla apples, the IDCs are parallel straight lines, (Figure 9.41) for which the MRS remains constant. In such cases, the consumer's equilibrium will be at one corner of the IDC, implying corner solution. The analysis shown through the Figure 9.41 reveals that, for the given level of income of the farmer and for the prices of two commodities PL_1 is the price line and it touches the IDC I_1 at Q_1 (consumer's equilibrium), where the consumers consumes only OM_1 number of Kashmir apples (because of corner equilibrium). Assume, the price of Kashmir apple is increased in the market and thereby, the new price line is PL_2 and it touches the new IDC I_2 at Q_2 (consumer's equilibrium), where the consumer consumes only ON_2 number of Shimla apples (because of corner equilibrium).

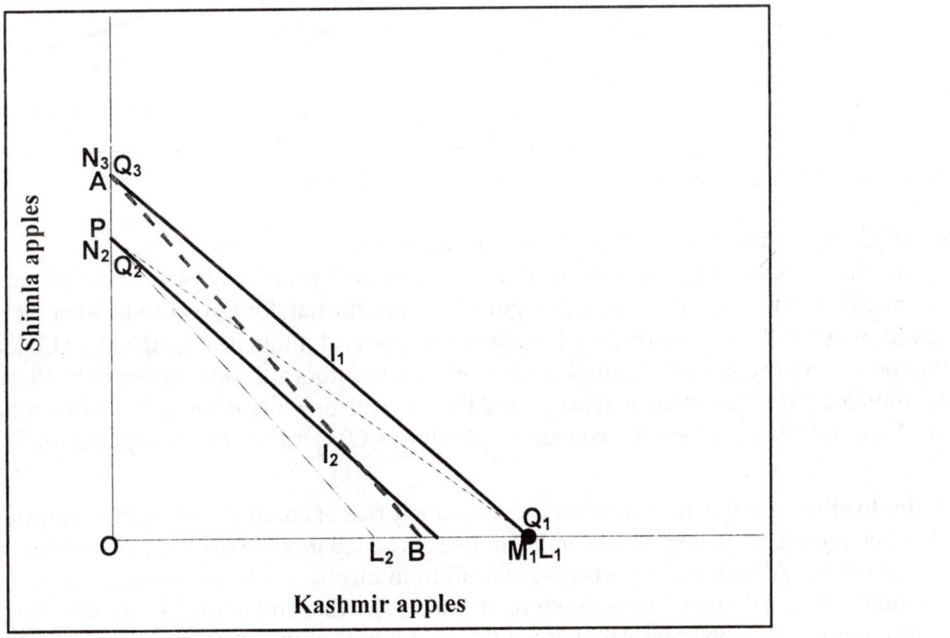

Figure 9.41: Price effect in perfect substitutes.

In order to eliminate the income effect due to rise in price of Kashmir apple, compensatory variation in income has been made by giving some amount of money to the consumer, so as to maintain his real income constant. So, BL_2 is the compensatory variation in income with reference to Kashmir apples and PA is the compensatory variation in income with reference to Shimla apples, thereby, AB is the compensated price line and it touches the original IDC I_1 at Q_3 (consumer's equilibrium), where the consumer purchases only ON_3 number of Shimla apples (due to rise in price of Kashmir apples and they are prefect substitutes). So, the substitution effect in case of perfect substitutes is very large *i.e.*, from Q_1 to Q_3.

Analysis of Substitutes in Case of Three-Commodities

Hicks considered three commodities *i.e.*, X,Y and money (M) (money as a composite good) to explain the substitution relationship between the commodities, as two commodities invariably have a substitution relationship, because of convexity of IDC. He defined the substitution relationship between the goods as *'good Y is a substitute for X, if the MRS of Y for M falls, as good X is substituted for M such that, the consumer is no better than before'*. That means, if the MRS between Y and M decreases, as more of X is purchased, then X and Y have substitution relationship. This is explained through Panel A of the Figure 9.42 where, the quantity of rice purchased by the consumer is increased from OR_1 to OR_2, it leads to fall in the MRS of money due to purchase of an additional unit of wheat. In other words, if the consumer purchases an additional unit of rice, then he will spend less money to purchase of an additional unit of wheat. This infers, the MRS_{RM} is more than MRS_{WM}. Hicks further concluded that, if the consumer divides his income in the purchases of two goods rice and wheat, then they have substitution relationship. The new IDC I_2 becomes more flatter, indicates fall in MRS_{WM} over MRS_{RM}.

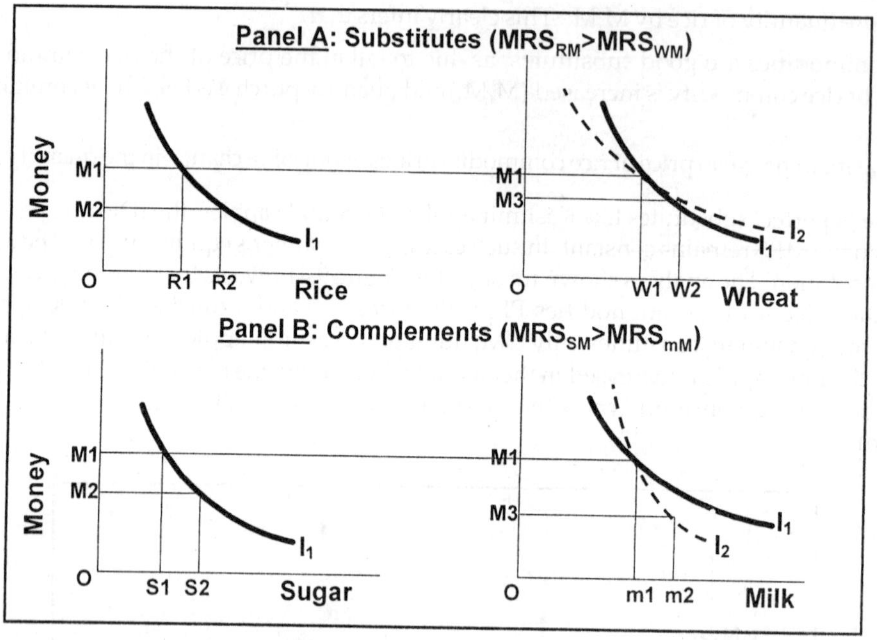

Figure 9.42: Substitutes and Complements – Three commodities case.

Analysis of Complements In case of Two-Commodities

To make the analysis simple, let us consider two commodities *viz.*, computer monitors and CPU's which are good complements. The analysis shown through Figure 9.43 reveals that, for the given level of income of the consumer and for the given prices of the commodities, PL_1 is the price line and it touches the IDC I_1 at Q_1 (consumer's equilibrium), where the consumer purchases OM_1 number of computer monitors and ON_1 number of CPUs. Assume, the price of the computer monitor is decreased in the market, and thereby, the new price line is PL_2 and it touches the new IDC I_2 at Q_2 (consumer's equilibrium), where the consumer purchases OM_2 number of computer monitors and ON_2 number of CPUs.

In order to eliminate the income effect due to fall in price of computer monitors, compensatory variation in real income has been made by taking away some amount of money income from the consumer, so as to maintain his real income constant. So, AP is the compensatory variation in income with reference to CPUs and BL_2 with reference to computer monitors and thereby, AB is the compensated price line and it touches the original IDC I_1 at Q_3 (consumer's equilibrium), where the consumer purchases OM_3 number of computer monitors and ON_3 number of CPUs. This movement along the same IDC I_1 between Q_1 to Q_3 (or) from OM_1 to OM_3 infers substitution effect. This analysis infers that,

- ☆ Due to the elimination of income effect by compensating the variation in real income, the substitution effect between Q_1 and Q_3 points on IDC I_1 infers that, though both the commodities are good complements, they behave like good substitutes too in some portion (Q_1Q_3) of the IDC I_1.
- ☆ In case of good complements, for a large change (decrease) in price of the commodity, (say, computer monitors), there is a small change (increase) in its quantity demanded, thereby, in the Q_1Q_3 portion of the IDC I_1, the two commodities will substitute each other. Thus, price effect includes substitution effect (M_1M_3) and Income effect (M_3M_2).

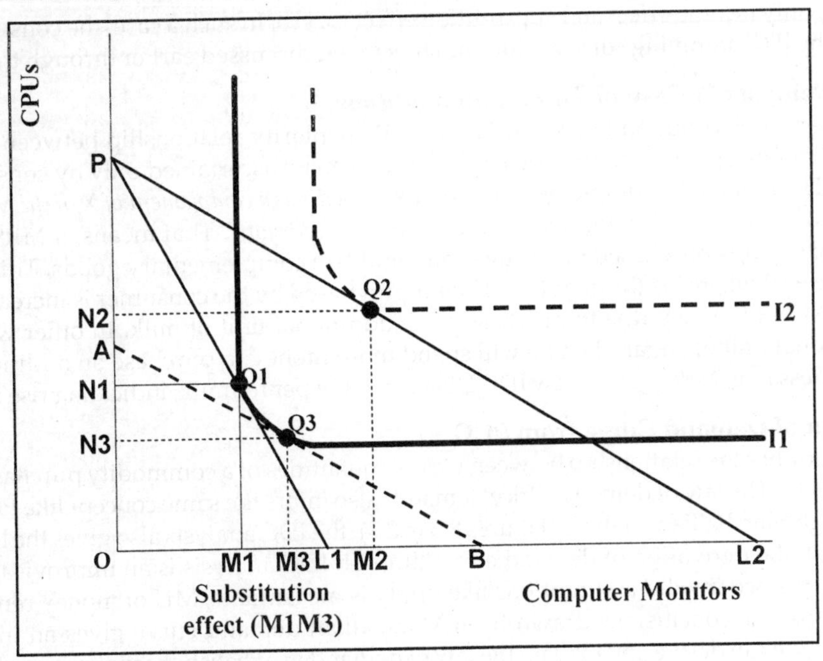

Figure 9.43: Substitution effect in good complements.

However, in case of perfect complements like rice and water in cooking rice, the IDCs are right angled (or) 'L' shaped and the substitution relationship between the commodities is completely absent *i.e.*, MRS is equal to zero. The analysis shown through the Figure 9.44 reveals that, for the given level of income of the consumer, PL_1 is the price line and it touches the IDC I_1 at Q_1 (consumer's equilibrium), where the consumer uses OM_1 quantity of rice and ON_1 volume of water. Assume, the price of rice is decreased in the market, and thereby, the new price line is PL_2 and it touches the new IDC I_2 at Q_2 (consumer's equilibrium), where the consumer employs OM_2 quantity of rice and ON_2 volume of water.

In order to eliminate the income effect due to fall in price of rice, compensatory variation in real income has been made by taking away some amount of money income from the consumer, so as to maintain his real income constant. So, AP is the compensatory variation in income with reference to volume of water and BL_2 with reference to quantity of rice and thereby, AB is the compensated price line and it touches the original IDC I_1 at Q_1 (consumer's equilibrium), where the consumer purchases the original OM_1 quantity of rice and ON_1 volume of water. So, the substitution effect in case of perfect complements is zero. The movement from Q_1 to Q_2 refers to income effect. So, in case of perfect complements,

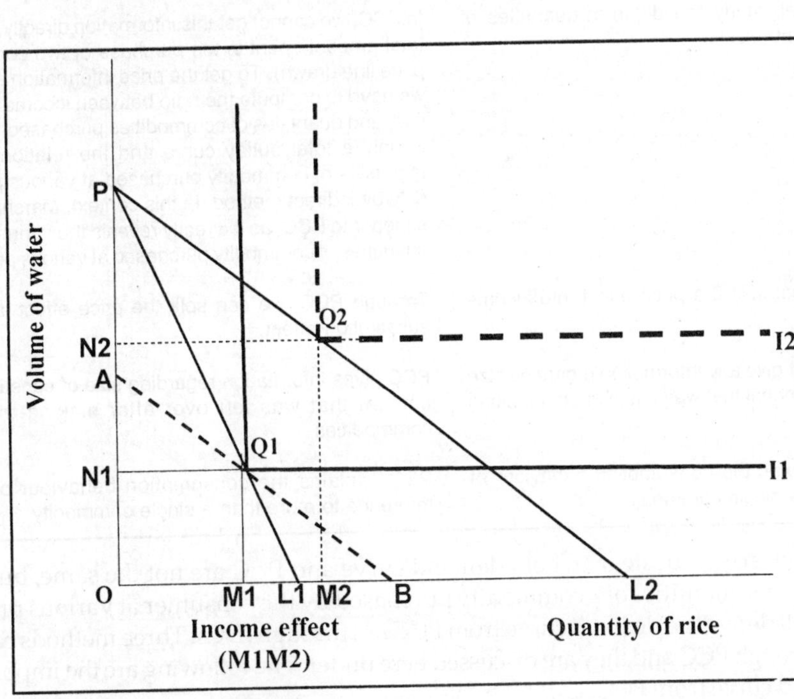

Figure 9.44: Price effect in perfect complements.

price effect includes only income effect and substitution effect is zero. In such a case, the consumer's equilibrium will be at one corner of the IDC, implying corner solution. The same is discussed earlier through the Figure 9.23.

Analysis of Complements in Case of Three Commodities

Hicks considered three commodities to explain complementarity relationship between the commodities. The complementarity relationship between the two commodities can be ascertained only by considering a minimum of three commodities. He defined the complementary good as, '*good Y is a complement of X, if the MRS of Y for M increases, as good X is substituted for M such that, the consumer is no better than before*'. That means, if MRS between milk and M increases, as more of sugar is purchased, then sugar and milk are complementary goods. This is explained through Panel B of the Figure 9.42 where, as the quantity of sugar purchased by the consumer is increased from OS_1 to OS_2, it leads to rise in the MRS of money due to purchase of an additional unit of milk. In other words, if the consumer purchases an additional unit of sugar, then he will spend more money to purchase an additional unit of milk. This infers that, MRS_{SM} is less than MRS_{mM}. The new IDC I_2 becomes steeper for milk, indicating rise in MRS_{mM} over MRS_{SM}.

9.5.3.6. Derivation of Demand Curve from PCC

We know, PCC studies the relationship between various quantities of a commodity purchased at various levels of prices of the commodity. The law of demand (Price demand) also infers the same concept like PCC. As we derived the demand curve through Marshallian analysis (Figures 8.3 to 8.7), the IDC analysis also gives the basis for the derivation of demand curve. But, the derivation of demand curve through IDC analysis is an improvisation over Marshallian analysis, as the latter framed invalid assumptions like utility is measurable, MU of money remains constant etc. So, with such assumptions, the conclusions drawn from Marshallian demand curve gives an invalid picture. Hence, through IDC analysis, we can derive and explain the law of demand in a realistic manner, as such invalid assumptions are not framed here. So, in IDC analysis, since the PCC curve infers the same concept like law of demand, it can be employed to derive and explain the law of demand. However, before deriving the demand curve from PCC, it is essential to study the differences between Marshallian demand curve and PCC (Table 9.10).

Table 9.10: Differences between Marshallian Demand curve and PCC

Marshallian Demand Curve	*PCC*
Demand curve is drawn by taking quantity of commodity demanded on X-axis and price of the commodity on Y-axis.	PCC is drawn through IDCs taking two commodities, one each on X-axis and Y-axis or with money income on Y-axis and one commodity on X-axis.
In demand curve, price of the commodity is directly mentioned on the Y-axis.	In PCC, price of the commodity is not directly mentioned, but a price line is drawn indicating the price ratio of two commodities or we can derive price information about the commodities through computing the ratio between money income and quantities of the commodities purchased by the consumer.
Through demand curve, we can study the different quantities of commodity demanded at different prices.	In PCC, we cannot get this information directly, as it indicates only the total amount spent in the purchase of two commodities (through the price line drawn). To get the price information about the commodities, we have to compute the ratio between income of the consumer (price line) and quantities of commodities purchased. This implies that, PCC is only a total outlay curve and the relationship between various quantities of commodity purchased at various prices can be deduced only by indirect method. In this context, Marshallian demand curve is superior to PCC, as it directly reveals the information between various quantities of commodity purchased at various prices of the commodity.
Marshallian demand curve do not split the price effect into income effect and substitution effect.	Through PCC, we can split the price effect into income effect and substitution effect.
Marshallian demand curve do not give any information regarding size of consumer's income and the amount that was left over after making the purchase of a commodity.	PCC gives information regarding size of consumer's income and the amount that was left over after making the purchases of two commodities.
Marshallian demand curve explains the consumption behaviour of the consumer with reference to a single commodity.	PCC explains the consumption behaviour of the consumer with reference to more than a single commodity.

The above differences clearly indicate that, both demand curve and PCC are not the same, but they are alike in explaining the aspect of various quantities of a commodity purchased by the consumer at various prices. So, inspite of the above differences, we still derive the demand curve from PCC *i.e.*, through IDCs. Three methods can be employed to derive the demand curve through PCC and they are discussed here under. The following are the important assumptions formulated to derive demand curve from PCC.

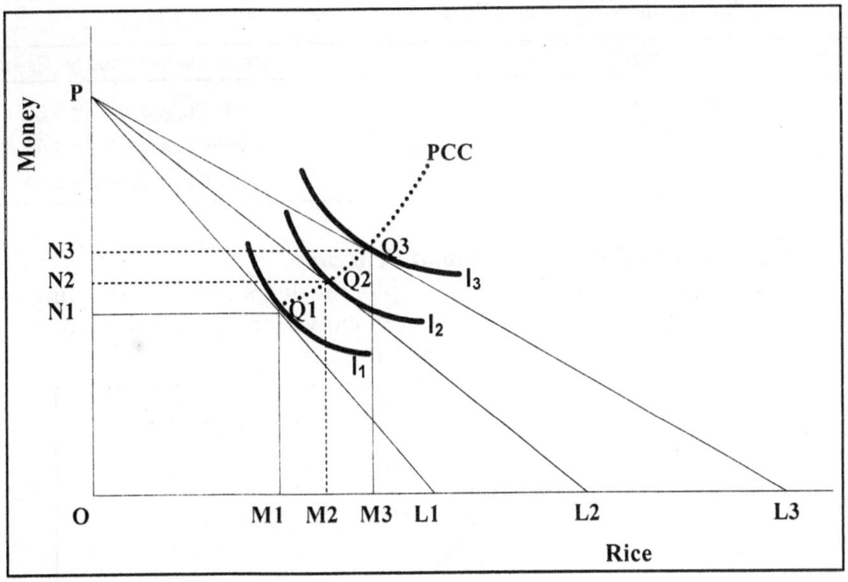

Figure 9.45: Method 1– Derivation of PCC for commodity rice (Normal good).

- The money spent on the commodity rice is constant.
- The price of the commodity under consideration decreases in the market and the prices of other related commodities remains same.
- The tastes and preferences of the consumer remains same.

1. *Method 1*

This method involves the following steps:

- Derivation of PCC for the commodity (rice).
- From the PCC drawn, we can prepare a demand schedule for the commodity.
- Based on the demand schedule, we can derive the demand curve for the commodity.

Derivation of PCC for the Commodity

On X-axis, commodity rice and on Y-axis money income of the consumer are taken (Figure 9.45). For a given level of income of the consumer and given price of the commodity rice, PL_1 is the price line drawn and it touches the IDC I_1 at Q_1, where he purchases OM_1 quantity of rice commodity by spending PN_1 amount of money. When price of the commodity rice is decreased in the market, PL_2 is the new price line and it touches the higher IDC I_2 at Q_2, where the consumer purchases OM_2 quantity of rice commodity by spending PN_2 amount of money. If the price of rice commodity is still decreased in the market, PL_3 is the new price line and it touches still higher IDC I_3 at Q_3, where the consumer purchases OM_3 quantity of commodity by spending PN_3 amount of money. If we connect Q_1, Q_2 and Q_3 points of consumer's equilibrium, it will give the upward sloping PCC.

Preparation of Demand Schedule for Rice

From the PCC drawn through the Figure 9.45, we can derive the demand schedule for the commodity. When the consumer is on the IDC I_1, at consumer's equilibrium Q_1, he purcnased OM_1 quantity of commodity rice by spending PN_1 amount of money. That means, the price of the commodity can be given by PN_1/OM_1 or we can also get the same information through computing the ratio between OP/OL_1. Similarly, when the consumer is on the IDC I_2, at consumer's equilibrium Q_2, he purchased OM_2 quantity of commodity rice by spending PN_2 amount of money. That means, the price of the commodity can be given by PN_2/OM_2 or we can also get the same information through computing the ratio between OP/OL_2. When the consumer is on the IDC I_3, at consumer's equilibrium Q_3, he purchased OM_3 quantity of commodity rice by spending PN_3 amount of money. That means, the price of the commodity can be given by PN_3/OM_3 or we can also get the same information through computing the ratio between OP/OL_3. So, now we have the information regarding various quantities of rice purchased by the consumer (OM_1, OM_2 and OM_3) at various prices (PN_1/OM_1 or OP/OL_1, PN_2/OM_2 or OP/OL_2 and PN_3/OM_3 or OP/OL_3 respectively). Let us suppose, money income of the consumer (OP) is Rs. 300, OL_1=50, OL_2=70 and OL_3=90 and at these prices, he purchased OM_1=20kgs, OM_2=30kgs and OM_3=40kgs. From this assumed data, we can prepare the demand schedule for rice and it is shown through Table 9.11.

Table 9.11: Demand schedule for rice derived through PCC.

Quantity Demanded (kg)	*Price of the Commodity (Rs/kg)*
OM_1=20	=OP/OL_1 = 300/50 = 6.00
OM_2=30	=OP/OL_2 = 300/70 = 4.50
OM_3=40	=OP/OL_3 = 300/90 = 3.33

Derivation of Price Demand Curve from the Demand Schedule

From the demand schedule (Table 9.11), we can easily plot the demand curve, showing an inverse relationship between price and quantity demanded for rice and the same is shown through Figure 9.46.

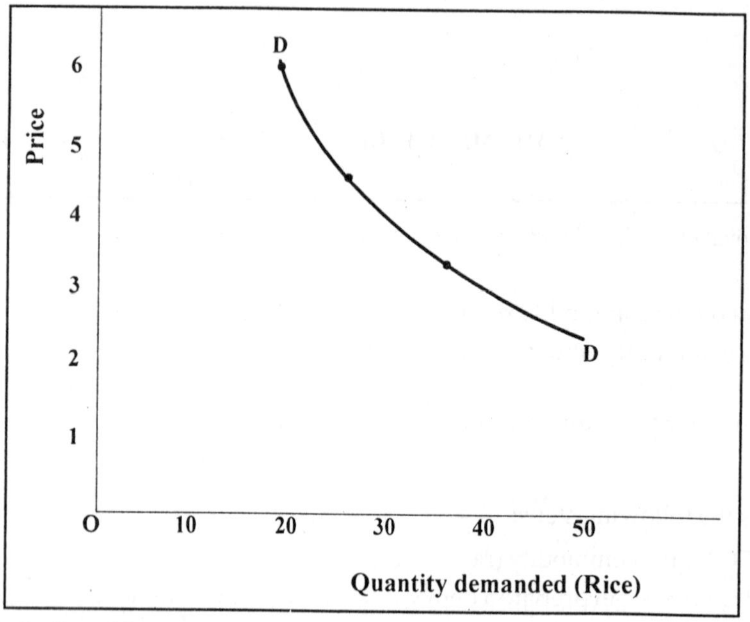

Figure 9.46: Method 1– Derivation of Demand curve from PCC (Normal good).

2. Method 2

This method is much more simple compared to the earlier method, as here, we need not have separate figures for PCC and demand curve and they both are combined to deduce the demand curve from the PCC. From the PCC drawn as in the above method (Figure 9.45), we extend the equilibrium points Q_1, Q_2 and Q_3 to the lower Panel B as Q_4, Q_5 and Q_6 equilibrium points respectively, thereby, we arrive at the different quantities of rice commodity purchased at different prices of the rice commodity, which is nothing but a demand curve. A close examination of the Panel A and Panel B of the Figure 9.47 reveals that, in Panel B, on Y-axis, the price information of rice derived from Panel A (discussed in Method 1) are taken and this price information is related to different quantities of rice commodity purchased by the consumer and this helps to derive the demand curve directly.

3. Method 3

In this method, instead of drawing the demand curve in the lower panel below the PCC, here the demand curve is shown within the Panel drawn for PCC. That means, in this method, the two panels drawn in Figure 9.47 are combined and shown as a single Panel (Figure 9.48).

As explained through Figure 9.45, the PCC is derived for the commodity rice. Now, to derive the demand curve for the commodity, consider equilibrium position Q_1. We know, at this equilibrium position Q_1, the consumer purchases OM_1 quantity of rice by paying PN_1 amount of money. So, we can get the unit price of rice commodity as PN_1/OM_1 or OP/OL_1. The computation of unit price of the commodity is shown through the Figure 9.48, by considering a point 'A' on the X-axis at 'a distance of one unit' from M_1. From this point 'A' draw a line parallel to price line PL_1, so that it will meet the perpendicular line Q_1M_1 at point B. It is clear from the Figure 9.48 that, the triangles OPL_1 and M_1BA are similar. So, we can say, $OP/OL_1 = M_1B/M_1A$. Since, OP is the price of OL_1 quantity, then M_1B is the price of M_1A quantity. Since, M_1A represent a unit quantity of the rice commodity, M_1B represent unit price of rice. Similarly, the triangles OPL_2 and M_2DC corresponding to equilibrium Q_2 are similar. So, we can say, $OP/OL_2 = M_2D/M_2C$. Since, OP is the price of OL_2 quantity, then M_2D is the price of M_2C quantity. Since, M_2C represent a unit quantity of the rice

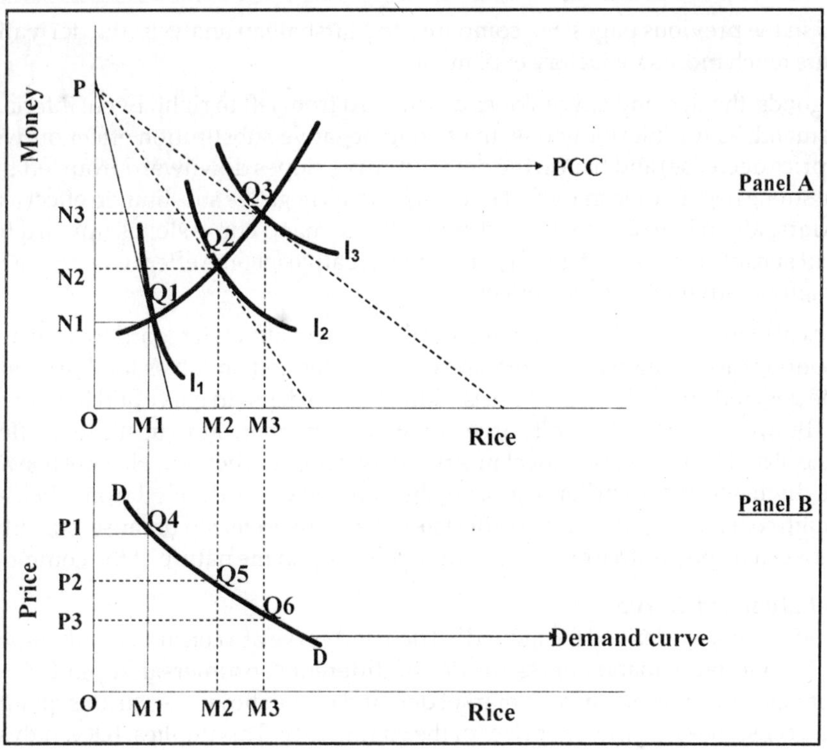

Figure 9.47: Method 2 – Derivation of demand curve from PCC.

commodity, M_2D represent unit price of rice. Likewise, corresponding to equilibrium Q_3, the triangles OPL_3 and M_3FE are similar. So, we can say $OP/OL_3 = M_3F/M_3E$. Since, OP is the price of OL_3 quantity, then M_3F is the price of M_3E quantity. Since, M_3E represent a unit quantity of the rice commodity, M_3F represent unit price of rice. Since, all the points B, D and F represent the different quantities of rice purchased at different prices, then by joining B, D and F, we derive a downward sloping demand curve D_1D_1, indicating more quantities of rice purchased, as the price falls from M_1B to M_3F.

So, these three methods explain that, demand curve can be easily derived from PCC and it is a more convincing explanation in the sense that, the total price effect can be further divided into income and substitution effects, which

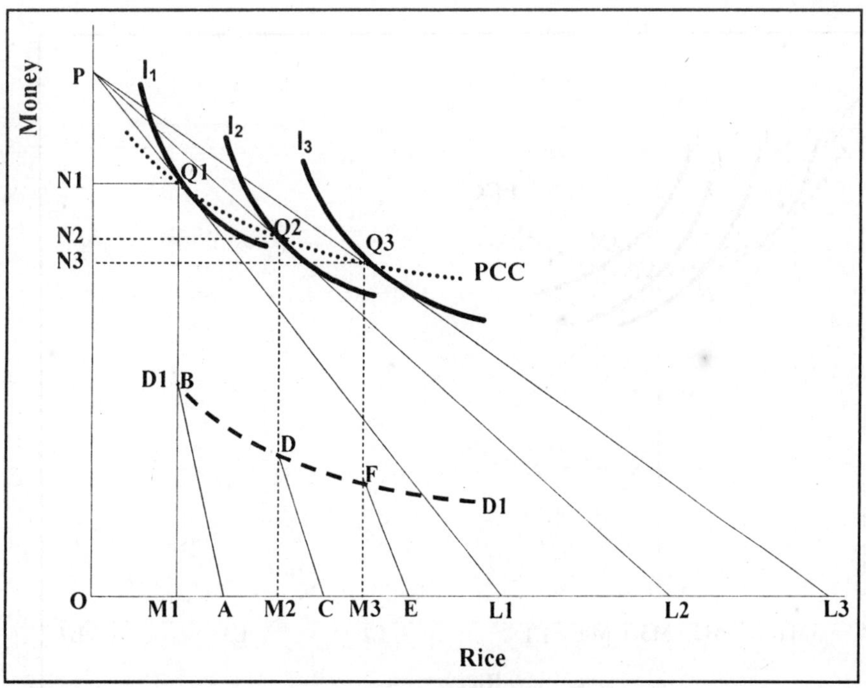

Figure 9.48: Method 3 – Derivation of demand curve from PCC of rice (normal good).

was already discussed in the previous pages. So, compared to Marshallian analysis, the derivation of demand curve from IDC analysis offers much more satisfactory explanation.

In case of normal goods, the demand curve slopes downward from left to right. Even, inferior goods also obey the Marshallian law of demand, as for inferior goods, the strong negative substitution effect outweighs weak negative income effect (case of price decrease) and hence, the demand curve slopes downward from left to right. But, in case of giffen goods, where the strong negative income effect outweighs weak negative substitution effect (case of price decrease), they do not obey the Marshallian law of demand and hence, the demand curve slopes upward from left to right. But, due to the availability of substitutes even for giffen goods, the negative income effect will not be that much strong and hence, it cannot outweigh negative substitution effect.

The three geometrical methods employed above gives the same result with regard to derivation of demand curve from the PCC. The important aspect here is, the demand curve so derived, involves total price effect, as discussed by Alfred Marshall. So, the demand curve derived above is a Marshallian demand curve and it involves both income effect and substitution effect. But, to derive the demand curve in a realistic manner, only substitution effect is to be considered and hence, it should be isolated from the price effect in terms of two approaches *viz.*, Hicks-Allen approach and Slutsky approach and they are discussed in the earlier pages. So, the demand curve derived from Hicks-Allen approach and Slutsky approach through considering substitution effect alone is more preferred because, substitution effect is always negative, but income effect may be positive or negative depending upon the nature of the commodity.

Derivation of Market Demand Curve

Through Figures 9.45 to 9.48, we derived the individual demand curve of a consumer with respect to rice commodity. If we summate all the individual demand curves for rice of different consumers and plotted on the same graph, it represents a market demand curve. That means, a market demand curve shows the various quantities of a commodity rice purchased by all the consumers at different prices of the commodity. This implies, IDC analysis guides in deriving the market demand curve for a commodity and this is a meritorious aspect over Revealed Preference Theory.

9.5.3.7. Computation of E_D from PCC

It is possible to compute the E_D from the PCC (Figure 9.49) in terms of total outlay method. As explained through the Figure 9.49, PCC is drawn for rice commodity. Assume that, income of the consumer is Rs. 120, OL_1=20, OL_2=30, OL_3=40 and OL_4=50 and quantities demanded at different prices are, OM_1=5.0, OM_2=10.0, OM_3=13.3, OM_4=16.0. Now, the total outlay spent on rice at different prices is shown through the Table 9.12 and a close look at the Figure 9.49 reveals the following interesting points:

- ✰ If the price of rice is decreased from Rs. 6.00 to Rs. 2.40, the quantity of rice demanded is increased from 5.0 kgs to 16.0 kgs.
- ✰ When the price of rice falls from Rs. 6.00 to Rs. 4.00, the quantity demanded increases from 5kgs to 10 kgs, and thereby, total outlay on rice increases from Rs. 30 to Rs. 40. This indicates the E_D for rice is more than unity.

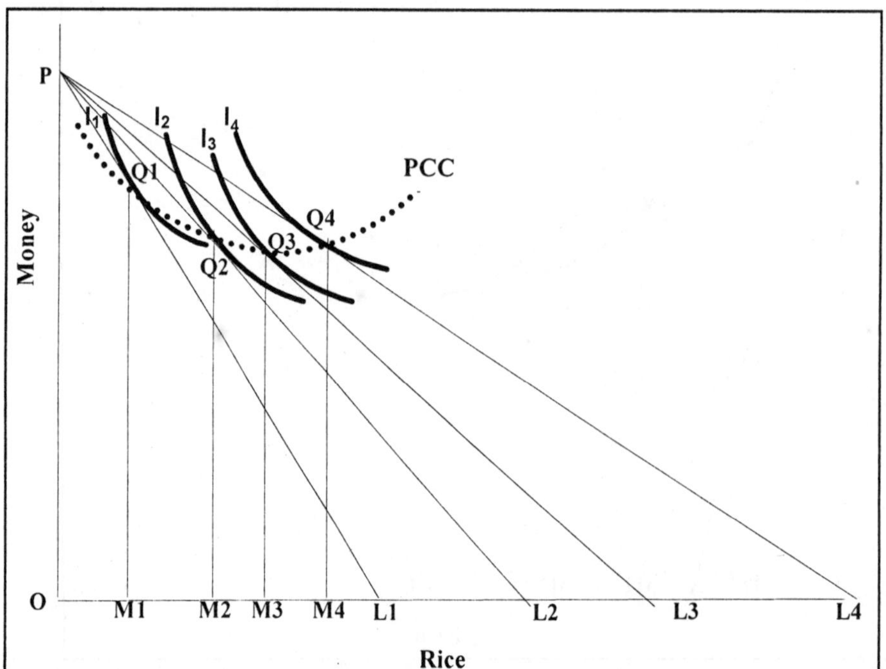

Figure 9.49: Estimation of E_D from PCC.

That means, in the Figure 9.49, between Q_1 and Q_2 of the PCC, the E_D is more than unity. Rather, we can say, when the PCC slopes downward, the E_D for rice is more than unity.

☆ When the price of rice falls from Rs. 4.00 to Rs. 3.00, the quantity demanded increases from 10.0kgs to 13.3kgs, but total outlay on rice remains same at Rs. 40. This indicates the E_D for rice is unity. That means, in the Figure 9.49 between Q_2 and Q_3 of the PCC, the E_D is unity and between these points, the PCC is a straight line parallel to X-axis.

☆ When the price of rice falls from Rs. 3.00 to Rs. 2.40, the quantity demanded increases from 13.3kgs to 16.0 kgs, but the total outlay on rice decreases from Rs. 40.00 to Rs. 38.40. This indicates the E_D for rice is less than unity. That means, in the Figure 9.49, between Q_3 and Q_4 of the PCC, the E_D is less than unity. Rather, we can say, when the PCC slopes upward, the E_D for rice is less than unity.

Table 9.12: Total outlay on rice commodity at different prices.

Price	*Qty Demanded*	*Total Outlay (Rs)*
(OP/OL1)=6.00	OM1=5.0	30.00
(OP/OL2)=4.00	OM2=10.0	40.00
(OP/OL3)=3.00	OM3=13.3	40.00
(OP/OL4)=2.4C	OM4=16.0	38.40

9.5.3.8. Marshallian Uncompensated Demand Curve Vs Compensated Demand Curves

The demand curve indicates various quantities of a commodity purchased at different levels of prices of the commodity. This concept is explained by the Marshallian law of demand. The demerit of this Marshallian version of explaining law of demand is that, here the total price effect was studied and it was not decomposed into substitution effect and income effect. However, in true sense, as explained earlier, when price of the commodity is decreased, more quantity of that commodity is purchased, because of both substitution effect and income effect. So, to isolate the true substitution effect, we have to eliminate the income effect and this was studied in IDC analysis through Hicks-Allen and Slutsky approaches. So, when the income effect is eliminated by the said approaches, the resultant demand curve is called compensated demand curve. That means, the compensated demand curve shows the various quantities of a commodity the consumer purchases at various prices, but by nullifying the income effect. To put it simply, a compensated demand curve shows only the substitution effect and no income effect is considered. In this aspect, the compensated demand curve differs from the Marshallian demand curve, where in Marshallian demand curve, we study the total price effect *i.e.*, both substitution effect and income were studied together and income effect is not separated from the substitution effect. So, Marshallian demand curve is otherwise called as Marshallian uncompensated demand curve, as income effect is not eliminated.

As explained earlier, in case of Hicks-Allen method, the income effect is eliminated through compensating the indirect increase or decrease in real income due to decrease or increase in price of the commodity. In case of Slutsky method, cost difference is worked out to eliminate the income effect due to decrease or increase in price of the commodity. This elimination of income effect will ensure to study only the substitution effect and this helps to deduce compensated demand curve. The derivation and comparative picture about these uncompensated Marshallian demand curve and compensated demand curves of both Hicks-Allen and Slutsky approaches is studied through the Panels A and B of the Figure 9.50.

In Panel A of the Figure 9.50, the total price effect and components of price effect *i.e.*, substitution and income effects derived through Hicks-Allen and Slutsky approaches are shown. In Panel B, all the three types of demand curves derived *viz.*, Marshallian uncompensated demand curve, Hicks-Allen compensated demand curve and Slutsky compensated demand curve are shown.

Marshallian Uncompensated Demand Curve

In Panel A of the Figure 9.50, for a given level of income of the consumer and at a given price of rice commodity (P_1), PL_1 is the price line drawn and it touches the IDC I_1 at Q_1, where the consumer purchases OM_1 quantity of rice. When the price of rice commodity is decreased to P_2, PL_2 is the new price line and it touches the higher IDC I_2 at Q_2, where the consumer purchases OM_2 quantity of rice. So, this increase in quantity of rice purchased from OM_1 to OM_2 is due to fall in the price of the commodity and hence, the movement from Q_1 of IDC I_1 to Q_2 of IDC I_2 refers to Price effect (which includes both Substitution and Income effects).

The same is represented in the Panel B in the form of Marshallian uncompensated demand curve D_1, which shows that, at OP_1 level of price, OM_1 is the quantity of rice demanded and at OP_2 price, OM_2 is the quantity of rice demanded.

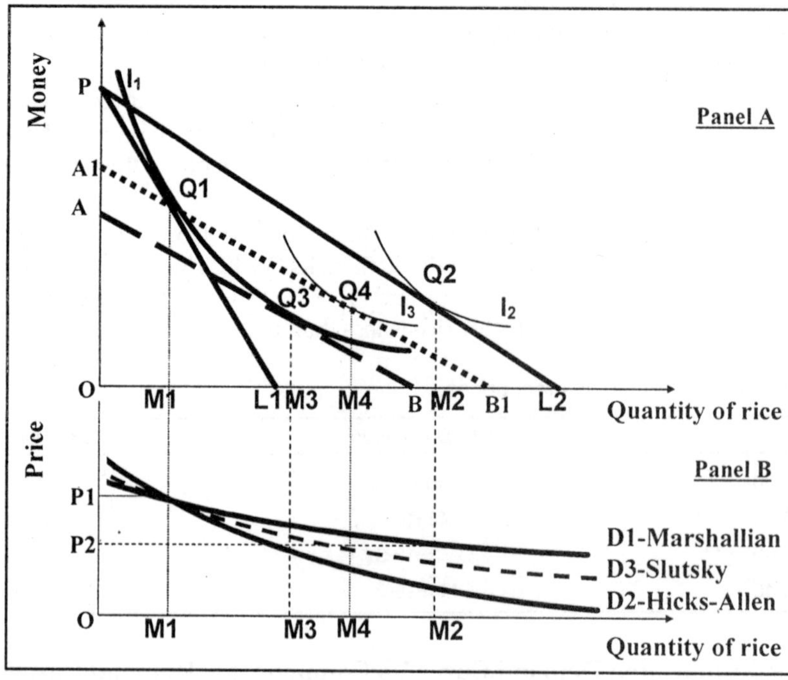

Figure 9.50: Derivation of Marshallian and Compensated demand curves.

Hicks-Allen Compensated Demand Curve

In Hicks-Allen method, the income effect is eliminated through compensating the indirect increase or decrease in real income due to decrease or increase in price of the commodity. In this case, the indirect increase in real income of the consumer is nullified by taking away some amount of money from the consumer due to fall in price of the rice commodity and this helps to trace out the substitution effect alone. Hence, the indirect increase in real income PA or BL_2 was removed and AB is the new compensated price line drawn parallel to PL_2. This compensated price line AB will touch the original IDC I_1 at Q_3, where the consumer purchases OM_3 quantity of rice. This movement from Q_1 to Q_3 on the same IDC I_1 refers to Substitution effect. So, in Hicks-Allen method, to study the substitution effect alone, the consumer was brought back on to the same IDC I_1. (The movement from Q_3 of IDC I_1 to Q_2 of IDC I_2 refers to income effect).

The same is explained in the Panel B with reference to the demand curve D_2. The movement from Q_1 to Q_3 or M_1 to M_3 on the demand curve D_2 refers to substitution effect, where the consumer purchased more quantity of rice (OM_3>OM_1) due to fall in price of rice from OP_1 to OP_2.

Slutsky Compensated Demand Curve

In this method, indirect increase in real income of the consumer due to fall in price of rice commodity was studied through computing the cost difference. So, this cost difference was removed from the increased real income of the consumer, so as to get the Slutsky compensated price line given by A_1B_1, drawn parallel to PL_2. This new price line A_1B_1, passes through original combination Q_1 of IDC_1, where the consumer, if desires, he purchases OM_1 quantity of rice. This infers that, PA_1 or B_1L_2 is the cost difference removed from the PL_2 price line, so as to trace out only substitution effect or to eliminate income effect. But, with the fall in price of rice commodity, the consumer will definitely purchase more of rice and hence, he moves to higher IDC I_3 and this price line A_1B_1 touches IDC I_3 at Q_4, where the consumer purchases OM_4 quantity of rice commodity. This movement from Q_1 to Q_4 refers to substitution effect and movement from Q_4 to Q_2 refers to income effect.

The same is explained in the Panel B with reference to the demand curve D_3. The movement from Q_1 to Q_4 or M_1 to M_4 on the demand curve D_3 refers to substitution effect, where the consumer purchased more quantity of rice (OM_4>OM_1) due to fall in price of rice from OP_1 to OP_2.

Comparison between Marshallian Uncompensated Demand Curve and Compensated Demand Curves

A close look at the three demand curves *viz.*, D_1 (Marshallian uncompensated demand curve), D_2 (Hicks–Allen compensated demand curve) and D_3 (Slutsky compensated demand curve) reveals that, D_1 curve is more elastic compared to D_2 and D_3 curves. This is because, D_1 curve has studied the total price effect and it has not divided the price effect into substitution and income effects. So, D_1 curve represents total price effect unlike D_2 and D_3 curves, as they represent only substitution effect and income effect was eliminated. Hence, D_1 curve infers that, the consumer purchases more quantity of rice (OM_2>OM_1) due to fall in price of rice from OP_1 to OP_2, when compared to D_2 and D_3 demand curves.

Among the two compensated demand curves *viz.*, D_2 (Hicks–Allen compensated demand curve) and D_3 (Slutsky compensated demand curve), D_3 is more elastic compared to D_2. This is because, in Slutsky method, the indirect increase in real income is measured through studying the cost-difference and this is removed from the PL_2 price line, so as to trace out only substitution effect or to eliminate income effect. As a result, the consumer moved to a higher IDC I_4 and he purchased more quantity of rice ($OM_4 > OM_1$) when compared to D_2 curve ($OM_3 > OM_1$). In case of D_2 curve, the consumer was brought back on to the same IDC I_1 through compensating the indirect increase in real income and hence, he purchases OM_3 quantity of rice due to fall in price of rice from OP_1 to OP_2 and hence, it is less elastic compare to D_3 (because $OM_3 < OM_4$).

Another important difference between these three demand curves is that, the compensated demand curves D_2 and D_3 always slopes downward from left to right because, in these two curves only substitution effect was studied and income effect is eliminated. However, in case of Marshallian demand curve (D_1), the total price effect was studied, which includes both income and substitution effects and hence, depending upon the relative strengths of income and substitution effects, the shape of demand curve changes and hence, it may or may not slope downwards. This is because, for a price decrease,

- ☆ In case of normal good, the demand curve always slopes downward from left to right because, both negative substitution effect and positive income effect act in the opposite direction of price (*i.e.*, these two effects complement each other).
- ☆ In case of inferior goods (say, jowar), the demand curve slopes downward from left to right because strong negative substitution effect outweighs weak negative income effect.
- ☆ In case of giffen goods (say, bread), the demand curve slopes upwards from left to right *i.e.*, it has a positive slope because, strong negative income effect outweighs weak negative substitution effect.

9.6. Similarities between Marshallian Approach and IDC Approach

Though the concepts of Marshallian and IDC analyses differ in their approaches to arriver at consumer's equilibrium, there are some similarities between them and they are discussed here under.

- ☆ Both in Marshallian analysis and IDC analysis, we assume that, the consumer is rational in his consumption behaviour and he tries to achieve maximum satisfaction in the consumption process. In Marshallian analysis, we arrive at the point of maximum satisfaction in quantitative terms ie, in LDMU, when MU is equal to zero, TU is maximum or MU of expenditure is equal to MU of money and in case of LEMU, the point of maximum satisfaction is when MU of expenditure is same across all the commodities. In IDC analysis, we arrive at the point of maximum satisfaction in qualitative terms, as this approach is based on ordinalism. This approach states that, if the consumer moves from lower IDC to higher IDC, it indicates the consumer is deriving higher satisfaction and *vice versa*. The consumer is said to be at the point of maximum satisfaction, where the price line is tangent from below to the highest attainable (convex) IDC. So, both the approaches aim at maximizing the consumer's satisfaction in the consumption of commodities, of course, Marshallian approach through the concept of cardinalism and IDC approach in terms of ordinalism.
- ☆ In both Marshallian analysis and IDC analysis, the method of explaining the consumer's equilibrium is same. In case of Marshallian analysis, the point of consumer's equilibrium in terms of MU of expenditure with respect to rice and wheat commodities (LEMU) is given by,

  $(MU_R/P_R) = (MU_W/P_W)$ ***Equation 9.12***

  The above equation, can be written as, $(MU_R/ MU_W)=(P_R/P_W)$ ***Equation 9.13***

  In IDC analysis, the consumer's equilibrium with respect to rice and wheat is given by

  $MRS_{RW} = P_R/P_W$ or $(\Delta W)/(\Delta R) = (P_R/P_W)$ ***As per Equation 9.11***

  So, from the above two equations 9.13 and 9.11, we can infer that,

  $(MU_R/ MU_W) = MRS_{RW}$ ***Equation 9.14***

  Considering Equation 9.11, we can write the above equation 9.14 as,

  $(MU_R/ MU_W) = (P_R/P_W)$ ***Equation 9.15***

  By cross multiplication of the terms in Equation 9.15, we get

  $(MU_R/P_R) = (MU_W/P_W)$ ***Equation 9.16***

  So, the above Equation 9.16 is similar to Equation 9.12 and this implies that, through IDC analysis, we can arrive at the consumer's equilibrium as in Marshallian analysis, but without invalid assumptions.
- ☆ Marshallian analysis states the assumption that, MU of a commodity diminishes as the consumer goes on consuming a commodity, while in IDC analysis, the MRS between the related commodities diminishes, if the

consumer goes on substituting one commodity for the other. That means, the concept of diminishing MRS of IDC analysis is considered as parallel to the concept of diminishing MU of Marshallian analysis

- Both the approaches apply psychological and introspective (self experienced) methods of consumer behaviour. Marshallian analysis is based on psychological introspection and IDC analysis is based on the hypothetical experimentation of the consumer behaviour in the consumption of related commodities.
- Both the approaches arrive at same conclusions regarding the consumer behaviour in terms of deriving additional satisfaction gained in the consumption of commodities, in arriving at the point of consumer's equilibrium and in derivation of the demand curve.

The similarities discussed above between Marshallian analysis and IDC analysis and the same conclusions revealed by both the approaches led the Prof. D.H. Robertson to criticize the IDC analysis as, *'old wine in new bottle'*. He opined that, IDC analysis simply substituted the concepts of Marshallian analysis and there is nothing like new in IDC approach. But, other Modern Economists are of the opinion that, since MRS_{RW} can be written as MU_R/MU_W (as per the equation 9.14), if the MUs are quantifiable (as per the cardinality approach), then their ratio will give MRS_{PW}. If they are not quantifiable, (IDC approach), the MRS_{RW} can still explain the concept of consumer's equilibrium. Hence, IDC analysis is like a new wine and it is definitely an improvisation made over Marshallian analysis. This can be further emphasized through analyzing the superiority of IDC analysis over Marshallian analysis.

9.7. Superiority of IDC Analysis over Marshallian Analysis

Though both Marshallian analysis and IDC analysis have some similarities in dealing with the consumer behaviour, the IDC approach is certainly more meritorious in explaining the consumption behaviour of the consumer and its superiority is discussed here under.

- ***More realistic ordinality approach:*** The IDC analysis is based on ordinalism, which is more realistic compared to cardinalim concept followed in Utility analysis. This is because, quantification of utility is purely imaginary, as utility derived by the consumer is subjective and a psychological feeling. So, it is meaningful to rank the satisfaction as low, medium and high rather than quantifying in the form of utils. Hence, IDC analysis discarded the Marshallian approach on the cardinality grounds.
- ***Fewer assumptions:*** The IDC analysis arrives at the same conclusions regarding the consumer's behaviour like Marshallian utility analysis, but with less number of assumptions and that too not considering invalid assumptions like, utility is measurable, constancy of MU of money, goods have independent utilities etc., and thereby, it sounds more realistic and superior in its approach. It framed realistic assumptions compared to Marshallian utility analysis. The assumptions such as ordinality measurement of utility, consumer have scale of preferences for the commodities, consumer deals with related commodities etc., are far superior to assumptions of Marshallian analysis and hence, the conclusions drawn from the IDC analysis are more scientific and practically acceptable.
- ***Studies the combination of two goods rather than one good:*** In Marshallian analysis, we deal with only one commodity and we assume that, the utility of this commodity is independent of the other commodity. So, in the Marshallian utility analysis, we measure the utility of a commodity in isolation from other commodities, *i.e.*, it confines itself to a single commodity model. Hence, the discussion pertaining to substitution and complementarity relationships among the commodities is avoided. But, in reality, consumer prefers a related combination of goods (complements) and generally substitutes one good for another good (substitutes). On this ground, IDC analysis is superior over Marshallian approach, as it deals with two commodities *i.e.*, either substitutes, complements, inferior goods, giffen goods etc. This further widened the scope of IDC analysis over Marshallian analysis.
- ***Offers better explanation for substitutes and complementary goods:*** Marshallian analysis deals with only price effect and it assumed the commodities have independent utilities. Hence, it could not offer much explanation regarding the consumption behaviour of substitutes and complementary goods. The earlier Economists explained the substitution and complementarity relationships among the goods in terms of E_C. However, in IDC analysis, the concept of independent utilities of commodities is not assumed and further it guides the splitting up of price effect into income effect and substitution effect and this facilitates to analyze the consumption behaviour with reference to related commodities. The splliting up of price effect into substitution effect and income effect helps to explain the Marshallian law of demand with reference to inferior goods and giffen goods. Since, Marshasllian analysis deals with only price effect, it could not offer explanation, why the giffen goods disobey the law of demand.
- ***Replacement of diminishing MU:*** The concept of diminishing MRS of IDC analysis is parallel to diminishing MU concept of LDMU. This is really a positive change and not a mere translation because, in case of IDC

analysis, we will study the consumer behaviour with respect to two commodities, whereas, in Marshallian analysis (LDMU approach), we will deal with only one commodity. This concept of diminishing MRS is more scientific and realistic and is well applicable in the fields of consumption, production and distribution. Moreover, it is based on comparability of utilities of the two related commodities and not on the quantification of utilities. So, this concept of diminishing MRS has explained the theory of consumer's demand more realistically.

☆ ***No assumption of constancy of MU of money:*** The Marshallian approach has based on unrealistic assumption *viz.*, constancy of MU of money. However, such assumption is not framed in IDC analysis because, if the consumer goes on spending the money, the MU of money increases, rather than remaining constant. Hence, IDC analysis has not assumed constancy of MU of money because, MU of money changes with the change in price of the good. In IDC analysis, we assumed that, there is indirect increase in the real income of the consumer due to fall in price of the good and hence, he purchases either more of normal good or less of inferior good and this is called as income effect and this is separated from substitution effect. Since the income effect is considered, the concept of IDC analysis has not assumed constancy of MU of money.

☆ ***Explains proportionality rule in a better way:*** In both Marshallian analysis and IDC analysis, the method of explaining the consumer's equilibrium is same, but IDC analysis explained the concept without assuming invalid assumptions like utility is measurable, constancy of MU of money, goods have independent utilities etc., and the same is proved through Equations 9.12 to 9.16.

☆ ***Explain the dual effect of price effect:*** IDC analysis explains the dual effect of price effect *i.e.*, it explains both income effect and substitution effect due to change in price of the commodity. On the other hand, in Marshallian analysis, we study total price effect of the commodity and it is not categorized into income and substitution effects. Since, MU of money is assumed constant, the income effect is not considered in Marshallian analysis. In this aspect, IDC analysis is far superior to Marshallian analysis, as dual effect of price effect helps in explaining the concept of law of demand for inferior and giffen goods. Further, MU of money is not assumed constant and hence, it explained the law of demand in a realistic manner.

☆ ***Explain 'Giffen paradox':*** Marshallian theory does not break up the price effect into income effect and substitution effect and thereby, it does not explain why the giffen goods does not obey law of demand. Since, IDC analysis explains the break up of the price effect into income effect and substitution effect, the concept infers that, the strong negative income effect outweighs the weak negative substitution effect (case of price decrease) and this is responsible for disobeyance of law of demand by the giffen goods.

☆ ***Helps to derive a general 'demand theorem':*** In IDC analysis, the constancy of MU of money and independent utilities of commodities are not assumed and further the analysis guides to decompose the price effect into income effect and substitution effect. These meritorious aspects of IDC analysis helps to derive a general 'demand theorem' based on the relative strengths of income effect and substitution effect. If E_Y of a commodity is positive and if both income effect and substitution effect act in the same direction (*i.e.*, substitution effect is negative and income effect is positive), then that commodity will obey the law of demand (Normal goods). If the E_Y for a commodity is negative and if the negative substitution effect outweighs the negative income effect (case of price decrease), then that commodity will obey the law of demand (Inferior goods). If the E_Y for a commodity is negative and if the negative income effect outweighs the negative substitution effect (case of price decrease), then that commodity will not obey the law of demand (Giffen goods).

☆ ***Concept of real income considered:*** In Marshallian analysis, the concept of real income is not considered. Hence, only price effect is considered to explain the Marshallian law of demand. Since, no income effect is isolated from the price effect, the concept of rise in real income of the consumer due to fall in price of the commodity is not studied. But, in IDC analysis, the real income is measured by means of 'compensating variation' of income (Hicks-Allen approach) and 'cost difference' (Slutsky approach) methods and this helps to isolate the income effect on the consumption of the commodity. So, tracing of income effect is one of the major discoveries of the IDC analysis.

☆ ***More objective in approach:*** The concept of IDC analysis is more objective compared to Marshallian utility analysis. We know, the Marshallian utility analysis is subjective in its approach.

☆ ***Applicability for the analysis of Welfare Economics:*** The IDC technique is more popular, as it is mostly used for the analysis of Welfare Economics. For example, it is employed to explain that, the direct tax imposes a lesser burden than an indirect tax upon the consumer. Further, it explains that, the goods should not be distributed equally (rationing), as the income and preferences of consumers differ. It also helps for constructing the supply curve of labour in the economy.

9.8. Applications of IDC Analysis

The following are the important applications of IDC analysis:

1. Consumption

The explanation offered, so far, in this chapter, highlights the importance of IDC analysis in the field of consumption of commodities. An IDC indicates all possible combinations of two commodities that yield same level of satisfaction to the consumer. To derive maximum satisfaction, the consumer has to select one combination from different possible combinations and this is possible by drawing a price line taking into consideration his income and prices of the two commodities. The point of tangency between the highest attainable IDC and price line will indicate the point of maximum satisfaction with reference to two commodities. But, in practical situation, the consumer is faced with more than two commodities. In such cases also, this IDC analysis is applicable by taking money income on Y-axis (it indicates total purchasing power of the consumer) and a commodity on X-axis. So, by this analysis, we can draw the PCC and from which we can derive demand curve for each commodity and this represents the consumer's purchasing behaviour in the market.

2. Production

The concept of IDC analysis of consumption theory is contemporary or synonymous to isoquant of Production Economics. An isoquant is a curve, which indicates all the possible combinations of two resources that are capable of producing same level of output. In IDC analysis, at any point on the same IDC, the level of satisfaction remains same, but on the same isoquant, the level of output remains same. An isoquant will simply indicates different possible combinations of two resources to produce a given level of output, but it won't indicate which combination is best suitable to minimize the cost in the production programme. So, to select the best or least cost combination of resources, we draw a price line taking the income of the farmer and prices of the resources into consideration and the point of tangency between isoquant and price line indicates the point of cost minimization or least cost combination of resources.

3. Exchange

1DC analysis is very useful in the field of exchange of commodities, when two persons are entering the market and exchange the commodities between them. That means, IDC analysis is useful in 'Barter system' of trade. Edgeworth-Bowley box diagram employing the 1DC analysis helps to explain the barter system of trade between two persons.

Suppose, there are two farmers X and Y producing two commodities rice and wheat and they enter into barter system of trade. The process of exchange between the two farmers is explained through the Figure 9.51. In the Figure 9.51, IDCs from IDC I_1 to IDC I_9 are shown for farmer X taking rice on X-axis and wheat on Y-axis. Similarly, the IDCs from IDC I_1 to IDC I_9 are shown for farmer Y by turning the map to 180 degrees angle. So, for farmer Y, rice is taken on X-axis and wheat on Y-axis, so that, the axes of both the farmers X and Y will be in opposite direction and the entire graph is in the form of a rectangle. The IDCs drawn for farmers X and Y are tangent to each other indicating that, the

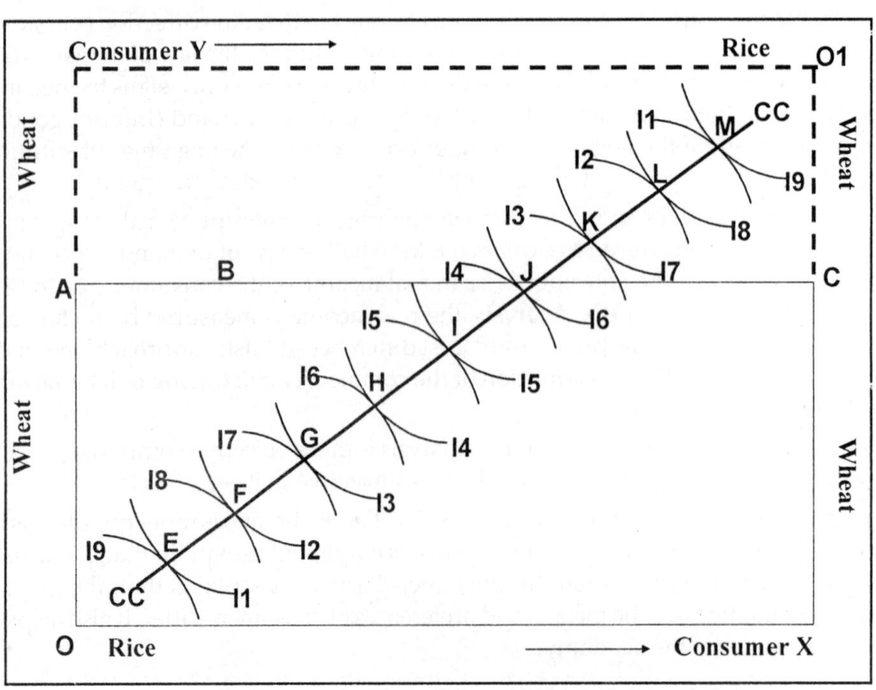

Figure 9.51: Application of IDC analysis in the field of trade (Edgeworth–Bowley Box).

MRS of two commodities is same between both the farmers and the line connecting these tangent points is called Contract curve, CC and along this curve, the trade takes place between the farmers.

It is clear from the Figure 9.51 that, farmer X posses OA units of wheat and AB or OB units of rice. Obviously, farmer Y posses BC units of rice and O_1C or O_1A units of wheat. So, in terms of exchange, farmer X posses more of wheat OA and less of rice AB or OB. Conversely, farmer Y posses more of rice BC and less of wheat O_1C or O_1A. So, farmer X wishes to exchange wheat with farmer Y to get more of rice and farmer Y wishes to exchange rice with farmer X to get wheat, so that, both derive maximum satisfaction, as for farmer X, rice is having more economic significance and for farmer Y, wheat is having more economic significance. So, to derive maximum satisfaction, farmer X will go on exchanging wheat to get more of rice from farmer Y. Similarly, farmer Y will go on exchanging rice to get more of wheat from farmer X to derive maximum satisfaction. So, in the process of exchange and to derive maximum satisfaction, both the farmers moves to higher IDCs. The exchange will continue between farmer X and farmer Y, till the MRS between the two commodities becomes equal to the price ratio of the two commodities for each farmer. Since the price ratio of two commodities remains same (assuming prices remains constant for rice and wheat), the MRS of two commodities between the two farmers will be the same. This is indicated by the same slopes of different IDCs *i.e.*, from IDC_1 to IDC I_9 of the two farmers, as they are touching each other at E, F, G, H, I, J, K, L, M points. On these points, *i.e.*, from E to M, the two farmers will try to exchange the commodities and it represents the final contract between them. Hence, the line connecting the equal slopes of IDCs of two farmers from E to M indicates a Contract curve. Now, the important question is, at which point on the contract curve, both the farmers will derive maximum satisfaction, so that, they will get mutually benefited in the trade mechanism for which the exchange of commodities is done. The ideal point of exchange of two commodities between the two farmers is at point 'I' on the contract curve because, both the farmers move to the highest possible IDC *i.e.*, IDC_5, so that, both the farmers can get mutually benefited.

Let us suppose, if both the farmers settled at point 'K' on the contract curve. This implies that, farmer X derives more satisfaction because, he moves to higher IDC I_7, indicating that, he is getting more of rice for exchanging wheat. Conversely, for farmer Y, at the point 'K' on the CC curve, he is on IDC I_3, and he derives less satisfaction compared to farmer X. On the other side, if the farmers settle at 'F', the farmer X is the loser, as he is on the lower IDC I_2, but farmer Y is the gainer, as he is on the higher IDC I_8. However, both the farmers will be at an equal position of trade advantage *i.e.*, at point I and they can settle at this point I only by mutual agreement. Otherwise, the gain in the trade depends upon the income or bargaining power. That means, there is no guarantee that, both the farmers settle down at point 'I' on the CC curve because, it depends upon the income level of the two farmers. Say, if farmer X is having more money, he can purchase more of rice from farmer Y and he moves to higher IDCs (say I_7) and he derives more satisfaction compared to farmer Y. On the other hand, if the other farmer Y possesses more money, he will settle at higher IDC (say, I_8) and he derives more satisfaction compared to farmer X. However, to have mutual benefit, both the farmers will exchange the commodities at point I on contract curve.

This exchange mechanism can be equally applicable in the field of international trade, where countries exchange different goods and services for their mutual benefit.

4. Rationing Vs Liberal Scheme of Distribution of Commodities

We know, Government resort for rationing at times of shortage of production of commodities. In rationing, the Government will distribute specific quantities of specific commodities to the consumers of similar income groups. Besides rationing, the Government also announces the liberal scheme, where the consumers are allowed to purchase the commodities as per their likes or dislikes in requisite quantities. In this context, the IDC analysis is useful to analyze and prove that, liberal scheme is better to the consumers than rationing system.

Suppose, there are two commodities rice and wheat that are rationed by the Government, because of shortage of production and supply of these two commodities. Assuming prices of these two commodities and the income level of the consumers remains constant (in rationing, commodities are offered in specific quantities to the consumers of similar income group). So, for the given level of income of the consumers and given of prices of the commodities, PL_1 is the price line drawn (Figure 9.52). In this system of rationing, suppose, there are two consumers A and B purchasing these two commodities. IDC I_A is the IDC for consumer A and IDC I_B is the IDC for consumer B. These two consumers will be given equal quantities of rice and wheat *i.e.*, OR_1+ OW_1 respectively and hence, they derive same level of satisfaction.

Let us suppose, the Government announces the liberal scheme, under which, the consumers can purchase the commodities as per their likes and dislikes in requisite quantities. Suppose, consumer A prefers more of rice and less of wheat, as per his tastes and preferences and thereby, he will be on IDC I_{A1} and it touches the same price line PL_1 (because their income levels are kept constant) at Q_2. So, the consumer A purchases OR_2 of rice and OW_2 of wheat *i.e.*, he purchases more of rice and less of wheat as per his tastes and preferences.

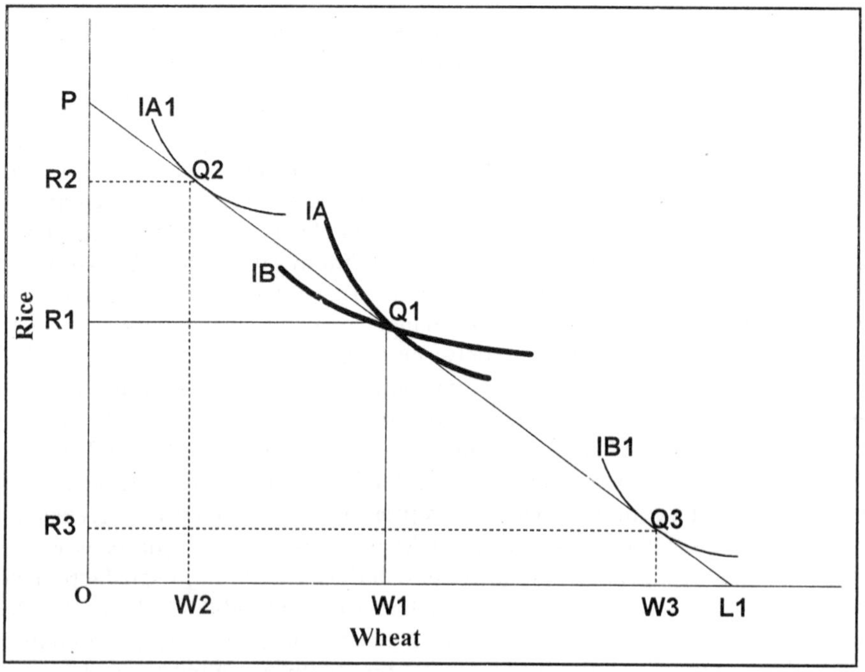

Figure 9.52: Application of IDC technique in rationing of commodities.

Suppose, consumer B prefers more of wheat and less of rice, as per his taste and preferences and thereby, he will be on IDC I_{B1} and it touches the same price line PL_1 (because their income levels are kept constant) at Q_3. So, the consumer B purchases OR_3 of rice and OW_3 of wheat *i.e.*, he purchases more of wheat and less of rice as per his tastes and preferences.

A close examination of the Figure 9.52 and analysis of the two systems of having the commodities reveals that, in rationing system, the consumers gets fixed quantities of both the commodities, whereas, under liberal system, the consumer can purchase the requisite quantities of both the commodities as per his tastes and preferences. The comparative analysis of the two systems reveals that, liberal system is more favourable to the consumers compared to rationing of commodities across the consumers, as liberal system ensures better distribution of commodities among the consumers as per their tastes and preferences. This is because, in both the cases of rationing and liberal schemes, the total quantity of rice and wheat sold in the market remains same *i.e.*, POL_1. Among POL_1 quantities of rice and wheat supplied and sold in the market, in liberal scheme, the consumers purchase these two quantities as per their tastes and preferences and hence, they derive higher satisfaction compared to rationing system. However, in rationing system, the consumers must necessarily purchase both the commodities and atleast one commodity they have to purchase more against their interest. So, a liberal scheme ensures better distribution of commodities and hence, higher satisfaction to the consumers compared to rationing system. Further, the Government's aim of controlled distribution of commodities is not disturbed under liberal scheme. This analysis further infers that, as the tastes and preferences of the consumers differ, the goods should not be distributed equally (rationing) and they should be made available on liberal basis.

5. In Taxation

The principle of IDC technique can be applied in the field of taxation or public finance. So, the concept of IDC analysis helps to analyze the welfare effects of different taxation policies implemented by the Government. We are aware that, the Government imposes two types of taxes *viz.*, direct tax (income tax) and indirect tax (sales tax on a commodity). This IDC concept helps to analyze, whether paying income tax by the consumer is more burdensome or paying the tax indirectly through sales tax of the commodity is more burdensome. This is important because, irrespective of the mode of taxation policy followed by the Government *i.e.*, either direct tax or indirect tax, the Government can get the same revenue, but the impact of these taxes on the consumers will be different and the same is explained through the Figure 9.53. For a given level of income of the consumer and for a given price of commodity wheat, PL_1 is the price line. Initially, the consumer will be at equilibrium, when price line PL_1 touches the IDC I_1 at Q_1, where the consumer purchases OM_1 quantity of wheat by spending PN_1 amount of money or R_1N_1 amount of money.

Case1: Government Imposed Sales Tax on the Commodity

Let us suppose, the Government imposed sales tax on wheat and this leads to increase in the price of wheat in the market. So, the price line PL_1 pivots towards origin as PL_2 and it touches the new IDC I_2 at Q_2 (consumer's equilibrium), where the consumer purchases OM_2 quantity of wheat by spending PN_2 or R_1Q_2 amount of money. Out of this R_1Q_2

amount of money spent by the consumer in purchasing OM_2 or PR_1 quantity of commodity, S_1Q_2 is the amount of money goes to the Government as sales tax revenue. This infers that, due to imposition of sales tax by the Government, the consumer purchases less quantity of wheat OM_2 ($OM_2<OM_1$) and hence, he derives less amount of satisfaction, which is also indicated by his movement from higher IDC I_1 to lower IDC I_2.

Case 2: Government Imposed Direct Tax

Let us assume another case, where Government imposed direct tax (income tax) on the consumer because, the Government decided to collect the money (revenue) directly from the consumer, instead of collecting the same through imposing sales tax on the commodity. Due to imposition of income tax, the income of the consumer will fall and thereby, the price line comes close to origin as AB. Since, the amount of revenue earned by the Government in the form of direct tax is same as that of indirect tax, the price line AB lies below the original price line PL_1 at a distance of S_1Q_2 or passes through the point Q_2 and touches the higher IDC I_3 at Q_3. This indicates that, $PA = S_1Q_2$ is the amount of direct tax revenue goes to the Government. On this price line AB, the consumer will move to a higher IDC I_3 and he will not remain on IDC I_2, as I_2 is an IDC drawn because of rise in price of commodity due to imposition of sales tax. But, with reference to AB price line, since the price of the commodity remains same, the consumer definitely moves to higher IDC I_3 and touches the I_3 at Q_3 indicating that, the consumer purchases OM_3 quantity of wheat by spending PN_3 amount of money.

The comparative analysis of the above two cases of taxation reveals that, even though the Government is getting same amount of revenue (S_1Q_2=PA), imposition of direct tax is beneficial compared to imposition of indirect tax on the part of the consumers. This is because, in case of direct tax, the consumer moves to a higher IDC I_3 and he purchases more quantity of wheat OM_3 compared to imposition of indirect tax with reference to IDC I_2, where he purchases only OM_2 quantity of wheat ($OM_3>OM_2$). So, the consumer derives more satisfaction, if direct tax is imposed when compared to indirect tax imposed on the commodity. So, we can infer that, direct tax is more beneficial than indirect tax on the part of the consumers, though the revenue earned by the Government from either of the sources remains same.

6. Effect of Subsidy on Consumers

IDC technique helps to analyze and compare the effects of different types of subsidies offered by the Government to the beneficiaries. Subsidy can be given by the Government in two ways *viz.*, by offering the commodities to consumers at concessional prices and by giving money value of subsidy as cash directly to the consumers. So the IDC technique helps to analyze and compare, which form of subsidy is advantageous or meritorious to the beneficiaries and the same is explained through the Figure 9.54. For a given level of income and given price of rice, PL_1 is the price line and it touches the IDC I_1 at Q_1, where the consumer purchases OM_1 quantity of rice by spending PN_1 amount of money.

Case 1: Supply of Rice at Concessional Rate

Suppose, the Government is offering rice to the beneficiaries at concessional price through fair price shops. So, the original price line PL_1 pivots away from the origin as PL_2 and it touches the new IDC I_2 at Q_2, where the consumer purchases OM_2 quantity of rice by spending only PN_2 amount of money due to supply of rice at a concessional price. So,

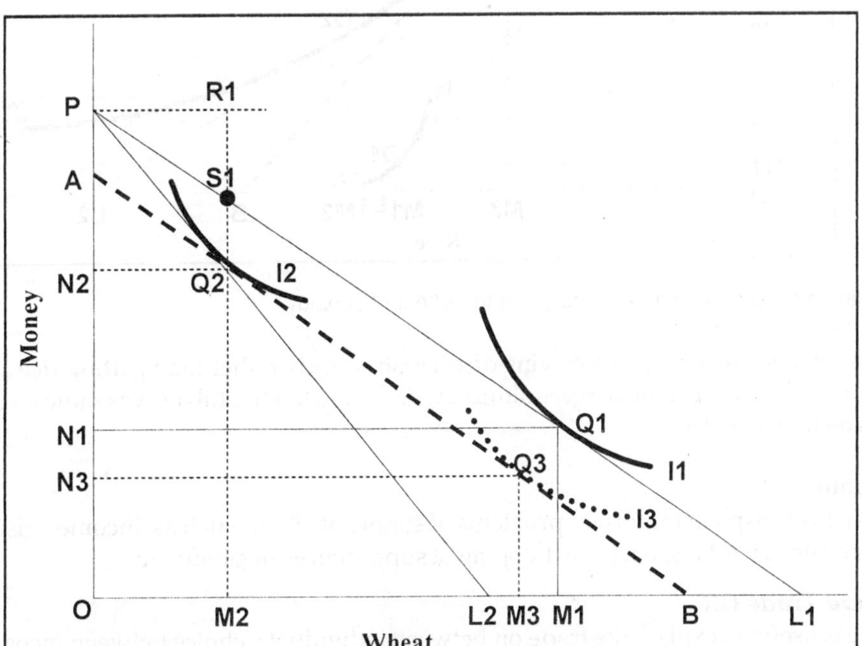

Figure 9.53: Application of IDC technique in taxation policy.

when the rice supplied at concessional price, the consumer purchases more quantity *i.e.*, $OM_2>OM_1$. This also infers that, N_2N_1 is the subsidy offered by the Government or availed by the consumers (in the form of subsidised price) in purchasing rice and this amount goes to the cereal suppliers and not to the consumers.

Case 2: Offering Money Value of Subsidy as Cash to the Beneficiaries

Let us suppose, Government offered liquid cash to the consumers directly, instead of supplying rice at concessional price. So, when money cash is offered to the consumer as a subsidy, his income will increase and hence, the price line will move away from the origin as AB. This new price line AB will touch the IDC I_2 at Q_3, where the consumer purchases OM_3 quantity of rice by spending AN_3 amount of money. Out of AN_3 amount of money spent by the consumer, PA is the cash subsidy offered by the Government and this will go to the consumers directly and the consumer's actual money incurred is only PN_3.

From the Figure 9.54, we can have the following interesting conclusions:

- When cash is paid to the consumer directly as subsidy, he will purchase less quantity of rice (OM_3), when compared to the situation, where rice is offered at concessional prices to him *i.e.*, OM_2 ($OM_3<OM_2$). So, there is less burden on the part of Government regarding the quantity of supply of rice into the market and this is actually what the Government wants.
- The value of cash subsidy given directly by the Government to the consumers *i.e.*, PA is less than the value of subsidy incurred by the Government when rice is supplied at concessional price to the consumers, N_1N_2 *i.e.*, $PA<N_1N_2$. This again reveals less burden on the part of the Government in offering subsidy in the form of cash, rather than offering the rice commodity at concessional price to the consumers.
- Whatever may be the form of subsidy offered by the Government to the consumers, the level of satisfaction derived by the consumer is same, as he lies on the same IDC I_2.

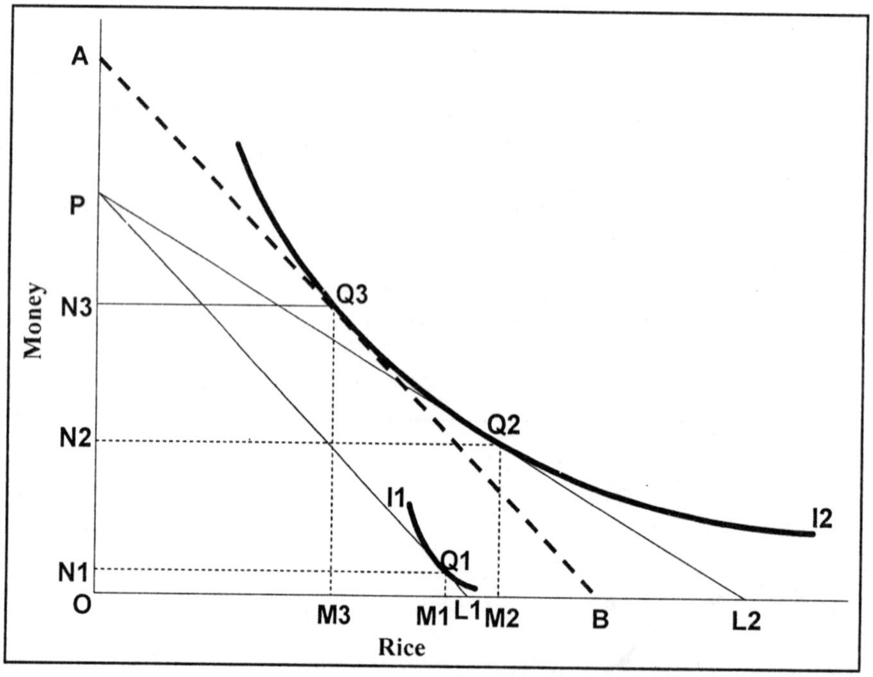

Figure 9.54: Application of IDC technique in analyzing the effect of subsidy.

- The consumer is also happy in receiving direct cash subsidy rather than getting rice at a concessional price because, he incurs less personal expenditure *i.e.*, $PN_3<PN_2$, but he still derives same level of satisfaction, as he remains on the same IDC I_2.

7. Supply of Labour

IDC analysis helps to explain important problems of supply of labour such as, income–leisure trade off, offering high overtime wage rate to workers, backward sloping of supply curve of labour etc.

(a) Income–Leisure Trade Off

IDC technique is useful to explain the trade off between individual's choice between income earned and leisure hours preferred. We know, there exists an inverse relationship between income earned and leisure hours enjoyed by the worker in a processing factory *i.e.*, if the worker enjoys more number of leisure hours (less number of working hours), his income will be low and *vice versa*. The worker offers his service in a factory taking into consideration various factors like nature of work, wage rate offered by the factory, number of working hours in a factory (number of leisure hours the worker prefers) etc. So, the decision between income earned by the worker and his preferences for leisure hours (working hours) can be explained by IDC technique. This analysis is based on the following assumptions:

- ☆ From the institution side or factory side, there should not be any constraint regarding the working for a fixed number of hours.
- ☆ The worker himself can freely select his working hours and leisure hours.
- ☆ Wage rate remains constant.

Let us suppose, a processing factory is offering employment to workers and they can perform the operations in a factory in three shifts *viz.*, 7AM to 3PM, 3PM to 11PM and 11PM to 7AM. The worker can select any shift in the factory and if he desires to work for more number of hours, he can increase his working hours by accepting to work in other shift continuously. So, total number of working hours offered by the factory is 24 hours and assume the worker has contributed his work for 'A' number of hours. So, the number of leisure hours (L) the worker enjoys is given by:

$$L = 24 - A \text{ or } A = 24 - L$$

In terms of income parameters, we can write the above equation as

$$A \times W = (24 \times W) - (L \times W),$$

where, W refers to the wage rate/hour in a factory, A x W refers to actual income earned by the worker through offering his labor service in the factory, 24 x W refers to the potential income or maximum possible income the worker can earn, if he works for all the working hours (in all three shifts) offered by the factory, L x W refers to the income missed by the worker by enjoying 'L' leisure hours.

The Figure 9.55 explains the Income-Leisure trade off of the worker in a factory. In the Figure 9.55, IL is the Income–Leisure line, which indicates all possible combinations of income and leisure hours enjoyed by the worker or working hours employed in the factory. That means, say, if the worker enjoys OL leisure hours, the income earned by the worker is zero. On the other hand, if the worker enjoys no leisure hours or if he works continuously in all the three shifts offered by the factory, he derives the OI level of (potential) income. The slope of Income–Leisure line indicates wage rate/hour and this is given by, (Δ total income/Δ total working hours).

As shown in the Figure 9.55, the worker derives maximum satisfaction both in terms of income earned and leisure preference, where the income leisure line IL is tangent to IDC I_1 at Q_1, which indicates that, the worker earns OI_1 level of income through enjoying OL_1 number of leisure hours or L_1L working hours.

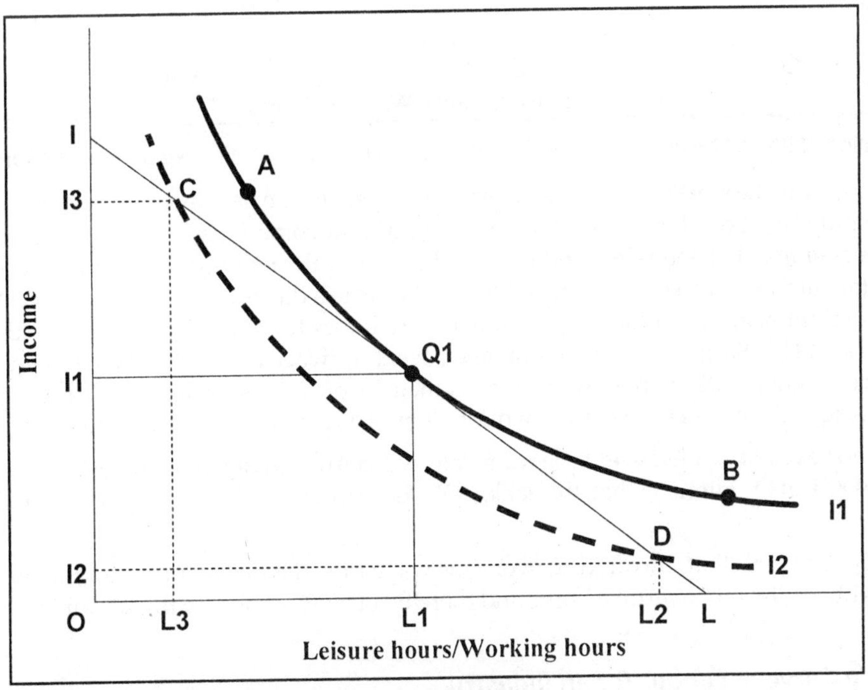

Figure 9.55: Application of IDC technique in Income-Leisure trade off.

Let us take two points on IDC I_1 *i.e.*, A and B, but on these two points, the worker cannot derive maximum satisfaction because, Income-Leisure line, IL will not touch these two points. This is because, IL line always remains constant, as number of working hours in a factory will not be above 24 hours a day and we also assumed wage rate as constant. So, deriving maximum satisfaction at A and B points on IDC I_1 is not possible.

Let us suppose, if the worker is offering his service on IDC I_2, then at both the points C and D, the labour can derive maximum satisfaction. This is because IL line touches these two points, C and D on IDC I_2. At point D on IDC I_2, the worker earns OI_2 level of income ($OI_2<OI_1$) by enjoying OL_2 leisure hours or working for L_2L ($L_2L<L_1L$) working hours. At point C on IDC I_2, the worker earns OI_3 level of income ($OI_3>OI_1$) by enjoying OL_3 leisure hours or for working L_3L ($L_3L>L_1L$) working hours. So, we can conclude that, by working for more number of hours in a factory or enjoying less number of leisure hours, the worker derives more income and *vice versa*. But, the worker prefers Q_1 combination of income and leisure on IDC I_1, as he derives more satisfaction on the highest IDC I_1 than IDC I_2. The IDCs so drawn between income and leisure preferences are called income-leisure trade off curves.

(b) Effect of Offering Over Time Wage Rate on Income Earned by the Labour

Let us suppose, the processing factory offered an higher over time wage rate to the worker. That means, if the worker avails the over time opportunity offered by the factory by working for more number of hours, the worker gets more income and hence, he moves to higher IDC I_2 (Figure 9.56) and realizes more satisfaction.

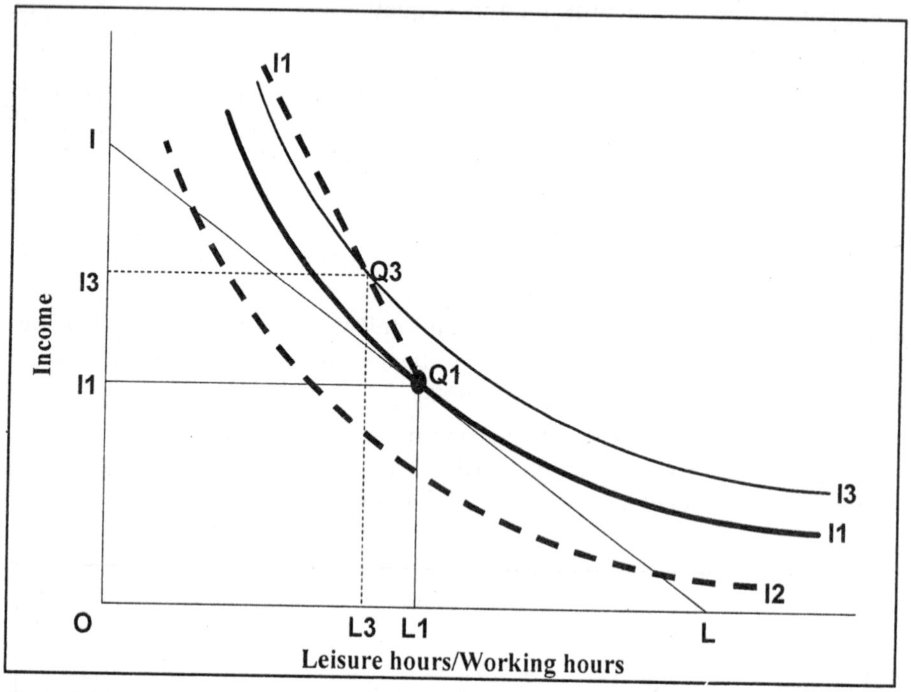

Figure 9.56: Application of IDC technique in Income-Leisure trade off when over time wage rate is offered.

Due to rise in income of the worker by availing over time wage rate, the new Income–Leisure line is I_1Q_1L and it will start at a higher position *i.e.*, above I on Y-axis compared to earlier Income-Leisure line IL. This new Income–Leisure line I_1Q_1L is more steeper than IL (steep slope indicates higher wage rate per hour *i.e.*, over time wage rate) and bears a kink at Q_1, as the total number of working hours and leisure hours in a day is equal to OL hours and it will not change. This constancy of total number of working hours plus leisure hours for a factory will give a kink at Q_1 to the new Income–Leisure line *i.e.*, I_1Q_1L. So, the workers, who availed over time wage rate, they will definitely move to higher IDC I_3 and enjoy more satisfaction with increased income, as their Income–Leisure line I_1Q_1L is tangent to IDC I_3 at Q_3, inferring that, they enjoy OL_3 leisure hours or L_3L working hours ($OL_3<OL_1$ or $LL_3>LL_1$) and earn OI_3 level of income.

The workers who have not availed over time wage rate, will derive satisfaction on IDC I_1, as their Income-Leisure line IL is tangent to IDC I_1 at Q_1, inferring that, they enjoy OL_1 leisure hours (LL_1 working hours) and earn only OI_1 level of income.

The above analysis reveals that, by offering over time wage rate, the number of working hours of a worker will increase from LL_1 to LL_3 *i.e.*, by L_3L_1 overtime hours and thereby, income earned by the labour increases from OI_1 to OI_3 *i.e.*, by I_1I_3.

(c) Deriving Supply Curve of Labour for an Industry

The IDC technique can be employed to derive supply curve for labour for an industry. We know, a labour will supply his service in an industry based on various factors like nature of the work, wage rate offered by the industry, number of working hours offered by the industry, leisure hours the labour prefers etc.

In the Figure 9.57, hours of leisure and work are taken on X-axis and income is taken on Y-axis. For a given working hours (or leisure hours enjoyed) and the income earned by the worker, I_1L is the Income-Leisure line. The slope of this I_1L line indicates wage rate/hour. This I_1L line touches the IDC I_1 at Q_1, where the labour earns ON_1 or Q_1L_1 income for performing LL_1 working hours or enjoying OL_1 leisure hours.

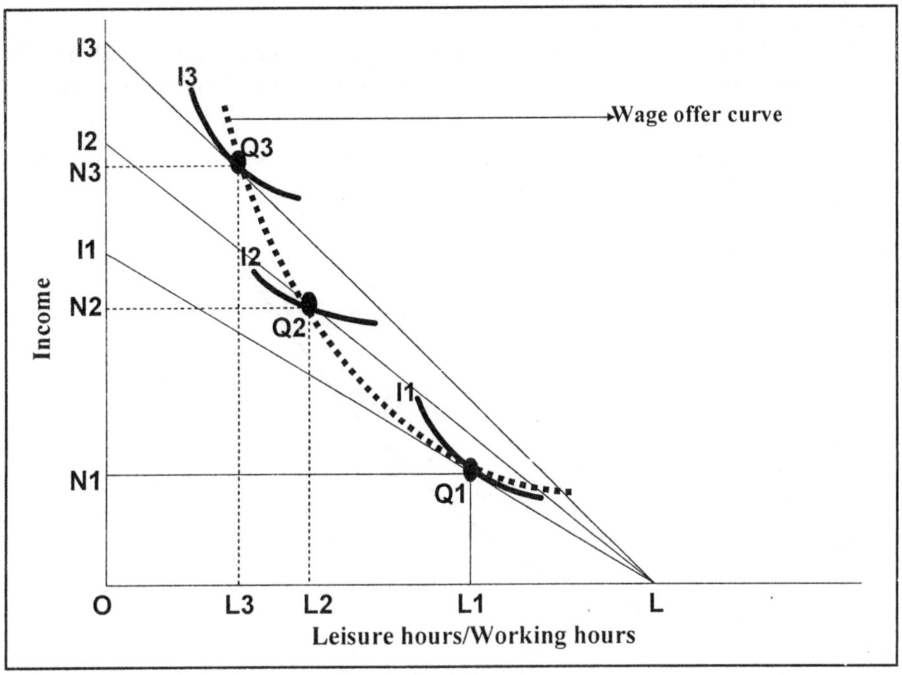

Figure 9.57: Application of IDC technique to derive wage-offer curve.

Let us suppose, the industry increased the wage rate and the labour shows the tendency to work for more number of working hours to earn more income. So, the new Income-Leisure line is I_2L. In the Figure 9.57, the new Income-Leisure line I_2L touches the point L on X-axis because, the total number of leisure hours and working hours is constant for a day. So, with the increase in income due to increase in wage rate, the labour derives more satisfaction and hence, he moves to higher IDC I_2, where I_2L line touches IDC I_2 at Q_2. So, it is clear that, when more wage rate is offered, the labour works for more number of hours *i.e.*, $LL_2>LL_1$ and thereby, his leisure hours are decreased *i.e.*, $OL_2<OL_1$ and thereby, he derived higher income $ON_2>ON_1$ or $L_2Q_2>L_1Q_1$.

If still the wage rate is increased, the labour contributes still more number of working hours *i.e.*, $LL_3>LL_2$ and thereby leisure hours are decreased *i.e.*, $OL_3<OL_2$ and the labour derives a higher income $ON_3>ON_2$ or $L_3Q_3>L_2Q_2$. Note that, if the wage rate is increased, the income-leisure line becomes more steeper and touches the higher IDC.

If we connect all the equilibrium points, Q_1, Q_2 and Q_3 of different incomes at different wage rates and working hours, it gives rise to a Wage–Offer curve. It is important that, this Wage-Offer curve is not synonymous to supply curve and we have to derive the supply curve from this Wage–Offer curve. For the Figure 9.57, the data is assumed and shown in Table 9.13.

Table 9.13: Supply schedule of labour for an industry

Equilibrium Point	*Wage Rate/Hour (Rs)*	*Working Hours*	*Leisure Hours*
Q_1	$OI_1/OL=15$	$L_1L=6$	$OL_1=18$
Q_2	$OI_2/OL=20$	$L_2L=8$	$OL_2=16$
Q_3	$OI_3/OL=25$	$L_3L=10$	$OL_3=14$

From the Figure 9.58, it is clear that, if wage rate is increased in an industry, the labour contributes more number of working hours and hence, the supply curve of labour will have a positive slope. This is mainly due to the following two reasons:

- ☆ ***Substitution effect:*** When wage rate is increased by the industry, the labour prefers to work for more number of hours and this indicates that, the labour has substituted working hours for leisure hours. So, due to substitution effect, working hours preference is more than leisure hours preference. Thus, due to substitution effect, the working hours to leisure hours ratio will increase.

☆ *Income effect:* When more wage rate is offered by the industry, the labour contributes more number of working hours and thereby, he earns more income and with this earned income he becomes potentially better-off and thereby, he prefers more number of leisure hours compared to working hours. This is due to income effect in which, leisure hours preference is more than working hours preference. Thus, due to income effect, the working hours to leisure hours ratio will decrease.

But, the supply curve will have a positive slope as in the Figure 26.7, indicating that, with increase in wage rate, the labour contributes more number of working hours and substitute leisure hours because, substitution effect is stronger than income effect.

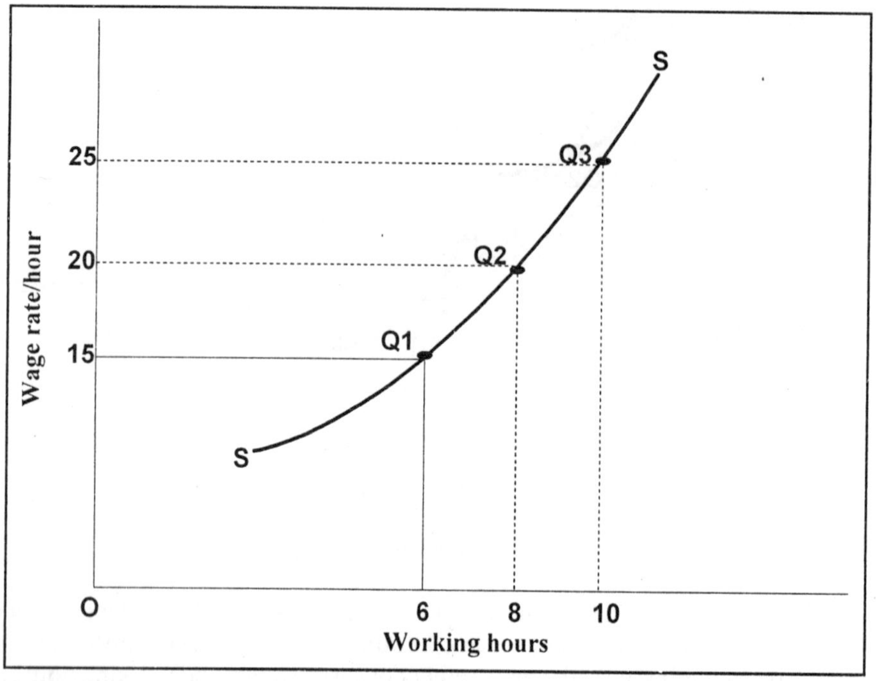

Figure 9.58: Application of IDC technique to derive supply curve for labour from wage offer curve.

(d) Backward Bending of Labour Supply Curve (Supply curve of labour for an economy as a whole)

Compared to the industry level, at economy level, the supply of labour depends upon still more factors *viz.*, economic and political factors, population growth, sex ratio, traditions, customs, society's attitude regarding women employment, geographical environment, age-structure, work preference, leisure preference, attractive pay norms, expertise, trainings received, presence of strong trade unions etc. At economy level, it is generally noted that, if the wage rate increases, the labour contributes more number of working hours to earn more income. But, it is interesting to say that, if wage rate is still increased, the number of working hours contributed by the labour will not increase proportionately, rather they decrease because, at one point of time, the labour prefers leisure hours to working hours.

In the Figure 9.59, when the wage rate is Rs.15/hr, the labour contributed six working hours, when the wage rate increased to Rs.20/hr, then the number of working hours are increased to eight, and when the wage rate is still increased to Rs.30/hr, the labour contributed 12 working hours. But, when the wage rate is further increased to Rs.32/hr, now the labour contributed less number of working hours *i.e.*, only 11.25 working hours. This infers that, at higher wage rates, the labour prefers more number of leisure hours compared to working hours. This is because, the labour will have inelastic cash requirements. That means, when the wage rate is Rs.30/hr, the labour contributed 12 working hours and thereby, he derived Rs. 360/- income. But, when the wage rate increased to Rs.32/hr, he contributed only 11.25 working hours and still he derives same amount of income *i.e.*, Rs. 360/-. So, the labour prefers more number of leisure hours and less number of working hours at very high wage rate, because of inelastic cash requirements.

From the Figure 9.59, it is clear that, up to AB portion of supply curve, substitution effect is greater than income effect and hence, the supply curve has positive slope. But, between B and C portion of supply curve, income effect is greater than substitution effect and hence, the supply curve has negative slope and thereby, at very high wage rate, the labour prefers leisure hours to working hours.

8. Saving Plan of an Individual

The level of savings of an individual depends upon various factors like present level of income, future needs, prices of the commodities in the present and in the future, tastes and habits of individual, rate of interest etc. For

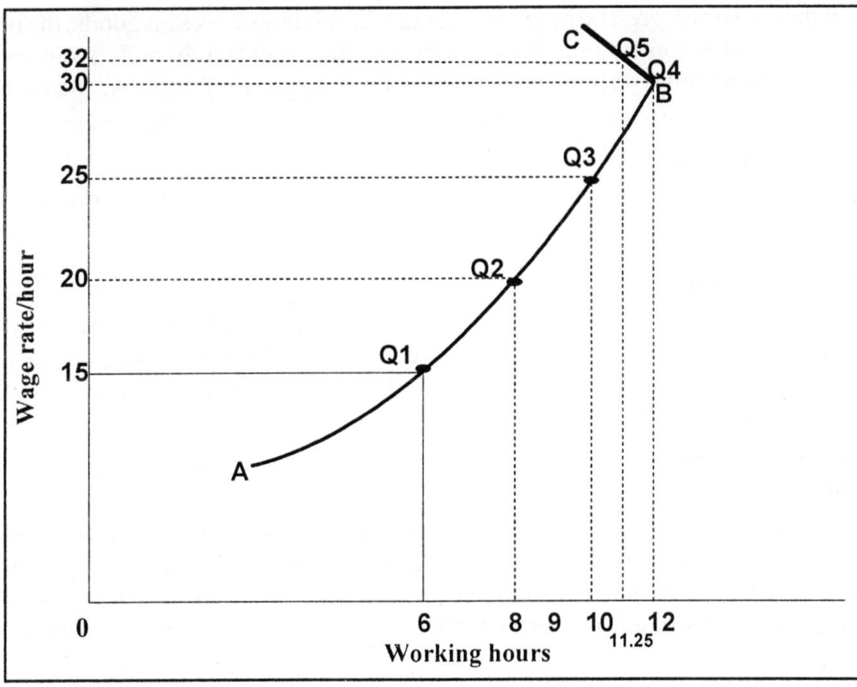

Figure 9.59: Supply curve of labour for an economy as a whole (Backward bending supply curve).

example, if the consumer prefers more of present goods to future goods, his present spending will be more and thereby, he saves less for the future and *vice versa*. This saving plan of the individual can be explained with the help of an IDC technique.

With the given level of income of the individual and given level of prices of both present goods and future goods P_1L_1 is the price line and it touches IDC I_1 at Q_1, where the consumer prefers OM_1 of present goods and ON_1 of future goods (Figure 9.60). If the consumer prefers to have more of present goods compared to future goods, the individual spends more on present goods and hence, the price line is P_2L_2 and it touches the IDC I_2 at Q_2, where the consumer purchases more of present goods *i.e.*, OM_2 ($OM_2>OM_1$) and purchases less of future goods *i.e.*, ON_2 ($ON_2<ON_1$). So, when the consumer prefers to have more of present goods, he substitutes present goods for future goods and hence, he saves less for the future.

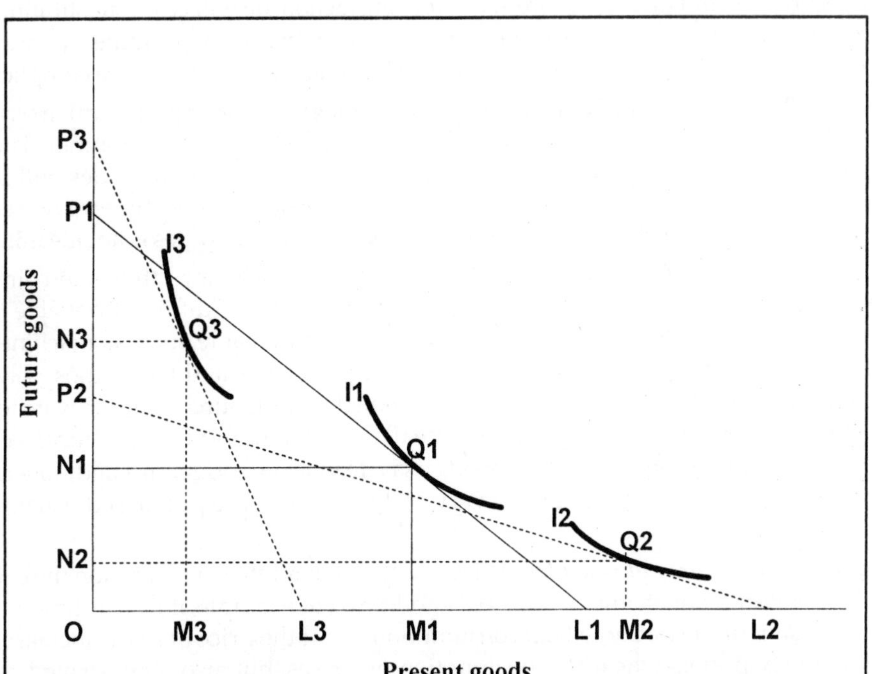

Figure 9.60: Application of IDC technique to plan investment on present goods Vs future goods.

On the other hand, if the consumer prefers more of future goods and less of present goods, the price line is P_3L_3 and it touches the IDC I_3 at Q_3, where the consumer purchases ON_3 of future goods ($ON_3>ON_1$) and less of present goods ($OM_3<OM_1$). So, the individual substitutes future goods to present goods and for this he has to save more income in the present period.

9. Measurement of consumer's surplus

The IDC technique is useful in explaining the concept of consumer's surplus in a better way compared to Marshallian analysis and the same is discussed in the Chapter 13.

9.9. Criticisms of IDC Analysis

From the above merits of IDC analysis over Marshallian utility analysis, we can conclude that, the IDC analysis offers more explanation to the consumer's behaviour compared to Marshallian analysis, as it has not framed invalid assumptions that are considered under Marshallian analysis. But, even the concept of IDC analysis is also subjected to severe criticisms and they are discussed here under.

- ☆ ***The consumer is not rational:*** The consumer is not rational, as he acts under various social, economic and legal disabilities.
- ☆ ***Complete knowledge of scale of preferences impossible:*** The IDC analysis is unrealistic because, the consumer may not have complete knowledge about his scale of preferences, different combinations of commodities that yield same level of satisfaction and a particular combination that can maximize his satisfaction etc. So, the assumption of 'perfect knowledge' on the consumer behaviour is not valid under ordinary market behaviour of the consumer.
- ☆ ***Weak ordering hypothesis:*** Prof. Samuelson criticized IDC technique as 'Weak ordering hypothesis'. This is because, on the same IDC, the consumer is indifferent between various combinations of two commodities, as any combination will yield him same level of satisfaction. Among these possible combinations, the consumer has to select one combination in relevance to his income and prices of both the commodities and the optimum combination of commodities is decided by the point of tangency between price line and highest attainable IDC in the indifference map. So, in IDC analysis, choice is made among the possible preferences of the consumer on the same IDC and this implies 'Weak ordering'. But, in the ordinary market behaviour, there is 'Strong ordering hypothesis' with respect to the consumer, as his choice about a particular combination of commodities directly reveals the preference. So, in IDC technique, choice is made among the possible preferences, but in the ordinary market behaviour, choice he makes directly reveals his preference and other possible combinations of commodities are completely ruled out.
- ☆ ***Deals with only two commodities:*** IDC technique helps to study the consumer's behaviour with respect to two commodities only and their possible combinations. But, in practical situation, the consumer will be faced with several commodities and to derive more and more satisfaction, he will go on substituting one commodity for another commodity. Further, the consumer buys large number of commodities to satisfy his unlimited wants. If we consider more than two commodities, the IDC analysis becomes more complicated.
- ☆ ***Purely imaginary:*** The concept of IDC analysis is purely imaginary or hypothetical in assuming different combinations of two commodities yielding same level of satisfaction to the consumer. The empirical data pertaining to income level of the consumer and market prices of the commodities will be available, but conducting experimentation on imaginary aspects of different scales of preferences of two commodities and getting same level of satisfaction from different combinations of commodities is not meaningful.
- ☆ ***All commodities are not perfectly divisible:*** The IDC analysis assumes that, the commodities under consideration are perfectly divisible, so that, the consumer can frame different possible combinations of commodities. But, in practicality, all the commodities are not divisible such as machinery, cell phones, watches, televisions etc. Say, for example, purchasing three watches and two radios or two watches and three radios may yield same level of satisfaction on the same IDC. But, purchasing 2.5 watches and 2.5 radios, which will yield same level of satisfaction on the same IDC, is absolutely absurd or incorrect. In some cases, the consumer prefers to purchase only one commodity say, television, fridge, air-conditioner, tractor etc., and not in combinations, as these commodities are very costly and they are purchased usually once in the life time.
- ☆ ***Other related commodities are not considered:*** IDC analysis takes into consideration the price behaviour of two commodities only and not of other related commodities. For example, if IDC technique is used to study the consumer's behaviour of rice and wheat consumption and if the price of rice is increased in the market, the consumer not only prefers wheat for consumption purposes, but also other related commodities like bajra, jowar, ragi etc. Further, the assumption of constancy of tastes and preferences of the consumer is also not valid.

☆ ***Demonstration effect is neglected:*** IDC technique has neglected the demonstration effect on the part of the consumer behaviour. That means, the concept of IDC analysis explain the consumer behaviour with reference to two commodities only. The concept explains that, if the income of the consumer is increased, he moves to higher IDCs and derives more satisfaction by consuming more quantities of both the commodities. But, in practicality, with the increase in income of the consumer, the consumer may not purchase the same commodities and will not move to higher IDCs, but he may altogether change the commodity preferences. For example, if the consumer is on the IDC with respect to bajra and jowar commodities and if the income of the consumer increases, he may prefer superior commodities like rice and wheat instead of the same jowar and bajra as considered earlier. Similarly, the influence of advertisement and other marketing techniques are not considered in IDC analysis, but they significantly influence the consumer's behaviour. Hence, IDC analysis is considered only as a static concept and not a dynamic concept.

☆ ***Perfect competition will not hold good:*** IDC technique assumes perfect competitive market situation *i.e.*, homogeneity of good and constancy of market prices of commodities. But in practical situation, heterogeneity of commodities will prevail in the market. For example, there may be inter-mixing of grains of different rice varieties. Similarly, prices will not remain constant in the market. For example, if consumers in the society prefer more of rice to wheat, the price of rice will increase, thereby, price line come close to origin and touches the lower IDC and consumer derives less satisfaction. But, assuming perfect competitive situation, where same price prevails for same commodity, the consumer may remain at higher IDC. But, this is not actually the case.

☆ ***Midway house:*** Schumpeter described IDC analysis as a '*midway house*', as it is particularly no better than the Utility analysis.

☆ ***Cannot explain uncertainity:*** This technique cannot explain the consumer's behaviour, when the element of risk or uncertainity is involved in making the choice among the preferences.

☆ ***IDCs are non-transitive*:** The concept of IDC technique was severely criticized by Prof. W.E. Armstrong. He criticized that, in IDC analysis, the meaning attached to the word 'indifference' is not clear. He opined that, there is no practical methodology to prove the satisfaction is indifferent at different points on the same IDC. This concept of 'indifference' was introduced as an important assumption in this technique. He opined that, the consumer is indifferent between various combinations of commodities not because, he knows all the preferences or scale of different combinations of commodities yielding same level of satisfaction, but because he is incapable of judging the differences between satisfaction levels at various combinations of commodities. He further opined that, any two points on the same IDC are indifferent not because they are of iso-utility, but because, they have zero utility difference. It is only because of zero utility difference, the consumer is indifferent among different combinations of two commodities on the same IDC. As per the concept of IDC, consumption of five apples and one mango (Combination A) yield same level of satisfaction as consumption of one apple and five mangoes (Combination B) on the same IDC. But, in practical situation, per unit consumption of apple and per unit consumption of mango will yield different levels of satisfaction to the consumer. In such cases, the total satisfaction realized with reference to Combination A is different compared to Combination B. This is because, in case of Combination A, the consumer is saturated with apples and in Combination B, the consumer is saturated with mangoes. As the per unit satisfaction levels contributed by apples and mangoes differ, the total satisfaction realized from Combinations A and B are different. If we admit these differences in the levels of satisfaction at different combinations of commodities, then we can infer that, the combinations on IDC are not transitive. So, if there is no zero utility difference, the concept of IDC has no meaning at all. In the Figure 9.61, the consumer is indifferent between A or B combinations and between C or D combinations because, the differences between satisfaction levels among each pair *i.e.*, between A and B or C and D is insignificant or imperceptible. But definitely, the consumer will have utility difference between A and D because, at A he prefers more of apples and less of mangoes and at D he prefers more of mangoes and less of apples. Since, the satisfaction derived from every unit of consumption of mangoes and apples are different at A and D combinations, the difference in total satisfaction between A and D combinations is not equal to zero. That means, the difference in total satisfaction between A and D points on IDC is significant or perceptible. Hence, the points A and D on IDC are not transitive. So, when there is non-transitivity, the concept of indifference utility will not exist.

☆ ***Micro in its scope*:** The concept of IDC technique is micro in its scope because, it deals with only one consumer and with two commodities only. It is not possible to deal the group behaviour of consumers and with more number of commodities.

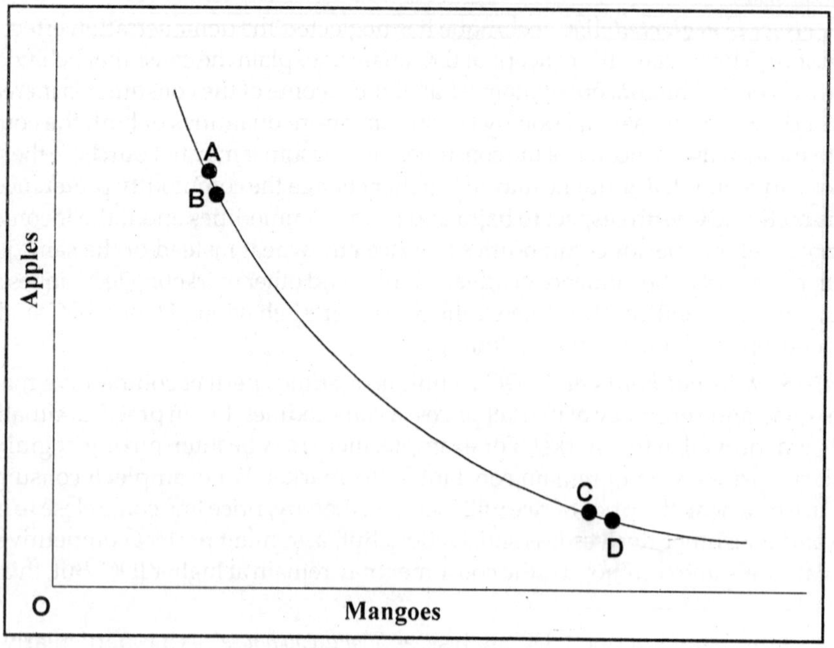

Figure 9.61: Non-transitivity of IDC curve.

From the above discussion, it is clear that, the IDC technique is not without defects, but when we take into consideration the concept as a whole, we find its approach is far superior to that of Marshallian utility analysis. Moreover, it explained the concepts of consumer behaviour in a more scientific way and they have wider practical application.

10

Revealed Preference Theory

Revealed Preference Theory of consumer's behaviour was proposed by Paul A. Samuelson. It is an improvised version over Marshallian utility analysis and IDC analysis and it is treated as the *'third root of the logical theory of demand'*. Marshallian analysis was based on cardinality approach or cardinal measurement of utility, where its assumed utility is measurable. IDC analysis was based on ordinality approach, where it assumed utility cannot be quantified or measured, but can be expressed in qualitative terms like high, medium and low levels of satisfaction. But, both these methods mainly depend upon introspective behaviour of the consumer. Among these two methods, IDC technique made a big assumption that, consumer can draw different combinations of two commodities and any combination of commodities on the same IDC will yield same level of satisfaction. But, in practicality, these two methods suffer from limitations because, utility is not measurable (Marshallian approach) and drawing different combinations of two commodities in advance and any combination yielding same level of satisfaction on the same IDC (IDC approach) are not valid. To overcome these limitations, Samuelson proposed Revealed Preference Theory for explaining the consumer behaviour or the theory of demand, but without the use of IDCs and with fewer assumptions. Since, this analysis of consumer behaviour in which, the consumer is asked to reveal his preferences for the commodities in any set of circumstances, it is known as Revealed Preference Theory. The one major advantage of Revealed Preference Theory is that, it is explicitly designed to allow *Econometricians* to make use of it. This theory was mainly based on studying actual or ordinary market behaviour of the consumer and no scope on his introspective and psychological reactions due to changes in income-price situations. Thus, with the introduction of the concept of Revealed Preference Theory, there is a shift from the psychological to behaviouristic explanation of the consumer behaviour. So, Revealed Preference Theory indicates that, choice made by the consumer directly reveals his preference in purchasing the commodities. This is where, Revealed Preference Theory differs from IDC technique because, in IDC technique, the consumer makes a choice among various preferences of combinations of commodities that are open to him; whereas, in case of Revealed Preference Theory, a single choice reveals his preference and other possible combinations are not considered at all. Further, in IDC analysis, the consumer is expected to give some subjective or hypothetical information such as, drawing up various combinations of commodities that yield same level of satisfaction and indicating his preference among different combinations. But, in Revealed Preference Theory, such subjective information is not needed and the consumption behaviour is studied based on the ordinary market behaviour. So, it is unnecessary to assume in Revealed Preference Theory that, the consumer draws different scale of preferences of the commodities. Revealed Preference Theory is also called as 'Behaviourist ordinal-utility theory', as it provides behaviouristic explanation of the consumer's demand. Tapas Majumdar also described Revealed Preference Theory as *'Behaviourist Ordinalist'* because, this theory employs two basic features to explain consumer theory *viz.*, ordinary market behaviour of the consumer and concept of ordinal utility.

10.1. Assumptions

This theory is based on the following assumptions:

- ☆ The consumer is rational in making the choice regarding the purchase of combination of commodities and the choice made by him reveals his preference.
- ☆ The consumer chooses only one combination of the commodities in any given income-price situation.
- ☆ No two observed market behaviour of the consumer can conflict with regard to combinations of commodities preferred by him in a given income-price situation.
- ☆ The consumer's tastes and preferences do not change.
- ☆ The consumer's choice is based on strong ordering.
- ☆ This theory assumes consistency of consumer behaviour. That means, if the consumer prefers combination A to combination B in one particular instance, he cannot prefer combination B over combination A in another instance, if both combination A and combination B are present. This is called two-term consistency.
- ☆ The theory assumes 'Transitivity', which also refers to three term consistency. For example, if the consumer prefers combination A to combination B and combination B to combination C, then he prefers combination A to combination C. This assumption is essential in Revealed Preference Theory to ensure that, the consumer is consistent in his consumption behaviour.
- ☆ E_Y must be positive.

10.2. Concepts in Revealed Preference Theory

Prof. Samuelson in his article '*Consumption theory in terms of revealed preference*' stated the hypothesis as '*through any observed equilibrium point A, draw the budget equation straight line with arithmetical slope given by the observed price ratio. Then all combinations of goods on or within the budget line could have been bought in preference to what was actually bought. But, they weren't. Hence, they are 'revealed' to be inferior to A*'. This hypothesis was explained through the following postulates framed under the theory:

1. Behaviourist-ordinalist Approach

The Revealed Preference Theory explains, how the consumer reacts regarding the purchase of a combination of commodities under different income-price situations. It deals with the consumer behaviour in the actual marketing situation and thereby, the choice he makes regarding the combination of commodities directly reveals his preference. Generally, a consumer prefers a combination of commodities, which will give him the highest satisfaction implying that, it is based on ordinality approach. Since, we study the consumer behaviour regarding the purchase of combination of commodities taking the actual market behaviour and through ordinality approach, the Revealed Preference Theory is otherwise termed as Behaviourist ordinalist approach. Thus, this theory was built mainly based on 'behaviour' and 'ordinalism'.

2. Preference Hypothesis

Samuelson explained the concept of Revealed Preference Theory taking Preference hypothesis as a base. As shown in the Figure 10.1, the consumer makes a choice of combination A of commodities, where he can purchase OM quantity of rice and ON quantity of wheat commodities for a given income-price situation. So, his choice reveals the preference for having combination A of rice and wheat commodities in the market economy. This infers that, choice made by the consumer reveals his preference for having that particular combination of commodities. Hence, the name 'Preference Hypothesis'. In the words of Samuelson, '*choice reveals preference*'.

In the Figure 10.1, PL_1 is the price line drawn for the consumer for a given income-price situation. With this price line, the consumer can purchase any combination of commodities say A, B, C on the price line or D, E, F, G below price line or any other combination with in the POL_1. But, here, the consumer made a choice of combination A indicating that,

- ☆ Combination A of commodities is considered to be superior when compared to other combinations possible and hence, considered them as inferior.
- ☆ Making the choice for combination A reveals his preference for having OM quantity of rice and ON quantity of wheat commodities and thereby, rejecting all other possible combinations.

The above explanation holds good, as long as the income-price situation remains constant. When there is change in income or change in prices of the commodities or both, the choice made by the consumer also varies regarding combinations of commodities preferred.

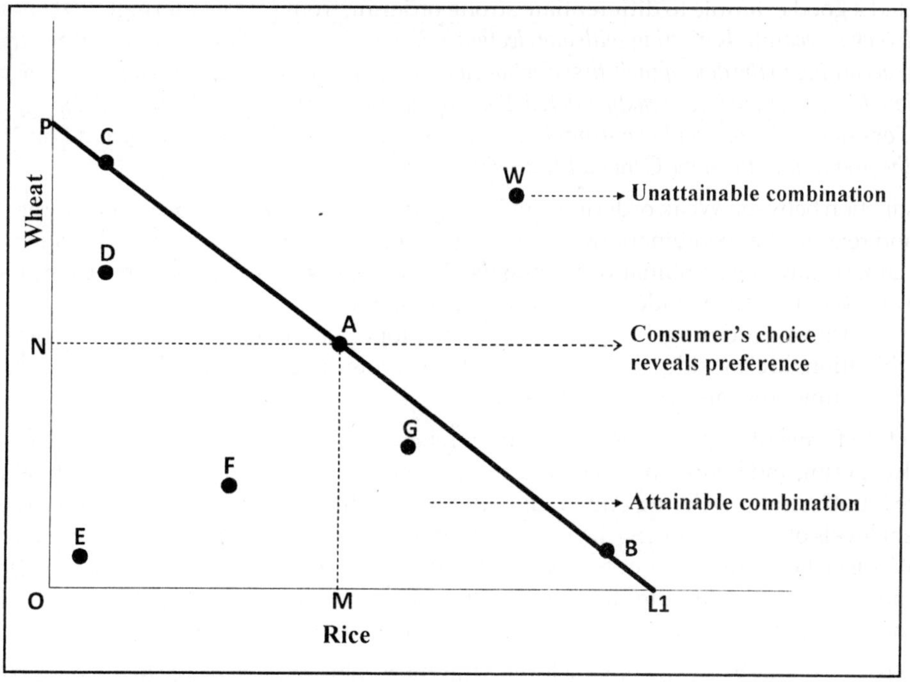

Figure 10.1: Consumer revealing his preference for the combination 'A' of commodities.

3. 'Preference' here Means Once-ever Choice

According to Prof. Samuelson, consumer's preference is a dynamic concept and not static. This is because, his choice reveals the preference in a given income-price situation. There is no way of revealing the indifferent behaviour by the consumer, as his single choice in the ordinary market behaviour reveals his final preference and indirectly other possible combinations are considered 'inferior' to the choice made.

4. Strong Ordering

In the Figure 10.1, it is clear that, the consumer made a choice for combination A of commodities, where he will purchase OM quantity of rice and ON quantity of wheat commodities and since, he prefers this definite combination of commodities over all other possible combinations, it implies 'strong ordering' of the consumer for combination A of commodities and other combinations are completely ruled out. So, strong ordering means definite ordering for a particular combination of commodities under a given income-price situation. That means, the consumer strongly ordered combination A of commodities and this combination is considered superior to other combinations. This strong ordering of commodities implies that, the consumer is not indifferent regarding various alternate combinations, but he is altogether different with reference to other combinations. This is, where, Revealed Preference Theory differs from IDC analysis, as in the latter, we assumed any combination of commodities on the same IDC will yield same level of satisfaction to the consumer, but in Revealed Preference Theory, due to strong ordering, combination A, which will give him maximum satisfaction was preferred and other combinations are completely ruled out thereby, the consumer is altogether different with respect to other combinations. So, under strong ordering, the concept of indifference between the choice he makes and other alternate combinations is completely ruled out. In the Figure 10.1, the consumer prefers combination A of commodities, which implies strong ordering for combination A when compared to other possible combinations B, C, D, E, F, G etc. This strong ordering also explains the consistency postulate or consistency behaviour of the consumer with reference to combination A of commodities. If the consumer prefers combination B over other combinations under the same income-price situation, it indicates that, the consumer is not behaving consistently in the market.

In case of strong ordering, each item of consumer's set of purchases is assigned with a definite place or number and each number reveals only one item (combination) thereby, it enable the consumer to reveal his preference on strong ordering. In the Figure 10.1, the consumer will assign different numbers to different combinations possible and he strongly ordered for combination A, thereby rejecting all other possible combinations *i.e.* B,C,D,E,F,G etc.

There is another kind of ordering called Weak ordering. In case of weak ordering, there may be some items in the consumer's set, which cannot be assigned with a definite number or preferential order, so that, the consumer is unable to indicate, which item he prefers over the other.

J.R.Hicks quoted a good example to differentiate strong ordering from weak ordering. *'Parliaments of Great Britain are strongly ordered; each is separate, beginning with one election and ending with the next. However, if one arranged all Members of Parliament (MPs) according to the date of their first election, a number of them would always have been elected for the first time at a particular election. MPs are, therefore, weakly ordered. Some of them are in cluster, with the same date of first election, though of course, some MPs or groups of them would be strongly ordered. Some individual MPs will be strongly ordered because, no one else was elected for the first time at the same General Election or at a particular by-election'.*

The major distinction between Weak ordering and Strong ordering is that, in Weak ordering, the consumer chooses one combination and rejects other combinations that are open to him, not because other combinations are 'inferior', but because, he has to choose only one combination among the different possible combinations that have the preference in terms of yielding satisfaction. Hence, under weak ordering, the choice made by the consumer fails to reveal a definite preference. However, in Strong ordering, the consumer's choice for a particular combination reveals his preference and other possible combinations are considered 'inferior' and hence, completely ruled out. So, under strong ordering, the choice made by the consumer reveals a definite preference.

Among these two forms of ordering, Revealed Preference Theory assumes strong ordering and rules out the possibility of weak ordering and hence, the concept of indifference is ruled out in Revealed Preference Theory. In IDC analysis also, there is both Strong ordering and Weak ordering. Strong ordering with reference to ranking of IDCs on the basis of different levels of satisfaction. So, between the IDCs, we can strongly order for a particular IDC that yields the highest level of satisfaction. But, Weak ordering is with reference to different combinations of two commodities on the same IDC, as any combination on the same IDC will yield same level of satisfaction. So, the consumer cannot strongly order among different combinations of commodities on the same IDC, as they are equally satisfactory.

So, by Preference hypothesis and Strong ordering, the consumer reveals his preference for combination A in a decisive manner, such that, other possible combinations are completely rejected. So, these two postulates help to obtain useful information and compare about various choices made by the consumer for a combination of commodities under different income-price situations and it is called as Consumer's preference scale. Thus, these two postulates *viz.*, Preference hypothesis (preferring or making a choice of combination A) and Strong ordering (rejecting all other combinations and making a choice for a particular combination) guides the consumer to make a choice of combination of commodities in a decisive manner and also helps to derive a definite information about the consumer behaviour in the actual market situation under a given income-price situation.

5. Consistency Postulate

This is one of the important assumptions of Revealed Preference Theory and this is implied in Strong ordering hypothesis. This consistency postulate is only the counter part of utility maximization concept of Marshallian utility analysis and IDC analysis. It can also be stated that, if a consumer chooses combination A to combination B in one particular instance, then he cannot choose combination B to combination A in any other instance, when both A and B are possible combinations. That means, if the consumer prefers combination A to combination B in one instance then, he cannot reveal his preference for combination B to combination A in another instance, but if he chooses combination B to combination A, he said to be behaving inconsistently in the market. This inconsistency behaviour of consumer is completely ruled out in Revealed Preference Theory, as it is based on Strong ordering. The consistency behaviour of consumer with reference to combination A to combination B under two different situations refers to two term consistency and it is coined by J.R.Hicks. This axiom of revealed preference, according to which the consumer behaviour is consistent is otherwise termed as 'Weak Axiom of Revealed Preference (WARP). It implies that, if a consumer strongly orders combination A to combination B, which also can be purchased by the given price line, then it cannot happen to strongly order combination B to combination A in some other situation in which the consumer can also have combination A. The same is explained through Panel A of the Figure 10.2, where for the given income of the consumer and for the given prices of the commodities *viz.*, rice and wheat, PL is the price line. On this price line PL, let us suppose, the consumer strongly ordered the combination A to combination B to have both rice and wheat commodities. If price of rice falls in the market, after making income adjustment by cost-difference method, P_1L_1 is the new price line. This price line P_1L_1 is more flatter compared to PL indicating that, now price of rice commodity is less in the market compared to the earlier situation PL. With this new income-price situation *i.e.*,P_1L_1, if the consumer prefers or strongly orders combination B to combination A, where still under this new P_1L_1 situation, he can have combination A, then the consumer is said to be behaving inconsistently in the market behaviour and thereby, he is violating the WARP. Such inconsistency is ruled out in the Revealed Preference Theory.

For interesting sake, let us have another situation (Panel B of the Figure 10.2), where for the given income of the consumer and given prices of the commodities *viz.*, rice and wheat, PL is the price line. On this price line, let us suppose the consumer strongly ordered for combination A of the commodities and thereby, all other combinations are completely ruled out. If the price of rice falls, the new price line is P_1L_1, after making the income adjustment through cost-difference method. If the consumer strongly orders combination B under new income-price situation P_1L_1, where for such new

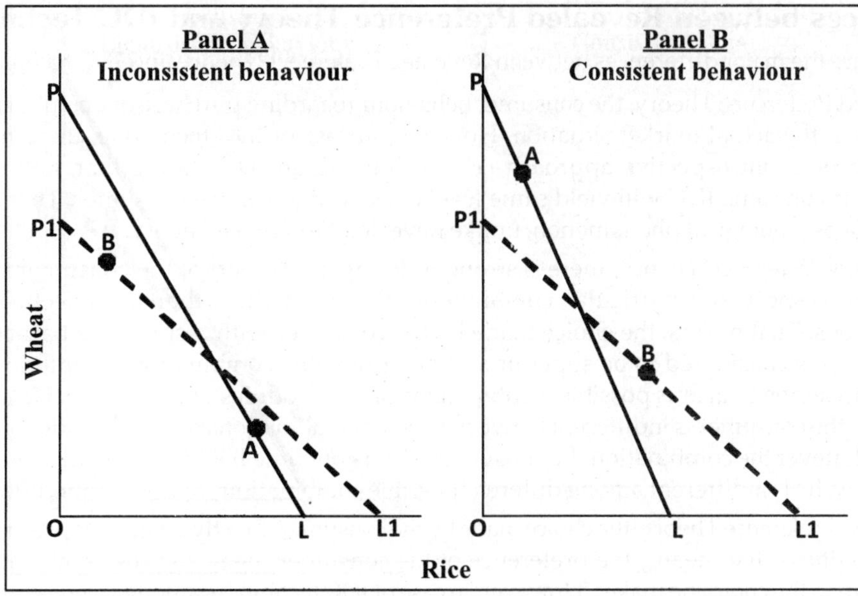

Figure 10.2: Consistency postulate of consumer behaviour.

situation, the earlier combination A is not possible, then the consumer is said to be behaving consistently, as earlier combination A is unattainable with P_1L_1. So, this strong ordering for commodity B under the new income-price situation, where the earlier preference for the combination A is not possible, indicates WARP.

6. Transitivity

This feature of Revealed Preference Theory infers three-term consistency. Say for example, if a rational consumer prefers combination A to combination B and combination B to combination C of the commodities, then he will also prefer combination A to combination C. So, we can say if A>B and B>C then, A>C. The same is explained through Figure 10.3, where for the given income level of the consumer and given prices of the commodities *viz.*, rice and wheat, PL is the price line. On the price line, the consumer strongly ordered combination A to combination B. Let us suppose P_1L_1 is another price line for which, the consumer strongly ordered combination B to combination C. So, from the transitivity assumption, we can infer that, in the entire shaded region of the Figure 10.3, the consumer prefers combination A to combination C. So, transitivity ensures 'circular' relationships are ruled out.

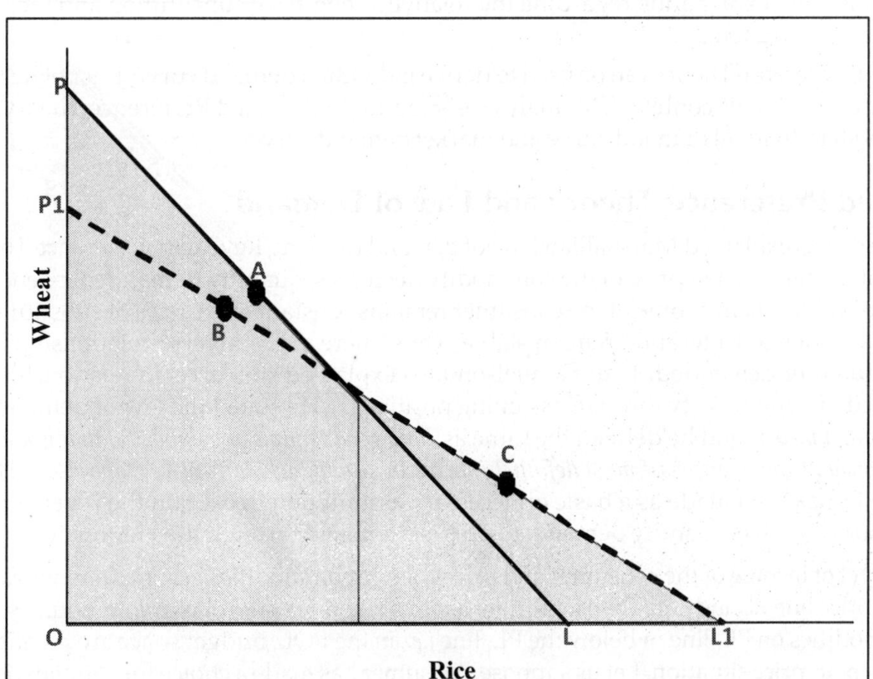

Figure 10.3: Transitivity assumption of Revealed Preference Theory.

10.3. Differences between Revealed Preference Theory and IDC Technique

The following are the major differences between Revealed Preference Theory and IDC technique:

- In Revealed Preference Theory, the consumer behaviour regarding purchase of a combination of commodities is studied in the actual market situation. However, in case of IDC technique, the consumer behaviour is studied through introspective approach *i.e.*, by drawing an IDC stating that, any combination of two commodities on same IDC will yield same level of satisfaction to the consumer. That means, IDC is more related to a psychological phenomenon, unlike Revealed Preference Theory.
- In Revealed Preference Theory, there is strong ordering on the part of the consumer such that, a choice is made with respect to a particular combination of commodities thereby, rejecting all other possible combinations. That means, the choice made by the consumer with respect to a particular combination of commodities is considered to be superior and other possible combinations as inferior. So, the concept of indifference among various possible combinations of commodities will not arise. However, in case of IDC technique, the consumer is indifferent between various combinations of commodities on the same IDC. That means, whatever the combination the consumer prefers on same IDC, it will yield same level of satisfaction and thereby, he is indifferent among different possible combinations of the commodities.
- In Revealed Preference Theory, the choice made by the consumer directly reveals his preference for combination of commodities. That means, the preference of the consumer regarding the combination of commodities depends upon the choice he makes. However, in case of IDC technique, among various preferences of different combinations of two commodities that are open to the consumer, he will make a choice to maximize the satisfaction.
- In IDC technique, the point of maximum satisfaction is where, the price line is tangent to the highest attainable IDC. That means, an IDC is drawn here. However, in Revealed Preference Theory, simply through drawing a price line, the strong ordering of consumer in making a choice for a particular combination of commodities is indicated and it is the point of maximum satisfaction. So, in Revealed Preference Theory, no IDC is drawn.
- The Revealed Preference Theory is based on strong ordering or an actual market behaviour of the consumer and hence, it is not possible to the make a distinction between income effect and substitution effect. So, by mere observation of the consumer behaviour, it is not possible to isolate substitution effect from the income effect. In this case, IDC analysis is more superior to the Revealed Preference Theory, as these two components of price effect are studied by drawing IDCs. Moreover, Revealed Preference Theory assumed positive E_Y and this further made difficult to establish substitution effect.
- Since, the Revealed Preference Theory assumed positive E_Y, it could not offer much explanation regarding the price effect of inferior goods and giffen goods. So, the Revealed Preference Theory could not explain giffen paradox. In this context, the IDC analysis is more meritorious compared to Revealed Preference Theory, as it offered the valid explanation regarding the relative strengths of substitution and income effects of inferior goods and giffen goods.
- Revealed Preference Theory can be used to derive individual demand curve, but not helpful to derive market demand curve. In this context, IDC analysis is superior to Revealed Preference Theory, as it is applicable to derive both individual demand curve and market demand curve.

10.4. Revealed Preference Theory and Law of Demand

Prof. Samuelson has explained Marshallian Law of demand by using Revealed Preference Theorem. According to Marshallian law of demand, when price of the commodity decreases, quantity demanded extends and *vice versa*. This relationship is applicable when income of the consumer remains constant and prices of substitutes and complements for the commodity under consideration remain stable. Thus, there exists inverse relationship between price of the commodity and quantity demanded. Paul Samuelson also explained this inverse relationship between price and quantity demanded of a commodity through assuming positive E_Y. He stated the law of demand as the '*Fundamental theorem of consumption theory*' and he defined the same as, '*any good (simple or composite) that is known always to increase in demand when money income alone rises must definitely shrink in demand when its price alone rises*'. This definition clearly indicates positive E_Y has been made as a basic or necessary assumption to explain the inverse relationship between price of the commodity and its quantity demanded. This is explained through the Figures 10.4 and 10.5.

For a given level of income of the consumer and prices of two commodities *viz.*, rice and wheat, PL_1 is the price line (Figure 10.4). This price line PL_1 indicates two important aspects *viz.*, price ratio of two commodities and any combination of these two commodities on PL_1 line or below the PL_1 line *i.e.*, in the POL_1 budget space are attainable by the consumer under the given income-price situation. Let us suppose, consumer has made a choice for combination A of commodities, which implies strong ordering for that combination A, where he can purchase OM_1 quantity of rice and ON_1 quantity

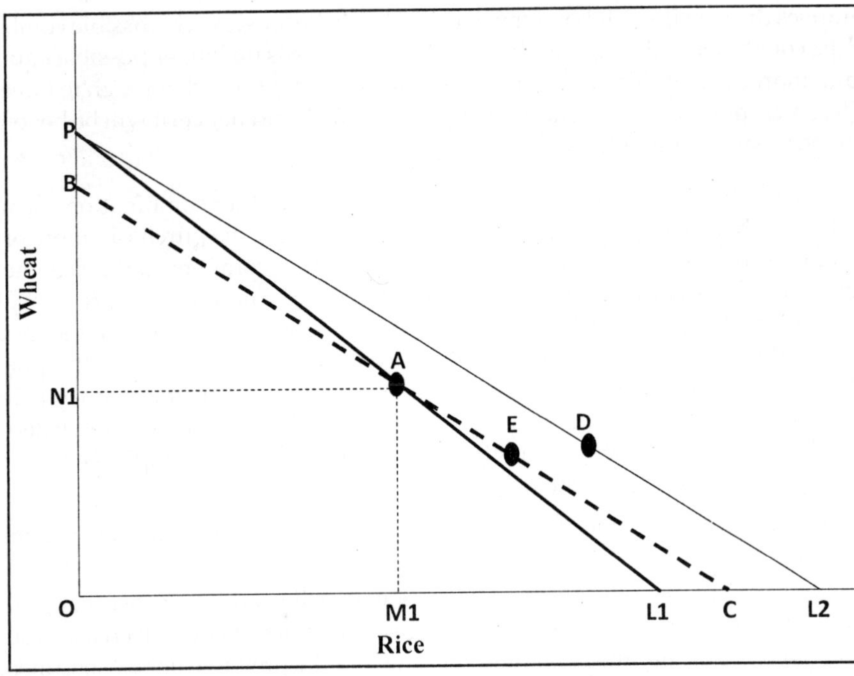

Figure 10.4: Revealed Preference Theory explaining law of demand – Case of decrease in price of rice.

of wheat commodities. This strong ordering for combination A implies that, the consumer has rejected all other possible combinations of commodities within the POL_1.

Case1: Fall in Price of Rice Commodity

Let us suppose, price of commodity rice is decreased in the market and price of wheat remains same. Due to the fall in price of rice, the consumer purchases more of rice and hence, PL_2 is the new price line (Figure 10.4). But, on this new price line PL_2, the consumer will not purchase his original choice for combination A, (which is the strong ordering for the consumer for PL_1 price line), as with the fall in price of rice, he prefers to purchase more of rice and less of wheat, thereby, he may choose combination D on PL_2 price line. This movement from A to D is called price effect, as due to fall in price of rice, more quantity of it is demanded by the consumer.

When price of the rice commodity is decreased, the consumer psychologically feels that, his real income is increased. So, to trace out the substitution effect alone due to fall in price of the rice commodity, this indirect increase in real income should be compensated. Samuelson employed cost-difference method to measure the extent of indirect increase in real income and this is removed from the consumer's money income for maintaining his real income constant. This is indicated by PB with reference to commodity wheat and by CL_2 with reference to rice commodity. So, BC is the compensated price line drawn parallel to PL_2 line, which passes through combination A *i.e.*, original combination of commodities chosen by the consumer when his price line is PL_1. This passing of BC compensated price line through combination A implies, the consumer's real income was brought back to original level. So, OBC triangle becomes the area of choice of the consumer. But, due to fall in price of rice, the consumer will definitely purchase more of rice. The consumer will not purchase any combination of commodities above A *i.e.*, between BA, below BA, between AL_1 and below AL_1. This is because of the following reasons:

- ☆ At low price, the consumer purchases more of rice (when compared to original price line PL_1 or even on BC).
- ☆ This theory assumes the consumer has strong ordering for combination A of commodities. If he purchases any combination of commodities above A or below B and A, he will purchase less of rice and more of wheat commodities when compared to PL_1 price line. But, this will not happen because, with the fall in price of rice, he will definitely purchase more of rice and not less of rice. If he chooses any combination of commodities above A or below B and A, his choice could be inconsistent, as all combinations above A are already rejected by him when he is on the price line PL_1. Since, we have assumed consistency of choice behaviour of the consumer, he will not choose any combination above A or below B and A in the new income-price situation. Similarly, the consumer will not purchase the commodities on AL_1 or below AL_1, as he already rejected these combinations with reference to PL_1 price line.
- ☆ Since the compensated price line BC passes through A, the combination D is unattainable. The consumer, therefore, moves from combination A to combination E. Since PL_1 and BC lines cross each other at A, the combination E cannot lie left to combination A. Say, if E lies left to A *i.e.*, between BA portion and below AL_1

portion, it implies the consumer is behaving inconsistently because, such possible combinations are already rejected by the consumer with PL_1 price line. So, PAB triangle is no longer possible equilibrium area for the consumer and moreover, with compensated price line BC, he is not rich enough to have the combination in PAB triangle. So, combination E will lie at the right to point A and hence, it will be beyond the price line PL_1, consequently the consumer is better-off at E compared to A.

So, the consumer will purchase either original combination A or any other combination of commodities between A and C. So, the triangle L_1AC becomes the area of choice of the consumer, as he purchases more of rice due to fall in its price. Say, if the consumer purchases combination A again on the new compensated price line BC, it implies that consumer has resorted for strong ordering for combination A even under the new income-price situation. If he purchases combination E on the compensated price line BC, it implies that, the consumer has purchased more of rice and less of wheat, when price of rice is decreased in the market. In IDC analysis, we can easily identify this point E, by bringing the consumer back onto the original IDC in case of Hicks–Allen method and the consumer moves to higher IDC in case of Slutsky method, where the compensated price line is tangent to that IDC. But, in case of Revealed Preference Theory, because of ordinary market behaviour, we cannot easily locate the point E on the compensated price line BC. From the Figure 10.4, the following are the important conclusions:

- For the given income-price situation on price line PL_1, the consumer made strong ordering for combination A of commodities.
- With the fall in price of rice commodity, PL_2 is the new price line and the consumer purchases combination D of commodities *i.e.*, more of rice and less of wheat because, price of rice is decreased (Substitution effect) on one hand and psychologically the consumer feels his real income was increased (income effect) and this movement from A to D indicates price effect.
- If the indirect increase in real income was brought back to normal level by cost-difference method, on the new compensated price line BC, the consumer purchases combination E of commodities and this movement from A to E indicates substitution effect alone because, he purchased more of rice due to its price fall and money income is compensated or real income is maintained constant.
- If the consumer is at point E and if money is given back to the consumer, he will move to new price line PL_2 and he prefers combination D of commodities and this movement from E to D indicates income effect. This movement from E to D also indicates positive E_Y *i.e.*, when income increases, quantity demanded of both rice and wheat increases.
- With the fall in price of rice commodity, the consumer psychologically feels his real income is increased and thereby, its quantity demanded also increases. It infers that, there exists inverse relationship between price of the commodity and quantity demanded and positive relationship between income of the consumer and quantity demanded. So, positive E_Y implies negative E_P.
- We can conclude that, price effect = income effect+substitution effect.
- So, RPT also divides the price effect into income effect and substitution effect, but it is a less fundamental division than the one found in IDC analysis, as it involves the ordinary market behaviour and thereby, it does not allow us to separate out a pure income effect and pure substitution effect. But, this division is important for two reasons *viz.*, it helps to derive Marshallian demand curve and it is purely based on mathematical calculations (cost difference method) without drawing any IDC.

Case 2: Rise in Price of Rice Commodity

Let us suppose, price of rice commodity is increased in the market and price of wheat remains same. Due to the rise in price of rice, the consumer purchases less of rice and hence, PL_2 is the new price line (Figure 10.5), which pivots towards the origin. But, on this new price line PL_2, the consumer cannot purchase his original choice for combination A, (which is the strong ordering for the consumer for PL_1 price line) because, combination A is not available on PL_2 line. But, he may choose combination D *i.e.*, he purchase less of rice and more of wheat with the increase in price of rice. This movement from A to D is called price effect, as due to rise in price of rice, less quantity of it is demanded by the consumer.

When price of the rice commodity is increased, the consumer psychologically feels his real income is decreased. So, to trace out the substitution effect alone due to rise in price of the rice commodity, this indirect decrease in real income should be compensated. Samuelson employed cost-difference method to measure the extent of indirect decrease in real income and this is added to the consumer's money income for maintaining the real income constant. This is indicated by PB with reference to commodity wheat and by CL_2 with reference to commodity rice. So, BC is the compensated price line drawn parallel to PL_2 line, which passes through combination A *i.e.*, original combination of commodities chosen by the consumer when his price line is PL_1. This passing of BC compensated price line through combination A implies, the real income was brought back to original level. The amount of money given to the consumer to buy the original

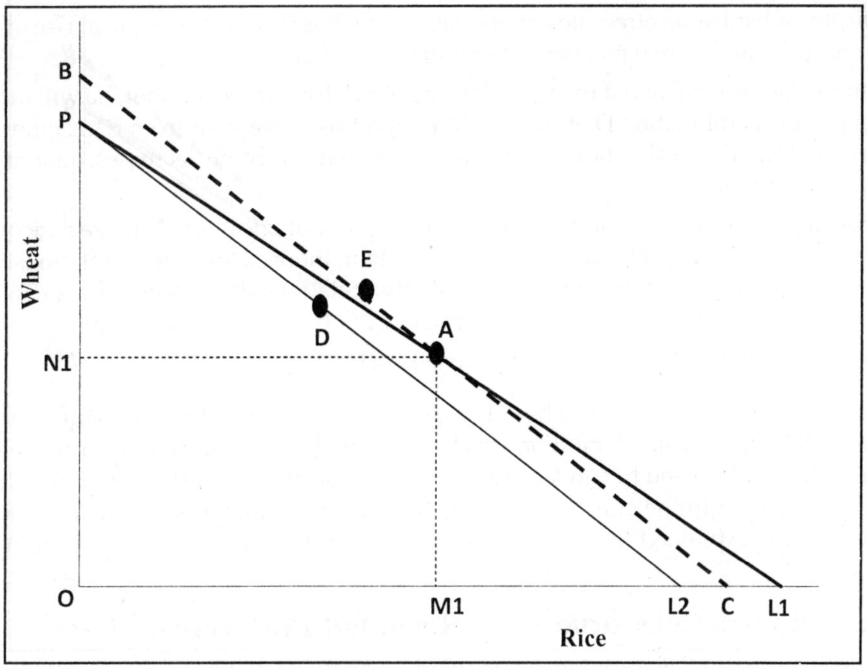

Figure 10.5: Revealed Preference Theory explaining law of demand – Case of increase in price of rice.

combination A is called as Over-compensation effect by Paul Samuelson. This has been called as 'Cost difference' by Slutsky. So, OBC triangle becomes the area of choice of the consumer. But, due to increase in price of rice, the consumer will definitely purchase less quantity of rice. The consumer will not purchase any combination of commodities below A *i.e.*, between A and C, below AC, between A and P and below AP. This is because of the following reasons:

- At high price, the consumer purchases less of rice (when compared to original price line PL_1 or even on BC).
- This theory, assumes the consumer has strong ordering for combination A of commodities. If he purchases any combination of commodities below A or below A and C, he will purchase more of rice and less of wheat commodities when compared to PL_1 price line. But, this will not happen because, with the rise in price of rice, he will definitely purchase less of rice and not more of rice. If he chooses any combination of commodities below A or below A and C, his choice could be inconsistent, as all combinations below A are already rejected by him when he is on the price line PL_1. Since, we have assumed consistency of choice behaviour of the consumer, he will not choose any combination below A or below A and C in the new income-price situation.
- Since, the compensated price line BC passes through A, the combination E is attainable. So, the consumer, therefore, moves from combination A to combination E. The consumer will not purchase any combination of commodities between A and P and below AP, as such combinations are already rejected by him (with reference to PL_1 price line) and if he still purchases them, then he is said to be behaving inconsistently. So, point E will not lie between AP and below AP. Considering the compensated price line BC, the consumer purchases combination E of the commodities *i.e.*, less of rice and more of wheat.

So, the consumer will purchase either original combination A or any other combination of commodities between A and B. So, the triangle PAB becomes the area of choice of the consumer, as he purchases less of rice due to rise in its price. Say, if the consumer purchases combination A again on the new price line BC, it implies that, consumer has resorted for strong ordering for combination A even under new income-price situation. If he purchases combination E on the compensated price line BC, it implies that, the consumer has purchased less of rice and more of wheat, when price of rice is increased in the market. From the Figure 10.5, the following are the important conclusions:

- For the given income-price situation on price line PL_1, the consumer made strong ordering for combination A of commodities.
- With the rise in price of rice commodity, PL_2 is the new price line and the consumer purchases combination D of commodities *i.e.*, less of rice and more of wheat because, price of rice is increased (Substitution effect) and psychologically the consumer feels his real income was decreased (income effect) and this movement from A to D indicates price effect.
- If the indirect decrease in real income was brought back to normal level by cost-difference method, on the new compensated price line BC, the consumer purchases combination E of commodities and the movement from

A to E indicates substitution effect alone because, he purchased less of rice due to rise in its price and money income is compensated or real income is maintained constant.

- If the consumer is at point E and if money is collected back from the consumer, he will move to new price line PL_2 and he prefers combination D of commodities and this movement from E to D indicates income effect. This movement from E to D also indicates positive E_Y *i.e.*, when income decreases, quantity demanded of both rice and wheat decreases.
- With the rise in price of rice commodity, the consumer psychologically feels his real income is decreased and thereby, quantity demanded also decreases. It infers that, there exists inverse relationship between price of the commodity and quantity demanded and positive relationship between income of the consumer and quantity demanded. So, positive E_Y implies negative E_P.
- We can conclude that, price effect = income effect+substitution effect.

Thus, the explanation of substitution effect based on Revealed Preference Theory is different when compared to IDC approach. In case of IDC approach, substitution effect is explained by bringing the consumer back onto the original IDC in case of Hicks–Allen method and he moves to higher IDC (case of price fall) or lower IDC (case of price rise) in case of Slutsky method. So, substitution effect is explained through drawing IDCs. However, in Revealed Preference Theory, substitution effect is explained through the movement from original price line to compensated price line without drawing any IDCs.

10.5. Derivation of Demand Curve from Revealed Preference Theory

In the IDC analysis, we justified in deriving the demand curve through drawing an indifference map. But, this is an unreasonable procedure, as in each case of varying price situations (say, price decrease), the consumer moves from one IDC to another IDC by drawing different scales of preferences and try to attain equilibrium on highest attainable IDC. However, it is more reasonable to suppose that, the consumer behaviour is consistent in making the purchases of the combination of commodities on the basis of observed market behaviour and thereby, this form the basis to derive the demand curve. This is because, the Modern Economists prefer to analyze the consumer behaviour in the ordinary market situations, as it facilitates to formulate and test the hypothesis.

Having studied the explanation offered by the Revealed Preference Theory for the law of demand, the same theory guides to derive the demand curve for the consumer. In Panel A of the Figure 10.6, for a given level of income of the consumer and price of rice commodity, PL_1 is the price-income line. Let us suppose, consumer has made a choice at point A on the price-income line PL_1, which implies strong ordering for OM_1 quantity of rice. This strong ordering for point A on PL_1 indicates that, the consumer has rejected all other possible quantities of rice that can be purchased within the POL_1 and on the PL_1 line.

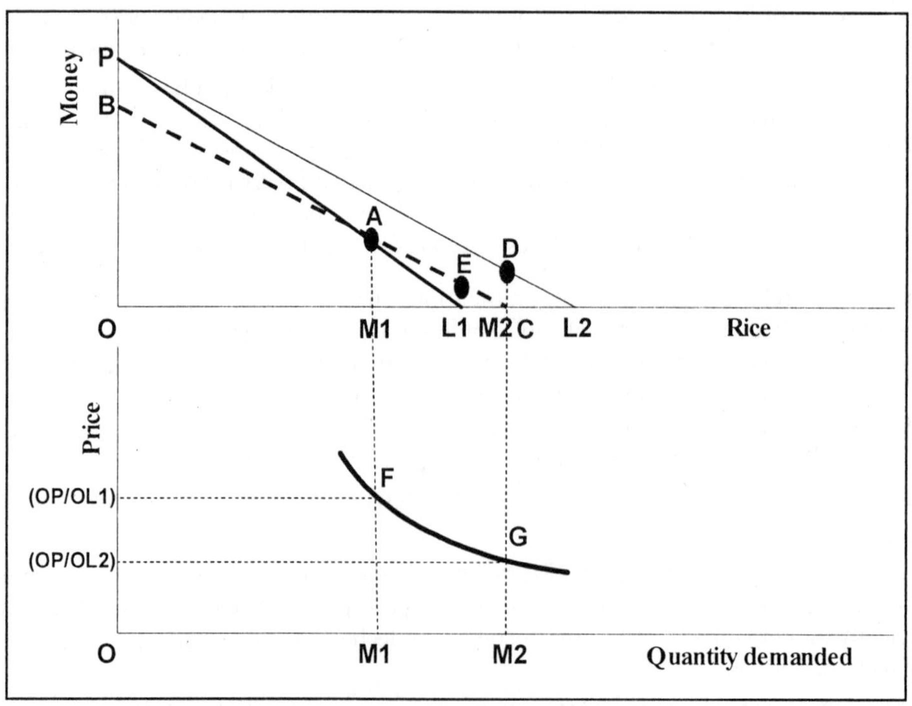

Figure 10.6: Revealed Preference Theory – Derivation of demand curve.

Let us suppose, price of commodity rice is decreased in the market. Due to the fall in price of rice, the consumer purchases more of rice and hence, PL_2 is the new price line, which lies away from the origin. But, on this new price line PL_2, the consumer will not purchase his original choice for combination A, (which is the strong ordering for the consumer for PL_1 price line) because, with the fall in price of rice, he will purchase more quantity of rice. So, he may choose combination D *i.e.*, he may purchase more of rice than earlier due to fall in the price of rice. This movement from A to D is called price effect, as due to fall in price of rice, more quantity of it is demanded by the consumer.

When price of the rice commodity is decreased, the consumer psychologically feels that, his real income is increased. So, to trace out the substitution effect alone due to fall in price of the rice commodity, this indirect increase in real income should be compensated. Samuelson employed cost-difference method to measure the extent of indirect increase in real income and this is removed from the consumer's money income for maintaining his real income constant. This is indicated by CL_2 with reference to rice commodity. So, BC is the compensated price line drawn which passes through combination A *i.e.*, original combination of commodities chosen by the consumer, when his price line is PL_1. This passing of BC compensated price line through combination A implies, the consumer's real income was brought to original level. So, OBC triangle becomes the area of choice of the consumer. But, due to fall in price of rice, the consumer will definitely purchase more of rice. The consumer will not purchase any combination of commodities above A. This is because of the following reasons:

- At low price, the consumer purchases more of rice (when compared to original price line PL_1 or even on BC).
- This theory assumes the consumer has strong ordering for combination A of commodities. If he purchases any combination of commodities above A or below B and A, he will purchase less of rice when compared to PL_1 price line. But, this will not happen because, with the fall in price of rice, he will definitely purchase more of rice and not less of rice. If he choose any combination of commodities above A or below B and A, his choice could be inconsistent, as all combinations above A are already rejected by him when he is on the price line PL_1. Since, we have assumed consistency of choice behaviour of the consumer, he will not chooses any combination above A or below B and A in the new income-price situation. Similarly, the consumer will not purchase the commodities on AL_1 or below AL_1, as he already rejected these combinations with reference to PL_1 price line.
- Since the compensated price line BC passes through A, the combination D is unattainable. The consumer, therefore, moves from combination A to combination E. Since PL_1 and BC lines cross each other at A, the combination E cannot lie left to combination A. Say, if E lies left to A *i.e.*, between BA portion and below AL_1 portion, it implies the consumer is behaving inconsistently because, such possible combinations are already rejected by the consumer with PL_1 price line. So, PAB triangle is no longer possible equilibrium area for the consumer and moreover, with compensated price line BC, he is not rich enough to have the combination in PAB triangle. So, combination E will lie at the right to point A and hence, it will be beyond the price line PL_1, consequently the consumer is better-off at E compared to A.

So, the consumer will purchase either original quantity of rice *i.e.*, OM_1 or any other quantity of rice between A and C. So, the triangle L_1AC becomes the area of choice of the consumer, as he purchases more of rice due to fall in its price. Say, if the consumer purchases combination A (OM_1 quantity of rice) again on the new compensated price line BC, it implies that consumer has resorted for strong ordering for original quantity of rice at A, even under the new income-price situation. If he purchases combination E on the compensated price line BC, it implies that, the consumer has purchased more of rice, when price of rice is decreased in the market. From the Figure 10.6, the following are the important conclusions:

- For the given income-price situation on price line PL_1, the consumer made strong ordering for combination A *i.e.*, OM_1 quantity of rice.
- With the fall in price of rice commodity, PL_2 is the new price line and the consumer purchases combination D of commodities *i.e.*, more of rice because, price of rice is decreased (Substitution effect) on one hand and psychologically the consumer feels his real income was increased (income effect) and this movement from A to D indicates price effect.
- If the indirect increase in real income was brought back to normal level by cost-difference method, on the new compensated price line BC, the consumer purchases E combination of commodities and this movement from A to E indicates substitution effect alone because, he purchased more of rice due to its price fall and money income is compensated or real income is maintained constant.
- If the consumer is at point E and if money is given back to the consumer, he will move to new price line PL_2 and he prefers combination D of commodities and this movement from E to D indicates income effect. This movement from E to D also indicates positive E_Y *i.e.*, when income increases, quantity demanded of both rice and wheat increases.

☆ With the fall in price of rice commodity, the consumer psychologically feels his real income is increased and thereby, its quantity demanded also increases. It infers that, there exists inverse relationship between price of the commodity and quantity demanded and positive relationship between income of the consumer and quantity demanded. So, positive E_Y implies negative E_P.

The movement from A to D *i.e.*, price effect helps to derive the demand curve for the consumer. From PL_1 price line, we can compute the price of the rice commodity as (OP/OL_1). At this price, the consumer purchased OM_1 quantity of rice. In the new income-price situation *i.e.*, PL_2 price line, the consumer purchased OM_2 quantity of rice at (OP/OL_2) price. So, to derive the demand curve for rice commodity, in Panel B of the Figure 10.6, on Y-axis, price of the commodity (as computed above) is taken and on X-axis, quantity of rice demanded is taken. The points A and D from panel A are extended downward into Panel B as F and G price-quantity combinations and by connecting these points, we get a downward sloping demand curve, which shows that, as price of the commodity decreases from OP/OL_1 to OP/OL_2, the quantity demanded of rice extends from OM_1 to OM_2. Thus, we proved the '*Fundamental theorem of consumption theory*' and thereby, proving the Marshallian law of demand through Revealed Preference Theory.

10.6. Derivation of IDC from Revealed Preference Theory

Revealed Preference Theory guides to derive an IDC and prove the convexity of an IDC. The methodology followed by the Revealed Preference Theory to derive the IDC is far superior compared to the IDC approach. In IDC analysis, we formulated certain invalid assumptions like consumer has to choose different combinations of two commodities, any combination of two commodities on a particular IDC yield same level of satisfaction and the consumer should act rationally to select the best combination (in relevance to his income, prices of the commodities, tastes and preferences) that gives him maximum satisfaction. Hence, Revealed Preference Theory is a superior approach to IDC approach, as it has not formulated such invalid assumptions and based purely on the ordinary market behaviour of the consumer and still guides us in deriving the IDC and further explains the convexity of the IDC through framing the following few assumptions:

☆ The consumer's choices are consistent (transitivity).
☆ The consumer's tastes and preferences remains constant.
☆ The consumer is rational in *Pareto* sense *i.e.*, he prefers more goods to less.
☆ The consumer deals with two commodities X and Y.

For the given level of income of the consumer and prices of the two commodities, PL is the price line and the consumer reveals his preference for combination A of the commodities on the price line and thereby, he considers all other possible combinations on PL price line and any other combination within POL as inferior to combination A (Figure 10.7). Let us draw perpendiculars through the preferred combination A *i.e.*, AB and AC and all the combinations of commodities on these lines AB and AC and in the area between them are preferred to A, as they contain more of both

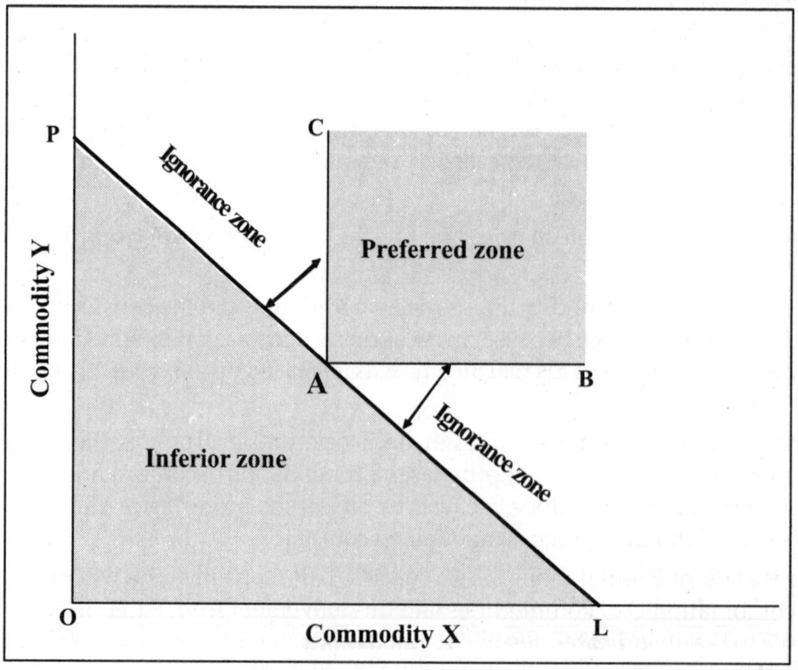

Figure 10.7: Consumer's choice for combination A indicating preferred, inferior and ignorance zones.

the commodities or atleast one commodity (as per the assumption). So, the zone ABC is considered as Preferred zone by the consumer. However, still there are two more zones, one between ABL and the other between ACP. These two zones are called Ignorance zones because, the combinations of the commodities in these two zones are still not ordered by the consumer. However, we may rank the combinations of the two commodities of the two ignorance zones relative to combination A by adopting the following procedure.

Assume that, price of commodity X falls in the market and the new (compensated) price line is P_1L_1 and it passes through original price line PL at point M (Figure 10.8). So, considering the new price line P_1L_1, the consumer chooses either combination M or any other combination in the triangle MLL_1, as all other possible combinations on P_1M are considered inferior, since it lies below the original price line (PL). Assume, the consumer revealed his preference to combination M on P_1L_1 price line. So, all other possible combinations on ML_1 price line and the area between MLL_4 are considered as inferior. So, considering the transitivity assumption, we can say that,

On original price line (PL) situation: A > M

On new price line (P_1L_1) situation: M > MLL_1 and hence, A>MLL_1.

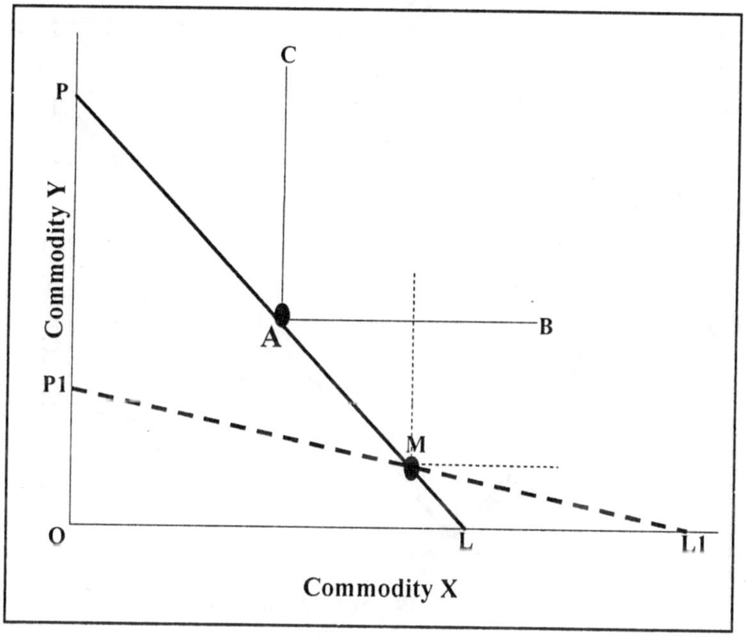

Figure 10.8: Narrowing the lower ignorance zone.

So, we considered the combinations of commodities in the area MLL_1 as inferior to A and this guides not to consider the area MLL_1 in the ignorance zone below ABL. So, by drawing several new price lines, we can narrow down the ignorance zone below the ABL.

By following the same procedure, we can also narrow the ignorance zone ACP. For this, assume the price of commodity X is increased in the market and thereby, the new (compensated) price line is P_2L_2 and it passes through original combination A, indicating the real income of P_2L_2 is equal to PL at point A (Figure 10.9).

Considering the new price line P_2L_2, the consumer chooses either original combination A or any other combination between P_2A of P_2L_2 price line, as all other possible combinations below A in POL area are already considered inferior. Assume, the consumer revealed his preference to combination N on P_2L_2 price line. So, from the Figure 10.9, using the rationality assumption we can say that,

- ☆ RNCS > N, as RNCS combinations of commodities contains more of both the commodities or at least one commodity than combination N.
- ☆ N > A, as per the revealed preference principle, and
- ☆ RNCS > A.

So, we ranked the combinations of the commodities in the zone RNCS as superior to combinations N and A due to price increase of commodity X and this facilitate to narrow down the upper ignorance zone ACP. We can repeat the same procedure by drawing several new price lines (with increase a price of X) and we can further narrow down the upper ignorance zone ACP. Thus, by narrowing both the areas below A *i.e.*, ABL and ACP, we can locate the IDC I_1 as shown through Figure 10.10.

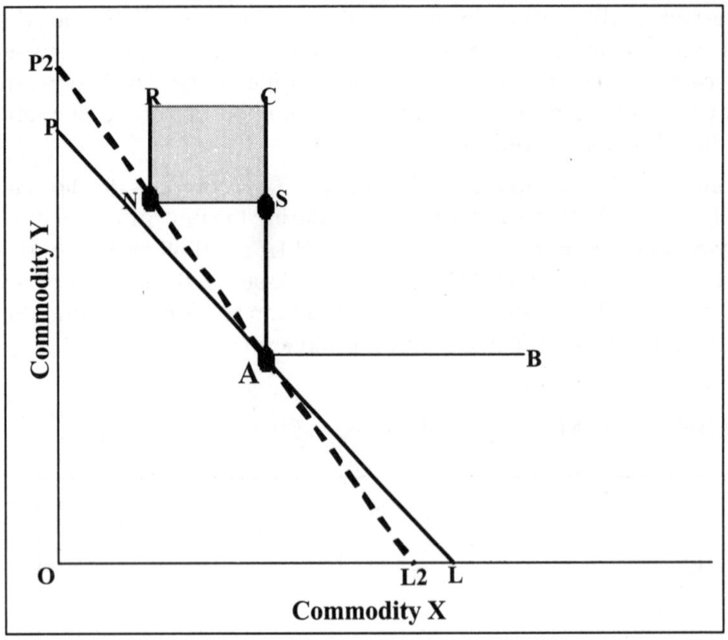

Figure 10.9: Narrowing the upper ignorance zone.

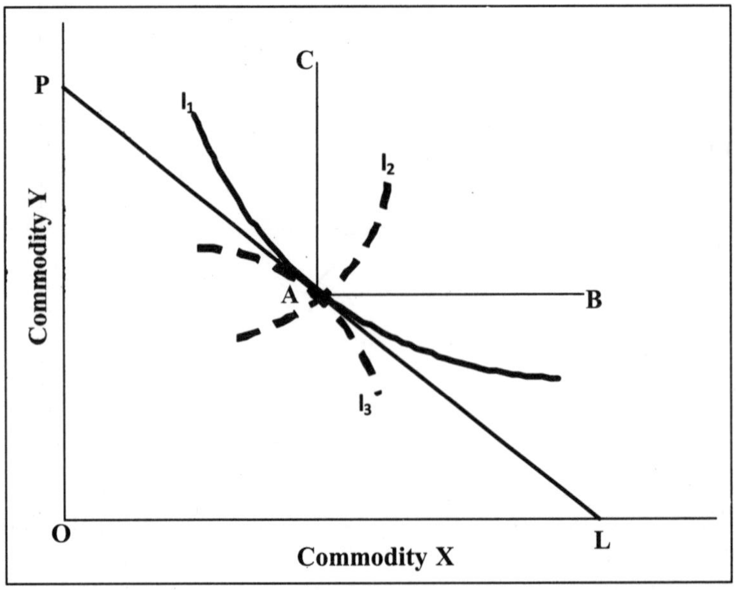

Figure 10.10: Convexity of IDC curve.

So, by assuming the price fall and price rise of commodity X, we could be able to locate the position of IDC I_1 highlighting that, on this IDC, the consumer revealed his preference for original combination A of the two commodities. So, the concept of Revealed Preference Theory helps to derive the IDC from the ordinary market behaviour of the consumer.

Establishing the Convexity of the Derived IDC I_1

Now, the important aspect is to establish the convexity of the derived IDC I_1. A close look at the Figure 10.10 reveals that, IDC I_1 must have a convex shape at point A, as it has to pass through the two ignorance zones ABL and ACP. This can be further proved from the following facts, why the IDC will not assume any other shape except convexity to the origin.

- ✰ The IDC cannot have the straight line shape PL, as all other combinations on PL are considered inferior by the consumer relative to A and thereby, consumer cannot be indifferent with other combinations on PL compared to A. But, the IDC assumes that, the consumer should be indifferent between all the possible combinations of the commodities on the same IDC. So, the IDC will not have a straight line shape, PL.

- ☆ The IDC cannot have a shape like I_2 cutting the original combination A on price line PL because, all the possible combinations of two commodities below A are considered inferior by the consumer with respect to PL price line and thereby, the consumer cannot be indifferent with them.
- ☆ The IDC cannot have concave shape like I_3, for the same reason mentioned above.

So, since all the other possible shapes of IDC are eliminated, the only possible shape of IDC is that, it is convex to origin, as indicated by IDC I_1. On this convex shaped IDC I_1, the consumer revealed his preference for combination A of the commodities and other possible combinations on the IDC are ignored by the consumer and they are not inferior.

10.7. Superiority of Revealed Preference Theory over Marshallian Utility Analysis and IDC analysis

The Revealed Preference Theory is superior to the earlier theories of consumer behaviour *viz.*, Marshallian utility analysis and IDC analysis on the following grounds:

- ☆ The Revealed Preference Theory is not based on introspective behaviour of the consumer, unlike Marshallian utility analysis and IDC analysis. It is based on behaviouristic approach of the consumer in the practical (market) situation. Among these two approaches *viz.*, behaviouristic approach and introspective or psychological approach, the earlier approach is more reliable, as it is based on the ordinary market behaviour of the consumer and cannot go wrong and hence, it gained more support from the Modern Economists.
- ☆ Both, Marshallian utility analysis and IDC analysis assume 'continuity' in the consumption behaviour of the consumer. But, Revealed Preference Theory avoids this 'continuity' assumption and it is based on the 'discontinuity', as it is the practical situation in the market behaviour of the consumer. Taking this 'discontinuity' concept into consideration, Hicks revised the assumption of 'continuity' by replacing it with 'strong ordering' assumption.
- ☆ The Revealed Preference Theory completely dispenses with the assumption that, the consumer always acts rationally in choosing the combination of commodities, so as to attain maximum satisfaction in the consumption process. This theory has not laid any hypothesis like diminishing MU of Marshallian utility analysis and diminishing MRS of IDC analysis. Instead, it formulated 'consistency postulate' to explain the consumer's demand theory.
- ☆ The Revealed Preference Theory assumes over-compensation effect to compensate the real income back to the normal or origin level and this is considered superior over Hicks-Allen method. This is because, in Hicks-Allen method, through compensating variation in income, the consumer is brought back onto the original IDC, whereas, in Revealed Preference Theory, the consumer is moved onto the compensated price line and he derives higher satisfaction like Slutsky approach.
- ☆ Marshallian utility analysis and IDC analysis guides to analyze the consumer's behaviour and thereby, gives the information to prove Marshallian law of demand. However, both the approaches depend upon unrealistic assumptions. Marshallian utility analysis assumed utility is measurable or quantified, constancy of MU of money, commodities have independent utilities etc. Similarly, IDC analysis also framed invalid assumptions like consumer draw a complete list of possible preferences and he is indifferent among the possible combinations on the same IDC. However, Revealed Preference Theory guides to derive the law of demand through analyzing the ordinary market behaviour of the consumer and that too without formulating such invalid assumptions considered in the earlier two approaches. Moreover, no IDCs are used to derive the Marshallian law of demand. If the tastes and preferences of the consumer do not change, Revealed Preference Theory guides the derivation of demand curve through observing the ordinary behaviour of the consumer in the market situation. Thus, Revealed Preference Theory assumes constancy of tastes and preferences of the consumer.
- ☆ Without drawing IDCs, the Revealed Preference Theory guides to study the price effect and to decompose the same into income effect and substitution effect through cost-difference method, which is more reliable compared to compensating variation in income of Hicks-Allen method. Further, this theory explains the income effect of Hicks in a much simpler way and it shows the superiority of this theorem. This notion gives the foundation for the Welfare Economics in stipulations of observable performance based on reliable choice.
- ☆ Revealed Preference Theory explains the derivation of IDC and also explains its important property of 'convexity'.

10.8. Criticisms of Revealed Preference Theory

Though Revealed Preference Theory was meritorious compared to Marshallian utility analysis and IDC analysis, it is also not free from limitations and they are discussed on the following grounds:

☆ The Revealed Preference Theory is based on the postulate of 'strong ordering'. But, since the consumer is often confronted with various alternatives due to the presence of close substitutes, the concept of indifference cannot be ruled out altogether even in the ordinary market behaviour of the consumer. According to W.E.Armstrong, around the point of chosen combination of commodities, there are points of indifference. So, the consumer in the process of attaining maximum satisfaction under a given income-price situation goes through these indifference points. So, the combination actually chosen by the consumer is one among the indifference combinations that the consumer faces in the consumption process. As shown in the Figure 10.11, the compensated price line 'BC' passes through original combination A and the consumer strongly ordered point E on BC line (under new income-price situation). The point E lies very close to the point A of PL_1 price line, thereby, the consumer is indifferent between E and A combinations of the commodities, as per the assertion of W.E.Armstrong. So, this breaks down the Samuelson's Revealed Preference Theory because,

- ❖ The consumer is indifferent between E and A
- ❖ With the fall in price of rice, the consumer strongly ordered or purchased more quantity of rice at E and still he is indifferent with combination A on PL_1 price line, as according to Armstrong, around the point of earlier chosen combination A (say, at combination E), the consumer is indifferent even in the new income-price situation.

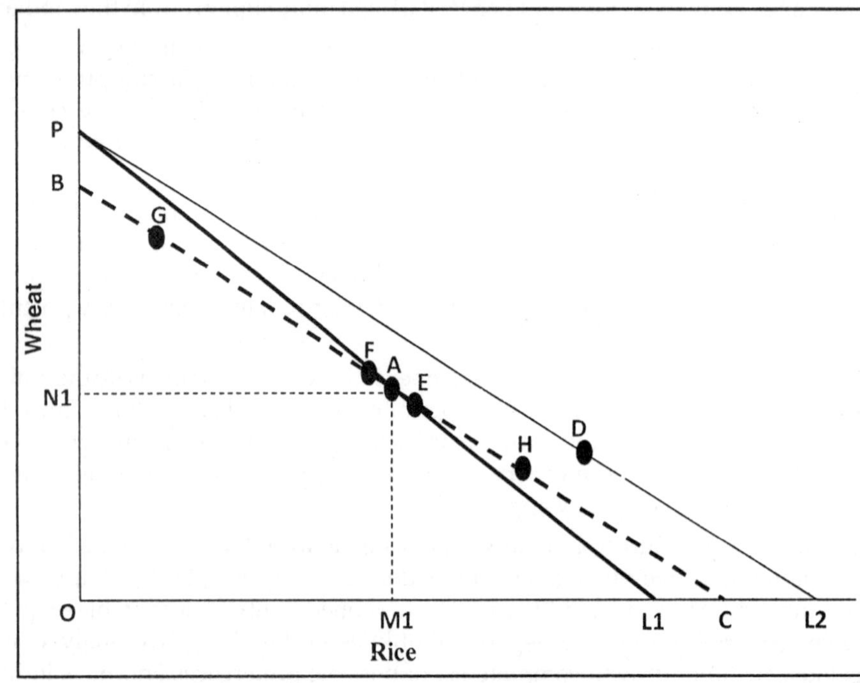

Figure 10.11: Breakdown of Samuelson's theorem – Consumer remains indifferent even in new income-price situation.

☆ The Revealed Preference Theory is based on strong ordering or ordinary market behaviour of the consumer and hence, it is not possible to make a distinction between income effect and substitution effect. So, by mere observation of the consumer behaviour, it is not possible to isolate substitution effect from the income effect. In this case, IDC analysis is more superior to the Revealed Preference Theory, as two components of price effect are studied through drawing IDCs. Moreover, Revealed Preference Theory assumed positive E_Y and this further made difficult to establish substitution effect. However, Samuelson made the distinction between income effect and substitution effect through using cost- difference method, similar to Slutsky method.

☆ Since, the Revealed Preference Theory assumed positive E_Y, it could not offer much explanation regarding the price effect of inferior goods and giffen goods, as for them, the income effect is negative. So, the Revealed Preference Theory could not explain giffen's paradox. In this context, the IDC analysis is more meritorious compared to Revealed Preference Theory, as it offered the valid explanation regarding the relative strengths of substitution and income effects of inferior goods and giffen goods.

☆ In Revealed Preference Theory, Samuelson emphasized 'choice reveals preference' but this is subjected to severe criticism because, the consumer does not act rationally at all times in the market and hence, the choice he reveals for the commodities may not be as per his preferences. So, according to Tapas Majumdar, *'the Revealed Preference Theory is invalid for situations, where the individual choosers are known to be capable of employing strategies of a game theory type of behaviour'.*

- ☆ Revealed Preference Theory can be used to derive individual demand curve, but not helpful to derive market demand curve. This is because, as explained through Figure 10.6, the demand curve of a consumer is drawn with reference to rice commodity. But, if the price of rice commodity is decreased in the market, the price of other related goods also gets affected and this will alter the distribution of real income in the economy. This alteration of real income of the community may lead to upward sloping demand curve, instead of downward sloping curve, thereby, the Revealed Preference Theory is not applicable to derive market demand curve. In this context, IDC analysis is superior to Revealed Preference Theory, as it is applicable to derive both individual demand curve and market demand curve.
- ☆ Through Revealed Preference Theory, it is difficult to analyze the consumer behaviour for the choice involving risk and uncertainity situations. Say for example, if combination A is preferred to combination B and combination B is preferred to combination C, as per the assumption of transitivity, combination A is preferred to combination C. But, if the probability of having combination A is very less and the consumer's preference for combination A over combination B and C cannot be said that, the choice is made on ordinary market behaviour.
- ☆ The customer always does not choose only one combination and hence, the theorem itself is incorrect.
- ☆ This theory is not valid for a game theory and it is unsuccessful during jeopardy or indecisive conditions.

Considering the above demerits, this theorem cannot prove to be better than the IDC concept, since the latter gives overall performance of many customers and this theorem fails in this aspect. Hence, this theorem is no way superior to the other concepts of demand analysis.

11

Theory of Demand

In Chapter 3, we studied that, business firms and consumers constitute the major economic players in an economic system. Consumers, in particular, play a very crucial role in the economy, as they direct the business activities based on their income and tastes and preferences. In economic sense, the income level of the consumer is crucial in purchasing the various goods and services and the process through which the consumer fulfills his consumption needs constitute 'demand'.. This chapter focuses on the consumer's demand for various goods and services in the ordinary market behaviour.

The word 'demand' is often confused with desire, wish, want etc., thereby, the need to define it precisely arises. Demand is more than a desire or wish or want, in the sense that, it involves the ability of the consumer to pay for having the commodity. So, 'demand' in Economics means, a desire or wish or want to have a good or service supported by the willingness and ability to pay for it. Say, if the consumer wishes to purchase a quintal of rice, but if he does not have the ability to pay for it, it will simply remain as a desire and it is not a demand from the consumer. So, demand is an effective desire, *i.e.*, a desire, which is backed up by the willingness and ability to pay by the consumer to have a commodity. The following are the popular definitions of demand given by the Economists,

'Demand means the various quantities of goods that would be purchased per time period at different prices in a given market'.

—(Hibdon)

'The demand for anything, at a given price, is the amount of it, which will be bought per unit of time at that price'

—(Benham)

'By demand, we mean the various quantities of a given commodity or service which consumers would buy in one market in a given period of time at various prices or at various incomes or at various prices of related goods'

—(Bobber)

From the above definitions, we can infer the following three important aspects of demand in an economic sense:

- ☆ To ensure a consumer demands for a commodity, his desire must be accompanied or backed up by willingness and ability to pay.
- ☆ Demand is always at a price. If we say, a consumer demanded for something with out referring the price at which the transaction had taken place, it will be meaningless.

☆ Demand is always with reference to a particular period of time, as time element plays a very vital role in bringing the changes in the demand and supply forces, thereby, influences the market price of the commodity.

From the point of view of seller, the demand price is synonymous with Average Revenue (AR). We know, AR is the Total Revenue (TR) divided by the quantity of output *i.e.*, AR = (TR/Quantity) = ((Quantity x Price)/Quantity)). So, AR = Price of the commodity. Hence, demand curve is also called AR curve.

11.1. Types of Demand

In economic sense, the demand for a particular commodity is mainly influenced by three important factors viz., price of the commodity (Price demand), income of the consumer (Income demand) and prices of the related commodities (Cross demand). These three types of demand are discussed in-detail in the ensuing pages.

11.1.1. Price Demand

In Economics, demand refers to Price demand. It implies that, various quantities of a commodity purchased by the consumer at its various prices. The concept of price demand can be well-explained through the Law of demand.

11.1.1.1. Law of Demand

The concept of law of demand was proposed by Alfred Marshall. This law expresses a relationship between the quantity demanded and its price. From the above definitions, we can infer that, the demand for a commodity is influenced by price of the commodity. It is the practical experience of every consumer that, with the fall in price of the commodity, the consumer purchases more of the commodity and *vice versa*. So, there lies inverse relationship between price and quantity demanded of the commodity and this inverse relationship is explained by the Law of demand. The various statements of the Law of demand given by the famous Economists are as under.

'The greater the amount sold, the smaller must be the price at which it is offered, in order that it may find purchasers; or in other words, the amount demanded increases with a fall in price and diminishes with rise in price'

—(Alfred Marshall)

'Under the same conditions of demand, the quantity of a commodity which will be purchased tends to vary inversely with its price'

—(Meyers)

'The quantity demanded varies inversely with price'

—(Ferguson)

'The law of demand states that, people will buy more at lower prices and buy less at higher prices, other things remaining the same'.

—(Samuelson)

'Other things remaining the same, the quantity demanded of a commodity will be smaller at higher market prices and larger at lower market prices'.

—(E. Miller)

To put it in simple words, 'when the price of the commodity decreases, the consumers prefer more of that commodity and when the price of the commodity increases, the consumers prefer less of that commodity, *ceteris paribus*'

The above definitions infer that, the quantity demanded for the commodity responds in the opposite direction of price or varies inversely with the price. Of course, besides price, there are several other factors that influence the quantity demanded for the commodity and this functional relationship between various factors and quantity demanded of a commodity is called Demand function. It is represented in the form of an equation given below.

$$Q_R = f(P_R / I, P_W, T, ...)$$ ***Equation 11.1***

Q_R = Quantity demanded of commodity rice

f = Functional relationship between quantity demanded and factors influencing it

P_R = Price of rice commodity

I = Income of the consumer

P_W = Price of the wheat (substitute)

T = Tastes and preferences of the consumer

The 'bar' in the above Equation 11.1 denotes that, the variables I, P_W, T etc., are kept constant and the influence of price of the rice commodity is studied on the quantity demanded for rice. We can re-write the above Equation 11.1 in terms of relationship between quantity demanded and price of rice commodity (Demand function) and this is shown below.

$$Q_R = f(P_R) \text{ } \textit{ceteris paribus}$$

The above general functional form does not show, how much quantity of rice is demanded by the consumer due to change in price of the rice. So, to estimate the quantity demanded in numerical form, we generally consider a linear demand function given below,

$$Q_R = a - bP_R \quad \textit{Equation 11.2}$$

In the above equation, 'a' implies intercept of the demand function, 'b' represents the regression coefficient or slope of the demand function. The negative sign in the above function implies negative relationship between the two parameters under consideration. To explain the linear demand function, assume the following function:

$$Q_R = 20 - 4P_R \quad \textit{Equation 11.3}$$

The above Equation 11.3 implies that, for every one unit increase in the price of the commodity, the quantity demanded of rice decreases by four units.

Features of Law of Demand

From the law of demand, we can ascertain the following features:

- ☆ There exists inverse relationship between price and quantity demanded of the commodity.
- ☆ Price of the commodity is the independent variable and quantity demanded is the dependent variable.
- ☆ It operates under *ceteris paribus* assumption.
- ☆ It helps to estimate the E_D for the commodity, based on which we can forecast the variables.

Assumptions of the Law of Demand

According to Stigler and Boulding, the demand function between quantity demanded and price of the commodity is valid, when the following assumptions holds good:

- ☆ The income of the consumer must remain constant.
- ☆ The prices of other related commodities *i.e.*, substitutes and complements must remain constant.
- ☆ No discovery of new substitutes.
- ☆ There is not change in the tastes and preferences of the consumers.
- ☆ Consumers do not feel that, the present fall in price is a prelude to a further decline in price *i.e.*, there should not any changes in the anticipations.
- ☆ No change in population.
- ☆ No change in total assets.
- ☆ No change in weather conditions.
- ☆ Consumer behaves rationally in the consumption of commodity.
- ☆ The commodity is of ordinary type and homogenous.
- ☆ No changes in psychology of the consumer.
- ☆ No changes in the credit availability to the consumer.
- ☆ No changes in the propensity to consume.
- ☆ No changes in trade cycles.
- ☆ No product differentiation.
- ☆ No changes in advertisement and propaganda about the commodity.
- ☆ The habits of the consumers should remain unchanged.

If the above assumptions do not hold good, the law of demand may not hold true. Say, for example, when the price of rice in the market is Rs. 25/kg, the consumer demands 30 kilograms. If the price of rice is increased in the market, say to Rs. 35/kg, now the consumer demands only 20 kilograms. This indicates that, with increase in price of the rice commodity, the quantity demanded for rice is decreased, thereby, the law of demand holds good, as long as other factors *viz.*, income of the consumer, price of the wheat (substitute good), tastes and preferences of the consumer etc., are held constant. If the increase in price of the rice commodity is accompanied with rise in income of the consumer, then the law of demand may not hold good, as the consumer is still capable of purchasing more quantity of rice, even though its price is increased in the market. Similarly, if price of the wheat commodity is decreased in the market, the consumer may switch over towards consumption of wheat, thereby, he demand less of rice, though its price is stable in the market.

If the consumer prefers to have more of wheat (say, as per Doctor's advise) than rice, now the demand for wheat increases, though the price of wheat is stable in the market. From these examples, we can infer that, the demand for the commodity responds inversely to the change in price of the commodity, only under *ceteris paribus* condition. If this *ceteris paribus* condition is not fulfilled, the law of demand may not hold true. Demand, thus, is a negative relationship between price and quantity, under *ceteris paribus* condition and thereby, the demand curve (between quantity demanded of a commodity and price) will have a negative slope. In the words of Bilas, *'other things being equal, the quantity demanded per unit of time will be greater, lower the price, and smaller, higher the price'*.

Exceptions to the Law of Demand

The Law of demand is a general statement explaining the inverse relationship between quantity demanded and price of the commodity. However, there are certain peculiar cases in which the law of demand will not hold good. In such cases, more quantity of commodity is demanded even at a higher price and *vice versa* and hence, the demand curve will have positive slope *i.e.*, it slopes upward. The following are the list of such exceptions of law of demand:

- ☆ ***Prestige goods:*** There are certain commodities like diamond, costly jewellery, sports bikes, sports cars etc., which are purchased as a mark of distinction in the society. These items are purchased by the wealthy people in the society and they purchase these goods to distinguish themselves from the average people, even if the prices of these goods increases. So, for such goods, the demand varies proportionately with the price of the commodity. Such type of consumption is called Conspicuous consumption. This concept was proposed by Theorstein Veblen and he stated in his doctrine, *'Conspicuous consumption'* that, these goods are purchased not for their intrinsic value, but for their snob appeal or prestige value. But, this is not really an exception because, with the fall in price of diamonds, the rich people may discard it, as it loses its distinctiveness in the market. However, when the diamonds are offered at lower prices, initially, the demand will shoot up from the middle-income group people, thereby, the law of demand operates.
- ☆ ***Conspicuous necessities:*** There are certain necessities in modern human life, which are purchased even at high price, as they attained the symbol of status. They include television, fridge, cell phone, internet connection etc., and they are purchased even at a high price.
- ☆ ***Price expectations:*** The people may purchase more of a commodity even at a high price in the present market because, they expect the price rise in the future. Of course, the violation of the law in this case is only temporary.
- ☆ ***Ignorance of the consumer:*** If the consumer is ignorant about the fall in price of a commodity in the future, he will purchase more of the same at a high price in the present market.
- ☆ ***Giffen goods:*** Sir Robert Giffen explained that, poor people will demand more for inferior goods, if their prices are high and they demand less, if the prices are low. In case of giffen goods, say due to price decrease, the income effect is strongly negative and it outweighs the weak negative substitution effect. If the prices of such goods increases, due to strong positive income effect, (as the poor consumers spend large proportion of their income) the demand also increases. So, in case of rise in price, the consumer economizes the expenditure on other goods and purchases more of giffen goods. If the price of giffen goods decreases, the consumers will divert their extra purchasing power on other (superior) goods and thereby, the consumer purchases less quantity of giffen goods. All giffen goods are inferior goods. The examples of giffen goods include, bread, potatoes, bajra etc. So, the quantity demanded varies proportionately with the price of the giffen good, thereby, they disobey the law of demand. But, this is not really an exception because, if the price of giffen good increases, by reducing expenditure on superior goods, the consumer is diverting the expenditure to giffen goods, which is nothing but an income effect. But, we assumed under the law of demand that, income of the consumer does not change.
- ☆ ***Speculative effect:*** In speculative market, the people purchase the stocks, shares etc., even at high prices with the anticipation that, their prices will still rise in the future. So, Marshall opined that, speculation is one of the important exceptions, where Marshallian law of demand is not obeyed. In the words of Alfred Marshall, *'the law of demand does not apply to the demand in a campaign between groups of speculators. A group, which desires to unload a great quantity of a thing on to the market, often begins by buying some of it openly. When it has thus raised the price of the thing, it arranges to sell a great deal quietly, and through unaccustomed channels'*. But, this is not really an exception because, if the prices continues to increase for a longer period, the demand will not rise further, as there is a certain limit, beyond which the demand will fall for any commodity. So, speculative effect is an exception only for a shorter period of time.
- ☆ ***Hyper inflation:*** At times of hyper inflation, the people purchase more of goods even at high prices with the anticipation that, the prices will still rise in the future.

- ☆ *Depression:* At times of depression, the purchasing power of the people in the economy is low and thereby, they demand less even at low prices. But, this is not really an exception because, under the law of demand, we assumed 'no change in consumer's income'.
- ☆ *Price delusion:* Some consumers will be deluded by the price under the notion that, high priced articles are superior in quality and low priced articles are inferior. So, due to price delusion, people will demand more for the commodity, if the price increases and they demand less, if the price decreases. But, this is not really an exception because, under the law of demand, we assumed there is no change in psychology of the consumer.
- ☆ *Out of fashion commodities:* If the commodity goes out of fashion, the demand for such commodity decreases, even if the price of the commodity decreases.
- ☆ *Lack of knowledge:* Due to lack of knowledge regarding the competitive price for the commodity, the consumer may purchase more quantity of commodity even at high price. For example, the farmers due to lack of knowledge, purchase more quantity of pesticides at higher prices from the dealers. Sometimes, the sellers will quote different names and different labels for the same commodity and fix a higher price. The consumers, out of ignorance, presume it is a new commodity and thereby, purchase the commodity at a higher price in the market. But, this is not really an exception because, once the consumer realizes the commodity is not a new one in the market, the demand will rapidly fall for it.
- ☆ *Abnormal situations:* At times of situations like wars, droughts, floods etc., the consumers purchase the commodities even at high price fearing their shortage in the near future.
- ☆ *Advertisement and propaganda(Demonstration effect):* Because of demonstration effect, the consumer gets motivated and stimulated to have the commodity even at a higher price. But, this is not really an exception because, under the law of demand, we assumed 'rational behaviour of the consumer'.

This discussion infers that, the above exceptions of law of demand are not really true exceptions, as in most of the cases, the assumptions under the law of demand are violated and if some exceptions are valid, their validity is only upto a certain point. So, the demand curve may rise upward only upto a part of its length. For such exceptional cases, the Marshallian law of demand is not obeyed and such demand curves are called Perverse demand curves.

Importance of Law of Demand

The concept of law of demand is very useful for the farmers, traders and Government and it is discussed under the following heads:

- ☆ *Importance to farmers:* The law of demand guides the farmers towards realization of better price for their crop produce. If a good crop fails to increase the demand, the prices will fall sharply. So, the farmers will realize that, high price for the produce depends upon the demand situation in the economy. At low price, the demand for the produce will increase, *ceteris paribus* and *vice versa*.
- ☆ *Importance to traders:* Through analyzing the demand situation for the commodity, the traders will fix the price for the commodity. If an increase in price adversely affects the quantity demanded for the commodity, then the trader will quote a low price for the commodity to boost its sales in the market. On the other hand, if a slight decrease in the price of the commodity brings up a significant increase in demand, then the trader will reduce the price of the commodity and increase his profits through more quantum of sales.
- ☆ *Importance to Government:* This law is extremely important to the Finance Minister of the country. If the Finance Minister imposes high tax on the commodity, which will adversely influence its demand, then there is no use of imposing higher tax, as it fails to increase the revenue on the part of Government due to low sales of the commodity in the market. So, he has to impose high tax only on those commodities for which the demand is not substantially reduced even after imposing the tax.
- ☆ *Guides the Monopolist:* The law of demand guides the monopolist in fixing the price of the commodity. He will draw the market demand schedule in advance and thereby, fixes the price of the product at which he can transact large quantities of output.
- ☆ *Planning the establishment of industries:* The first basic step in studying the feasibility analysis of establishment of an industry in an area is to assess the demand for the product. If the product to be produced in an area will fetch more demand from the society, it will indicate the entrepreneurs to establish an industry in an area.

11.1.1.2. Individual Demand

The individual's demand for a commodity is the quantity of a commodity, which the consumer is willing to purchase at a given price in the market at a given point of time. The demand of the individual consumer is called Individual Demand. The individual's demand for a commodity varies inversely with the price of the commodity, *ceteris paribus.*

Individual Demand Schedule

We know a consumer purchases different quantities of a commodity at different prices. So, a schedule or a table showing such information for a commodity at a given point of time refers to Individual demand schedule. The following Table 11.1 shows the hypothetical demand schedule of a consumer for a rice commodity at different prices. This infers that, as the price of the rice decreases in the market, more quantity of it is demanded by the consumer and *vice versa.*

Table 11.1: Individual demand schedule for rice in the market.

Price (Rs/kg)	*Quantity Demanded (kg)*
18	50
20	47
25	44
30	40
35	38

Individual Demand Curve

Demand curve is a graphical representation of the demand schedule. According to Lipsey, *'the curve, which shows the relation between the price of a commodity and the amount of that commodity the consumer wishes to purchase, is called demand curve'*. The individual demand schedule can be converted into an individual demand curve by measuring price on Y-axis and quantity of commodity demanded on X-axis, as shown in Figure 11.1. DD is the demand curve for rice, which slopes downwards from left to right showing that, when price decreases, the quantity demanded extends and *vice versa.* So, different points on the individual demand curve indicates different quantities of a commodity purchased by the consumer at its different prices. Here, it is often important to distinguish the 'demand' from the 'quantity demanded'. Typically, 'demand' refers to the entire demand curve and not just one point on the curve (the entire relationship between price and quantity demanded), while 'quantity demanded' is a point on the demand curve. So, when the price of the commodity changes, the quantity demanded changes, but demand does not change. But, when size of the family changes or cost of production changes, demand changes, thereby, the entire demand curve will shift.

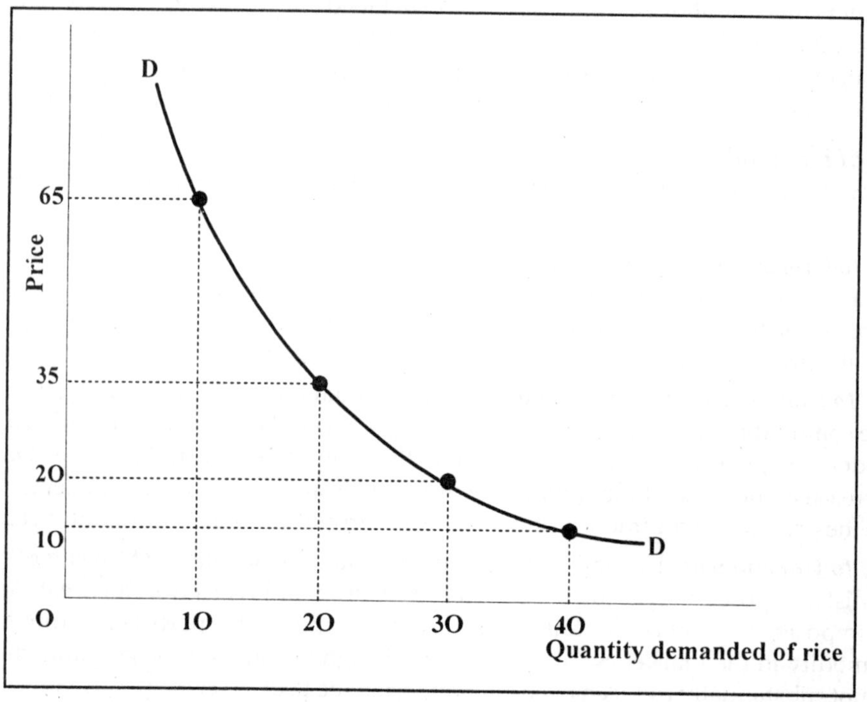

Figure 11.1: Downward (Negative) slope of demand curve.

Reasons for Downward Sloping of Demand Curve

In general, for normal goods, as the quantity demanded varies inversely with the price of the commodity, the demand curve will have a negative slope. That means, the two important variables *viz.*, price and quantity demanded work in opposite direction and this gives negative slope to the demand curve. So, downward sloping demand curve depicts the law of demand. According to Alfred Marshall, the downward sloping of demand curve is in accordance with the LDMU concept. But, according to J.R.Hicks and R.G.D.Allen, the direction and relative strengths of substitution effect and income effect is responsible for the downward sloping of demand curve. The following are the main reasons for the downward sloping demand curve.

- ☆ ***LDMU concept:*** According to Alfred Marshall, the downward sloping of demand curve is explained by the concept of LDMU. In LDMU, the consumer attains equilibrium when MU of expenditure is equal to the MU of money *i.e.*, ($MU_R/P_R = MU_M$). MU_R is the MU derived from rice and P_R is the price of rice / unit and MU_M is the MU of money. Say, if price of rice is decreased in the market, in the above equation, left hand side component will increase over MU_M. That means, MU of expenditure is more compared to MU of money. So, to ensure equality between these two or to attain consumer's equilibrium, MU of rice should be decreased, as MU_M is assumed constant in LDMU (Marshallian) approach. This MU_R will be decreased, only if the consumer consumes more of rice. So, the quantity demanded for rice increases with the fall in price of the rice.

- ☆ ***Direction and relative strengths of substitution effect and income effect:*** According to J.R.Hicks and R.G.D.Allen, the concept of IDC analysis helps to isolate both substitution effect and income effect and their direction and relative strengths will influence the obeyance of law of demand by different commodities. In case of normal goods, the substitution effect is negative and income effect is positive, thereby, they both act in the same direction to increase the quantity demanded of commodity due to fall in its price and hence, for normal goods, the demand curve slopes downward from left to right. In case of inferior goods, the strong negative substitution effect outweighs weak negative income effect, thereby, the quantity demanded for the commodity increases due to fall in its price and hence, for inferior goods also, the demand curve slopes downward from left to right. However, in case of giffen goods, strong negative income effect outweighs weak negative substitution effect, thereby, the quantity demanded of a commodity decreases due to fall in its price and hence, for giffen goods, the demand curve slopes upward from left to right.
- ☆ ***Change in number of consumers:*** If the price of the commodity falls, some new consumers will start using the commodity and the old consumers purchase more of the commodity. So, the total quantity demanded of the commodity will increase with the decrease in its price. On the other hand, if the price of the commodity increases in the market, new consumers withdraw from using the commodity and even the old consumers purchases less of the commodity. So, the total quantity demanded of the commodity will decrease with the increase in its price.
- ☆ ***Dominant low-income group:*** In general, a society is dominated by low income group compared to high income group. This low income group consumers show the tendency of purchasing more quantity of commodity at low price and *vice versa*. But, consumers from high income group are capable of purchasing the same commodity even at higher price. Since, low-income group dominate the society, this leads to downward (negative) sloping demand curve.
- ☆ ***Multiple uses of a commodity:*** When the commodity is offered at a cheaper price, the tendency of the consumers is to employ it in various uses. So, to employ the commodity in various uses, they purchase (demand) more of that commodity. On the other hand, if the price of the commodity is increased, the consumers uses it cautiously in important uses, thereby, they demand less of the commodity. This gives the negative slope to the demand curve. For example, when the electricity charges are more, the consumers may use it only for domestic lighting. But, when the same is offered at low prices, they use it for various purposes like heating, boiling, cooking etc.
- ☆ ***Differences in tastes:*** Generally, consumers differ in their tastes in having different commodities and different quantities of same commodity. A consumer who has low preference for a commodity, will buy more of it, only when it is quoted at less price in the market. Hence, sellers will have to quote less price for the commodity to boost its sales in the market.

11.1.1.3. Market Demand

It refers to the demand for the commodity in the entire market *i.e.*, various quantities of a commodity purchased by all the potential consumers at various prices in the market at a given point of time. The market demand for a commodity is obtained by summing up the quantity demanded at various prices by all the individual consumers at a given time in the market. It is, otherwise, described as the horizontal summation of the individual's demand for a commodity at various prices in market. Since, individual's demand varies inversely with the price of the commodity, the market demand for a commodity also varies inversely with the price of the commodity.

Market Demand Schedule

Since, all the potential consumers in the market purchase different quantities of a commodity at different prices, the schedule or a table showing such information for a commodity at a given point of time refers to Market demand schedule. So, we can arrive at the market demand schedule by summing up the individual demand schedules for a commodity. It is, otherwise, defined as the inverse relationship between price of the commodity and quantity demanded by all the potential consumers in the market. The market demand schedule for a commodity can be prepared in two ways *viz.*, Additive method and Representative method. In additive method, as explained above, the horizontal summation of individual demand schedule gives the market demand schedule for a commodity (Table 11.2). In Representative method, we consider one individual consumer as a representative of the population and his demand schedule will be multiplied by the number of consumers in that market. As shown in the Table 11.4, the number of consumers in the market is 500 and the total market demand is derived by multiplying the number of consumers with the quantity demanded by the representative consumer at different levels of prices of the commodity.

The following Table 11.2 shows the market demand schedule for rice commodity at different prices. It infers that,

- ✰ Market demand schedule is the horizontal summation of individual demand schedules of a commodity at different prices.
- ✰ Market demand schedule also obey the law of demand.

Market Demand Curve

The graphical representation of market demand schedule is the market demand curve for the commodity. Figure 11.2 shows the market demand curve for rice in the market. The market demand schedule can be converted into market demand curve by measuring price on Y-axis and quantity of commodity demanded by all the potential consumers on X-axis. D_MD_M is the market demand curve (Panel B of the Figure 11.2), which slopes downwards from left to right reveals that, when price decreases, the market demand for the commodity extends and *vice versa*. So, different points on the market demand curve indicates different quantities of rice commodity purchased by all the potential consumers at its different prices in the market. With the fall in price of the commodity, new buyers enter the market and this will further increase quantity demanded of the commodity. This will also contribute to downward falling of market demand curve from left to right with the fall in price of the commodity.

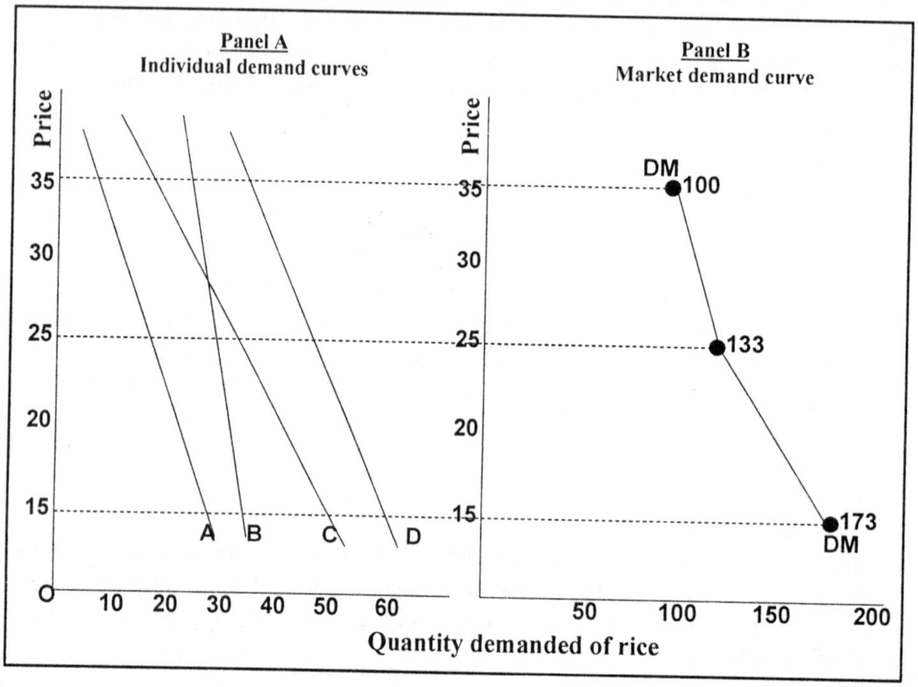

Figure 11.2: Derivation of Market demand curve from Individual demand curves for commodity rice.

Table 11.2: Market demand schedule for rice in the market

Price (Rs/kg)	*Quantity Demanded (kgs)*				
	Consumer A	*Consumer B*	*Consumer C*	*Consumer D*	*Market Demand*
15	29	34	50	60	173
25	19	30	36	48	133
35	8	27	29	36	100

Table 11.3 shows the market demand schedule for commodity X and please note the individual demand schedule of consumer C, as for him the commodity X is a giffen good, indicated by more quantity of commodity X purchased with increase in its price. But, the market demand curve (Figure 11.3) is a downward sloping curve from left to right indicating that, the commodity obeys Marshallian law of demand, as for majority of the consumers in the market, it may be either a normal good or an inferior good.

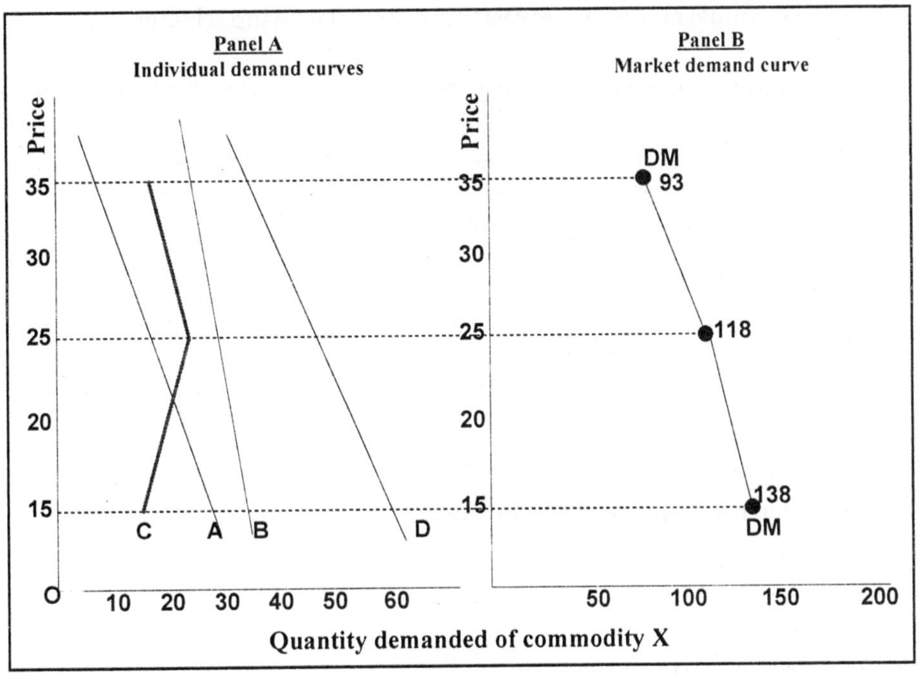

Figure 11.3: Derivation of Market demand curve from Individual demand curves for commodity X.

Table 11.3: Market demand schedule for commodity X in the market.

Price (Rs/kg)	*Quantity Demanded (kgs)*				
	Consumer A	*Consumer B*	*Consumer C*	*Consumer D*	*Market Demand*
15	29	34	15	60	138
25	19	30	21	48	118
35	8	27	22	36	93

However, the Additive approach of computation of market demand is more time consuming, as it difficult to collect and compile the individual demand schedules. Hence, Representative method is commonly followed to prepare market demand schedule (Table 11.4), where the total market demand is derived by multiplying the number of consumers with the quantity demanded by the representative consumer at different levels of prices of the commodity.

Table 11.4: Market demand schedule for rice in the market (Representative method)

Price (Rs/kg)	*Demand of Representative Consumer (kgs)*	*Total Number of Consumers*	*Market Demand (kgs)*
15	30	500	15000
25	20	500	10000
35	5	500	2500

So, the above explanation infers that, if we have the information pertaining to individual demand schedules for a commodity, we can easily prepare individual demand curves, market demand schedule and market demand curve for the commodity.

Individual Demand Schedule vs Market Demand Schedule

The important differences between individual demand schedule and market demand schedule are discussed here under.

- ☆ Individual demand schedule indicates the consumption behaviour with reference to single consumer, while market demand schedule indicates the consumption behaviour of all the consumers in the market.
- ☆ Individual demand schedule forms the basis for Market demand schedule, whereas market demand schedule indicates the individual demand schedule.
- ☆ Individual demand schedule may obey the law of demand, but Market demand schedule may not obey the law of demand and *vice versa*.

- ☆ Individual demand schedule may be more elastic, but Market demand schedule may be less elastic and *vice versa*.
- ☆ Individual demand may not influence the price of the commodity, but Market demand influences the price of the commodity
- ☆ The Individual demand schedule may involve discontinuity, as it is quite difficult to predict how the consumer behaves at different prices of the commodity. But, there is continuity with reference to Market demand schedule, as individual differences may disappear in the collective demand.

11.1.2. Income Demand

Income demand studies the relationship between income of the consumer and the quantity of the commodity demanded at a point of time. That means, it refers to various quantities of a commodity purchased by the consumer at different levels of his income, *ceteris paribus*. It is presented in the form of an equation given below,

$Q_R = f(I / P_R, P_W, T, \ldots\ldots)$ *Equation 11.4*

Q_R = Quantity demanded of commodity rice

f = Functional relationship between quantity demanded and factors influencing it

I = Income of the consumer

P_R = Price of rice commodity

P_W = Price of the wheat (substitute)

T = Tastes and preferences of the consumer

The 'bar' in the above equation denotes that, the variables P_R, P_W, T etc., are kept constant and the influence of income of the consumer is studied on the quantity demanded for rice. We can write the above Equation 11.4 in terms of relationship between quantity demanded and income of the consumer and this is shown below.

$Q_R = f(I)$ *ceteris paribus*

The income-demand relationship for the consumer is usually positive. That means, if the income of the consumer increases, the quantity demanded of the commodity increases and *vice versa*. However, there are some exceptions and they are discussed here under.

Normal Goods

For normal goods, the income-demand relationship is direct or positive. That means, if the income of the consumer increases, the quantity demanded of normal goods also increases and *vice versa*. So, the income-demand curve will have a positive slope for normal goods. If the income of the consumer increases from OY_1 to OY_2 (Panel A of the Figure 11.4), the quantity demanded for rice increases from OQ_1 to OQ_2. These goods are income positive and price negative. Examples include, rice, wheat, mangoes, apples etc.

Inferior Goods

For inferior goods, the income-demand relationship is negative. That means, if the income of the consumer increases, the quantity demanded of inferior goods decreases and *vice versa*. So, the income-demand curve will have a negative slope for inferior goods. If the income of the consumer increases from OY_1 to OY_2 (Panel B of the Figure 11.4), the quantity demanded for jowar decreases from OQ_1 to OQ_2. These goods are income negative and price positive. Examples include, jowar, bajra, ragi, vegetable ghee etc.

Giffen Goods

They are special type of inferior goods, for which the income-demand relationship is negative and price-demand relationship is positive. Hence, unlike inferior goods, giffen goods do not obey Marshallian law of demand (Table 11.5). This concept is already explained in-detail by employing the IDC technique in the Chapter 9. Since, Marshallian utility analysis considered only price effect in explaining the consumption behaviour, it could not offer valid explanation, why giffen goods are price positive or do not obey Marshallian law of demand. Since, they are income negative, the income demand curve slopes downward from left to right, as in case of inferior goods. Examples for giffen goods include, bread, potato etc.

11.1.3. Cross Demand

It refers to the various quantities of commodity rice purchased by the consumer not due to change in price of rice commodity, but due to change in price of any other related commodity like wheat etc. It is represented in the form of an equation given below,

$Q_R = f(P_W / P_R, I, T, \ldots\ldots)$ *Equation 11.5*

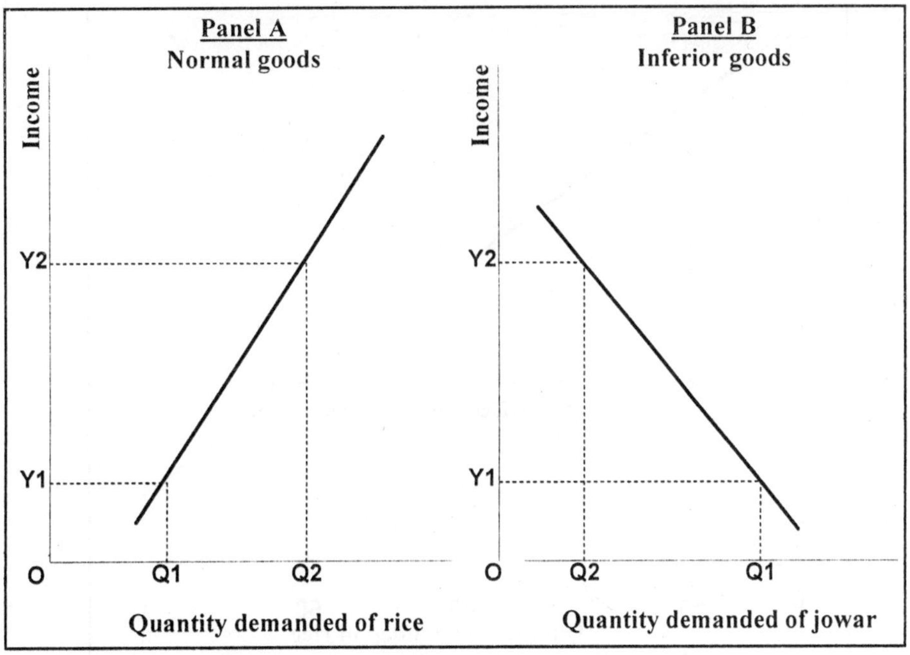

Figure 11.4: Income demand curve for different commodities.

Table 11.5: Differences between normal goods and inferior goods.

Basis of Difference	*Normal Goods*	*Inferior Goods*	*Giffen goods*
Meaning	These are the goods, whose demand increases with increase in income and *vice versa*	These are the goods, whose demand decreases with increase in income and *vice versa*	These are the goods, whose demand decreases with increase in income and *vice versa*
Income effect	Income effect is positive, thereby, act in the same direction as substitution effect (Figure 11.5). When price of rice decreases from Rs. 10 to Rs. 4, the quantity demanded increases from 10 kgs to 60 kgs due to movement of both substitution effect and income effect in the same direction or in the opposite direction of price.	Income effect is negative, thereby, act in opposite direction of substitution effect. But, the strong negative substitution effect outweighs weak negative income effect (case of price decrease). So, inferior goods also obey Marshallian law of demand (Panel A of the Figure 11.6). When price of jowar is decreases from Rs. 10 to Rs. 3, the quantity demanded increases from 20 kgs to 60 kgs due to strong negative substitution effect. But, since the income effect is negative, the quantity demanded decreases from 60 kgs to 40 kgs. However, strong negative substitution effect outweighs weak negative income effect and thereby, total price effect is positive ie., quantity demanded increases from 20 kgs to 40 kgs, and hence, the demand curve slope downward from left to right.	Income effect is negative, thereby, act in opposite direction of substitution effect. But, the strong negative income effect outweighs weak negative substitution effect (case of price decrease). So, giffen goods do not obey Marshallian law of demand. When price of potato decreases from Rs. 10 to Rs. 3, the quantity demanded increases from 40 kgs to 60 kgs due to negative substitution effect. But, since the income effect is strongly negative, the quantity demanded decreases from 60 kgs to 20 kgs. Since, strong negative income effect outweighs weak negative substitution effect, the total price effect is negative ie., quantity demanded decreases from 40 kgs to 20 kgs due to fall in price of potato. Thus, giffen goods do not obey Marshallian law of demand, thereby, demand curve slope upward from left to right (Panel B of the Figure 11.6).
Income-Demand curve	Positive slope	Negative slope	Negative slope
Marshallian demand curve	Negative slope	Negative slope	Positive slope
Movement of Income effect and Substitution effect with reference to price	Both substitution effect (negative) and income effect (positive) act in the opposite direction of price (say, case of price decrease).	Substitution effect moves in the opposite direction of price, but income effect is negative and acts in the direction of price (say, case of price decrease).	Substitution effect moves in the opposite direction of price, but income effect is negative and acts in the direction of price (say, case of price decrease).
Examples	Rice, wheat, apples, mangoes etc.	Jowar, bajra, ragi, vegetable ghee etc.	Bread, potato etc.

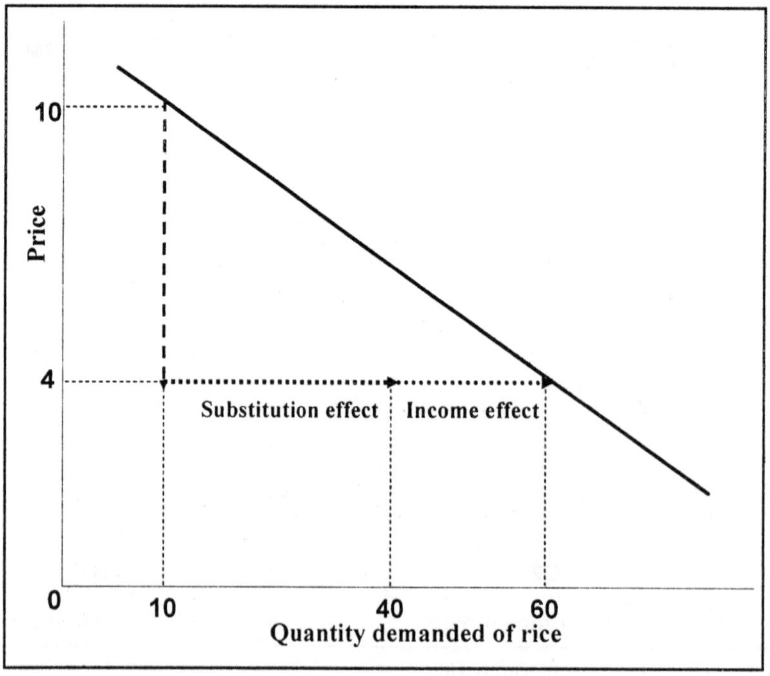

Figure 11.5: Strengths of income effect and substitution effect for a normal good.

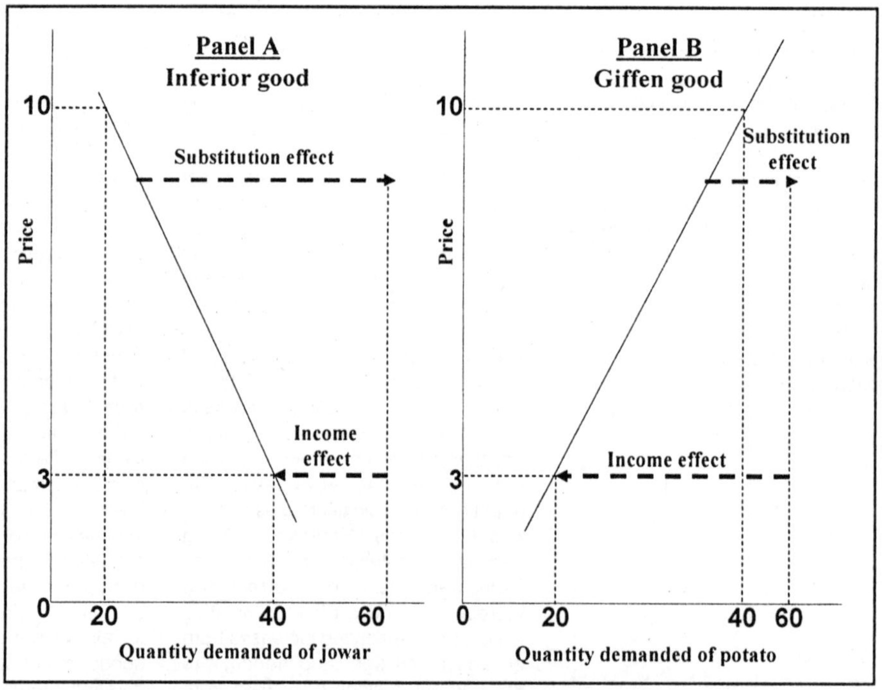

Figure 11.6: Relative strengths of income effect and substitution effect.

Q_R = Quantity demanded of commodity rice
f = Functional relationship between quantity demanded and factors influencing it
P_W = Price of the related commodity wheat (substitute)
P_R = Price of rice commodity
I = Income of the consumer
T = Tastes and preferences of the consumer

The 'bar' in the above equation denotes that, the variables P_R, I, T etc., are kept constant and the influence of change in price of wheat is studied on the quantity demanded for rice. We can write the above equation 11.5 in terms of relationship between quantity of rice demanded and price of wheat (substitute) and this is shown below.

$Q_R = f(P_W)$ *ceteris paribus*

It is important that, the related goods may be substitutes or complements for the commodity under consideration for studying the cross demand.

Cross Demand for Substitutes

The goods which can be used in place of other goods are called substitutes. In case of substitutes like rice and wheat, sugar and jiggery, tea and coffee, jam and butter etc., the cross demand is positive. That means, an increase price of a substitute commodity like wheat, the demand for rice will increase. This is because, with increase in price of wheat, the quantity demanded for wheat will decrease and thereby, the consumer demands more of rice. That means, when price of wheat increases, quantity demanded of rice increases, thereby, price (of rice) and quantity demanded (of wheat) will move in the same direction and hence, the cross demand is positive for substitute goods. So, the cross demand curve for substitutes will rise upward from left to right (concave shape). When price of wheat is increased in the market from OP_{W1} to OP_{W2}, the quantity demanded for its substitute good *i.e.*, rice is increased from OQ_{R1} to OQ_{R2} (Panel A of the Figure 11.7).

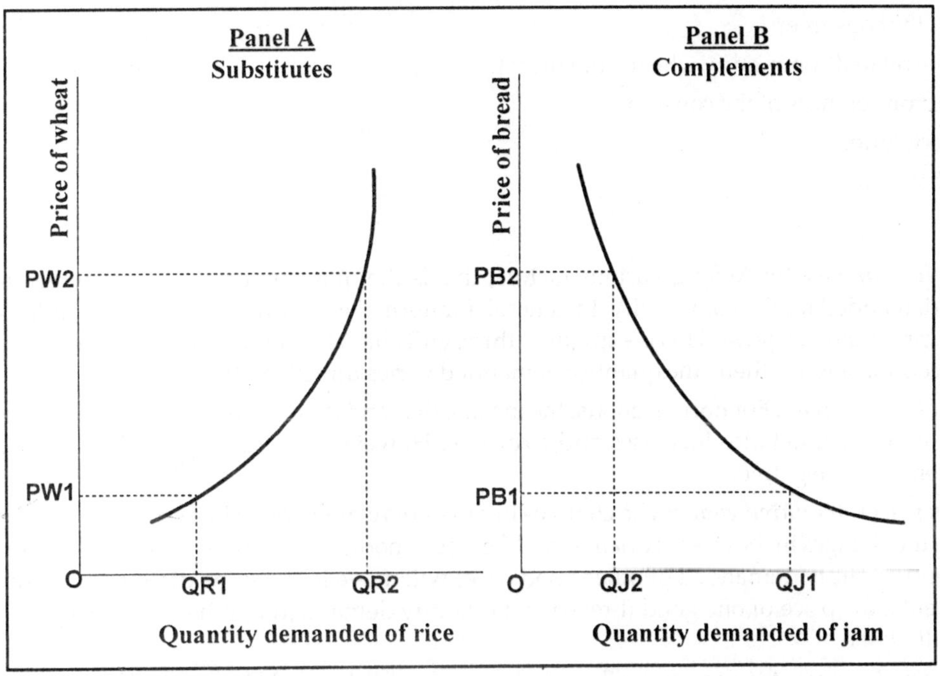

Figure 11.7: Cross demand curve for related goods.

Cross Demand for Complements

The goods which are required together to fulfill the consumption process constitute complements. In case of complements like milk and sugar, bread and jam, pen and ink etc., the cross demand is negative. That means, an increase price of a complementary good like bread, the demand for jam will decrease. This is because, with increase in price of bread, the quantity demanded for bread will decrease and thereby, the consumer demands less of jam. That means, when price of bread increases, quantity demanded of jam decreases, thereby, price (of bread) and quantity demanded (of jam) will move in the opposite direction and hence, the cross demand is negative for complementary goods. So, the cross demand curve for substitutes will fall downwards from left to right (convex shape). When price of bread is increased in the market from OP_{B1} to OP_{B2}, the quantity demanded for its complementary good *i.e.*, jam is decreased from OQ_{J1} to OQ_{J2} (Panel B of the Figure 11.7).

The above discussion reveals that, a change in price of one good will influence the quantity demanded of other good. But, the influence *i.e.*, whether positive or negative depends upon the relationship between the goods. In case of substitutes, the price of one good (wheat) and quantity demanded of other good (rice) will move in the same direction. However, in case of complements, the price of one good (bread) and quantity demanded of other good (jam) will move in the opposite direction.

However, if the commodities under consideration are independent, *i.e.*, a change in price of one commodity will not influence the quantity demanded of the other commodity such as, rice and computer, fertilizer and pen drive etc. So, there lies no meaning in drawing the cross demand curves for such examples, as we seldom study the relations between two unrelated goods. However, price-demand relations between substitutes and complementary goods assume more significance.

11.1.4. Demand Function

As discussed earlier, the functional relationship between quantity demanded of a commodity and the factors influencing it constitute demand function. Besides, price of the commodity, income of the consumer, prices of related goods (substitutes or complements), there are several other factors that influence the quantity demanded for a commodity. They are discussed here under through the following equation:

$$Q_{DR} = f(P_R, I, P_W, T, P, I_D, W \text{———})$$ *Equation 11.6*

where,

Q_R = Quantity demanded of commodity rice

f = Functional relationship between quantity demanded and factors influencing it

P_R = Price of rice commodity

I = Income of the consumer

P_W = Price of the related commodity wheat (substitute)

T = Tastes and preferences of the consumer

P = Size of population

I_D = Income distribution

W = Weather

- ☆ ***Price of the commodity:*** Among various factors, this is the most important factor, which influences the quantity demanded for the commodity. In general, for normal goods, the quantity demanded extends, with the fall in price and *vice versa*. This is explained through Table 11.1 and Figure 11.1. However, giffen goods are an exception and for them, the quantity demanded varies directly with the price.
- ☆ ***Income of the consumer:*** For normal goods, the income demand is positive. That means, as income increases, the quantity demanded also increases and *vice versa*. However, for inferior goods and giffen goods, the income demand is negative.
- ☆ ***Change in prices of related goods:*** The changes in prices of related goods like substitutes and complements also bring the changes in the quantity demanded for a commodity. In case of substitutes, the price of one good (wheat) and quantity demanded of other good (rice) will move in the same direction. However, in case of complements, the price of one good (bread) and quantity demanded of other good (jam) will move in the opposite direction.
- ☆ ***Tastes and preferences of the consumer:*** They influence the purchasing behaviour of the consumer. Generally, a consumer purchases the commodities suiting to his tastes and preferences. If his tastes and preferences changes, the nature and quantity of the commodities purchased by him also will change. For example, if the consumers developed a taste for coffee, then the demand for coffee increases in the market. So, the demand curve for coffee will shift towards right, thereby, the demand curve for tea will shift towards left side, being a substitute good.
- ☆ ***Size of population:*** The size of demand for commodities varies directly with the size of population. In case of large populated countries, the demand for necessaries will be more and less for luxuries. However, in less populated (developed) countries, the demand is more for luxuries compared to necessaries. If the population of a country increases due to immigration or through high birth rate, the demand for various types of commodities will increase even when the prices remains same. But, the commodity for which the demand is increased depends upon the majority or percentage of that age-group in the population. If there is high birth rate in the country, the demand for toys, children food etc., will increase. Similarly, if the per cent of aged people to the total population increases, the demand for walking sticks, artificial teeth etc., will increase.
- ☆ ***Fluctuations in trade cycle:*** The total quantity of commodities demanded is also affected by the cyclical fluctuations in economic activities. If the trade is in boom conditions, the demand for raw material, machinery etc., increases to encourage production activities. On the other hand, if the trade is under depression, the demand for producer's goods will fall sharply, as compared to the demand for consumer goods.
- ☆ ***Changes in technology:*** Advances in technology ensure cost-effective production of commodities and new products and this exerts positive influence on the quantity demanded of the commodities.
- ☆ ***Consumer's innovativeness:*** When the price of wheat flour or price of electricity falls, the consumer identifies new uses for the product. So, it creates new demand for the product.

- ☆ ***Seasonality factor:*** For some commodities like mangoes, grapes, apples etc., seasonality factor assumes vital role in influencing the quantity demanded for the commodities.
- ☆ ***Income distribution:*** If there is proper income distribution in the country, then the demand for all the commodities will remain in more or less same proportion. But, if there is uneven distribution of income and if more people are poor, the demand for necessaries will increase compared to luxuries. But, if there are more number of rich people in the country, the demand for luxury goods will be more.
- ☆ ***Changes in savings:*** The level of savings varies inversely with the quantity demanded for the commodity. If the level of savings in the economy increases, the consumer's demand for the commodities decreases and *vice versa*. On the other hand, if the liquidity preference increases with the consumer, the quantity demanded for the commodities increases.
- ☆ ***Propensity to consume:*** Increase in propensity to consumer increases the quantity demanded for a commodity and *vice versa*.
- ☆ ***Government policy:*** If the Government imposes higher tax on the commodity, then the demand contracts due to rise in price of the commodity and *vice versa*. However, the subsidies offered by the Government lowers the price of the commodity, thereby, demand will extend for the commodity.
- ☆ ***Weather:*** Changes in weather conditions brings about changes in quantity demanded for the commodities. For example, consumer's demand will be less for ice creams during winter compared to summer. Similarly, the consumer's demand for woolen clothes increases in winter and decreases in summer.
- ☆ ***Educational standards of the consumers:*** Improvement in educational standards of the consumers will influence both the nature and quantities of commodities demanded by the consumer. Examples include, demand for laptops, pen drives, statistical packages etc.
- ☆ ***Price differentiation and Product differentiation:*** Both influence the quantity demanded of a commodity on positive side, as the commodities are offered to the consumers at affordable prices and as per their tastes and preferences.
- ☆ ***Credit facilities:*** If credit facilities are extended to the consumers, the demand for the commodities will increase and *vice versa*.

Importance of Demand Function

The demand function guides us in the following aspects:

- ☆ It helps us to study the E_D for the commodity
- ☆ Based on the quantities demanded of the commodity at various prices, it guides the farmer to plan for cost-effective technologies in the context of inverse relationship between quantity demanded and price of the commodity.
- ☆ It helps in forecasting of prices and quantity demanded of the commodity and thus, guides the entrepreneurs in taking output-pricing decisions.
- ☆ It further guides in analyzing the influence of various factors on the quantity demanded of a commodity.

11.1.5. Movement versus Shifts in the Demand Curve

So far, we have discussed individual demand curve, market demand curve, law of demand etc., and it is noted that, the quantity demanded of a commodity varies inversely with the price of the commodity. However, the law of demand operates only under *ceteris paribus* assumption. If other factors are not held constant, they also exert influence on the demand of the commodity. In this context, it is essential to differentiate between demand and quantity demanded of a commodity. 'Demand' refers to the entire demand schedule or demand curve for the commodity, which indicates various quantities of a commodity purchased at various prices of the commodity. The 'changes in demand' for a commodity is influenced by several factors (discussed earlier) other than price. But, 'quantity demanded' infers only the quantity of a commodity purchased at a particular price. That means, it explains about a particular point on the demand curve. So, the 'changes in quantity demanded' of a commodity are influenced by price of the same commodity only and other factors are held constant. So, both 'changes in quantity demanded' and 'changes in demand' for a commodity can be studied in two ways *viz.*, Movement along the demand curve and Shifts in the demand curve respectively.

1. Movement along the Demand Curve

As explained through demand function, the quantity demanded for the commodity is influenced by various factors, implying that, demand is a multivariable function. If the influence of the price alone was studied on the quantity demanded for the same commodity, keeping other influential factors like income, changes in prices of related goods, tastes and preferences of the consumers, income distribution etc., constant, then we move along the same demand curve. This movement along the same demand curve represents various price-quantity combinations and it has a negative slope. Thus, whenever there is a change in the quantity demanded of a commodity due to change in its price, there is a movement from one point price-quantity combination to another on the same demand curve. Such a movement from one point price-quantity combination to another along the same demand curve is known as Movement along the demand curve. But, the movement along the same demand curve implies that, the level of demand remains consistent. So, in case of movement along the same demand curve, there is no shift in the demand curve either left side or right side. So, due to the movement along the same demand curve in accordance with the influence of the price of the same commodity, technically, there are two categories of demand called as Extension of demand and Contraction of demand.

☆ ***Extension of demand:*** Other things remaining equal, when more quantity of commodity is demanded due to fall in its price, it is called Extension of demand. So, the consumer moves from top to bottom along the same demand curve. The decline in price of rice from OP_{R1} to OP_{R2} leads to increase in quantity demanded from OQ_{R1} to OQ_{R2}. As a result of price fall, the consumer moves from point A to point B along the same demand curve (Figure 11.8).

☆ ***Contraction of demand:*** Other things remaining equal, when less quantity of commodity is demanded due to rise in its price, it is called Contraction of demand. So, the consumer moves from bottom to top along the same demand curve. The increase in price from OP_{R1} to OP_{R3} leads to decrease in its quantity demanded from OQ_{R1} to OQ_{R3}. As a result of price rise, the consumer moves from point A to point C along the same demand curve (Figure 11.8).

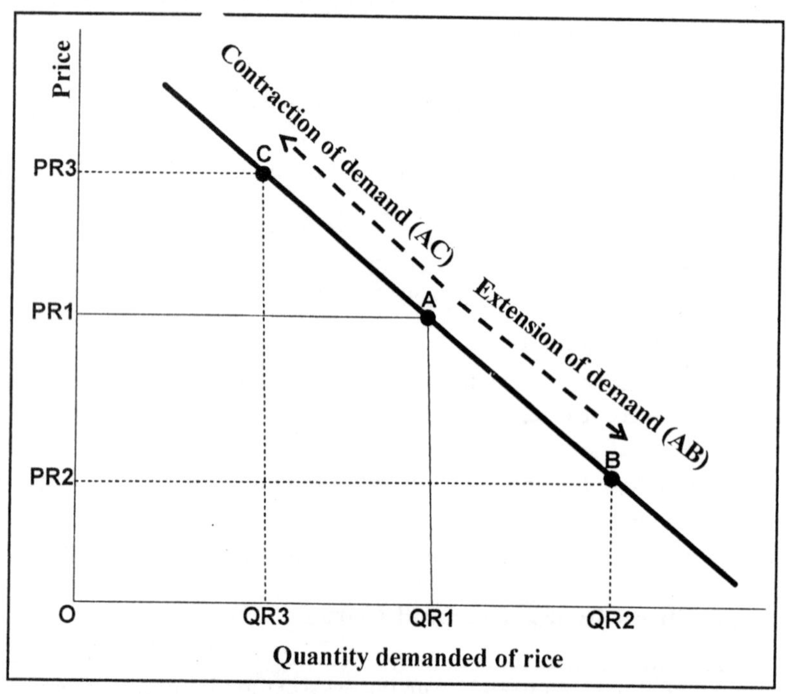

Figure 11.8: Movement along the demand curve – Extension and Contraction of demand.

The Figure 11.8 further reveals that, due to fall or rise in price of the commodity, the quantity demanded of the same commodity extends and contracts respectively and the important aspect is that, the consumer moves along the same demand curve, but there is no shift in the position of demand curve. This movement along the demand curve is designated as 'change in quantity demanded'.

2. Change or Shift in the Demand Curve

If we study the influence of one or more factors other than price factor of the same commodity (*i.e.*, price of the commodity remains same) on the quantity demanded for the commodity, then there is shift in demand curve from its original position. For example, if the level of income of a consumer increases or rise in price of close substitute or availability of commodity on credit basis etc., there is an increase in demand for the commodity. This increase in demand shifts the demand curve from its original position to the right side or away from the origin. On the other hand, if there is a fall in income of the consumer or fall in price of close substitute or change in consumer's taste or non-availability of commodity on credit basis etc., there is a reduction or decrease in demand for the commodity. This fall or decrease in demand for the commodity shifts the demand curve from its original position to the left side or towards origin. So, the shift in demand curve implies that, the original demand relationship has changed for the commodity. That means, in this case, at original price of the commodity, the demand curve shifts its position either left side or right side depending upon the influence of other factor(s) on the quantity of commodity demanded. Since, we study the relationship between various quantities of a commodity demanded at the same price, but due to change in other factors, the demand curve has a negative slope. So, such a movement in demand curve from one position to another position is known as Shift in the demand curve or Change in demand. In this case, there are two categories of demand called as Increase in demand and Decrease in demand.

☆ ***Increase in demand:*** When due to factors other than the price of the same commodity, more quantity of commodity is demanded at the same price, it is called as increase in demand. Since, more quantity of the commodity is demanded at the same price, the demand curve shifts towards right side or away from the origin. The right side shift in demand curve may be due to:

- Increase in consumer's income
- Increase in the prices of substitutes
- Consumer's preferences are more towards this commodity
- Increase in stock of assets
- Changes in climatic conditions favouring the purchase of this commodity
- Increase in family size
- Proper distribution of wealth in the society
- No government's restrictions in having that commodity.

As shown in the Figure 11.9, say, due to increase in family size, the consumer demanded more quantity of rice (OQ_{R2}>OQ_{R1}) at the same price OP_1. Hence, the demand curve D_1D_1 will move away from the origin as D_2D_2, indicating more quantity of rice is purchased by the consumer at the same price, due to increase in the family size. The higher demand curve shows that, consumers are able and willing to buy more of the commodity at each price than before. It is

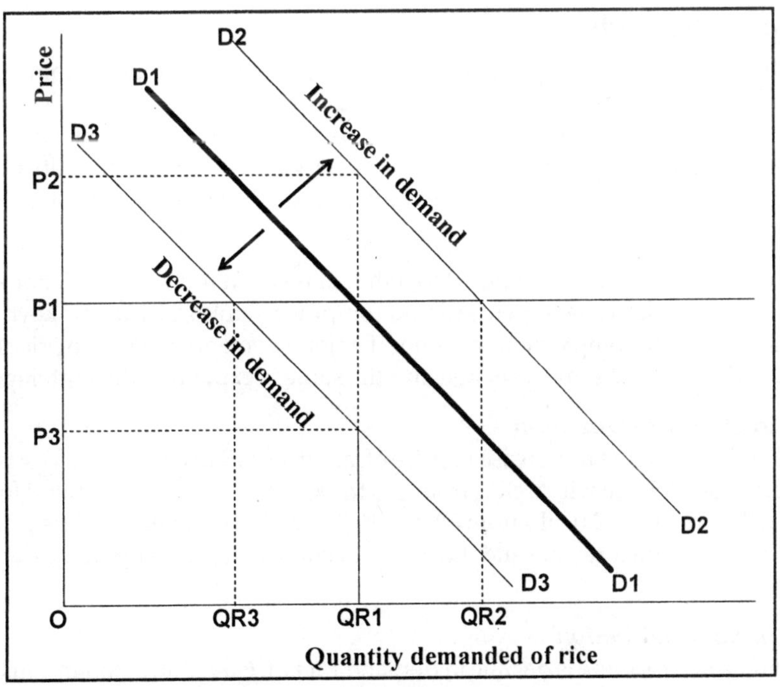

Figure 11.9: Shifts in the demand curve – Increase and Decrease in demand.

clear from the Figure 11.9 that, increase in demand refers to '*more quantity demanded* (OQ_{R2}) *at the same price* (OP_1) *or same quantity* (OQ_{R1}) *purchased at higher price* (OP_2)'.

☆ ***Decrease in demand:*** When due to factors other than the price of the same commodity, less quantity of commodity is demanded at the same price, it is called as decrease in demand. Since, less quantity of the commodity is demanded at the same price, the demand curve shifts left side or towards the origin. The left side shift in demand curve may be due to:

- ❖ Decrease in consumer's income
- ❖ Decrease in the prices of substitutes
- ❖ Change in consumer's tastes and preferences towards the related good
- ❖ Decrease in stock of assets
- ❖ Changes in climatic conditions disfavoring the purchase of this commodity
- ❖ Decrease in family size
- ❖ Improper distribution of wealth in the society
- ❖ Government imposed restrictions in the consumption of this commodity.

As shown in the Figure 11.9, say, due to decrease in family size, the consumer demanded less quantity of rice ($OQ_{R3}<OQ_{R1}$) at the same price OP_1. Hence, the demand curve D_1D_1 will move towards the origin as D_3D_3, indicating less quantity of rice is purchased by the consumer at the same price due to decrease in the family size. The lower demand curve shows that, consumers are able and willing to buy less of the commodity at each price than before. It is clear from the Figure 11.9 that, decrease in demand refers to *'less quantity demanded* (OQ_{R3}) *at the same price* (OP_1) *or same quantity* (OQ_{R1}) *purchased at lower price* (OP_3)*'*.

From the above discussion, we can infer the following:

- ☆ 'Changes in the quantity demanded' are represented by the movement along the same demand curve.
- ☆ Extension of demand refers to more quantity demanded due to reduction in price of the same commodity.
- ☆ Contraction of demand refers to less quantity demanded due to rise in the price of the same commodity
- ☆ 'Changes in demand' are represented by the shifts in the demand curve.
- ☆ Increase in demand refers to more quantity demanded due to changes in factors other than price of the same commodity.
- ☆ Decrease in demand refers to less quantity demanded due to changes in factors other than price of the same commodity.

11.1.6. Inter-related Demands

So far, we have discussed different types of demand *viz.*, price demand and income demand, where quantity demanded for a commodity is with reference to that commodity only. However, we observed in cross demand that, quantity demand of a particular commodity depends upon its relationship (either substitution or complementarity) with other commodity. Besides cross demand, there are also some cases, where the quantity demanded of a commodity is in connection with other commodity or commodities and they are discussed here under.

1. Joint Demand

When two or more commodities are demanded together, it is called Joint demand. For example, the demand for bread and jam, milk, tea powder and sugar, pen and ink, computer monitor, CPU and keyboard etc., refer to as joint demand. So, joint demand refers to complementary goods. For joint demand goods, the price of the goods will move in the opposite direction and also in the same direction and the same is explained through Figures 11.10 and 11.11.

(a) Price Movement in Opposite Direction

In Panel A of the Figure 11.10, due to increase in price of jam from OP to OP_1 (since, its supply is decreased from SS to S_1S_1), the quantity demanded for jam will decline from OQ to OQ_1. As a result, the demand for bread also will fall from DD to D_1D_1 and thereby, the price of bread falls from OP to OP_1 (Panel B). This infers that, the prices of the complementary goods tend to move in the opposite direction and the extent of change depends upon the degree of E_D for jam and the supply of bread.

(b) Price Movement in Same Direction

In Panel A of the Figure 11.11, there is decline in demand for jam (say, due to fall in consumers' income) thereby, demand curve shifts towards left side from DD to D_1D_1 and thereby, the price of the jam decline from OP to OP_1. Due to decline in demand for jam, the demand for bread also decreases in the market thereby, the demand curve for bread also shifts towards left side as D_1D_1 (Panel B). This will leads to fall in price of the bread from OP to OP_1. The extent of decline in price of these two goods depends upon the degree of E_D of two goods and their supply.

2. Composite Demand

If one commodity is demanded that can be put to several alternative uses, then that commodity is said to have composite demand. That means, there is competition among several alternative uses for the commodity and thereby, composite demand is also called as Rival demand. For example, coal is demanded by railways, factories and households.

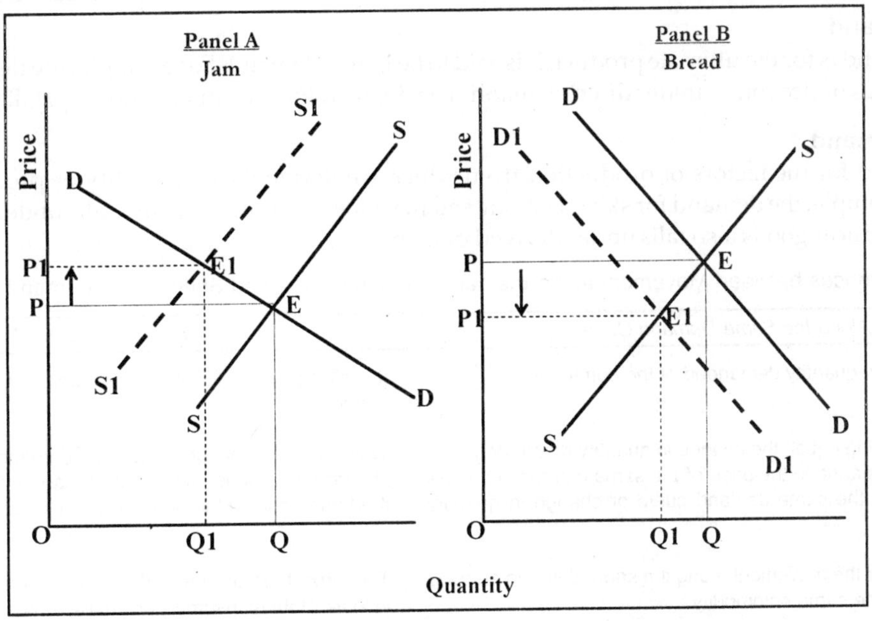

Figure 11.10: Joint demand for complementary goods – Price movement in opposite direction.

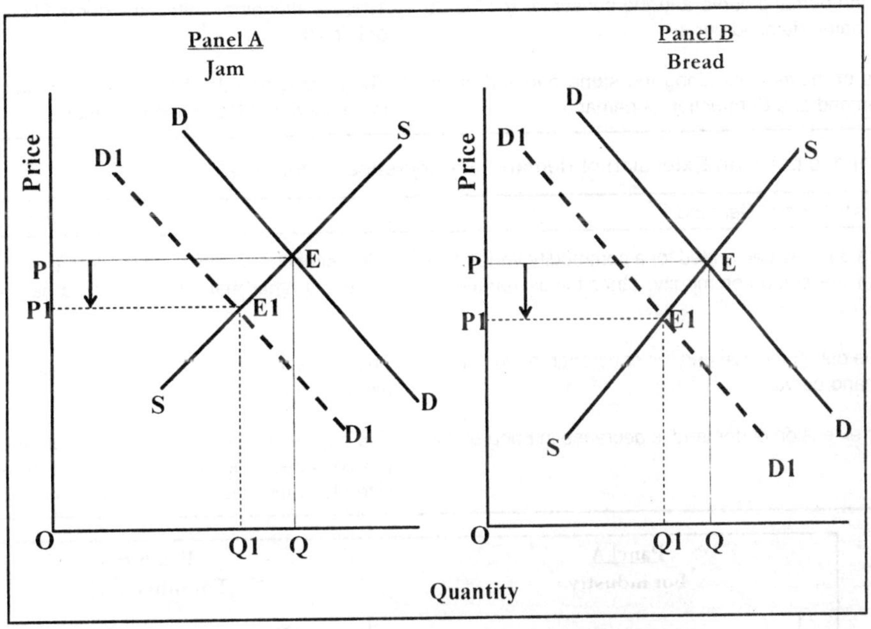

Figure 11.11: Joint demand for complementary goods – Price movement in same direction.

Similarly, land can be demanded for constructing a house, an industry, crop cultivation etc. Likewise, the demand for human labour, capital also falls under composite demand. When a commodity is demanded among several uses, its price will be determined by the Law of Substitution or LEMU.

Effect of Changes in Demand for the Composite Good

When the demand for the composite good is increased in one use, it will affect its supply to other alternative uses. The same is explained through the Figure 11.12. In Panel A, assume that, the demand for land increases for starting an industry thereby, the demand curve will shift towards right on D_1D_1 and hence, price (rent) for land will increase from OP to OP_1. Due to more supply of land for industrial use (increased from OQ to OQ_1 in Panel A), its availability for other alternative uses like crop cultivation, construction of houses etc., will decline, thereby, supply curve of land shifts towards left side as S_1S_1 (Panel B). So, due to decline in supply of land for other alternative uses, the price (rent) for land increases from OP to OP_1. The degree to which the extent of price rises for land in alternative uses depends upon the degree of the intensity of demand for land.

3. Direct Demand

If the demand is for the ultimate product, it is said to be Direct demand. For example, the demand of the consumer to purchase 50 kgs of rice comes under direct demand. The demand for consumer goods also falls under direct demand.

4. Derived Demand

The demand for the factors of production from which we derive the commodity or product is called Derived demand. For example, the demand for skilled labour and machinery is a sugar factory falls under derived demand. The demand for producer goods also falls under derived demand.

Table 11.6: Differences between Movement along the same demand curve and Shifts in the demand curve.

Movement Along the Same Demand Curve	*Shifts in the Demand Curve*
It implies changes in quantity demanded of the commodity.	It implies changes in the level of demand itself for the commodity in the economy.
Other things remaining equal, the change in quantity demanded for a commodity due to change in the price of the same commodity refers to Movement along the same demand curve or change in quantity demanded.	When there is change in quantity demanded of a commodity due to changes in various factors other than price of the same commodity, it refers to shift in the demand curve or change in the demand.
The causal factor for the movement along the same demand curve is change in price of the same commodity.	The causal factors for shifts in the demand curve are changes in income of the consumer, prices of substitutes, family size etc., other than price of the same commodity.
There is no shift in the demand curve and the consumer moves up and down along the same demand curve.	There is altogether shift in the demand curve either towards right side or left side.
There are two kinds of movements along the same demand curve *i.e.*, Extension of demand and Contraction of demand.	There are two kinds of shifts in the same demand curve *i.e.*, Increase in demand and Decrease in demand.

Table 11.7: Differences between Extension of demand and Increase in demand.

Extension of Demand	*Increase in Demand*
It is a situation, where quantity demanded for a commodity varies due to change in price of the same commodity, other things remaining equal.	It is a situation, where quantity demanded for a commodity varies due to factors other than price of the same commodity.
There is no shift in the demand curve and the consumer moves down along the same demand curve.	There is altogether shift in the demand curve towards right side or away from the origin.
The causal factor for extension of demand is decrease in price of the same commodity.	The causal factors for increase in demand are increase in income of the consumer, increase in prices of substitutes, increase in family size etc., other than price of the same commodity.

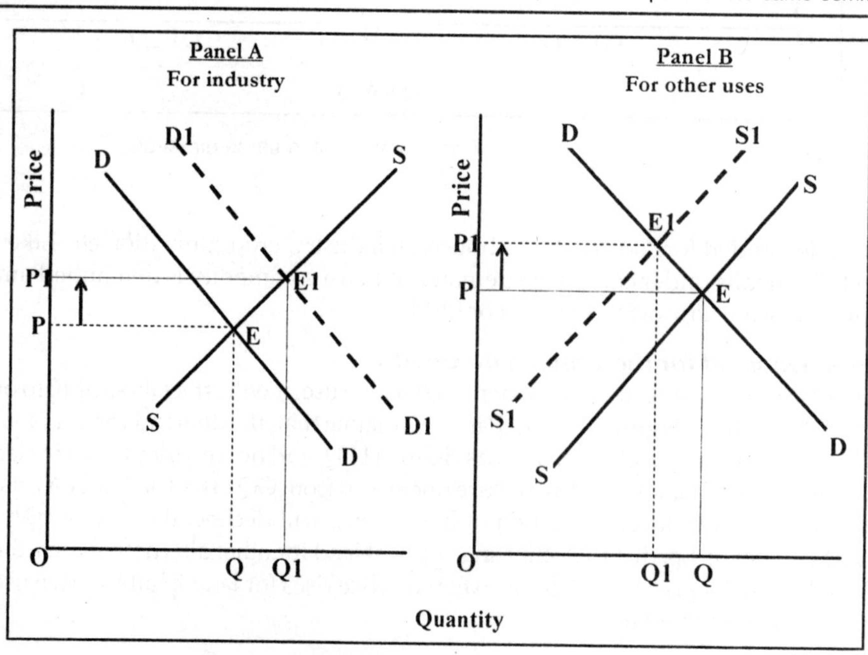

Figure 11.12: Effects of changes in demand and supply of composite good (Land).

Table 11.8: Differences between Contraction of demand and Decrease in demand.

Contraction of Demand	*Decrease in Demand*
It is a situation, where quantity demanded for a commodity varies due to change in price of the same commodity, other things remaining equal.	It is a situation, where quantity demanded for a commodity varies due to factors other than price of the same commodity.
There is no shift in the demand curve and the consumer moves up along the same demand curve.	There is altogether shift in the demand curve towards left side or towards the origin.
The causal factor for contraction of demand is increase in price of the same commodity.	The causal factors for decrease in demand are decrease in income of the consumer, decrease in prices of substitutes, decrease in family size etc., other than price of the same commodity.

However, the demand for a commodity may be direct demand or derived demand depending upon the purpose for which it is demanded. For example, the demand for paddy for seed purpose falls under derived demand, as it represents a producer good, while the demand for paddy for the final consumption comes under direct demand, as it represents a consumer good. Compared to direct demand, derived demand is less price-elastic or more inelastic.

12

Elasticity of Demand

From the earlier discussion regarding the theory and basic concepts of demand, it is evident that, the law of demand simply explains the inverse relationship between quantity demanded and price of the commodity. However, it does not explain the rate at which the quantity demanded for a commodity changes due to change in price of the commodity. So, the law of demand is straight forward, as it does not indicate, how much responsive the demand is to a change in price of the commodity. The impact of the price change is always not the same. Sometimes, the demand for a commodity changes considerably even for small price changes. On the other hand, there are some commodities for which, the demand is not affected much by price changes. The concept of elasticity has a very great importance in economic theory as well as applied economics. Demands for some commodities are very responsive to price changes, while demands for certain others are not so responsive to price changes. So, the concept of E_D gains significance in measuring the rate of change in demand in response to change in price of the commodity. This concept of E_D was introduced by Alfred Marshall. The following are the popular definitions of E_D:

'The elasticity (or responsiveness) of demand in a market is great or small according as the amount demanded increases much or little for a given fall in price, and diminishes much or little for a given rise in price'.

—Alfred Marshall

'Elasticity of demand is a measure of the relative change in the amount purchased in response to any change in price or a given demand curve'

—Meyers

'Elasticity of demand measures the responsiveness of the quantity demanded to change in the price'

—Boulding

'The elasticity of demand at any price is the proportional change of amount purchased in response to a small change in price divided by the proportional change of price'

—Robinson

'The elasticity of demand may be defined as the ratio of the percentage change in demand to the percentage change in price'

—Lipsey

12.1. Types of E_D

The quantity of a commodity demanded by the consumer per unit of time depends upon various factors *viz.*, price of a commodity, money income of the consumer, prices of related goods, the tastes and preferences of the consumer etc. While studying the E_D, we consider one of these factors (keeping other factors constant) and its influence will be analyzed on the responsiveness of various quantities of a commodity demanded by the consumer. So, when the relative responsiveness or sensitiveness of the quantity demanded of a commodity is measured due to changes in its price, the elasticity is said be E_P. When the change in demand for the commodity is the result of the given change in income of the consumer, it is referred as E_Y. When the change in demand of one commodity is due to change in price of other related commodity, it is referred as E_C. There is another important concept of elasticity called Substitution elasticity of demand (E_S), which is a measure of the ease or difficulty with which one commodity is substituted for another. These four types of E_D are discussed in-detail in the ensuing pages.

12.1.1. Price Elasticity of Demand (E_P)

The concept of E_P is commonly used in economic literature. Price elasticity means the degree of responsiveness or sensitiveness of quality demanded of a good to changes in its prices. In Economics, E_D refers to price elasticity. It is the measure of degree of responsiveness of the quantity demanded of a commodity to changes in its price, *ceteris paribus*. Precisely, it is the ratio between proportionate or percentage change in quantity demanded of a commodity to a given proportionate or percentage change in the price of the same commodity. Symbolically, the E_P with reference to rice commodity is given by

E_P = (per cent change in quantity demanded of rice/per cent change in price of rice)

So, to calculate the E_P, we need to know, what the percentage change in quantity demanded is and what the percentage change in price of the same commodity is. It is best to calculate these one at a time.

Percentage Change in Quantity Demanded

The formula used to calculate the percentage change in quantity demanded is:

[Quantity demanded(NEW) – Quantity demanded (OLD)] / Quantity demanded (OLD)

= Change in quantity demanded/Quantity demanded (old)

Percentage Change in Price

The formula used to calculate the percentage change in price of the same commodity is:

[Price(NEW) - Price(OLD)] / Price(OLD)

= Change in price/Price (old)

So, $E_P = (\Delta Q_R / Q_R) \div (\Delta P_R / P_R) = (\Delta Q_R / \Delta P_R) \times (P_R / Q_R)$ *Equation 12.1*

where,

Q_R = Original quantity of rice demanded

P_R = Original price of rice

ΔQ_R = Change in quantity demanded

ΔP_R = Change in price of rice.

Interpretation of E_P

The computation of E_P is helpful to see, how sensitive the demand for a commodity is to its price change. The higher the price elasticity, the more sensitive consumers are to price changes. A very high price elasticity suggests that, when the price of a commodity increases, the consumer will buy a great deal less of it and when the price of that commodity goes down, the consumers will buy a great deal more. A very low price elasticity implies just the opposite that, changes in price have little influence on demand. The following are the rules of thumb to interpret the E_P:

- ☆ If $E_P > 1$ then, demand for the commodity is price elastic (Demand is more sensitive to price changes)
- ☆ If $E_P = 1$ then, demand for the commodity is unit elastic
- ☆ If $E_P < 1$ then, demand for the commodity is price inelastic (Demand is less sensitive to price changes)

In computing the E_P, we use percentages rather than absolute amounts with reference to both quantity and price changes. If we use absolute changes, the choice of units of both quantity and price will arbitrarily affect our impression of consumer's responsiveness. For example, if the price of rice is reduced from Rs. 30 to Rs. 29 and consumers increase their purchases from 50 to 80 units, it will seem that, consumers are quite sensitive to price changes and therefore, that demand is elastic. This is because, after all, a price change of Rs.1 has caused a change (increase) in the amount demanded of 30 units. But, by changing the monetary unit from rupees to paisa, we find that, a price change of 100 paise causes a quantity change of only 30 units. This may falsely lead us to believe that, demand is inelastic. So, to avoid this problem, we employ percentage changes, as this particular price decline is the same whether we measure it in terms of rupees or paise. Further, by using percentages, we can correctly compare the consumer responsiveness to changes in the prices of different products. It makes little sense to compare the effects on quantity demanded due to (a) a Rs.1 increase in the price of a Rs.10000 farm implement with (b) a Rs.1 increase in the price of a Rs.1 chocolate. This is because, here, the price of the farm implement has increased by 0.01 per cent, while the price of the chocolate is up by 100 per cent. So, we can more sensibly compare the consumer responsiveness to price increases by using some common percentage increase in price for both.

It is essential to note that, in computing the E_P, we get a negative sign because, the change in quantity demanded of a commodity is in opposite direction to the change in its price. If price declines, then quantity demanded of a commodity will increase. This means that, the numerator in the Equation 12.1 will be positive and the denominator will be negative, yielding a negative E_P. For an increase in price, the numerator will be negative, but the denominator is positive, again yielding a negative E_P. Economists usually ignore the minus sign and simply present the absolute value of the elasticity coefficient to avoid an ambiguity that might otherwise arise. This is because, we are more concerned with the magnitude of response of change in quantity demanded due to the change in the price of the same commodity, thereby, E_P is taken as positive. It can be confusing to say that, an E_P of -5 is greater than -2. This possible confusion is avoided when we say an E_P of 5 reveals greater elasticity than 2. So, in what follows, we ignore the minus sign in the E_P and show only the absolute value. Being ratio in terms of percentages, the E_P will have no units for measurement. (This ambiguity does not arise with Elasticity of Supply (e_S) because, price and quantity supplied are positively related).

Example 1

Let us suppose that, price of a mango falls from Rs. 15/unit to Rs.10/unit in a day. The decline in price leads to increase in number of mangoes demanded by the consumer from 20 to 35. The E_P is computed using the above formula will be:

$E_P = (\Delta Q/\Delta P) \times (P/Q)$

Given, original quantity = 20, original price = Rs.15

$\Delta Q = 35-20 = 15$, $\Delta P = 10-15= -5$

$E_P = (15/-5) \times (15/20)= -2.25 =2.25$

So, E_P is >1 and this implies, the commodity is price elastic and thus, demand is very sensitive to price changes.

Example 2

Suppose that, price of rice falls from Rs. 35/kg to Rs.20/kg in a day. The decline in price leads to increase in the quantity demanded for rice from 20kgs to 25kgs. The E_P is computed using the above formula will be:

$E_P = (\Delta Q/\Delta P) \times (P/Q)$

Given, original quantity = 20, original price = Rs.35

$\Delta Q = 25-20 = 5$, $\Delta P = 20-35= -15$

$E_P = (5/-15) \times (35/20)= 0.58$

So, E_P is <1 and this implies, the commodity is price inelastic or less elastic and thus, demand is not very sensitive to price changes.

12.1.1.1. Degrees of E_P

As explained earlier, the demand for a commodity is sensitive or responsive to its price change. The variation in demand is, however, not uniform with a change in price across the commodities. For some commodities, a small change in price leads to a relatively larger change in quantity demanded. For example, a decline in price of the commodity by one per cent may extend the quantity demanded by 10 per cent. In some commodities, a greater change in price may bring about a very small change in the quantity of commodity demanded. For example, a decline in price of the commodity by 10 per cent may increase the quantity demanded by only one per cent. So, depending upon the degree of responsiveness and unresponsiveness of the quantity of a commodity demanded to a change in its price, the Economists categorized the degrees of E_P into five categories *viz.,* Perfect elastic or Infinite elastic demand, Perfect inelastic or Infinite inelastic demand, More elastic demand or More than unitary elastic demand, Unitary elastic demand and Less elastic demand or Less than unitary elastic or More inelastic demand and they are discussed here under.

- ☆ ***Perfect elastic or Infinite elastic demand***: When a very small change in price of the commodity brings about infinite change in quantity demanded for it, it refers to infinite elastic demand. This is an extreme degree of price elastic demand. But, this degree of price elasticity is not practically seen in the economy. The demand curve lies parallel to X-axis on which the quantity demanded for the commodity is taken, thereby, the E_P is equal to infinity (Panel A of the Figure 12.1). When we say, demand is 'perfectly elastic', it do not mean that, consumers are completely responsive to a price change. In that extreme situation, where a small price reduction causes buyers to increase their purchases from zero to all they can obtain, the E_P coefficient is infinite and Economists say demand is perfectly elastic. The perfect elastic demand for a commodity implies that, the seller can sell all his output at the ruling price. If the seller quotes a higher price, the demand for his commodity will drop to zero, as the buyers prefer to purchase the same commodity at the ruling price from other sellers.

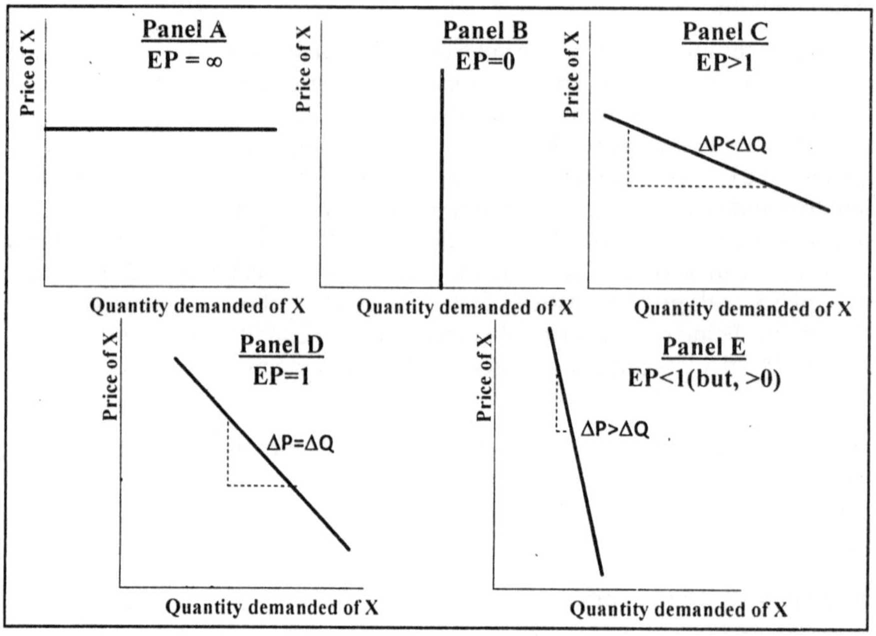

Figure 12.1: Different degrees of price elasticity of demand.

- ☆ ***Perfect inelastic or Infinite inelastic demand***: When an infinite change in price of the commodity brings about no absolute change in quantity demanded for it, it refers to infinite inelastic demand. That means, say, a 50 per cent decrease or increase in the price leads to no change in the quantity demanded of the commodity. This is an extreme degree of price inelastic demand. But, this degree of elasticity is also not practically seen in the economy. The demand curve lies parallel to Y-axis on which the price of the commodity is considered, thereby, the E_P is equal to zero (Panel B of the Figure 12.1). When we say, demand is 'perfectly inelastic', it do not mean that, consumers are completely unresponsive to a price change. In that extreme situation, where a price change results in no change whatsoever in the quantity demanded, Economists say that demand is perfectly inelastic. The E_P coefficient is zero because there is no response to a change in price. Approximate examples include an acute diabetic's demand for insulin or an addict's demand for heroin, cigarettes, alcohol etc.
- ☆ ***More elastic demand or More than unitary elastic demand:*** When a small change in price of the commodity brings about a greater or substantial change in quantity demanded, it implies, the demand for the commodity is more elastic. That means, when the percent change in quantity demanded for a commodity is greater than the percent change in its price, the demand is said to be more elastic. For example, one per cent change in price causes more than a one per cent change in the quantity demanded of a commodity, thereby, $E_P = (\%\Delta Q/\%\Delta P) > 1$. In general, all luxurious goods will have more elastic demand and the demand curve slopes downward with a flat slope indicating, $\Delta Q > \Delta P$ (Panel C of the Figure 12.1). When E_P is greater than one, a fall in price, increases the total expenditure and a rise in price, lowers the total expenditure on the commodity.
- ☆ ***Unitary elastic demand:*** When the percentage change in the quantity demanded of a commodity equals the percentage in its price, the E_P is said to have unitary elasticity. That means, $E_P = (\%\Delta Q/\%\Delta P) = 1$. For example, a one per cent change in price causes a response of exactly a one percent change in the quantity demanded of a commodity. The demand curve slopes downward with a uniform slope, thereby $\Delta Q = \Delta P$. When E_P is equal to one or unitary, a rise or fall in price leaves total expenditure incurred on the commodity unchanged. So, the demand curve with unitary E_P are called Constant total outlay curve. As shown in the Panel D of the Figure 12.1, the percentage change in price (ΔP) brings about an exactly equal percentage change in quantity demanded (ΔQ) at any point on the demand curve. The demand curve of unitary elasticity is, therefore, a

rectangular hyperbola. However, in practicality, it is difficult to find the commodities having unitary E_P. In general, a part of the demand curve will have a unitary E_P.

- ☆ ***Less elastic demand or Less than unitary elastic or More inelastic demand:*** When a greater or substantial change in price of the commodity brings only a small change in the quantity demanded, it implies, the demand for the commodity is less elastic or more inelastic. That means, when the per cent change in quantity demanded for a commodity is less than the per cent change in its price, the demand is said to be less elastic. For example, one per cent change in price causes less than a one per cent change in the quantity demanded of a commodity, thereby, $E_P = (\%\Delta Q/\%\Delta P) < 1$ (but, greater than zero). In general, all basic necessary and agricultural commodities will have less elastic demand and the demand curve slopes downward with a steep slope, thereby $\Delta Q < \Delta P$ (Panel E of the Figure 12.1). When E_P is less than one, a fall in price, decreases the total expenditure and a rise in price, increases the total expenditure incurred on the commodity.

So, the above discussion on different degrees of E_P infers that, the E_P ranges from zero to infinity across the commodities. For clarity purpose, different degrees of E_P and the degree of relationship of demand with price of the same commodity are explained through the Figure 12.2.

Degree of elasticity	E_P	Relationship of demand with price
Perfect elastic demand	Infinity	Infinite increase or decrease in demand for a very small change in price of the same commodity.
Perfect inelastic demand	Zero	No increase or decrease in demand even for an infinite change in price of the same commodity.
More elastic demand	Greater than one	% change in quantity demanded is more than % change in price of the same commodity. Ex: Cashew
Unitary elastic demand	Equal to one	% change in quantity demanded is equal to % change in price of the same commodity. Ex: Cloth
Less elastic demand	Less than one, but greater than zero	% change in quantity demanded is less than % change in price of the same commodity. Ex: rice

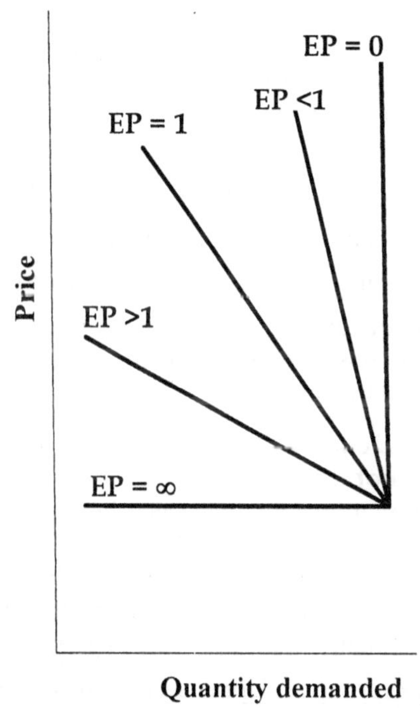

Figure 12.2: Different degrees of E_P and relationship of demand with price of the commodity.

It is essential to note that, the slope of a demand curve is not a reliable indicator of degree of E_P. That means, a demand curve with a flat slope does not necessarily mean a more elastic demand and similarly, a demand curve with a steep slope does not necessarily mean less elastic (inelastic) demand. This is because, the slope of the demand curve is expressed in terms of units of the commodity under consideration. So, if we change the units of the commodity, we can get a different slope of the demand curve. But, the E_P, on the other hand, is the percentage change in quantity demanded to the corresponding percentage change in price of the same commodity and hence, it has no units. The distinction between slope and E_P is discussed through the following cases:

Case 1: Two Demand Curves Originating from the Same Point on the Y-axis

Let us assume AB and AC are the two demand curves originating from the same point A on Y-axis (Panel A of the Figure 12.3). A glance at the two demand curves implies that, the demand curve AC is more elastic (as it is more flatter) than AB curve (as it is more steeper compared to AC). But this is not the case. Let us draw a horizontal line parallel to X-axis considering the price of the commodity as OP. This line touches the demand curve AB at D and it touches the demand curve AC at E. As per the Point elasticity method of measuring the E_P (discussed under section 12.2), at point D, the E_P is given by

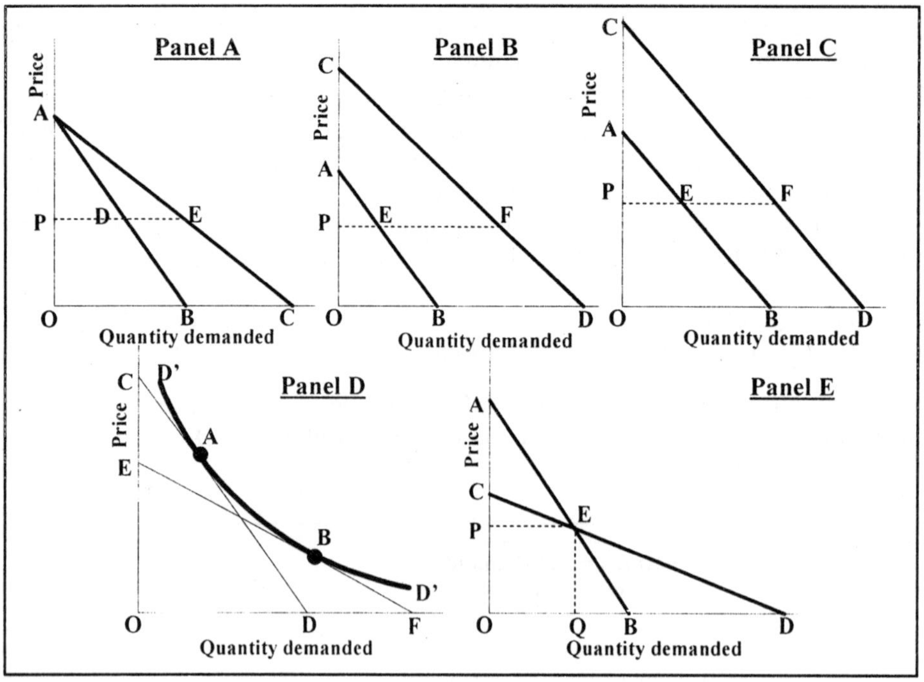

Figure 12.3: Slope of the demand curve and E_D.

E_P = Lower sector/Upper sector = BD/AD = OP/AP

Similarly, at point E, the elasticity is given by

E_P = Lower sector/Upper sector = CE/AE = OP/AP

So, the E_P at both the points D and E is same and thereby, the inference drawn earlier by seeing the slopes of the demand curves is wrong. So, we can now conclude that, if the linear demand curves originate from the same point, they have same E_P at the same price.

Case 2: Two Demand Curves Originating at Different Points on the Y-axis, but have Different Slopes, not Parallel and not Intersecting

Let us draw two demand curves AB and CD originating at different points on the Y-axis (Panel B of the Figure 12.3). A glance at the two curves implies the demand curve CD is more elastic (as it is more flatter) and demand curve AB is less elastic (as it is to more steeper). But, this is wrong. To prove this, let us consider a price OP and draw a horizontal line parallel to X-axis and it touches AB demand curve at point E and CD demand cure at point F. As per the point elasticity method, the E_P at point E is given by,

E_P = Lower sector/Upper sector = BE/AE = OP/PA = (>1)

Similarly, at point F, the E_P is given by

E_P = Lower sector/Upper sector = DF/CF = OP/PC = (<1)

So, (OP/PA) > (OP/PC) and this infers that, at point E, the demand curve AB is more elastic compared to point F on the demand curve CD at the same price.

Case 3: Two Demand Curves Originate at Different Points on the Y-axis, but Remain Parallel

Let us consider two demand curves AB and CD originating at different points on the Y-axis and they are parallel to each other (Panel C of the Figure 12.3). A glance at the two demand curves imply that, both the curves have same slope and hence, the same elasticity. But, this is wrong. To prove this, let us consider a price OP and draw a horizontal line parallel to X-axis and it touches the AB demand curve at E and CD demand curve at F. As per the point elasticity method, the E_P at point E is given by

E_P = Lower sector/Upper sector = BE/AE = OP/PA = (> 1)

Similarly, at point F, the E_P is given by

E_P = Lower sector/Upper sector = DF/CF = OP/PC = (< 1)

So, (OP/PA) > (OP/PC) and this infers that, at point E, the demand curve AB is more elastic compared to point F on the demand curve, CD at the same price. This gives the impression that, as the demand curve shifts away from the origin, the E_p at a given price will decrease and if the demand curve shifts towards the origin, the E_p at a given price will increase.

Case 4: Two Points on the Convex Shaped Demand Curve

Let us consider two points A and B on the convex shaped demand curve, D'D' (Panel D of the Figure 12.3). To study the E_p at these two points A and B, draw tangents CD and EF respectively. So, as per the point elasticity method, the E_p at point A on the demand curve is given by,

E_p = Lower sector/Upper sector = DA/CA = (>1)

Similarly, at point B on the demand curve, the E_p is given by

E_p = Lower sector/Upper sector = FB/EB = (<1)

So, the E_p at point A on the demand curve is more compared to point B.

However, there are some exceptions, where we have to depend upon the slope of the demand curve to ascertain the E_p. In the Panel E of the Figure 12.3, both the demand curves AB and CD intersect at point E, where OQ is the quantity demanded at OP price of the commodity. So, taking the quantity demanded and price of the commodity at point E, we get the same E_p for both the demand curves. In such a case, we have to necessarily depend upon the slope of the demand curve to ascertain the degree of E_p. A glance at the two demand curves AB and CD reveals that, demand curve AB is more steeper than demand curve CD, (or demand curve CD is more flatter than demand curve AB) and hence, as per the point elasticity method,

E_p at point E with reference to demand curve AB = BE/AE = OP/PA = (<1)

E_p at point E with reference to demand curve CD = DE/CE = OP/PC = (>1)

So, the demand curve CD having a flat slope will have more E_p at point E compared to demand curve AB with a steep slope at the same price. So, the demand curve with a flat slope will have more price elasticity compared to the demand curve with a steep slope.

However, as explained through the cases above, slope of the demand curve will not form the basis to judge the price elasticity, as the curves with a steeper slope may have high price elasticity compared to the demand curve with a flat slope. So, it is better to judge the E_p based on the point elasticity method rather than based on the slope of the demand curve. Further, the E_p is not the same throughout the slope of the demand curve and it is discussed in the ensuing pages (Figure 12.12).

Differences Between Slope of Demand Curve and E_p

The slope of the demand curve differs from E_p in the following aspects:

- The slope of demand curve is given by the ratio between change in price of the commodity to change in quantity demanded of the commodity, given by,

  Slope = $(\Delta P)/(\Delta Q)$.

  But, the E_p is the ratio between percentage change in quantity demanded of a commodity (paddy) to percentage change in price of the same commodity (paddy), given by,

  $E_p = (\Delta Q_p / Q_p) \div (\Delta P_p / P_p) = (\Delta Q_p/\Delta P_p) \times (P_p/Q_p) = [(\Delta Q_p/\Delta P_p)]/[(Q_p / P_p)]$

  = Marginal change/Average change

- Price is in the numerator and quantity is in the denominator for slope. In contrast, quantity is in the numerator and price is in the denominator for E_p. So, slope is the inverse of E_p. When one is bigger, the other is smaller.
- Slope of the demand curve is calculated using the measurement units (*i.e.*, absolute changes) for price and quantity. In contrast, E_p is calculated using percentage changes in price and quantity. As such, slope includes the measurement units (such as Rs/kg or Rs/quintal etc), whereas, E_p is just a number with no measurement units. So, the value of slope changes, if the measurement units change (such as, Rupees versus Dollars, kilograms versus pounds). However, the value of E_p remains same, as it is the ratio of percentages.

12.1.1.2. Factors Influencing E_p

The E_p is not the same for all the commodities. It may be high or low depending upon the number of factors influencing the quantity demanded of the commodity. The important factors which influence the E_p are discussed here under:

- *Availability of substitutes*: This is one of the most important factors that influence the E_p of a commodity. In general, the more the number of substitutes for a commodity, the more elastic the demand will be for the commodity under consideration. For example, if the price of wheat is increased in the market, the consumers could substitute rice or jowar or ragi for wheat. So, it means that, wheat is price elastic because, a rise in price will lead to large reduction in its quantity demanded by the consumers. However, if the commodity does not have substitutes like salt, that commodity will have more inelastic demand in the market.

- *Proportion of income available to spend on the good:* The greater the proportion of total amount of money income the consumer spends on the commodity, the greater is the E_P. For example, the consumer spends a large proportion of money income on luxury goods like television, fridge, car etc., and hence, for these goods, the demand is more elastic. Conversely, if the consumer spends a less proportion of total amount of money income on the commodity, the demand for that commodity is less elastic or more inelastic. For example, the consumer spends a small proportion of his money income on vegetables, salt, chilly powder, shampoo etc., and hence, the demand for these goods is less elastic.
- *Urgency:* If there is urgency in having the commodity, the demand will be more inelastic, as more will be purchased even at a high price. If that commodity does not have substitutes, the inelasticity will be further high.
- *Time element:* We know, the demand for a commodity is always related to some period of time *i.e.*, say, a day or a week or a month etc. So, the period of time plays an important role in shaping the demand curve. In the short run, when the consumption of a good cannot be postponed, its demand will be less elastic. In the long run, if the rise in price persists, people will find out methods to reduce the consumption of that good. Thus, the E_D varies with the length of the time period and the greater the time period, the greater will be the E_D and *vice versa*. For example, if the price of the cigarette is increased by Rs.5/- in the market, then the addicted smoker will continue smoking the cigarettes and if he is brand loyal, he will not change the brand. So, the demand for the cigarettes is inelastic. On the other hand, if the smoker decides to give up smoking cigarettes as per the doctor's advice or he is not affordable to pay high price for cigarettes, he will stop smoking the cigarettes over a period of time, thereby, the demand for cigarettes becomes elastic in the long run. Similarly, if the electricity charges goes up, it is very difficult to cut back its consumption in the short run. So, the demand for electricity is inelastic in the short run. However, if the same higher electric charges are imposed on the consumers for a long period of time, then the consumer will plan the strategies like using gas stoves in place of electric stoves, less use of air-conditioners etc., so that, they consume less electricity. So, the demand for electricity is more elastic in the long run than in the short run. Thus, in general, demand for the commodity is inelastic in the short run and more elastic in the long run. In the short run, there will be inelastic demand for the change in price, because of the following reasons:
    - It will take some time on the part of the consumers to know about the change in price of the commodity.
    - Even after they are aware of change in price of the commodity (say, decrease in price), they will wait and watch for further fall in the price of the commodity.
    - Consumer cannot change their habits or patterns of expenditure in the short run.
    - Some goods are durable goods and hence, they continue to use them even the prices changes in favour of them (say, further decrease in price).
    - The consumers may not find suitable substitutes immediately.

So, all the above factors will contribute to inelastic demand for the commodities in the short run. However, in the long run, the consumers plan for suitable adjustments in accordance with the price changes and thereby, the demand for the commodities becomes more elastic.

- *Multiple uses of the commodity:* If a commodity can be put to a several uses, the demand for that commodity is more elastic. For example, if electricity charges are reduced by the Government, then the demand increases greatly from the farming, household and industrial sectors. Similarly, if the price of coal decreases, the demand rapidly increases from households, industries, railways etc.
- *Durability of goods:* The more the durability of goods, the less is the E_D. This is because, when the prices of the goods rises, the people may prefer to get the old things repaired than buy new things.
- *Postponement of consumption:* If the purchase or consumption of a commodity can be temporarily postponed by the consumer, then for that commodity, the demand will be more elastic. For example, the consumer can postpone the purchases of television, fridge etc., if they are not offered at affordable prices now and for such commodities the demand will be more elastic. On the other hand, say, for medicines, the consumer cannot postpone their purchase and hence, the demand for medicines is inelastic.
- *Habits:* If the consumer is habituated or addicted for a commodity (say, cigarette), the demand will be inelastic for that commodity, as he is ready to pay any amount of money to have that commodity.
- *Joint demand:* If two commodities are required jointly for a common purpose, then the E_D of a particular commodity depends upon the E_D of other commodity. For example, if the demand for bread is less elastic, the demand for jam is also less elastic.

- *Frequency of purchases:* If the consumer purchases the commodity very frequently, he buys more when the price falls and buys less when the price rises, thereby, the demand for the commodity is more elastic. On the other hand, if the commodity is purchased rarely, then the demand for the commodity is less elastic, as the response for the price changes is little.
- *Range of prices:* The E_P also depends upon the range of the prices at which the commodities are offered. At very high range of prices and at very low range of prices, the commodities have less elastic demand. This is because, at very high price range of luxury goods like Audi car, Plasma TV, tractor etc., a small decline in the price of the commodities will not extend the demand significantly, as such goods are preferred by few rich class in the society. Similarly, at very low range of prices, generally in case of necessaries, the decline in prices will not extend the demand significantly, as any body can easily have the same commodities even below the price fall. So, at very high price range and at very low price range, the demand for the commodities is inelastic. But, the commodities with middle range of prices have more elastic demand.
- *Income group:* For the same commodity, the demand for the poor people may be more elastic and for the rich people, the demand is less elastic. If price of rice is increased in the market, the poor people tend to shift their consumption pattern towards jowar or bajra, whereas, high income group will still purchase and consume rice.
- *Nature of the commodities:* The E_P depends upon the nature of the commodity *i.e.*, whether the commodity is a necessary or comfort or luxury. In general, for necessaries like rice, wheat etc., the demand is less elastic, as everyone will demand a certain quantity, whatever may be the price. However, even in necessities, for the commodities having good substitutes (rice and wheat, sugar and jaggery, tea and coffee etc.,) the demand is more elastic and for the necessities having no substitutes (salt, chilly powder etc.) the demand is less elastic. That means, among necessities, rice has more elastic demand compared to salt.

In general for comforts and luxuries, the demand is more elastic. Further, in case of luxuries, it should be noted that, for all commodities the demand is not more elastic. This is because, in case of diamonds, which are preferred by few rich class in the society, any fall in the price of diamonds will not extend the demand and even rich people may discard this commodity, as it loses its distinctiveness.

- *Unequal distribution of income and wealth:* Distribution of income influences the consumption pattern and hence, the demand for various commodities. According to Taussig, unequal distribution of income and wealth makes the demand, in general, more elastic. If the government attempts redistribution of income to make it equitable, the demand for luxuries will decline and the demand for necessities of life will increase.

12.1.1.3 E_P and PCC

As discussed in Chapter 9, PCC indicates the various quantities of a commodity purchased by the consumer at various levels of prices of that commodity. So, PCC also indicates the E_P.

- If the PCC slopes downward (Figure 12.4), it implies the E_P for the commodity is more elastic.
- If the PCC is horizontal (Figure 12.4), it implies, the E_P for the commodity is unitary elastic.
- If the PCC is upward sloping (Figure 12.4), the E_P for the commodity is less elastic.
- So, we can have different types of E_P on the same PCC (Figure 12.4). Between Q_1 and Q_2, the PCC is downward sloping, indicating $E_P>1$ (more elastic), between Q_2 and Q_3, the PCC is almost a horizontal line, indicating $E_P=1$ (unitary elastic) and between Q_3 and Q_4, the PCC is upward sloping, indicating $E_P<1$ (less elastic).

12.1.1.4. Importance of E_P

The concept of price elasticity of demand has great practical significance, as indicated by the following examples:

(a) Large Crop Yields

The E_P for most farm products is highly inelastic ranging between 0.20 to 0.25. So, due to increase in farm output arising from a good crop season or from increased productivity tend to depress both the prices of farm products and the TR of farmers. So, considering the inelastic E_P, a large farm output is undesirable at the farmers' level. It further suggests the policy makers that, to achieve higher income, it requires that, farm output must be restricted.

(b) Imposition of Excise Taxes

The concept of E_P guides the Government in the imposition of sales taxes on the commodities. For example, if Rs. 10 tax was imposed on the commodity and 50,000 units are sold, then tax revenue will be Rs. 500000 (Rs.10 × 50000 units sold). If the Government raises the tax to Rs. 15, but this higher price results in reduction in sales of the commodity to 30000 units, because of elastic demand for the commodity, then tax revenue will be reduced to Rs. 450000 (Rs. 15 x 30000 units sold). So, the imposition of tax on the commodity having higher E_P, will bring down the tax revenue to the

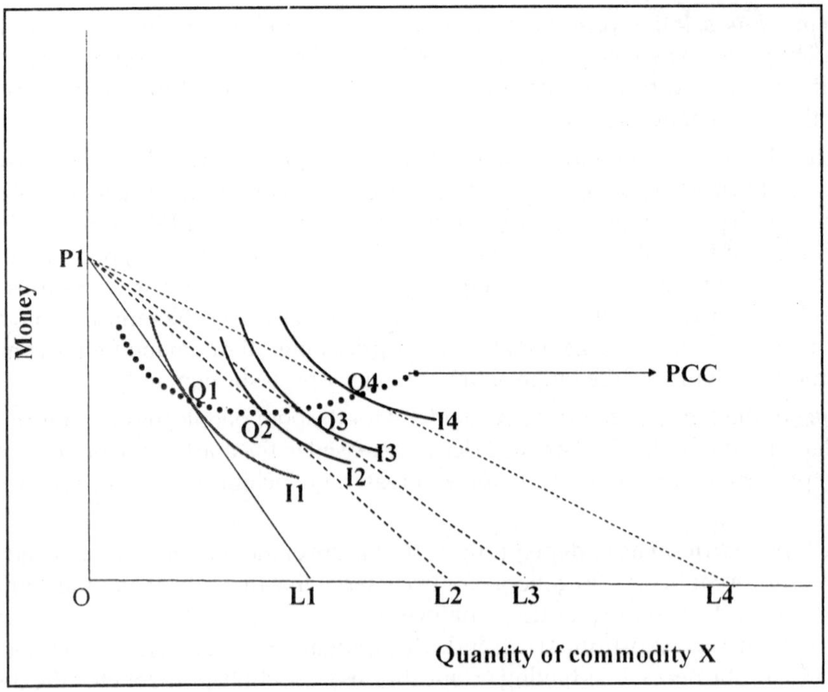

Figure 12.4: Varying price elasticities of demand shown by PCC.

Government. So, the Finance Minister tend to seek out for the products that have inelastic demand such as liquor, cigarettes etc., for levying taxes.

12.1.2. Income Elasticity of Demand (E_Y)

In the above discussion, we studied, if price of the commodity changes, keeping money income of the consumer and prices of other related goods constant, the demand for the commodity will change inversely to its price change. But, if there is change in income of the consumer keeping price of the commodity, prices of other related goods and other factors constant, the demand tends to change directly. The degree to which a change in income of the consumer leads to change in quantity demanded of a commodity refers to E_Y. It is computed as the ratio between proportionate or percentage change in quantity demanded for the commodity to a proportionate or percentage change in the income of the consumer. Symbolically, the E_Y is given by,

E_Y = (per cent change in quantity demanded of a commodity / per cent change in income)

$E_Y = (\Delta Q/\Delta Y) x(Y/Q)$ ***Equation 12.2***

where,

Q = Original quantity demanded

Y = Original income of the consumer

ΔQ = Change in quantity demanded

ΔY = Change in income.

So, to calculate the E_Y, we need to know what the percentage change in quantity demanded is and what the percentage change in income of the consumer is. It is best to calculate these one at a time.

Percentage Change in Quantity Demanded

The formula used to calculate the percentage change in quantity demanded is:

[Quantity demanded(NEW) – Quantity demanded (OLD)] / Quantity demanded (OLD)

Percentage Change in Income

The formula used to calculate the percentage change in income is:

[Income(NEW) - Income(OLD)] / Income(OLD)

So, $E_Y = (\Delta Q/Q) \div (\Delta Y/Y) = (\Delta Q/\Delta Y) \times (Y/Q)$

12.1.2.1. Degrees of E_Y

The computation of E_Y is helpful to see how sensitive the demand for a commodity is to an income change on the part of the consumer. The higher the E_Y, the more sensitive consumers are to income changes. Depending upon the degree of responsiveness of quantity of commodity demanded due to changes in income of the consumer, the E_Y is categorized into the following categories:

- ☆ ***E_Y more than unity or High income elasticity:*** E_Y will be more than unity, when the proportionate change in quantity demanded of a commodity is more than proportionate change in income of the consumer. That means, say, one per cent increase in income of the consumer causes greater than a one per cent increase in the quantity demanded of a commodity, thereby, $E_Y = (\%\Delta Q/\%\Delta Y) > 1$ (Figure 12.5). Luxury items usually have higher income elasticity because, when people have a higher income, they don't have to forfeit as much to buy these luxury items.
- ☆ ***E_Y equal to unity:*** E_Y will be equal to unity, when the proportionate change in quantity demanded of a commodity is equal to the proportionate change in income of the consumer. That means, say, one per cent increase in income of the consumer leads to one per cent increase in the quantity demanded of a commodity, thereby, $E_Y = (\%\Delta Q/\%\Delta Y) = 1$(Figure 12.5).

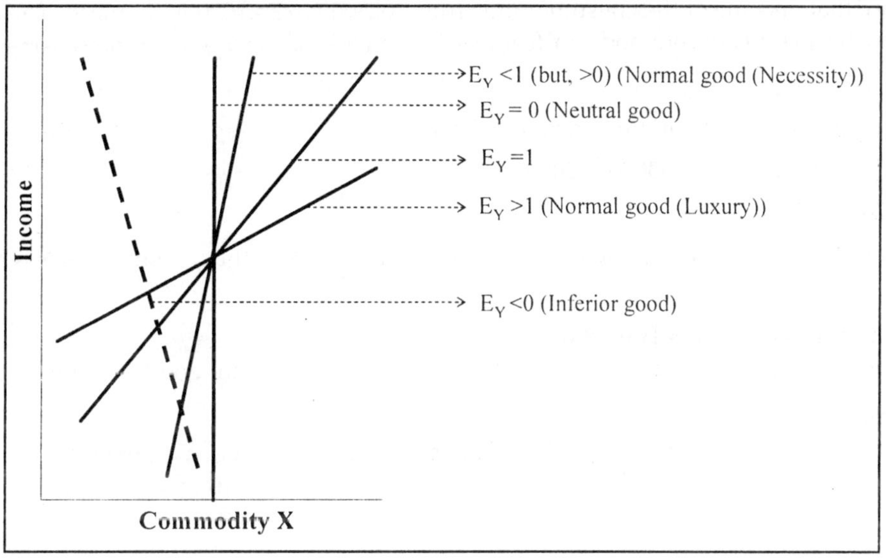

Figure 12.5: Degrees of E_Y.

- ☆ ***E_Y less than unity but greater than zero or Less income elasticity or Income inelasticity:*** E_Y will be less than unity (but greater than zero), when the proportionate change in quantity demanded of a commodity is less (quantity demanded increased and not decreased) than proportionate change in income. That means, say, one per cent increase in income of the consumer causes less than a one per cent increase (but, greater than zero) in the quantity demanded of a commodity, thereby, $E_Y = (\%\Delta Q/\%\Delta Y) < 1$, but >0 (Figure 12.5). Necessary goods like rice, wheat, edible oil etc., have less income elasticity.
- ☆ ***E_Y equal to zero:*** E_Y will be equal to zero, when the increase in income makes no change in the quantity demanded of a commodity purchased by the consumer. That means, say, one per cent increase in income of the consumer causes no change in the quantity demanded of a commodity, thereby, $E_Y = (\%\Delta Q/\%\Delta Y) = 0$ (Figure 12.5). So, the commodities that witness no change in quantity demanded despite a change in income usually have an income elasticity of zero. Examples include, salt, chilly powder, hair oil, shampoo, tooth paste etc., and these goods are called Neutral goods.
- ☆ ***E_Y less than zero or Negative income elasticity:*** E_Y will be less than zero or negative when the increase in income leads to less quantity of a commodity purchased. So, the commodities for which, the demand decreases as income increases have an income elasticity of less than zero or negative. This is more common in case of inferior goods, thereby, $E_Y = (\%\Delta Q/\%\Delta Y) < 0$(Figure 12.5). So, the inferior goods will be dropped by the consumer, who receives an income increase. For example, coarse grains like jowar, bajra, ragi etc., will have $E_Y < 0$, as they are inferior goods.

The following are the rules of thumb to interpret the E_Y:

- ☆ If $E_Y > 1$ then, demand for the commodity is income elastic (Demand is sensitive to income changes)
- ☆ If $E_Y = 1$ then, demand for the commodity is unit income elastic
- ☆ If $E_Y < 1$ then, demand for the commodity is income inelastic (Demand is not sensitive to income changes)

It is important to note that, for an individual consumer, the commodity may have income elastic demand, but if we consider the income elasticity of commodity in the entire market, its demand may be inelastic. E_Y being the ratio of percentages, it has no units.

Example 1

Let us suppose that, income of the consumer rises from Rs.10000 to Rs.12000 in a month. The increase in income leads to increase in the quantity of commodity X from 50kgs to 100kgs. The E_Y is computed using the formula will be:

$$E_Y = (\Delta Q/\Delta Y) \times (Y/Q)$$

Given, original quantity = 50, original income = Rs.10000

$$\Delta Q = 100\text{-}50 = 50,\ \Delta Y = 12000\text{-}10000 = 2000$$

$$E_Y = (50/2000)\times(10000/50) = 5.$$

So, E_Y is >1 and this implies, the commodity is income elastic and thus, demand is very sensitive to income changes.

Example 2

Let us suppose that, income of the consumer falls from Rs.3000 to Rs.1000 in a month. The decrease in income leads to decrease in the quantity of commodity Y from 150kgs to 100kgs. The E_Y is computed using the formula will be:

$$E_Y = (\Delta Q/\Delta Y) \times (Y/Q)$$

Given, original quantity = 150, original income = Rs.3000

$$\Delta Q = 100\text{-}150 = -50,\ \Delta Y = 1000\text{-}3000 = -2000$$

$$E_Y = (-50/-2000)\times(3000/150) = 0.50.$$

So, E_Y is <1 and this implies, the commodity is income inelastic and thus, demand is less sensitive to income changes.

12.1.2.2. Classification of Goods Based on E_Y

The concept of E_Y guides in classifying the goods as normal goods, inferior goods and giffen goods.

(a) Normal Goods

For normal goods, the E_Y coefficient is positive. That means, more of them are demanded as income rises. Such goods are also called as Superior goods. But, the value of E_Y varies greatly among normal goods. For example, E_Y for automobiles is about +3.0, while E_Y for most farm products is only about +0.20.

(b) Inferior Goods

For inferior goods, the E_Y coefficient is negative. That means, less of them are demanded as income rises. Examples include jowar, bajra, cabbage, used clothing etc.

(c) Giffen Goods

These are special type of inferior goods for which the E_Y is strongly negative compared to inferior goods. Examples include bread, potato etc.

12.1.2.3. Measurement of E_Y

In Chapter 9, we derived the Engel's curve from the ICC (Figure 9.30). As ICC indicates the E_Y for the commodity, the same can be measured from the Engel's curve and it is discussed through the following cases:

- ✰ *Concave shaped Engel curve:* As shown in the Panel A of the Figure 12.6, the Engel curve is concave to the origin. To measure the E_Y at point E on the Engel curve, a tangent is drawn with respect to point E and it touches the X-axis on the negative side. We use the E_Y formula as below,

$$E_Y = (\Delta Q/\Delta Y) \times (Y/Q) = (Q_1Q/QE) \times (QE/OQ) = (Q_1Q/OQ) > 1$$

This infers that, the E_Y is positive over much of its range and hence, it is more than one. This case is common in luxury goods.

- ✰ *Convex shaped Engel curve:* As shown in the Panel B of the Figure 12.6, the Engel curve is convex to the origin. To measure the E_Y at point E on the Engel curve, a tangent is drawn with respect to point E and it touches the X-axis on the positive side. We use the E_Y formula as below,

$$E_Y = (\Delta Q/\Delta Y) \times (Y/Q) = (Q_1Q/QE) \times (QE/OQ) = (Q_1Q/OQ) < 1$$

This infers that, the E_Y is less than one, but positive, (*i.e.,* ranges between zero to one). This case is common in necessary goods.

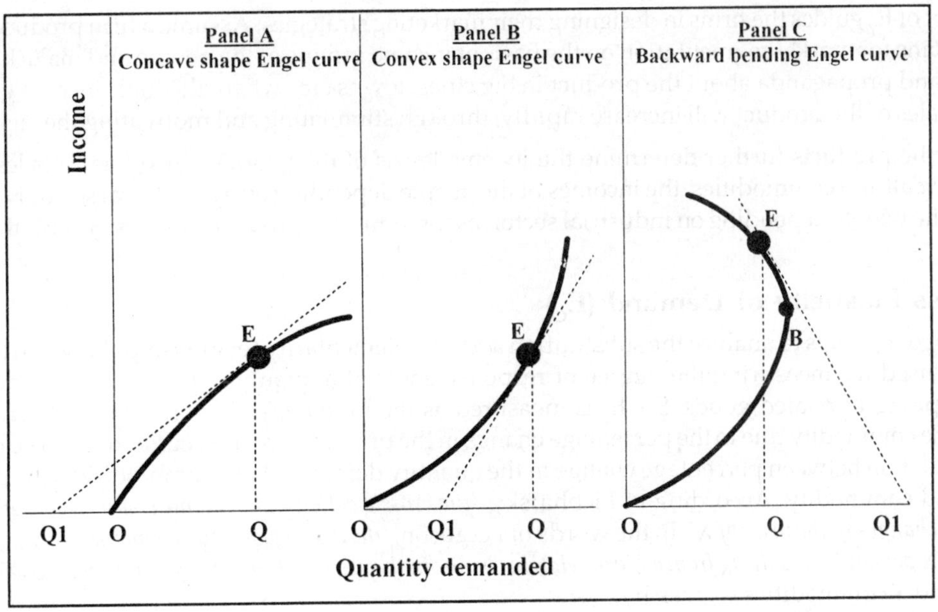

Figure 12.6: Measuring E_Y from Engel's curves.

- *Backward bending or sloping Engel curve:* As shown in the Panel C of the Figure 12.6, the Engel curve is backward sloping *i.e.*, with a negative slope. To measure the E_Y at point E on the Engel curve, a tangent is drawn with respect to point E and it touches the X-axis on the positive side. We use the E_Y formula as below,

  $E_Y = (\Delta Q/\Delta Y) \times (Y/Q) = (-Q_1Q/QE) \times (QE/OQ) = (-Q_1Q/OQ) < 0$

This infers that, the E_Y is negative for backward bending Engel curve and this is the case for inferior goods. However, before backward bending of the curve *i.e.*, before B, the Engel curve show the E_Y with respect to a necessary good (E_Y is <1, but positive).

12.1.2.4. Determinants of E_Y

The degree of E_Y depends upon the following factors:

- *Nature of the commodity*: The E_Y will be more than unity for luxuries, equal to unity for comforts, less than unity for necessaries and less than zero for inferior goods. However, in the practical world, a good which is a luxury for one person may become a necessity for some one.
- *Time period*: In the long run, due to inventions and innovations, the luxury good today may become a necessity in the future and hence, the E_Y changes. For example, cell phone is a luxury good during late-ninetees in India, but now, it became a necessary good among most of the customers.
- *Advertisement and Propaganda:* Due to more advertisement and propaganda about the products in the market, the consumers' tastes and preferences are likely to change and accordingly, the E_Y changes.
- *Income level of the consumer:* The degree of E_Y varies from one person to another person depending upon the level of income of the consumer. If the income level of the consumer increases, the quantity demanded for luxuries and comforts increases more than proportionately, the quantity demanded for necessaries increases less than proportionately and for inferior goods, the quantity demanded decreases.

12.1.2.5. Importance of E_Y

The concept of E_Y guides the producers in producing the goods and services. If the national income of the country is increasing at significant note, the concept of E_Y will guide the producers to produce more of luxury goods, as for such goods, E_Y is more than one. That means, during the boom periods in the economy, more of luxury goods are produced. On the other hand, at times of depression, the firms aim at producing the necessary goods (say, farm products, tooth paste, soaps etc), as they are demanded even at low income levels of the people. So, this gives the impression that, during boom periods of trade cycle, the firms wishes to expand their activities in producing luxury goods and at times of recession, the firms will not show much interest in expanding the existing business on luxury goods (like computers, plasma screen televisions etc). Products with relatively high E_Y coefficients such as automobiles (E_Y+3), housing (E_Y+1.5) and restaurant meals (E_Y+1.4) are generally hit hardest by recession. Those commodities with low or negative E_Y coefficients like food products prepared at home (E_Y+0.20) are much less affected.

The concept of E_Y guides the firms in designing their marketing strategies. Assume a firm produces a product with high E_Y. So, it plans for market segmentation on the basis of income groups of the people and make huge selling costs, advertisement and propaganda about the product in big cities, towns etc., where the high-income group people will reside. So, the sales of the product will increase rapidly, through stimulating and motivating the customers.

The E_Y for the products further determine the income levels of the people. Since, the E_Y is less than one (but, positive) for agricultural commodities, the incomes of the people depending on agricultural sector is low compared to the incomes of the people depending on industrial sector, as for industrial products (luxury goods), the E_Y is more than unity.

12.1.3. Cross Elasticity of Demand (E_C)

The concept of E_C helps to analyze the substitution and complementarity relationships between the commodities. This concept is used for measuring the degree of responsiveness of a quantity of a commodity demanded due to changes in the prices of related goods. So, E_C is measured as the ratio between percentage change in the quantity demanded of one commodity due to the percentage change in the price of its related commodity. The cross elasticity of demand is the relation between percentage change in the quantity demanded of a commodity to the percentage in the price of a related commodity. According to Liebhafsky, *'the cross elasticity of demand is a measure of responsiveness of purchases of Y to changes in the price of X'*. In the words of Ferguson, *'the cross elasticity of demand is the proportional change in the quantity of X demanded resulting from a given relative change in the price of the related good Y'*. Symbolically, the E_C with reference to X and Y commodities is given by,

$$\text{Cross elasticity of demand } (E_{XY}) = \frac{\text{\% change in quantity demanded of commodity X}}{\text{\% change in price of commodity Y}}$$

$$E_{XY} = (\Delta Q_X/\Delta P_Y) \times (P_Y/Q_X) \qquad \textit{Equation 12.3}$$

So, to calculate the E_C, we need to know what the percentage change in quantity demanded of commodity X is and what the percentage change in price of its related commodity Y is. It is best to calculate these one at a time.

Percentage Change in Quantity Demanded of Commodity X

The formula used to calculate the percentage change in quantity demanded of commodity X is:

[Quantity demanded(NEW) – Quantity demanded (OLD)]/Quantity demanded (OLD)

Percentage Change in Price of Commodity Y

The formula used to calculate the percentage change in price is:

[Price(NEW) - Price(OLD)]/Price(OLD)

So, $E_{XY} = (\Delta Q_X / Q_X) \div (\Delta P_Y / P_Y) = (\Delta Q_X / \Delta P_Y) \times (P_Y/Q_X)$

where, E_{XY} stands for cross elasticity of demand of commodity X with respect to commodity Y, Q_X = Original quantity demanded of commodity X, P_Y = Original price of commodity Y, ΔQ_X = Change in quantity demanded of commodity X, ΔP_Y = Change in price of commodity Y

12.1.3.1. Interpretation of E_C

The computation of E_C is important to study how sensitive the demand for a commodity X is to a price change of related commodity Y. Unlike E_P, we allow the coefficient of E_C to be either positive or negative. The numerical value of E_C depends on whether the two commodities in question are substitutes or complements or unrelated and the same is discussed here under.

1. E_C, if the Commodities Under Consideration are Substitutes

Substitutes are the commodities, which compete each other in a consumption process. That means, if the consumer prefers one commodity, say rice, he has to sacrifice the consumption of competing commodity, say wheat. So, rice and wheat are good substitutes. An increase in price of wheat, will reduce the demand for wheat, thereby, the consumer prefers and demands more of rice. Conversely, a decrease in price of wheat will increase the demand for wheat and thereby, the consumer prefers and demands less of rice. So, price of one commodity and quantity demanded of other commodity will move in the same direction. So, E_C between the substitute goods is positive, thereby, E_C curve will have a positive slope *i.e.*, slopes upward from left to right. In the above example, the greater the positive coefficient of E_C between rice and wheat (E_{RW}), the greater the substitution relationship between the two commodities, and so, the commodities are good substitutes. The other examples for good substitutes include jam and butter, sugar and jaggery, tea and coffee etc. Depending upon the degree of substitution between the two commodities, the E_C can be categorized as,

- ☆ E_C more than unity (E_{xy}>1)

- ☆ E_C equal to unity (E_{xy}=1)
- ☆ E_C less than unity, but positive (E_{xy}<1)
- ☆ E_C equal to zero (E_{xy} = 0)
- ☆ E_C less than zero (E_{xy} < 0)
- ☆ Infinite E_C (E_{xy} < ∞)

(a) E_C more than Unity

When a change (increase) in price of commodity Y, brings about more than proportionate change (increase) in the quantity demanded of other commodity X, then the E_C is referred as E_{xy} more than unity. As shown in the Panel A of the Figure 12.7, ΔQX > ΔPY, thereby, E_{xy} > 1. In this case, the commodities under consideration are good substitutes for each other and the E_C curve will have a flat positive slope.

(b) E_C Equal to Unity

When a change (increase) in price of commodity Y, brings about same proportionate change (increase) in the quantity demanded of other commodity X, then the E_C is referred as E_{xy} is unity. As shown in the Panel B of the Figure 12.7, ΔQX = ΔPY, thereby, E_{xy} = 1. In this case, the E_C curve will have a uniform positive slope.

(c) E_C less than Unity, but Positive

When a change (increase) in price of commodity Y, brings about less than proportionate change (increase) in the quantity demanded of other commodity X, then the E_C is referred as E_{xy} is less than unity (but, positive). As shown in the Panel C of the Figure 12.7, ΔQX < ΔPY, thereby, E_{xy} < 1. In this case, the commodities under consideration are poor substitutes for each other, thereby, the E_C curve will have a steep positive slope.

(d) E_C Equal to Zero

When a change in price of the commodity Y leads to no change in the quantity demanded of commodity X, then the E_C is equal to zero. As shown in the Panel D of the Figure 12.7, though the price of the commodity Y is decreased from OP_2 to OP_1, there is no change in the quantity demanded of commodity X, *i.e.*, ΔQX = 0. In this case, the commodities under consideration are unrelated to each other. Examples include, rice and pen drive, mango and computer etc, thereby, the E_C curve will lie parallel to Y-axis.

(e) E_C is less than Zero

When an increase in price of the commodity Y leads to decrease in the purchase of other commodity X, then the E_C is negative. As shown in the Panel E of the Figure 12.7, the increase in price of commodity Y from OP_1 to OP_2 leads to decrease in the quantity demanded of commodity X from OQ_1 to OQ_2. This is because, the demand for commodity Y is inelastic and so, whatever be the price of the commodity Y, the same quantity of commodity Y is purchased and thereby,

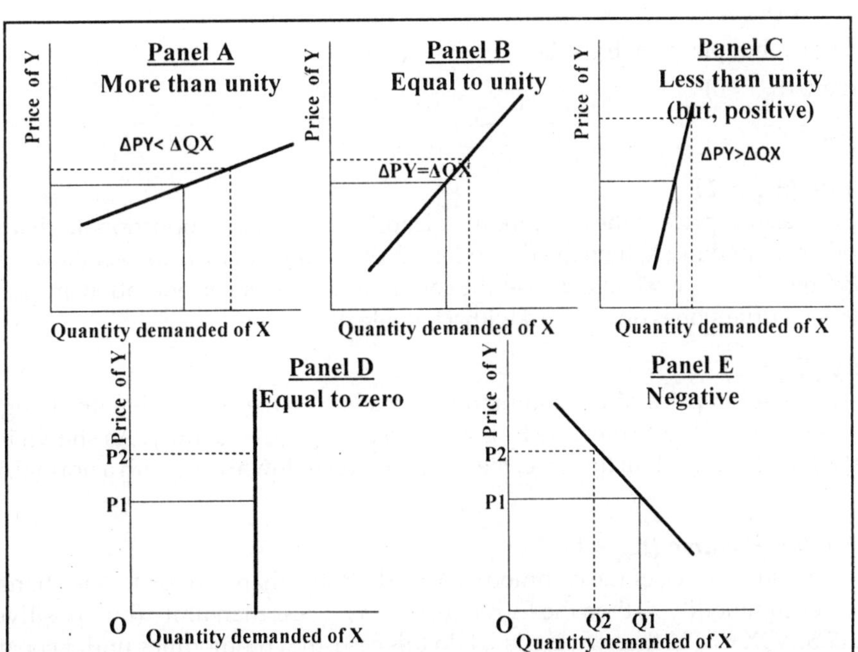

Figure 12.7: Different degrees of E_C for substitutes.

less of commodity X is demanded with the remaining money income of the consumer. So, the slope of demand curve shows negative E_C. But this is only an exceptional case.

(f) E_C is Infinity

The E_C between the two commodities will be infinite, when the two commodities are perfect substitutes. That means, when the price of one commodity Y is decreased, it leads to nothing of commodity X is purchased. Then the cross demand curve for commodity X will coincide with the Y- axis. Examples include, two brands of coffee, two brands of tea etc.

So, the above explanation infers that, the E_C for substitutes ranges between negative to infinity. It is important that,

- ☆ The E_C between rice and wheat (E_{RW}) is not the same as E_C between wheat and rice (E_{WR}). That means, say, if five per cent increase in price of wheat leads to 10 per cent increase in quantity demanded of rice, but a five per cent increase in price of rice may lead to only seven per cent increase in quantity demanded of wheat. In the first case, the E_{RW} is 2.0 and in the latter case, the E_{WR} is 1.4. So, the interpretation of E_C for substitutes is one way.
- ☆ The commodities, which are good substitutes, will have high E_C and the commodities, which are poor substitutes, will have low E_C. So, if different firms A, B, C, D and E produce the commodities A_1, B_1, C_1, D_1 and E_1 and if they are close substitutes (*i.e.*, the E_C between them is high), then we can say, these firms represent a particular group or industry, as they produce very close substitutes. On the other hand, if the commodities produced by these firms differ (heterogenous) and if the E_C between them is small/low, then each firm is considered as an independent firm or attain the monopoly status, as these commodities cannot be easily substituted.

2. E_C, if the Commodities under Consideration are Complements

Complements are the commodities, which are required together in the consumption process. So, a change in price of one of the commodities will influence the demand of both the commodities, as both commodities are required together in the consumption process. For example, bread and jam are good complements. An increase in the price of the jam, will reduce its demand and thereby, the demand for bread also decreases. So, price of one commodity and quantity demanded of other commodity will move in the opposite direction, thereby, the E_C between the complements is negative and hence, E_C curve for complements slopes downward from left to right (*i.e.*, negative slope). In the above example, the greater the negative coefficient of E_C between bread and jam, the stronger the complementarity relationship between the two commodities and hence, the commodities are good complements. The other examples for good complements include, milk and sugar, pen and ink etc. Depending upon the degree of complementarity between the two commodities, the E_C can be categorized as:

- ☆ E_C more than unity ($E_{XY} > 1$)
- ☆ E_C equal to unity ($E_{XY} = 1$)
- ☆ E_C less than unity, but positive ($E_{XY} < 1$)
- ☆ E_C equal to zero ($E_{XY} = 0$)
- ☆ Infinite E_C ($E_{XY} = \infty$)

(a) E_C more than Unity ($E_{XY} > 1$)

When a change (increase) in price of the commodity Y leads to more than proportionate change (decrease) in the quantity demanded other commodity X, then the E_C is referred as E_{XY} more than unity. As shown in the Panel A of the Figure 12.8, $\Delta QX > \Delta PY$, thereby, E_{XY} is >1. In this case, the commodities under consideration are good complements for each other, thereby, the E_C curve slopes downward with a flat negative slope.

(b) E_C Equal to Unity ($E_{XY} = 1$)

When a change (increase) in price of the commodity Y leads to same proportionate change (decrease) in the quantity demanded other commodity X, then the E_C is referred as E_{XY} equal to unity. As shown in the Panel B of the Figure 12.8, $\Delta QX = \Delta PY$, thereby, $E_{XY} = 1$. In this case, the E_C curve slopes dowmward with a normal or uniform negative slope.

(c) E_C less than Unity, but Positive ($E_{XY} <1$)

When a change (increase) in price of the commodity Y leads to less than proportionate change (decrease) in the quantity demanded other commodity X, then the E_C is referred as E_{XY} less than unity (but, positive). As shown in the Panel C of the Figure 12.8, $\Delta QX < \Delta PY$, thereby, E_{XY} is < 1. In this case, the commodities under consideration show less complementarity to each other and the E_C curve slopes downward with a steep negative slope.

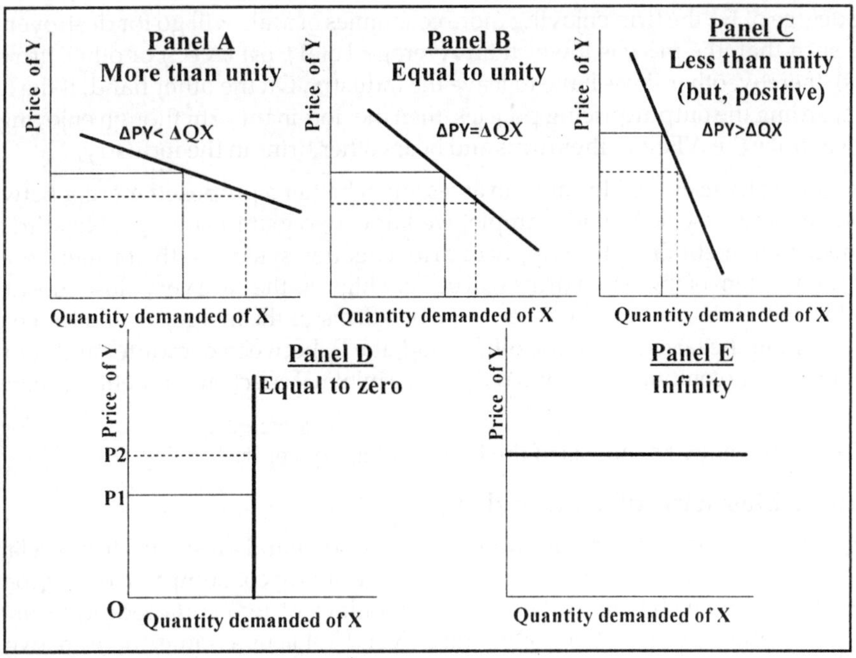

Figure 12.8: Different degrees of E_C for complements.

(d) E_C Equal to Zero ($E_{XY} = 0$)

When a change in price of the commodity Y leads to no change in the quantity demanded of commodity X, then the E_C is referred as E_{XY} equal to zero. As shown in the Panel D of the Figure 12.8, though the price of the commodity Y decreased from OP_2 to OP_1, there is no change in the quantity demanded of commodity X, *i.e.*, $\Delta X = 0$. So, the cross elasticity demand curve will be parallel to Y-axis.

(e) Infinite E_C

The E_C between the two commodities is infinity, when the two commodities are perfect complements. That means, when there is small decrease in price of one commodity, it leads to infinitesimal increase in the quantity demanded of other commodity. So, the cross elasticity demand cure will be parallel to X-axis. As shown in the Panel E of the Figure 12.8, for a very minute change in the price of commodity Y, there is infinite change in the quantity demanded of commodity X.

The above explanation infers that, the E_C for complements ranges between zero and infinity. It is important that, like substitutes, the interpretation of E_C for complements is also one way. The E_C between bread and jam (E_{BJ}) is not the same as E_C between jam and bread (E_{JB}). That means, say, if five per cent increase in the price of jam leads to 10 per cent decrease in quantity demanded for bread, but a five per cent increase in price of bread leads to only seven per cent decrease in quantity demanded of jam, then in the first case, the E_{BJ} is 2.0 and in the latter case, the E_{JB} is only 1.4.

However, the interpretation of different relationships between two commodities as close substitutes or poor substitutes or good complements or poor complements etc., as explained above is based on total price effect and hence, this concept of E_C may not give the clear picture about the true relationship between the commodities. This is because, in Marshallian analysis, the price effect is not decomposed into income effect and substitution effect. However, the same concept is discussed earlier under IDC analysis (Chapter 9) to analyze the relationship between the related commodities through decomposition of price effect into income effect and substitution effect and this offers more reliable explanation than Marshallian analysis.

3. E_C, if the Commodities are Independent

A zero or near-zero E_C suggests that, the two goods being considered are unrelated or independent goods. An example is rice and computer. We would not expect a change in the price of rice to have any effect on purchases of computers and *vice versa*.

12.1.3.2. Importance of E_C

The concept of E_C helps in formulating the pricing strategies for the products produced by the firm. We know, if the firm produces close substitutes, the E_C will be high. Say, if a firm produces A, B, C and D products, which are close substitutes and if the firm reduces the price of product A, it will adversely influence the demand for other products B, C and D. Similarly, if a firm produces A and B products, which are good complements and if the firm increases the price of product A, it will also adversely influence the demand for other product B.

Considering the degree of E_C, the firm enjoying more economies of scale will go for destroyer pricing in fixing the price of the product, such that, the price is lower than Average Total Cost (ATC) of other firms producing similar products (substitutes), thereby, other firms have to leave the industry. On the other hand, if the firms in the industry (oligopoly) collude regarding the output-pricing policies, then the dominant firm though enjoying economies of large scale, will quote a price above the ATC of other firms and bears other firms in the industry.

Government also implicitly uses the idea of E_C in assessing whether a proposed merger between two large firms will benefit the consumers in the society. For example, we know two giant firms *viz.*, Netafim and Finolex Plastro Plassion Ltd., have intense competition in the supply of drip irrigation system to the farmers. So, the E_C between the products of drip irrigation system of these two firms will be very high, as they are very close substitutes. Consequently, the Government would likely block a merger between these two firms, as the merger would lessen competition and the firms on collusion will exploit the farmers. On the other hand, the E_C between computer and car is zero. So, a merger between the firms manufacturing computer and car would have a minimal effect on competition, thereby, the Government would let that merger happen.

Table 12.1 provides a convenient synopsis of the E_P, E_Y and E_C concepts.

12.1.4. Substitution Elasticity of Demand (E_S)

The elasticity of substitution is another important concept of demand elasticity. It is a relative measure of the degree of substitution possibility between the two commodities meant for consumption by the consumer. It is defined as the ratio between proportionate change in the ratio of quantities of two commodities X and Y consumed and proportionate change in the ratio of prices of two commodities X and Y. Let us assume, there are two commodities X and Y, the quantities of two commodities consumed by the consumer are QX and QY and the prices of the two commodities are P_X and P_Y respectively, then the E_S is given by,

$$E_S = \frac{\text{Proportionate change in the ratio of quantities of two commodities X and Y consumed}}{\text{Proportionate change in the ratio of prices of two commodities X and Y}}$$

Proportionate change in the ratio of prices of two commodities X and Y

$$E_S = [(\Delta Q_X/Q_X)/(\Delta Q_Y/Q_Y)] \,/\, [(\Delta P_X/P_X)/(\Delta P_Y/P_Y)] \qquad \textit{Equation 12.4}$$

The E_S value ranges from zero to infinity. The Figure 12.9 shows that,

- ☆ In case of close or good substitutes like rice and wheat, tea and coffee, jaggery and sugar etc., the E_S will be more than one. Then, E_S curve will have flat slope. This is shown as curve A in Figure 12.9.
- ☆ In case of poor substitutes like tea and sugar, the E_S will be less than one. Then, E_S curve will have steep slope. This is shown as curve B in Figure 12.9.
- ☆ In case of perfect substitutes like two brands of coffee, two brands of tea etc., a fall in price of one commodity, will make the consumer to purchase that commodity only and none of the other commodity. So, the E_S will be infinity and the E_S curve will be parallel to X-axis. This is shown as curve C in Figure 12.9.

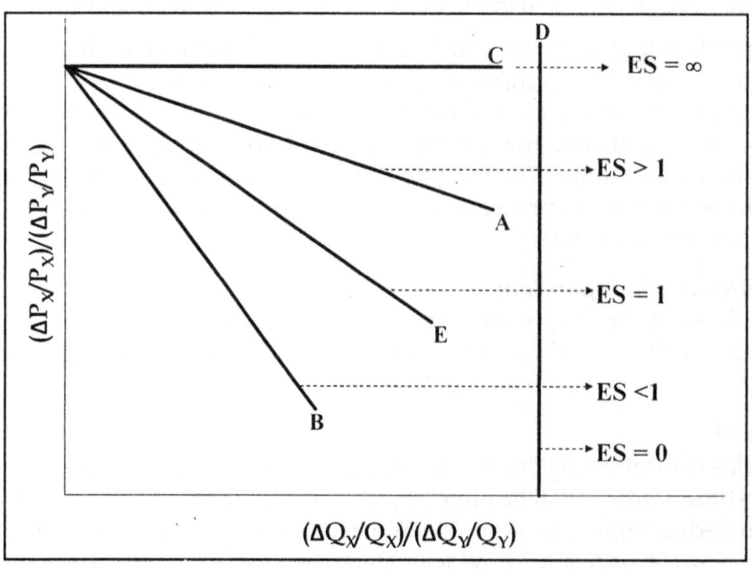

Figure 12.9: Degrees of E_S.

Table 12.1: Price, income and cross elasticities of demand

Elasticity Coefficient	*Description*	*Type of Good(s)*	*Shape of the Demand Curve*
1. Price elasticity			
Elastic or Relatively elastic ($E_P>1$)	Percentage change in quantity demanded of a commodity is greater than percentage change in its price	Luxuries, Comforts	Price demand curve is downward sloping from left to right with a flat slope
Inelastic or Relatively inelastic ($E_P<1$, but positive)	Percentage change in quantity demanded of a commodity is less than percentage change in its price	Necessaries (farm products)	Price demand curve is downward sloping from left to right with a steep slope
2. Income elasticity			
Positive ($E_Y>0$)	Quantity demanded of a commodity changes in the same direction of change in income of the consumer.	Normal or Superior good	Income demand curve is upward sloping from left to right with a flat slope
Negative ($E_Y<0$)	Quantity demanded of a commodity changes in the opposite direction of change in income of the consumer.	Inferior and Giffen goods	Income demand curve is downward sloping from left to right
3. Cross elasticity			
Positive ($E_C>0$)	Quantity demanded of a commodity X (rice) changes in the same direction as change in price of commodity Y (wheat).	Substitutes	Cross demand curve is upward sloping from left to right (concave)
Negative ($E_C<0$)	Quantity demanded of a commodity X (milk) changes in the opposite direction as change in price of commodity Y (sugar).	Complements	Cross demand curve is downward sloping from left to right (convex)

- ☆ In case of perfect complements, since the substitution between the commodities is zero, the E_S will be zero and the E_S curve will be parallel to Y-axis. This indicates that, what ever may be the change in price ratio of the commodities, the two commodities are preferred and the change in demand ratio is zero. This is shown as curve D in Figure 12.9.
- ☆ If the proportionate change in the ratio of quantities of two commodities X and Y consumed is equal to the proportionate change in the ratio of prices of two commodities X and Y, then E_S will be unity. Then, E_S curve will have a uniform slope. This is shown as curve E in Figure 12.9.

E_S in Terms of IDC Analysis

In IDC analysis, as we discussed earlier, if the consumer moves down the IDC curve, he will substitute one commodity for the other, so as to derive same level of satisfaction. The E_S in terms of IDC analysis is given by,

$$E_S = \frac{\text{Proportionate change in the ratio of quantities of two commodities X and Y consumed}}{\text{Proportionate change in the } MRS_{XY}}$$

$$E_S = [(\Delta Q_X/Q_X)/(\Delta Q_Y/Q_Y)]/[(\Delta MRS_{XY})/(MRS_{XY})] \qquad \textit{Equation 12.5}$$

When the two commodities are good substitutes, as the consumer moves down the IDC (Figure 9.2), the proportionate change in the ratio of quantities of two commodities X and Y consumed is more compared to the proportionate change in the MRS_{XY} of two commodities, thereby, E_S is high. Examples include, coffee and tea, rice and wheat, jam and butter etc. In case of perfect substitutes, where IDC is a straight line connecting the two axes (Panel A of the Figure 9.3), we can easily infer from the above explanation that, E_S is infinity, as infinitely large possibilities of substitution between the two commodities is possible and the proportionate change in the MRS_{XY} is zero (because, ΔMRS_{XY} is zero). Examples include, two brands of coffee, two brands of tea etc.

When the two commodities are good complements, the proportionate change in the ratio of quantities of two commodities X and Y consumed is little, as it is difficult to substitute one commodity for the other. Hence, the IDC is highly convex (the convexity of the IDC increases compared to good substitutes), thereby, the proportionate change in the MRS_{XY} of two commodities is more compared to the proportionate change in the ratio of quantities of two commodities X and Y consumed, thereby, E_S is very low. Examples include, milk and sugar, bread and jam etc. However, in case of perfect complements, where IDC is perfect L shaped, the proportionate change in the ratio of quantities of two commodities X and Y consumed is zero, as the commodities cannot be substituted and they are perfect complements in the consumption process, thereby, E_S is zero.

12.2. Methods of Measuring E_P

There are four methods of measuring the E_P *viz.*, Percentage method, Point method or Geometric method, Arc or Proportional method and Total outlay (expenditure) method. They are discussed in-detail here under.

1. Percentage Method

According to this method, the E_P is measured by the coefficient. The higher the coefficient, the greater the degree of responsiveness of quantity demanded to the change in price of the commodity (more elastic) and *vice versa*. The E_P is computed by the formula:

$$E_P = [(Q_2-Q_1)/Q_1]/[(P_2-P_1)/P_1] = (\Delta Q/\Delta P) \times (P/Q) = (\Delta Q/\Delta P)/(Q/P)$$
$$= \text{Marginal change/Average change}$$

Equation 12.6

Based on the value of E_P, we can categorize the demand for the commodity as more elastic ($E_P > 1$), unitary elastic ($E_P = 1$) and less elastic ($E_P < 1$). This method can be well-explained with the help of the following example (Table 12.2).

Table 12.2: Demand schedule for rice commodity.

Price of Rice (Rs/kg)	*Quantity Demanded (kgs)*
50	100
45	110
43	117
40	124
35	130
30	140
27	142

Case 1

Let us assume, the price of rice falls from Rs.50 to Rs.40 in the market and accordingly, the quantity demanded for rice increases from 100 kgs. to 124 kgs. So, the E_P is given by, 1.20 ($E_P > 1$).

Case 2

Let us assume the above case in the reverse direction. That means, if the price of rice commodity increases from Rs. 40 to Rs.50, then the quantity demanded for rice decreases from 124 kgs to 100 kgs. So, the E_P is given by, 0.77 ($E_P < 1$).

Case 3

From the above table, let us assume another case, where if the price of rice falls from Rs. 50 to Rs.30, then the quantity demanded for rice increases from 100 kgs to 140 kgs. So, the E_P is given by, 1.00 ($E_P = 1$).

Case 4

Assume case 3 in the reverse direction. That means, when the price of rice increases from Rs. 30 to Rs.50, then the quantity demanded for rice decreases from 140 kg to 100 kg. So, the E_P is given by 0.43 ($E_P < 1$).

So, from the above discussion, we can infer that, the value of E_P mainly depends upon the base values we use (with reference to original quantity and original price) in computing the ratio between the percentage change in quantity demanded and percentage change in price. If we observe case (1) vs case (2) and case (3) vs case (4), as discussed above, due to taking different base values with respect to original price and original quantity, the value of E_P is changed. Since, the ratio of base values of initial price and initial quantity are high in case (1) and case (3) compared to case (2) and case (4) respectively, we secured higher E_P values in case (1) and case (3) compared to their corresponding cases, but in the opposite direction.

2. Point Method or Geometric Method

This method was suggested by Alfred Marshall to measure the E_P, when relatively small changes in price and quantity demanded are experienced. As shown in the Figure 12.10, assume point A on the demand curve, where we have to estimate the E_P. So, consider another point A_1, which lies very close to the point A on the same demand curve DD. The assumption is that, the points A and A_1 lie very close to each other, such that, they both can be joined by a straight line and it touches the Y-axis and X-axis at B and C respectively. So, from the Figure 12.10 it is clear that, when the price of the commodity X is OP_1 the quantity demanded is OQ_1 and when the price decreases to OP_2, the quantity demanded extends to OQ_2. So, the E_P is given by,

$$E_P = (\Delta Q/Q)/(\Delta P/P) = (Q_2Q_1/OQ_1)/(P_2P_1/OP_1) = (Q_2Q_1/OQ_1) \times (OP_1/P_2P_1)$$
$$= (Q_2Q_1/P_2P_1) \times (OP_1/OQ_1)$$

Equation 12.7

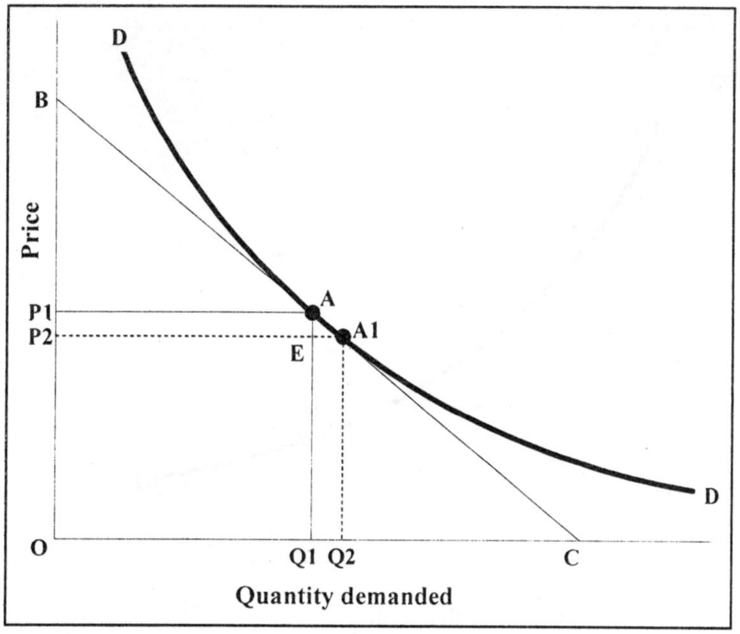

Figure 12.10: Measuring E_p at a point (Geometric method).

Considering EAA triangle, we can write E_p as

$E_p = (\Delta Q/Q)/(\Delta P/P) = (EA_1/OQ_1)/(EA/Q_1A) = (EA_1/OQ_1) \times (Q_1A/EA)$

$= (EA_1/EA) \times (Q_1A/OQ_1)$ *Equation 12.8*

In the Figure 12.10, EAA_1 triangle is similar to Q_1AC and therefore, we can substitute EA_1/EA by Q_1C/Q_1A in the above Equation 12.8 and we get,

$E_p = (Q_1C/Q_1A) \times (Q_1A/OQ_1) = Q_1C/OQ_1$ *Equation 12.9*

Initially, we framed an assumption that, both the points A and A_1 lie very close to each other, because of very small change in price and quantity demanded of the commodity and thereby, the points A and A_1 on BC line almost coincide. So, Q_1A coincide with Q_2A_1 and hence, in the right angled triangle OBC, the relation Q_1C/OQ_1 can be expressed as CA/BA. So,

$E_p = Q_1C/OQ_1 = CA/BA =$ Lower sector/Upper sector *Equation 12.10*

So, to estimate price elasticity at any point on the convex demand curve, draw a tangent to that point on the demand curve and it will touch the Y-axis and X-axis. Then, the price elasticity can be computed by finding the ratio between lower segment and upper segment of the tangent with reference to the point considered on the demand curve. As shown in the Figure 12.11, D_1D_1 is the convex shaped demand curve and A and A_1 are the two points on the demand curve, where the E_p should be estimated. So, two tangents BC and DE are drawn at these two points A and A_1 respectively connecting the Y-axis and X-axis. So, E_p at point A is given by,

$E_p =$ Lower sector/Upper sector $= CA/BA$

Similarly, E_p at point A_1 is given by

$E_p =$ Lower sector/Upper sector $= EA_1/DA_1$

Thus, E_p at point A is more compared to E_p at point A_1 on the demand curve D_1D_1.

From the above analysis, we can infer that, if the demand curve is a straight line (Figure 12.12),

- ☆ At mid point on the demand curve, the E_p is unity.
- ☆ If we move upward along the demand curve (at any point to the left of the midpoint), the ratio between lower sector and upper sector increases and thereby, E_p is more than unity.
- ☆ At the point, where the demand curve touches the Y-axis, the E_p is infinity (but, slope is zero).
- ☆ If we move down the straight line demand curve below mid point (at any point to the right of the midpoint), the ratio between lower sector and upper sector decreases and thereby, E_p is less than unity.
- ☆ At the point, where the demand curve touches the X-axis, the E_p is zero (but, slope is infinity).

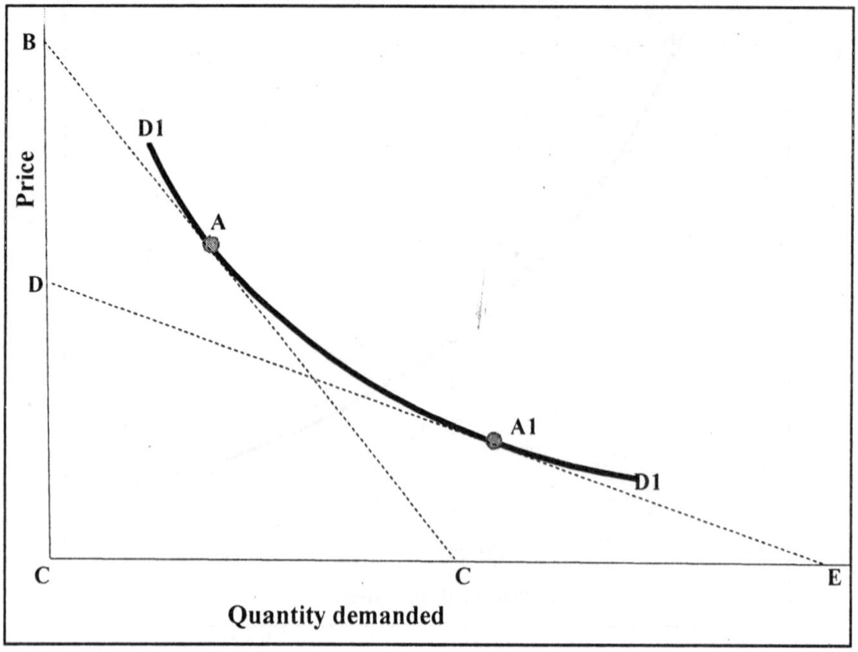

Figure 12.11: Measuring E_P at two points on a convex shaped demand curve.

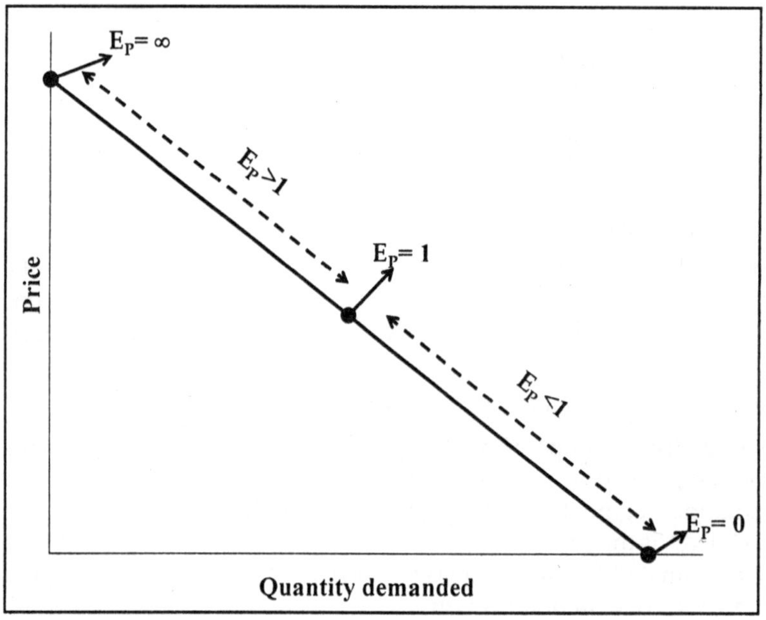

Figure 12.12: Different degrees of E_P on the same demand curve.

- ☆ Considering the straight line demand curve, Q = a – bP (Q = Quantity demanded, P = Price of the commodity), along the horizontal axis P = 0 and along the vertical axis Q = 0, thereby, at the midpoint of the demand curve P = (a/2b).
- ☆ At P = 0, the elasticity is 0, at Q = 0, elasticity is ∞. At P = (a/2b), the elasticity is one, at any price greater than 0 and less than (a/2b), elasticity is less than one, and at any price greater than (a/2b), elasticity is greater than one.

This again proves that, slope and E_P are not the same. Further, E_P is also not the same through out the demand curve. So, the E_P at different points on a linear demand curve varies between 0 to ∞. But, sometimes, the demand curves can be such that the E_P remains constant throughout. For example, in case of perfectly inelastic demand curve (Panel B of the Figure 12.1), whatever may be the change in price, the quantity demanded of the commodity remains same. So, a price change never leads to a change in the quantity demanded of a commodity and for such a demand curve E_P is always zero.

3. Arc or Proportional Method

In Percentage method, we measured the E_p taking the original quantity and original price as base and hence, the explanation of E_p is only one way. In case of Point method, we measured the E_p considering the price change and quantity change in demand are very small, thereby, it guides to measure the elasticity at a point on the demand curve. In both the above methods, since price and quantity changes are very small, we consider only original quantity and original price as a base to measure the E_p. But, if the changes in quantity and price are very large, the original quantity and original price cannot be taken as the basis of measurement to estimate the E_p. Even the final quantity and final price also cannot be taken as base for measuring the E_p. This can be exemplified from the case given through Table 12.3.

Table 12.3: Demand schedule of commodity X.

Price of Commodity X (Rs/kg)	*Quantity Demanded (kgs)*
500	1000
450	1100
420	1180
390	1280
350	1375
300	1400
250	1510

As shown through Table 12.3, the initial price of the commodity is Rs.500 and initial quantity demanded is 1000 kg of commodity. The price of the commodity is declined to Rs. 300, thereby, total quantity demanded extended to 1400 kg. The E_p, if initial quantity and initial price of commodity are considered as base is given by

$E_p = \Delta Q/\Delta P \times P/Q = 400/\text{-}200 \times 500/1000 = \text{-}1.00 = 1.00$

The E_p, if final quantity and final price of the commodity are taken as base, is given by

$E_p = \Delta Q/\Delta P \times P/Q = \text{-}400/200 \times 300/1400 = \text{-}0.43 = 0.43$

So, there exists a wide difference between the price elasticities of the commodity depending upon considering the base quantity and base price. This problem arises because, the fluctuations in the price and hence, the quantity demanded of the commodity are very large or significant. So, Economists proposed to measure the E_p over an arc of the demand curve, in such cases, where the fluctuations in price and quantity demanded are very large. Since, we have to study the price elasticity over an arc, AB (Figure 12.13), the initial and final data pertaining to both quantity demanded and price of the commodity should be taken into consideration. For then, we have to take the average of initial and final data of both quantity and price to compute the E_p and it is shown through the following equation.

$$E_P = \frac{(\text{New quantity} - \text{Old quantity})/[(\text{New quantity} + \text{Old quantity})/2]}{(\text{New Price} - \text{Old price})/[(\text{New price} + \text{Old price})/2]}$$

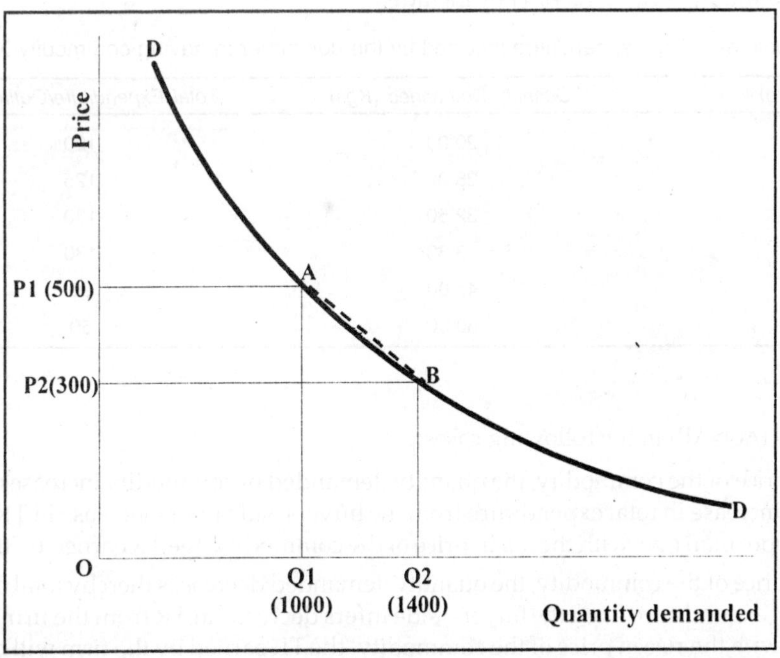

Figure 12.13: Measurement of E_P over an arc of the demand curve.

$$E_P = \frac{(\Delta Q)/[(Q_2 + Q_1)/2]}{(\Delta P)/[(P_2 + P_1)/2]}$$

Rearranging the terms, we get, $E_P = (\Delta Q/\Delta P) \times [(P_2+P_1)/(Q_2+Q_1)]$ ***Equation 12.11***

So, now substitute both the initial and final values of quantities and prices of the commodity in the above formula to compute the E_P. Between points A to B on the demand curve, the E_P is given by,

$E_P = (\Delta Q/\Delta P) \times [(P_2+P_1)/(Q_2+Q_1)] = [(1400-1000)/(300-500)] \times [(500+300)/(1400+1000)]$

$= 0.67$

Similarly, between B to A on the demand curve, the E_P is given by

$E_P = (\Delta Q/\Delta P) \times [(P_1+P_2)/(Q_1+Q_2)] = [(1000-1400)/(500-300)] \times [(300+500)/(1000+1400)]$

$= 0.67$

The above analysis infer that, when the price and quantity changes are significant, the averages of initial and final data of both quantity demanded and price of the commodity taken as base values reveal the same price elasticity, if we move up or down along the arc (AB) of the demand curve.

4. Total Outlay or Expenditure Method

This method was suggested by Alfred Marshall. According to this method, the E_P is measured based on the total expenditure incurred by the consumer in purchasing the commodity at different prices. In this method, we study the total expenditure incurred by the consumer at different prices of the commodity and the comparison of changes in total expenditure due to changes in prices of the commodity reveals the degree of E_P. We know, total expenditure incurred by the consumer is the product between price of the commodity and quantity of commodity purchased by the consumer. The decrease in price of the commodity may lead to increase in total expenditure or constancy of total expenditure or decrease in total expenditure depending upon the quantity demanded by the consumer. If the percentage increase in quantity is greater than the percentage decline in the price, the expenditure on the commodity will go up. On the other hand, if the percentage increase in quantity is less than the percentage decline in the price, the expenditure on the commodity will go down. And if the percentage increase in quantity is equal to the percentage decline in the price, the expenditure on the commodity will remain unchanged. Now, consider an increase in the price of a commodity. If the percentage decline in quantity is greater than the percentage increase in the price, the expenditure on the commodity will go down. On the other hand, if the percentage decline in quantity is less than the percentage increase in the price, the expenditure on the commodity will go up. And if the percentage decline in quantity is equal to the percentage increase in the price, the expenditure on the commodity will remain unchanged. So, accordingly, three degrees or propositions of E_P are possible in accordance with the changes in total expenditure incurred by the consumer due to changes in price of the commodity (Table 12.4). They include,

Table 12.4: Demand schedule and Total expenditure incurred by the consumer in having commodity X.

Price(Rs/kg)	*Quantity Demanded (Kgs)*	*Total Expenditure/Outlay (Rs)*
6	20.00	120
5	25.00	125
4	32.50	130
3	43.33	130
2	45.00	90
1	50.00	50

(a) $E_P > 1$

E_P is more than unity ($\Delta Q > \Delta P$) in the following cases:

- ☆ with the fall in price of the commodity, the quantity demanded of commodity increases, thereby, total outlay increases. (The increase in total expenditure from the buyers' side infers increase in TR from the firm's side. So, when E_P is more than one, with the fall in price of the commodity, the TR earned by the firm will increase).
- ☆ with the rise in price of the commodity, the quantity demanded decreases thereby total outlay decreases. (The decrease in total expenditure from the buyers' side infers decrease in TR from the firm's side. So, when E_P is more than one, with the rise in price of the commodity, the TR earned by the firm will decrease).

(b) $E_P = 1$

E_p is equal to unity ($\Delta Q = \Delta P$) in the following cases:

- ✰ with the fall in price of the commodity, the quantity demanded increases, but the total outlay remains same. (When there is no change in total expenditure from the buyers' side, it implies, there is no change in TR from the firm's side. So, when E_p is equal to one, with the fall in price of the commodity, the TR earned by the firm will remains same).
- ✰ with the rise in price of the commodity, the quantity demanded decreases, but the total outlay remains same. (When there is no change in total expenditure from the buyers side, it implies, there is no change in TR from the firm's side. So, when E_p is equal to one, with the rise in price of the commodity, the TR earned by the firm will remains same).

(c) $E_P < 1$

E_p is less than unity ($\Delta Q < \Delta P$) in the following cases:

- ✰ with the fall in price of the commodity, the quantity demand increases, but the total outlay decreases. (The decrease in total expenditure from the buyers' side infers decrease in TR from the firm's side. So, when E_p is less than one, with the fall in price of the commodity, the TR earned by the firm will decrease).
- ✰ with the rise in price of the commedity, the quantity demanded decreases, but the total outlay increases. (The increase in total expenditure from the buyers' side infers increase in TR from the firm's side. So, when E_p is less than one, with the rise in price of the commodity, the TR earned by the firm will increase).

So, the above explanation with reference to three possible types of E_p infers that, the total outlay or expenditure determines the responsiveness of demand due to change in price of the commodity. If the fall (rise) in price of the commodity leads to increase (decrease) in the total outlay on the commodity, it implies E_p is more than one ($E_p > 1$). If the fall or rise in price of the commodity, leads to no change in the total outlay on the commodity, it implies E_p is equal to one ($E_p = 1$). If the fall (rise) in price of the commodity, leads to decrease (increase) in the total outlay on the commodity, it implies E_p is less than one ($E_p < 1$).

It is clear from the Table 12.4 that,

- ✰ If the price of the commodity X is decreased from Rs. 6 to Rs. 5, the quantity demanded increases (20 kg to 25 kg), thereby, total outlay also increases from Rs.120 to Rs.125. This indicates that, the E_p is >1, as per the first proposition of this method.
- ✰ If the price of the commodity X is decreased from Rs.4 to Rs.3, the quantity demanded increases (32.50kg to 43.33kg), but the total outlay incurred by the consumer remains same *i.e.*, Rs.130. This indicates the $E_p = 1$, as per the second proposition of the method.
- ✰ If the price of commodity X is decreased from Rs.2 to Rs.1, the quantity demanded increases (45kg to 50 kg), but the total outlay incurred by the consumer decreases from Rs.90 to Rs.50. This indicates that, $E_p < 1$, as per the third proposition of this method.

The following Table 12.5 summarizes the three propositions and data of the Table 12.4.

Table 12.5: Three propositions of total outlay method.

Price	*Total Expenditure*	E_p
Decrease	Increase	More than unity
Increase	Decrease	(Price elastic)
Decrease	Constant	Unity
Increase	Constant	(Unitary elastic)
Decrease	Decrease	Less than unity
Increase	Increase	(Price inelastic)

The above discussion reveals that, the expenditure on the commodity would change in the opposite direction as the price change, if and only if the percentage change in quantity is greater than the percentage change in price, *i.e.*, if the commodity is price-elastic. The expenditure on the commodity would change in the same direction as the price change, if and only if the percentage change in quantity is less than the percentage change in price, *i.e.*, if the commodity is price inelastic. The expenditure on the commodity would remain unchanged, if and only if the percentage change in quantity is equal to the percentage change in price, *i.e.*, if the commodity is unit-elastic. If we plot the data portrayed in

Table 12.4 as a graph (Figure 12.14) taking total expenditure made on the commodity along X-axis and prices of the commodity along Y-axis, it gives a backward bending curve ABCD. It is clear from the Figure 12.14 that,

- If the price of the commodity decreases from Rs. 6 to Rs. 5, the total outlay increases from Rs. 120 to Rs. 125 and this is represented along AB portion of the curve and the $E_p > 1$.
- If the price of the commodity decreases from Rs. 4 to Rs. 3, the total outlay remains same at Rs. 130 and this represented along BC portion of the curve and the $E_p = 1$. This portion includes highest total expenditure incurred by the consumer in having the commodity.
- If the price of the commodity decreases from Rs. 2 to Rs. 1, the total outlay decreases from Rs. 90 to Rs. 50 and this represented along CD portion of the curve and the $E_p < 1$.

So, the Figure 12.14 infers the following interesting points:

- When $E_p > 1$, the total expenditure incurred on the commodity is inversely proportional to the change in its price.
- When $E_p = 1$, the total expenditure incurred on the commodity remains same irrespective of change in its price.
- When $E_p < 1$, the total expenditure incurred on the commodity is directly proportional to the change in its price.

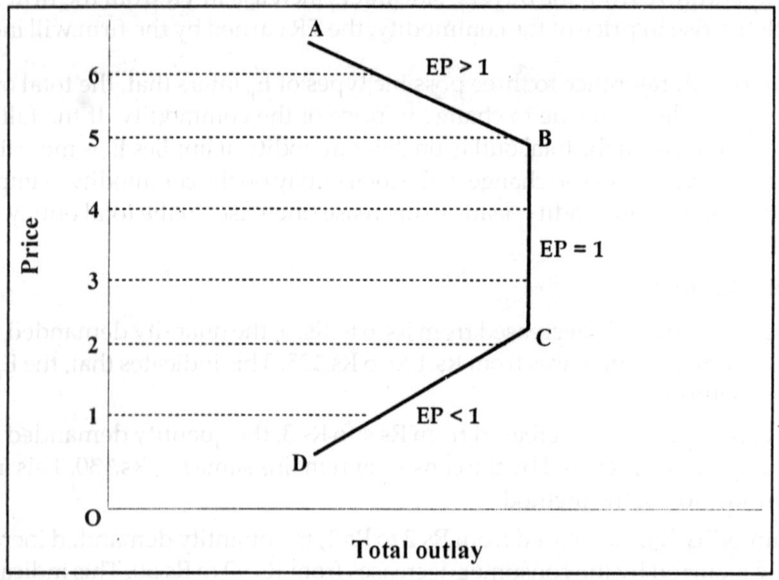

Figure 12.14: Total outlay curve to measure E_p.

We can also plot the same data of the Table 12.4 in deriving the demand curve for the commodity by considering quantity demanded of the commodity on X-axis and price of the commodity on Y-axis. The Figure 12.15 reveals that,

- The demand curve D_1D_1 slopes downward from left to right.
- The rectangles drawn below the demand curve implies total expenditure incurred by the consumer to purchase the desired quantity of commodity at a give price.
- The AB portion of the demand curve represents the $E_p > 1$, as indicated by the increase in total outlay (6x20=Rs.120, 5x25=Rs.125) with decrease in price of the commodity.
- The portion CD on the demand curve represents $E_p = 1$ as indicated by the constancy of total outlay with the fall price of the commodity (4x32.50=Rs.130, 3x43.33=Rs.130). This portion of the demand curve indicates the highest total outlay incurred by the consumer.
- The portion EF on the demand curve represents $E_p < 1$, as indicated by the fall in total outlay with the fall in price of the commodity (2x45.00=Rs.90, 1x50.00=Rs.50).

The above inferences further reveal the following interesting points:

- The movement along the same demand curve implies E_p.

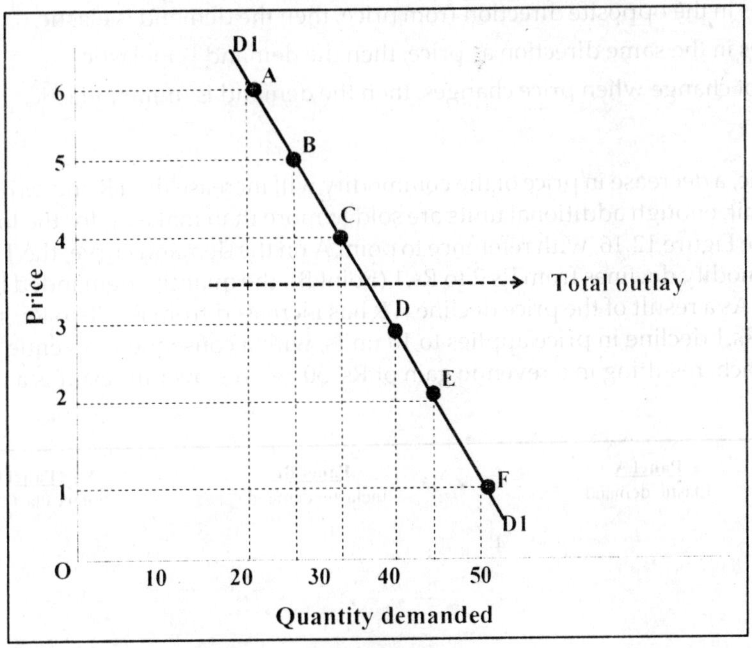

Figure 12.15: Demand curve showing total outlay of the consumer.

- It is important for the readers that, the E_p along the same demand curve varies. So, to say, the more elastic demand curve will have a flat slope or the less elastic demand curve will have a steep slope is wrong because, throughout the slope, the elasticity will not remains same, but it varies.
- The slope of the demand curve will not indicate E_p for the commodity, as they both are different. We generally infer that, the flatter the slope of the demand curve, the more is the E_p and the steeper the slope of the demand curve, the less is the E_p. But, this is wrong, as indicated by the following equations:

  Slope of the demand curve = $\Delta P / \Delta Q$

  $E_p = (\Delta Q/\Delta P) \times (P/Q)$

  So, $E_p = (1/\text{Slope}) \times (P/Q)$

This once again proved that, the slope of the demand curve and E_p are not the same. Further, the slope of the demand curve (straight line) remains same throughout its length, but the E_p along the straight line demand curve ranges between zero to infinity (*i.e.*, perfect inelastic demand to perfect elastic demand).

But, the main draw back of this method of analysis of E_p is that, we cannot estimate the value of E_p in numerical terms. Rather, it will simply indicate, whether the elasticity is more or equal or less than unity.

Choice among the Four Methods

The choice of selecting one among the four methods discussed above depends upon the data available with the investigator. If the investigator is provided with the data pertaining to demand schedule of the commodity and if the changes in price of the commodity and its quantity demanded are significant, Arc method of computing E_p is more reliable compared to Percentage method or Total outlay method. On the other hand, if the investigator need not compute E_p in terms of value, he prefers to choose Total outlay method. But, given a demand curve, to compute elasticity at a particular point, Geometric or Point method will be employed.

12.3. The Total-Revenue (TR) Test

The importance of elasticity for firms relates to the effect of price changes on TR and thus on profits (TR minus Total Costs (TC)). We know, TR is price times the quantity of output sold by the firm. So, a decrease in price or increase in quantity demanded will increase the TR for the firm. Graphically, TR is represented by the P x Q rectangle lying below a point on a demand curve. As shown in the Figure 11.1, when price is Rs.65, 10 units of the commodity is demanded. So, TR to the firm is Rs. 650 shown by the rectangle below the demand curve. We know from basic geometry that, the area of a rectangle is found by multiplying one side by the other. Here, one side is 'price' (Rs.65) and the other is 'quantity demanded' (10 units).

TR and the E_p are related. This Total-Revenue test is the easiest way to find out, whether the demand for the commodity is elastic or inelastic or unitary elastic. The rules of thumb include,

- If TR changes in the opposite direction from price, then the demand is elastic.
- If TR changes in the same direction as price, then the demand is inelastic.
- If TR does not change when price changes, then the demand is unitary elastic.

Elastic Demand

If demand is elastic, a decrease in price of the commodity will increase the TR and *vice versa*. Even though a lesser price is received per unit, enough additional units are sold to more than make up for the lower price. For example, as shown in Panel A of the Figure 12.16, with reference to point A on the demand curve, the TR is Rs. 20 (Rs.2 x 10 units). If the price of the commodity declines from Rs.2 to Rs.1 (Point B), the quantity demanded becomes 40 units and TR is Rs.40 (Rs.1 x 40 units). As a result of the price decline, TR has increased from Rs. 20 to Rs. 40. The TR has increased in this case because, the Rs.1 decline in price applies to 10 units, with a consequent revenue loss of Rs.10. But, 30 more units are sold at Rs.1 each, resulting in a revenue gain of Rs. 30. So, the overall result is a net increase in TR of Rs. 20 (=Rs.30 – Rs.10).

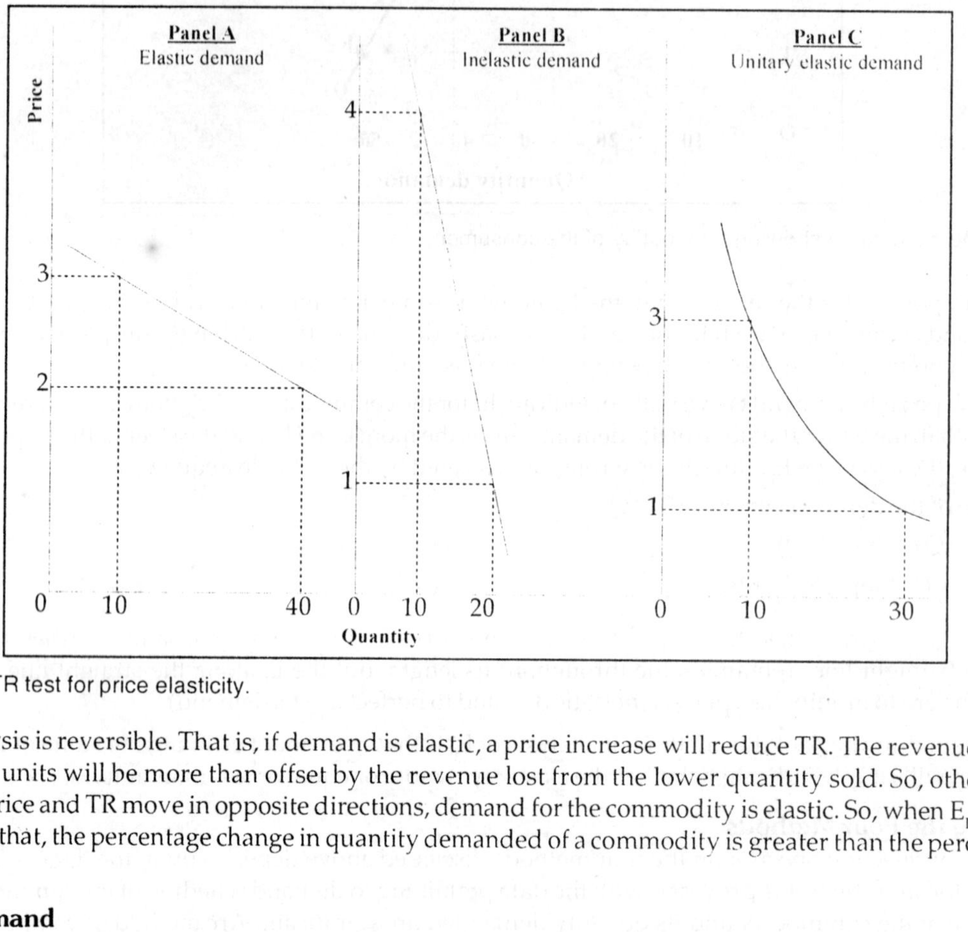

Figure 12.16: TR test for price elasticity.

The analysis is reversible. That is, if demand is elastic, a price increase will reduce TR. The revenue gained on the higher-priced units will be more than offset by the revenue lost from the lower quantity sold. So, other things being equal, when price and TR move in opposite directions, demand for the commodity is elastic. So, when E_p is greater than one, it implies that, the percentage change in quantity demanded of a commodity is greater than the percentage change in its price.

Inelastic Demand

If demand is inelastic, a price decrease will reduce TR and *vice versa*. That means, the increase in sales due to price decrease will not fully offset the decline in revenue per unit, thereby TR will decline. The same is explained through Panel B of the Figure 12.16. At point A on the demand curve, price is Rs.4 and quantity demanded is 10 units. Thus, TR is Rs. 40, shown as the rectangular area below the demand curve. If the price of the commodity drops to Rs.1 (Point B), TR declines to Rs.20, which obviously is less than Rs.40. The TR has declined because, the loss of revenue from the lower unit price is larger than the gain in revenue from the accompanying increase in quantum of sale of the commodity. So, when price of the commodity decreases, TR also decreases.

This analysis is again reversible. If demand is inelastic, a price increase will increase TR. So, other things being equal, when price and TR move in the same direction, demand for the commodity is inelastic. So, when E_p is less than one, it implies, the percentage change in quantity demanded of a commodity is less than the percentage change in its price.

Unitary Elastic Demand

In the special case of unit elasticity, an increase or a decrease in price of the commodity leaves TR unchanged. That means, the loss in revenue from a lower unit price is exactly offset by the gain in revenue from the accompanying increase in quantum of sales of the commodity. Conversely, the gain in revenue from a higher unit price is exactly offset by the revenue loss associated with the accompanying decline in the amount demanded. The same is explained through Panel C of the Figure 12.16. With reference to point A on the demand curve, at the price of Rs.3, 10 units will be demanded, thereby, the TR is Rs.30. At the lower price (Rs.1), a total of 30 units will be demanded, again resulting in Rs.30 of TR. The Rs.2 price reduction causes the loss of revenue is exactly offset by the revenue gain due to increase in number of units of commodity sold, thereby, TR does not change. In fact, that would be true for all price changes along this particular curve. So, other things being equal, when price of the commodity changes and TR remains constant, then the demand is unit-elastic. So, when E_p is equal to one, it implies, the percentage change in quantity demanded of a commodity equals the percentage change in price.

E_p and the TR Curve (TRC)

Although the demand curves depicted in Figure 12.16 explain the TR test for E_p, two of the graphs (Panel A and Panel B) involve specific movements along linear (straight-line) demand curves. That presents no problem for explaining the TR test. However, we know that, E_p typically varies over the different price ranges on the same demand curve (Figure 12.12) The exception is the curve in Panel C of the Figure 12.16, where E_p=1 along the entire curve. Through the following Table 12.6 and Figure 12.17, we can illustrate different E_p at different points on the same demand curve and this guides to formulate an interesting relationship between E_p and TR. Plotting the hypothetical data for milk shown in columns 1 and 2 of Table 12.6 yields a linear demand curve D′D′ in Panel A of the Figure 12.17. We see from column 3 of the table that, the E_p for this demand curve declines as we move from higher to lower prices. So, demand is more price-elastic towards the upper left (here, the Rs.5 to Rs.8 price range of D′D′ curve) than towards the lower right (here, the Rs.4 to Rs.1 price range of D′D′ curve). This is because, in the upper left segment of the demand curve, the percentage change in quantity is large because, the original reference quantity is small. Similarly, the percentage change in price is small in that segment because, the original reference price is large. The relatively large percentage change in quantity demanded divided by the relatively small change in price yields a large E_p *i.e.*, an elastic demand.

Table 12.6: E_p for milk as measured by the elasticity coefficient and TR test.

Price of Milk (Rs/unit)	*Quantity of Milk Demanded (Units)*	E_p	*TR*	*TR Test*
8	1	5.00	8	Elastic
7	2	2.60	14	
6	3	1.57	18	
5	4	1.00	20	Unitary elastic
4	5	0.64	20	
3	6	0.38	18	Inelastic
2	7	0.20	14	
1	8	–	8	

The reverse holds true for the lower-right segment of the demand curve. Here, the percentage change in quantity is small because, the original reference quantity is large. But, the percentage change in price is large because, the original reference price is small. The relatively small percentage change in quantity demanded divided by the relatively large percentage change in price results in a small E_p *i.e.*, an inelastic demand. So, as discussed earlier, the demand curve whether flat or steep will not form a basis for judging the E_p. This is because, the slope of the curve is computed from absolute changes in price and quantity, while elasticity involves relative or percentage changes in price and quantity. In our analysis, the demand curve in Panel A of the Figure 12.17 is linear, which by definition means that, the slope is constant throughout. But we have demonstrated that, such a curve is elastic in its high-price (Rs.5 to Rs.8) range and inelastic in its low-price (Rs.1 to Rs.4) range.

In Panel B of the Figure 12.17, we plotted the TR that corresponds to each price-quantity combination indicated along the demand curve D′D′ in Panel A. The price–quantity demanded combination represented by point A on the demand curve yields TR of Rs.8 (Rs.8 x 1 unit of milk). In Panel B, we graphed this Rs.8 amount vertically at 1 unit quantity demanded. Similarly, the price–quantity demanded combination represented by point B in the upper panel yields TR of Rs.14 (Rs7 x 2 units of milk). This amount is graphed vertically at Rs.14 TR amount with reference to 2 units of milk demanded in the lower panel. The ultimate result of such graphing is TRC, which first slopes upward, then reaches a maximum and finally turns downward. So, the TRC reaches a peak at a certain price and quantity demanded of a commodity and drops off on either side of it. It is evident that, TR reaches a maximum value at the same point that the E_p for the commodity is unitary elastic. At prices above Rs. 5/unit, demand is more price elastic, as a rise in price of the commodity will reduce the quantity demanded and TR will fall. At prices below Rs. 4/unit, demand is more price

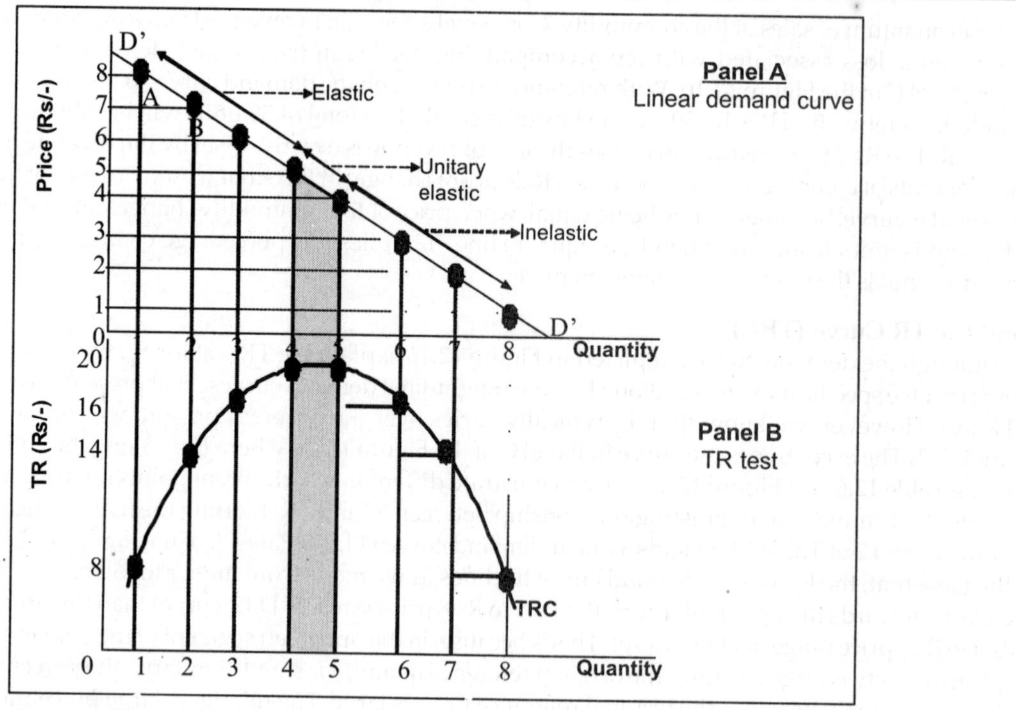

Figure 12.17: Relation between E_P and TR.

inelastic, as the fall in price of the commodity will increase the quantity demanded and TR will fall. Comparison of curves D'D' and TR sharply focuses the relationship between E_P and TR. Lowering the price of milk in the elastic range of demand *i.e.*, from Rs.8 to Rs.5, will increase TR. Conversely, increasing the price of milk in that range reduces TR. In both cases, price and TR change in opposite directions, confirming that the demand is elastic. The Rs.5 to Rs.4 price range of demand curve D'D' reflects unitary elasticity. When price either decreases from Rs.5 to Rs.4 or increases from Rs.4 to Rs.5, TR remains same at Rs.20. In both cases, price has changed, but TR has remained constant, confirming that demand is unitary elastic. In the inelastic range of demand curve D'D', lowering the price from Rs.4 to Rs.1, decreases TR and raising the price from Rs.1 to Rs.4 boosts the TR, as shown in Panel B. So, in both the cases, price and TR move in the same direction, confirming that demand is inelastic. The above discussion is summarized and presented below through Table 12.7.

Table 12.7: E_P and TR test – A summary

E_P	*Description*	*Impact on TR of a*	
		Price increase	*Price decrease*
Elastic or Relatively elastic	Percentage change in quantity demanded of a commodity is greater than percentage change in its price	TR decreases	TR increases
Unitary elastic	Percentage change in quantity demanded of a commodity is same as percentage change in its price	TR is unchanged	TR is unchanged
Inelastic or Relatively inelastic	Percentage change in quantity demanded of a commodity is less than percentage change in its price	TR increases	TR decreases

12.4. Significance of E_D

In the words of Keynes, '*the concept of elasticity is so important that, in the provision of terminology and apparatus to aid thought, I do not think, Marshall did any greater service than by the explicit introduction of the idea of the elasticity*'. This statement signifies the importance of E_D, as it offers the following theoretical and practical advantages:

☆ ***Production:*** The concept of E_D guides the producers in the production of commodities. For the commodities, which have more elastic demand, the producers should take the correct decision regarding the level of output. For example, due to imperfect knowledge, if the output is less from the firm than the demand, then the prices will rise drastically in the market, thereby, the firm receives less profits, as the consumers prefers substitute goods instead of this dearer good. On the other hand, if the commodity is having inelastic demand in the market, even the decline in production will not affect the firm, as the commodity will be purchased even at a high price.

☆ ***Price fixation:*** The monopolist, being the price maker, uses this concept of E_D to fix the price of the commodity. If the commodity produced by the monopolist have elastic demand, he will quote a low price, but produce more quantity of output. So, the low price of the commodity extends the demand for the commodity and thereby, he maximizes the revenue. On the other hand, if the commodity enjoys inelastic demand, he can quote a high price for the commodity and thereby, he maximizes the revenue. In the case of price discrimination, the monopolist will divide the market into different sub-markets based on the differences in the price elasticities of demand. He quote a low price for the commodity in sub-market with high elastic demand and high price in the sub-market with less elastic demand and thereby, he leaves no consumer's surplus.

☆ ***Distribution:*** The concept of E_D is widely useful in the distribution of rewards to different factors of production. For example, if the demand for labour is very elastic, the strike or efforts made by the trade unions will not meet the success. On the other hand, if the demand for labour is inelastic, even the threat of the strike called by the trade union will induce the employers to raise the wages of the labour. Similarly, the rewards of other factors of production are also influenced by the concept of E_D.

☆ ***In international trade:*** The concept of E_D forms the basis for international trade. In making exports and imports of commodities across the countries, we have to consider the elasticities of demand for different commodities exported and imported. Generally, a country will gain in international trade if,

- ❖ The country exports the commodities that have inelastic demand for a rise in price. This because, even at high price, the commodity gets exported and demanded in the international market.
- ❖ The country exports the commodities that have elastic demand for a fall in price. This because, by slight decline in price of the commodity, the demand in the international market increases significantly.
- ❖ The country imports the commodities that have elastic demand for a rise in price. This is because at high price, less quantity of commodities will enter the domestic market.
- ❖ The country imports the commodities that have inelastic demand for the fall in price. So, at low price, more quantity of commodities will enter the domestic market.

So, by these four ways, the terms of trade position of the country improves. Thus, the terms of trade depends upon the mutual elasticities of demand of the commodities that are exported and imported across the countries. This concept of E_P for exports and imports forms the basis for fixation of appropriate exchange rate between the two currencies of the two countries. That means, this concept also guides in devaluation of a currency. If a country devaluate its currency, exports become cheaper and imports become dearer. So, a country benefits from devaluation of its currency, when the demand for both exports and imports is price elastic. This concept is equally applicable to address other important issues like balance of payments, effects of tariffs, net trade of the countries etc.

☆ ***Fixation of tariffs:*** In international trade, tariffs will be imposed on the imported commodities, so as to safeguard the domestic producers from the import surge. The imposition of tariffs on the imports depends upon the E_D of the imported goods. Say, if a foreign country is dumping a commodity into India that have more elastic demand, a high tariff will be imposed, such that, its price will rise in the domestic market and thereby, domestic producers will be safe guarded. Here, the main intention of imposing high tariff is to safeguard the domestic producers from dumping. On the other hand, if a foreign country is dumping a commodity into India that have inelastic demand, even after imposing a high tariff, the commodity will enter the domestic market. Here, the main intention of imposing a high tariff is to earn more revenue by the Indian Government.

☆ ***Explain the 'paradox of poverty in the midst of potential plenty':*** It implies that, even if the producers produce more of the commodity, it will not bring prosperity to them, but might ruin them. This is possible, when the demand for that commodity is less elastic in the market.

Suppose, in an area, there is a bumper harvest of rice crop. Since the demand for rice is inelastic, the decline in prices of rice will ruin the farmers and thereby, they get less amount of revenue, even when they had a bumper harvest. Had the farmers received a normal crop, their revenue might have been more compared to bumper harvest. The same concept is explained through Figure 12.18. DD is the demand curve for rice and it is more steep indicating inelastic demand for rice. S_1S_1 is the supply curve for rice when the farmers had a normal crop. S_2S_2 is the supply curve for rice when the farmer had a bumper harvest for rice and hence, the supply curve shifted towards right. Considering the S_1S_1 supply curve, the equilibrium price is OP_1 where OQ_1 quantity of rice is bought and sold in the market. The TR earned by the farmers is $OP_1E_1Q_1$ when the farmers had a normal crop. Considering the S_2S_2 supply curve, the equilibrium price is OP_2, where OQ_2

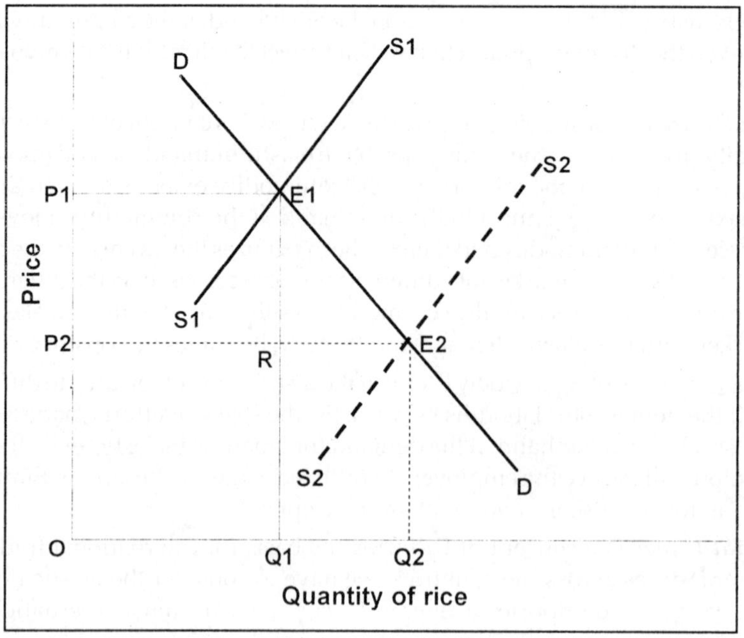

Figure 12.18: Price elasticity of demand explaining the 'Paradox of poverty in the midst of potential plenty'.

quantity of rice is bought and sold in market. This implies that, the demand for rice being less elastic, the increase in supply led to fall in price of the rice from OP_1 to OP_2. So, the TR earned by the farmers is $OP_2E_2Q_2$, when the farmers had a bumper harvest of rice. So, the loss in revenue to the farmers is the difference between the rectangles $P_1P_2E_1R$ - $RQ_1Q_2E_2$ even at times of bumper harvest of rice.

☆ ***For nationalizing an industry :*** The concept of E_D helps the Government to decide whether to nationalize an industry or not. Say, if a monopoly firm produces a commodity that has high inelastic demand, it is a clear case of declaring or bringing it under Government control, so as to safeguard the customers from higher prices.

☆ ***Helps in procuring buffer stocks:*** The concept of E_D guides the Government in planning for sizeable procurement and maintenance of buffer stocks. We know, the Government of India procures food grains *viz.*, rice and wheat to maintain sizeable buffer stocks to meet the emergency needs. At the point of equilibrium E (Figure 12.19) of demand and supply of rice *i.e.*, at OP_1 equilibrium price, OQ_1 is the equilibrium quantity of rice bought and sold in the market. If the Government decides to maintain sizeable buffer stocks in the coming season, it will slightly increase the Minimum Support Price (MSP) to OP_2 from OP_1. At this price OP_2, OQ_2 is the quantity of rice supplied, but only OQ_3 is the quantity demanded in the market. So Q_3Q_2 is the excess supply of rice in the market and the Government will procure this Q_3Q_2 supply at OP_2 price to maintain buffer stocks.

☆ ***In determination of prices of public utilities:*** The prices charged by the public utilities are fixed in accordance with the elasticities of demand for the product. If the product or service offered by the public authority has inelastic demand, then a high price is charged for it. On the other hand, if the demand is more elastic, then it fixes a low price for the product. For example, Indian Railways charge less price from the customers because, the authorities are aware of the presence of alternatives like air travel, road travel etc. On the other hand, for the transport of goods like coal, diesel, petrol etc., the demand being inelastic, they charge high price from the dealers.

☆ ***Distribution of National Income:*** If the price of the factor of production increases and if that factor has more E_S, its income will fall, thereby, its share in National Income will fall.

☆ ***Importance in taxation policy:*** The concept of E_P guides the Finance Minister in imposition of taxation policy. If the demand for the commodity is highly inelastic, a high rate of tax can be imposed on the commodity and the Government can succeed in getting higher revenue. However, the incidence of the tax depends upon the degree of E_P of the commodity. If the commodity under consideration is perfectly inelastic, then the tax imposed on the commodity will be borne by the consumers, as they have to purchase the commodity irrespective of the price level of the commodity. On the other hand, if the demand for the commodity is perfectly elastic, then the tax imposed on the commodity will be borne by the manufacturers or dealers and not by the consumers, as the increase in price of the commodity due to taxation at the consumer's level, will adversely affect the

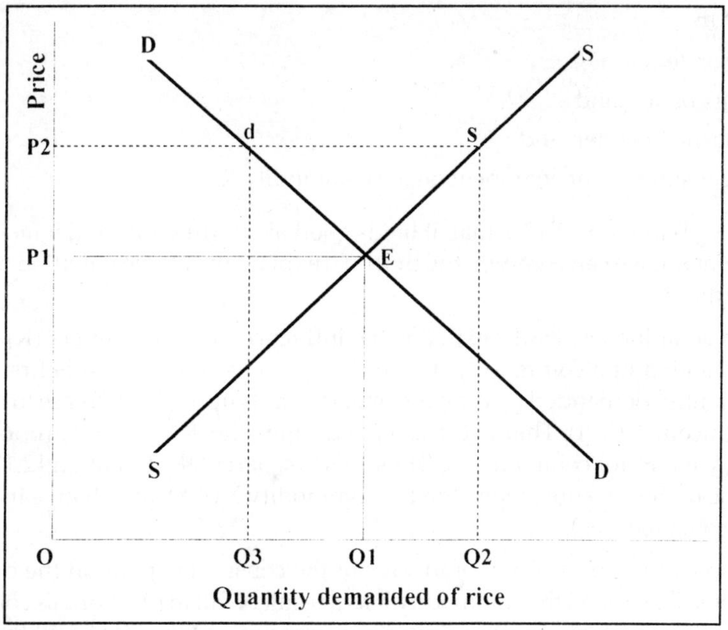

Figure 12.19: E_P guiding the procurement of rice.

demand for the commodity. This concept clearly indicates the importance of E_P in the imposition of taxes on the commodities.

☆ ***Prices of joint products***: In agriculture and allied sectors, the products are mostly joint products, such as paddy grain+straw, cotton seed+lint, sheep mutton+wool etc., as these commodities are produced simultaneously in the same production programme. But, it is difficult to fix the prices for these joint products, due to joint costs of production. In such cases, the concept of E_P guides to formulate the prices for the joint products. For example, since wool is having inelastic demand in the market, a high price is charged for wool and since, mutton is having elastic demand, it is charged low price in the market.

☆ ***Measurement of Monopoly power:*** The degree of control, which the monopolist can exercise on the price and output of his product is called Monopoly power. The extent of monopoly power is the reciprocal of E_P for the product. So, the lower the E_P for the product, the greater is the monopoly power and *vice versa*. This is because, the lower the E_P, the higher the price the monopolist can quote and hence, the higher the monopoly power.

☆ ***Mechanization in industries***: It is a misconception that, the use of machinery in an industry will displace the labour. However, it all depends upon the E_P for the product, the industry manufactures. If the demand for the product is price elastic, then machinery will be introduced in the industry, as it ensures cost effective production. This leads to extension of demand for the product in the market and this further calls for higher production, for which the employment of both workers and machinery will increase. On the other hand, if the demand for the commodity is inelastic, the introduction of machinery will ensure cost effective production (cost saving is profit making on the part of the entrepreneur), but it will greatly displaces labour force without affecting the demand for the product in the market.

☆ ***Sales policy in super markets***: We know that, in super bazaars, the commodities are offered at low prices compared to individual shops. These super markets to cover their establishment and maintenance costs, will quote a low price for the products, whose demand is elastic, thereby, the demand extends and the costs are recovered.

☆ ***Determination of AR and MR of a firm:*** The concept of E_P helps to estimate the AR and MR of a firm from the following relationship:

$$AR = MR\,[E_P/(E_P-1)] \qquad \textit{Equation 12.12}$$

12.5. Relationship Between Price Elasticity, Income Elasticity and Substitution Elasticity

As discussed in IDC analysis (Chapter 9), the price effect is the sum of income effect and substitution effect. Likewise, E_P, which is normally used in the economic analysis depends upon both income elasticity and substitution elasticity. So, we can formulate a mathematical relationship between these three types of elasticities of demand and it is given in the form of a Slutsky equation for two commodities X and Y, as given below,

$E_P = KX.E_Y + (1\text{-}KX)E_S$ *Equation 12.13*

E_P = Price elasticity of demand

E_Y = Income elasticity of demand

E_S = Substitution elasticity of demand

K = Proportion of consumer's income spent on the commodity X.

The merit of the above Equation 12.13 is that, it holds good in all situations and it facilitates that, if we know two types of elasticities of demand, we can compute the third. To explain the above equation, we can conveniently divide the equation into two parts.

The first part of the equation *i.e.*, $KX.E_Y$ explains the influence of income effect (K) on the E_P. That means, the increase in quantity demanded of a commodity due to change in its price depends first on the income effect. That means, the increase in quantity demanded by the consumer depends upon the increase in his real income due to fall in price of the commodity (income effect). This, in turn, depends upon the 'KX' *i.e.*, the proportion of consumer's income spent on the commodity X and E_Y for commodity X. Thus, this first part of the Equation 12.13, involves two components *viz.*, proportion of income of the consumer spent on the commodity X (KX) and change in quantity demanded due to change in income of the consumer (E_Y).

- ☆ The first component *i.e.*, proportion of income of the consumer spent on the commodity X refers to what amount of income is spent on the commodity X as a result of fall in price of this commodity. The income, thus, released due to fall in price of commodity X (increase in real income) will be used to purchase more of commodity X and other commodities as well. So, for the given E_Y, the greater the proportion of income spent on the commodity X (*i.e.*, KX) implies, greater the amount of income released (due to fall in price of commodity X) to purchase commodity X and other commodities as well and hence, more quantity of commodity X is demanded by the consumer.
- ☆ But, the income effect also depends upon the second component E_Y, which measures the amount of money released due to fall in price of commodity X and that will be spent on commodity X and on other commodities as well. Given the proportion of income spent on the commodity X (*i.e.*, KX), the greater the E_Y, the greater the amount of income released due to fall in price of commodity X and consequently, more of it is spent on commodity X.

The second part of the Equation 12.13 refers to the influence of substitution effect on the E_P of commodity X. This infers that, the E_P or the increase in quantity demanded of a commodity depends not only on the income effect, but also on the substitution effect. This is because, if the price of the commodity X falls, the consumer purchases more of commodity X at the expense of others, as it is relatively cheaper compared to other commodities. A close examination of the Equation 12.13 reveals that, the substitution effect depends upon two components *viz.*, the proportion of income spent by the consumer on the other commodities before the fall in price of the commodity X *i.e.*, (1-KX) and E_S. (Note that, KX in the first part of the equation represent income spent on commodity X and 1-KX in the second part of the equation represent income spent on other commodities *i.e.*, other than X).

- ☆ For the given E_S, the greater the income spent by the consumer on other commodities (1-KX) before the fall in price of commodity X, the greater the scope for substituting commodity X for other commodities due to fall in price of commodity X.
- ☆ Now given the 1-KX, the greater the E_S, the greater the income spent by the consumer on the commodity X due to fall in its price and thereby, more of commodity X is substituted for other commodities.

The above explanation offered with respect to influence of both income effect and substitution effect on the E_P infers that, E_P is determined by four important factors *viz.*, proportion of income spent on the commodity X (*i.e.*, KX), E_Y, proportion of income spent on other commodities *i.e.*, other than X (*i.e.*, 1-KX) and E_S.

13

Consumer's Surplus

13.1. Definition and Meaning of Consumer's Surplus

The concept of Consumer's surplus was first stated in crude form by Jules Dupuit (1844) to measure the benefits derived from public goods like bridges, irrigation projects, national highways etc. Later, Alfred Marshall refined this concept and popularized it as 'Consumer's surplus'. Prof. Boulding has renamed it as 'Buyer's Surplus'. This concept is based on cardinality approach of utility analysis and it gained significance in the Welfare Economics.

In the words of Marshall, *'excess of the price which a consumer would be willing to pay rather than go without a thing over that which he actually does pay is the economic measure of this surplus satisfaction ...it may be called consumer's surplus'*. According to Marshall, if the surplus increases for each and every individual then, the overall welfare of the economy increases. According to Penson, *'the difference between what we would pay and what we have to pay is called the consumer's surplus'*. In the words of Koutsoyannis, *'consumer's surplus is equal to the difference between the amount of money that consumer actually pays to buy a certain quantity rather than go without it'*. The essence of the concept of consumer's surplus is that, consumer, in general, derive more MU from the consumption of a commodity than the price he actually pay for it. So, he is prepared to pay more price to have the commodity, as he derives more MU compared to the actual price paid to have the commodity. For example, a person is anxious to see a movie and he is ready to pay Rs. 100 for the ticket. But, he actually paid Rs. 60 for getting the ticket. So, he get a surplus satisfaction of Rs. 40 over what he is willing to pay and what he actually pays, and this surplus satisfaction is called as Consumer's surplus. That means, the price a consumer pays to have a commodity or service does not always indicate the exact amount of satisfaction derived from it. So, by paying a lesser price to purchase the commodity than what he is prepared or willing to pay will give him a psychological feeling of amount being saved with him. This saving of money is in terms of mental satisfaction and it is derived by having the commodity at a lower price than the expected price. This psychological satisfaction on the part of the consumer constitute 'Consumer's surplus'. So, according to Marshall, consumer's surplus is a measure of the surplus satisfaction. It is, otherwise, defined as the difference between the 'price that one is willing to pay' and 'the price one actually pays' for the commodity under consideration. Alternatively, consumer's surplus is the benefit realized by the consumer by purchasing the commodity at a lesser price than he is willing to pay for it. The concept of consumer's surplus is the result of two important phenomena *viz.*, characteristic of consumer's behaviour and characteristic of market. The characteristic of consumer's behaviour is that, as he buys more and more of a particular commodity, the MU of the successive units begins to decrease. A rational buyer continuously purchases the commodity up to the unit, which equates his MU derived in terms of money of the commodity to the price he pays for it. The second phenomenon is that, there is perfect competition among sellers and a single price prevails in the market for a particular commodity at a particular time. The buyer is abie to get the first unit of the commodity at the same price as the second or pay any other unit thereafter.

13.2. Concept of a Consumer's Surplus

The concept of a consumer's surplus can be derived from the LDMU. As the consumer goes on consuming the commodity, the MU goes on diminishing for each and every successive unit of consumption. So, the consumer's willingness to pay for additional units of a commodity declines, because of the diminishing MU, as he possess more number of units of a commodity by that time. He then stops purchasing that commodity, when MU derived in terms of money equals with the price of the commodity. So, when MU starts diminishing, the gap between MU derived in terms of money and price paid for the commodity also diminishes, as the price paid for the commodity is given and constant at a point of time. So, this implies, the consumer's satisfaction both in terms of 'utils' (MU) and in terms of consumer's surplus (*i.e.*, margin between what he is willing to pay and what he actually pays) diminishes, if he goes on purchasing the commodity till the point of equilibrium.

Suppose, a consumer prefers to consume mangoes. Let the price of each mango is Rs. 2/unit and MU of money is constant at 12 utils/rupee. As discussed in the concept of LDMU, a person goes on consuming the commodity, as long as MU derived in terms of money is greater than the price paid for the commodity and he reaches equilibrium when equity is ensured between these two. Beyond this point, the consumer stops purchasing the commodity, as he derives less MU in terms of money compared to the price paid by him for the commodity. Since, the price of the commodity remains same at a given point of time, the decline in MU during the process of continuous consumption of commodity leads to continuous fall of surplus satisfaction realized by the consumer. This implies, the consumer's surplus fall continuously due to continuous decline in the MU derived from the additional units of commodity. That means, as more and more number of units of commodity are consumed, the consumer's surplus declines at the same rate as decline in MU of the commodity and finally it becomes zero, when MU derived in terms of money gets equal to price of the commodity. Beyond this equilibrium point, the consumer's surplus turns negative, as MU derived in terms of money is lower than the price paid for the commodity.

The analysis (Table 13.1) shows that, the consumer's surplus is maximum at the first unit of consumption of mango (1.33), as it gives the consumer more MU in terms of money (3.33), but he actually pays only Rs. 2/unit. So, he is prepared to purchase the commodity at Rs. 2/unit. This implies that, the consumer is prepared to pay Rs. 3.33 for the first unit of mango, but he actually pays Rs. 2.00 only. That means, the price the consumer is prepared to pay for the commodity is based on the MU derived from the commodity (40utils) or MU derived in terms of money (Rs. 3.33). This surplus satisfaction will decline gradually for the remaining units of the commodity due to the operation of the concept of LDMU. So, the consumer will not be prepared to pay Rs. 3.33 for the second mango, as it will give him lesser utility than the first. Let us imagine that, the second mango gives him lesser MU worth of Rs. 3.17, which the consumer is prepared to pay. The third mango gives still lesser MU, worth of Rs. 3.00 and the fourth mango gives MU worth of Rs. 2.33, the fifth mango gives MU worth of Rs. 2.17 and the sixth mango gives MU worth of Rs. 2.00. The consumer stops purchasing at the sixth mango because, the MU derived in terms of money at the sixth unit of mango is equal to the price of the mango/unit. So, the consumer's surplus becomes zero at the sixth unit of consumption of mango. So, the consumer will not purchase the seventh mango, as the MU derived in terms of money (Rs. 1.33) is less than price paid for the mango (Rs. 2.00). So, the consumer being rational, will not be prepared to pay Rs. 2.00 for the seventh mango from which he gets MU worth only Rs. 1.33.

Table 13.1: Marshallian approach of measurement of consumer's surplus for mango.

Number of Mangoes	*MU of Mangoes (Utils)*	*MU Derived in Terms of Money (MU of mangoes/MU of money) (Rs/unit)*	*Price/Mango (Rs/unit)*	*Consumer's Surplus (Rs/unit)*
1	40	3.33	2.00	1.33
2	38	3.17	2.00	1.17
3	36	3.00	2.00	1.00
4	28	2.33	2.00	0.33
5	26	2.17	2.00	0.17
6	**24**	**2.00**	**2.00**	**0.00**
7	16	1.33	2.00	-0.67
Total consumer's surplus (upto sixth unit of mango)		16.00 **(TU in terms of money)**	12.00 **(Total expenditure)**	**4.00**

The above analysis further indicate that, as the price of the mango is given and stable at a point of time, there is a decline in the surplus satisfaction or consumer's surplus due to continuous decline in the MU derived from the commodity due to the operation of LDMU concept. Accordingly, the consumer's surplus is maximum at the first unit of mango because, the consumer is prepared to pay Rs. 3.33 for that unit, as its consumption yields higher MU for the consumer compared to its price. Similarly, for the second mango, he is prepared to pay Rs. 3.17, but actually pays Rs. 2.00 and the extra benefit he gets is equivalent to Rs. 1.17. In the same way, a surplus benefit goes to the consumer in the purchase of the third, fourth and fifth mangoes to the tune of Rs. 1.00, Rs. 0.33 and Rs. 0.17 respectively, as the price he is prepared to pay for each unit is larger than the price he actually pays.

In the purchase of the sixth mango, the consumer derives no surplus satisfaction, as the price he is prepared to pay (Rs. 2.00) and the price he actually pays (Rs. 2.00) are the same. But, in purchase of the previous units of mangoes, there is surplus satisfaction on the part of consumer. The consumer definitely will not purchase the seventh mango, as the MU derived in terms of money (Rs. 1.33) is less than the price paid for seventh unit of mango (Rs. 2.00). Still, if he purchases seventh mango, the surplus satisfaction turns negative. So, invariably, the consumer stops consumption at sixth mango, as it is the point of consumer's equilibrium. This surplus satisfaction derived by the consumer in the consumption of successive units of a commodity is called as 'Consumer's surplus'. So, it can be defined as 'the excess what the consumer would be willing to pay for the commodity than the actual price paid for the commodity'. In the words of Taussig, '*the consumer's surplus is the difference between the potential price and actual price*'. The potential price is the price, what the consumer is prepared to pay to have the commodity, while the actual price is the price the consumer actually pays in the market to have the commodity. So, we can arrive at computing total consumer's surplus, by deducting the total expenditure spent on the commodity from TU gained from the commodity in terms of money. From the above analysis, we can arrive at consumer's surplus in the following ways:

Consumer's surplus = (What the consumer to willing to pay) – (What he actually pays) or
= (Price prepared to pay) – (Price paid) or
= (Σ MUs) – (Price x Number of units of a commodity purchased) or
= (Total utility) - (Total amount spent on the commodity) or
– Satisfaction - Sacrifice or
= Value to buyers - Amount paid by buyers

Diagrammatic Illustration

The concept of consumer's surplus is explained through the Figure 13.1. Along OX axis, the number of units of mangoes consumed and along OY axis, both price per unit of the mango and MU derived in terms of money are taken. It is evident that, MU curve (in terms of money) slopes downward due to continuous consumption of mangoes by the consumer. On the other hand, the price line PL remains parallel to X-axis, as the price of the commodity is given at a point of time. For the first unit of mango, the MU derived in terms of money is relatively higher (Rs. 3.33) than the price (Rs. 2.00) and hence, the consumer's surplus is Rs. 1.33. Similarly, for the second, third, fourth and fifth units of

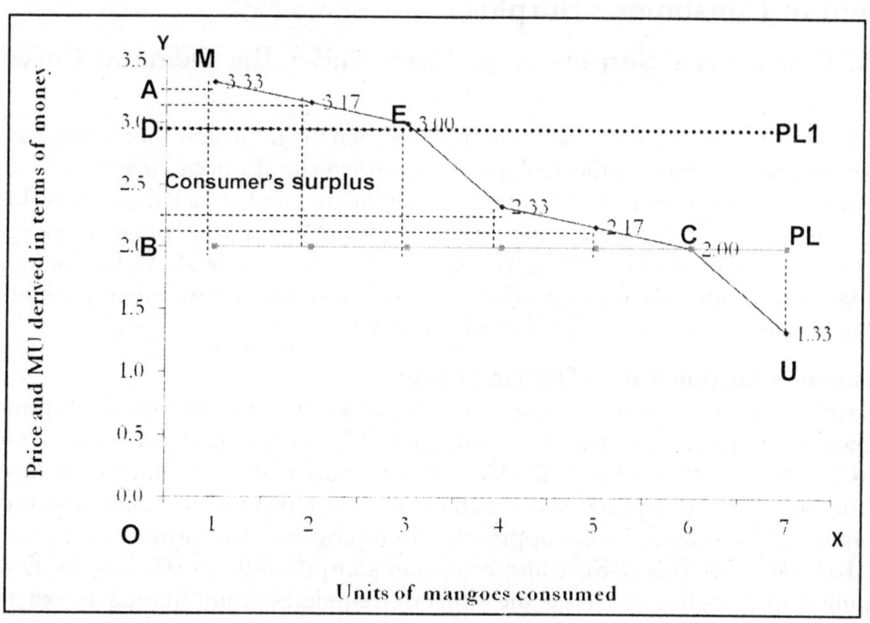

Figure 13.1: Marshallian approach of measurement of consumer's surplus.

mangoes, the consumer's surpluses are positive to the tune of Rs. 1.17, Rs. 1.00, Rs. 0.67 and Rs. 0.17 respectively. However, for sixth unit of mango, the consumer do not enjoy surplus satisfaction, as the MU derived in terms of money is equal to price paid for the commodity. So, the total surplus satisfaction enjoyed by the consumer is ABC. For interesting sake, even the consumption of seventh mango is shown though the figure, for which the surplus satisfaction turns negative (Rs. – 0.67). So, from this analysis, we can infer that, the concept of consumer's surplus is derived from the concepts of MU derived in terms of money and price of the commodity through the operation of the LDMU.

Changes in Consumer's Surplus due to Changes in Price and Quality of the Commodity

The consumer's surplus received from a commodity will change, if the price of the commodity and quality changes. For example, if price of the commodity increases, but people's willingness to pay remains the same, then the consumer's surplus will be less than before (Figure 13.1). If the price of the mango is increased to Rs. 3.00 (indicated by PL_1 price line), the total consumer's surplus is decreased to ADE. This is indicated by the fact that, for the first two units of mango, the consumer enjoys surplus satisfaction and he reaches equilibrium at third unit of mango. When compared to the earlier case (when price of mango is Rs. 2/unit), he misses the surplus satisfaction in the consumption of three more mangoes due to rise in price of the mango. This leads to decline in the total surplus satisfaction from ABC to ADE due to rise in price of the commodity. However, if quality of the commodity increases, but price remains the same, people's willingness to pay for the commodity may increase and this will increase the consumer's surplus.

13.3. Market Consumers' Surplus

The above analysis explains the concept of consumer's surplus with reference to a particular consumer in dealing with a particular commodity. So, individual consumer's surplus is the net gain from the purchase of a commodity and it is measured as the difference between the maximum price a consumer is willing to pay for a commodity and the actual price paid. On the same lines, we can extend the same concept to the entire market and it is referred as Market consumers' surplus. This gives the impression that, market consumers' surplus is the sum total of individual consumer's surpluses. However, it needs important assumptions like no changes in tastes, preferences, incomes of the consumers etc. Further, inter-personal differences among the consumers should also be ignored. Such assumptions will facilitate to calculate market demand and based on this, we can compute the market consumers' surplus. This can be calculated in the following two ways:

- By adding the consumer's surplus of each individual in the market
- By computing the consumer's surplus of a representative consumer and multiply his consumer's surplus with number of consumers in the market.

13.4. Assumptions of Consumer's Surplus

The concept of consumer's surplus involves the applications of two important concepts *viz.*, Marshallian utility analysis and Demand analysis of the commodity. So, the assumptions framed under these two concepts are equally applicable to the concept of consumer's surplus.

13.5. Measurement of Consumer's Surplus

1. Measurement of Consumer's Surplus as an Area Under the Demand Curve–Marshallian Approach

As discussed earlier, consumer's surplus is the difference between the total amount that the consumer is willing and able to pay for a commodity (indicated by the demand curve) and the total amount that he actually do pay. So, the ability of the consumer to pay for the commodity indicates his potential demand, which depends on his income and the actual amount the consumer pays for having the commodity depends on the market price of the commodity. So, the concept of consumer's surplus can be studied through demand curve analysis. In short, the triangular area formed between below the demand curve and above the equilibrium price line of the commodity represents the consumer's surplus. As shown in the Figure 13.2, the consumer's surplus is DAP.

Quantifying the Consumer's Surplus Under Demand Curve

Total consumer's surplus will be computed through summing up the benefits from each unit purchased under the MU curve (in terms of money) for the commodity. From the Figure 13.3, it is quite easy to compute consumer's surplus as 40+30+25+15+10+0 = Rs. 120. That means, below the MU curve, we add all the consumer's surpluses enjoyed by the consumer for different units of commodity purchased upto the price line. However, it is a labourious process. So, more commonly, Geometric approach is followed. In this approach, we employ the basic formula for estimating the area of a triangle *i.e.*, Area = 1/2 (Base)(Height). This is so because, consumer's surplus is the area below the demand curve (from above) and upto the equilibrium price line and it is in the form of a triangle. So, computation of area of the triangle will

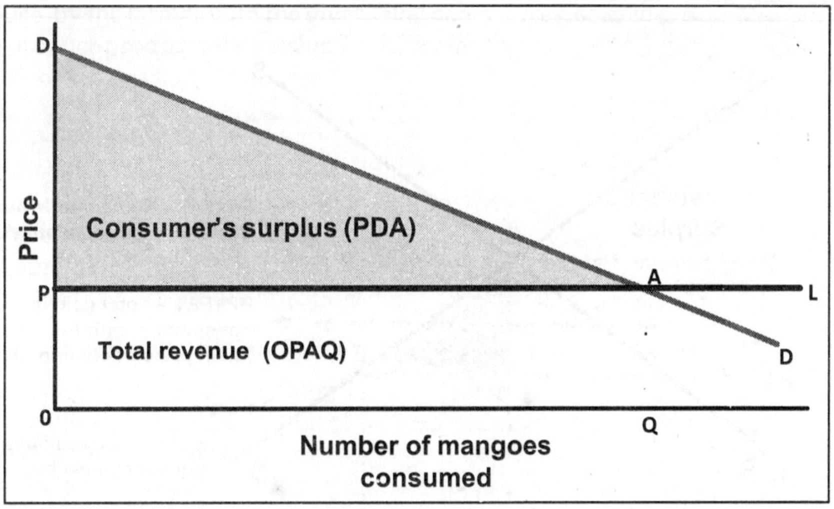

Figure 13.2: Consumer's surplus - Area below the demand curve and above the equilibrium price.

directly give the measurement of total consumers' surplus. In the Figure 13.3, base = 6, height = 40 and so, the total consumers' surplus is given by, 1/2 (6)(40) = Rs. 120. So, through Geometric approach, it is relatively easy to quantify the consumer's surplus. The major objection leveled against this Marshallian method of measuring the consumer surplus is that, the real income of the consumer does not remain constant along the demand curve even for unimportant commodities. Further, the demand curve is usually asymptotic to the price axis.

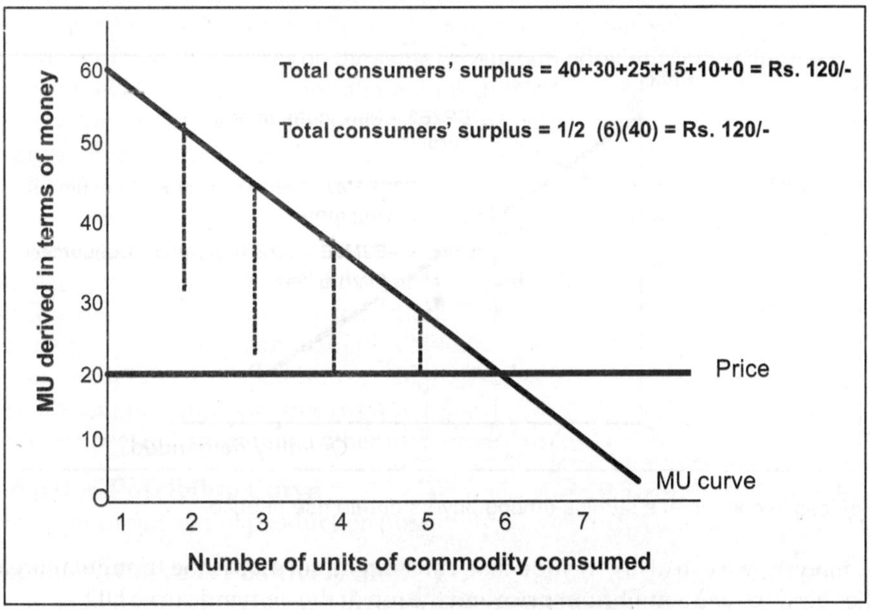

Figure 13.3: Measurement of consumer's surplus – Additive and Geometric approaches.

Share of Consumer's Surplus among the Consumers due to Change in Price of the Commodity

It is clear from the Figure 13.4 that, DD is the demand curve for the commodity under consideration, which is sloping downward from left to right in accordance with LDMU and SS is the supply curve for the commodity, which is sloping upward from left to right. Both demand and supply curves intersect at point E, which is the equilibrium point, thereby, indicating that, at OP level of equilibrium price, OQ is the quantity demanded and supplied in the market. So, the total amount the consumer paid to purchase the commodity is given by OQ x OP = area OQEP. But, the total amount of money that the consumer is capable or prepared to pay for OQ quantity of commodity is OQED. So, the consumer's surplus is OQED – OQEP = DEP. As explained earlier, the consumer's surplus increases with the fall in price of the commodity and *vice versa*. If the price of the commodity falls to OP_1, the consumer's surplus increases to DE_1P_1. So, the newly added consumer's surplus is PP_1EE_1. Of this newly added consumer's surplus *i.e.*, PP_1EE_1, PP_1EM_1 represents the increase in consumer's surplus to original buyers (consumers) and EE_1M_1 represent the consumer's surplus gained by the new consumers due to fall in price of the commodity.

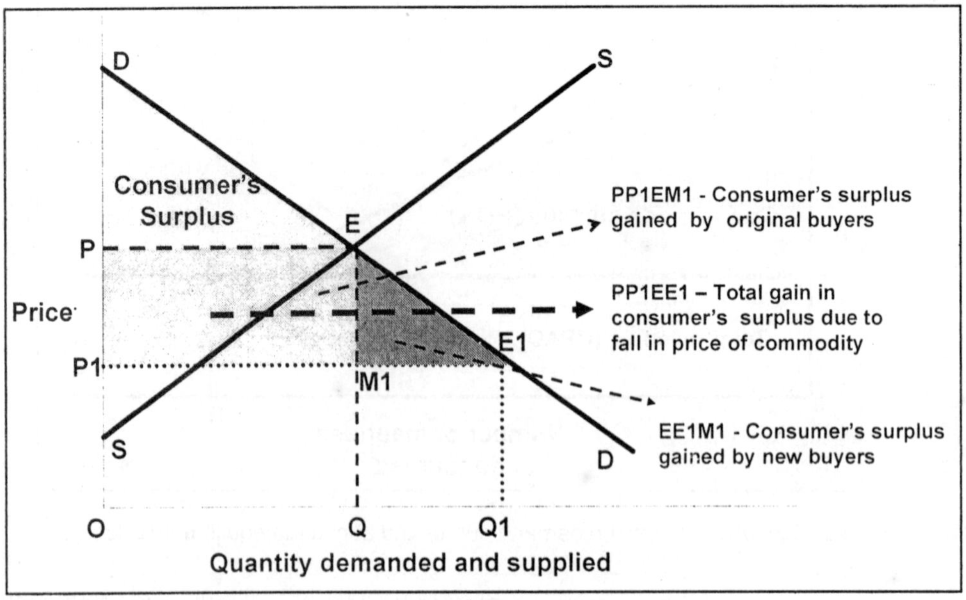

Figure 13.4: Splitting of gain in consumer's surplus among buyers due to fall in price.

On the other hand, due to rise in price of the commodity, the total loss in consumer's surplus is given by DPE – DP_2E_2 *i.e.*, P_2PE_2E (Figure 13.5). So, the new consumer's surplus is DP_2E_2. Of this total loss, *i.e.*, P_2PE_2E, the loss in consumer's surplus due paying more is $P_2PE_2M_2$ and loss in consumer's surplus due to buying less quantity of commodity is E_2M_2E.

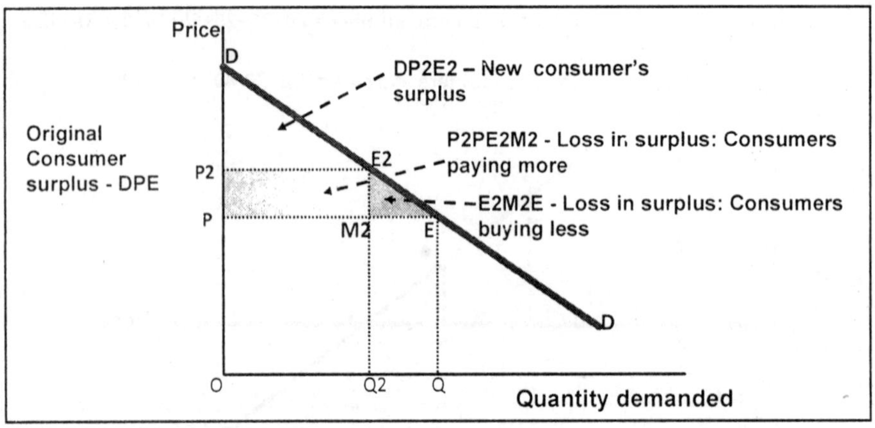

Figure 13.5: Splitting of loss in consumer's surplus among buyers due to rise in price.

From the above analysis, we can also conclude that, consumer's surplus is the triangular area formed below the demand curve and lies between the equilibrium price and the top of the demand curve DD.

Consumer's Surplus and E_D

From the above analysis, we concluded that, the consumer's surplus changes with change in price of the commodity. This helps to trace out the relationship between consumer's surplus and E_D, as demand is responsive to changes in price of the commodity. The following are the possible influences of different elasticities of demand on the consumer's surplus:

- When the demand for a commodity is perfectly elastic, the consumer's surplus is zero because, the consumers will actually pay for the commodity as much as they are prepared or willing to pay for that commodity. However, this is not so common, and this usually happen in highly competitive markets for luxurious goods.
- In contrast, when demand is perfectly inelastic, consumer surplus is infinite. This is because, whatever may be the price for the commodity, the consumer will purchase his requisite quantity (*i.e.*, there is no change in demand for the commodity) and so he is prepared to pay any amount of price for the commodity. This makes the consumer's surplus very large or infinite and this is most common in case of necessaries.

However, these two situations of perfect elastic and perfect inelastic demands are not common phenomena in the economy. In majority cases, the demand curve slopes downwards with relatively more elas icity or less elasticity and they are explained through the Figure 13.6. It is evident from the illustrations that, when the demand for the commodity is more elastic, the demand curve has flat slope and this leads to lower surplus satisfaction on the part of the consumer (Panel A). On the other hand, when the demand for the commodity is less elastic, the demand curve has steep slope and this leads to higher surplus satisfaction on the part of the consumer (Panel B). So, in the Figure 13.6, the consumer's surplus for less elastic commodities, DPE_1 is greater than consumer's surplus for more elastic commodities, DPE *i.e.,* $DPE_1 > DPE$.

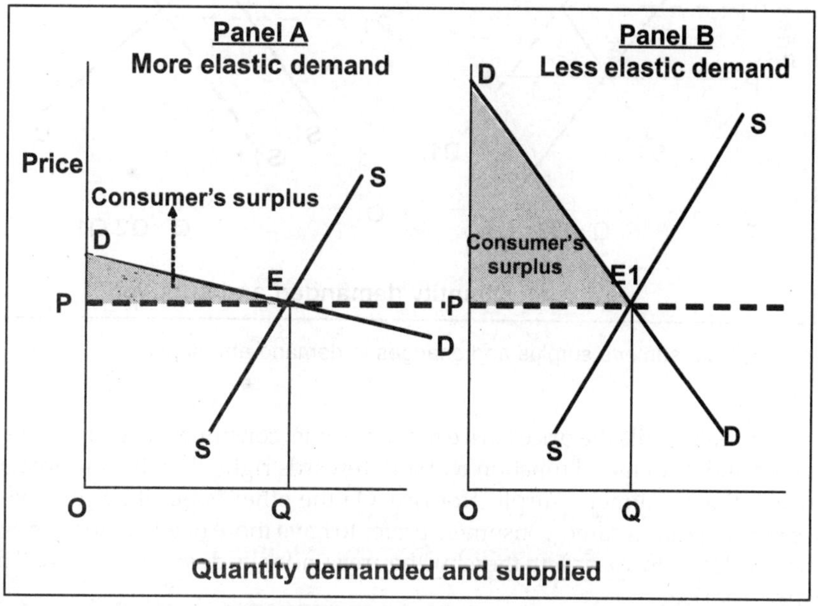

Figure 13.6: Relation between consumer's surplus and E_D.

Changes in Consumer's Surplus due to Changes in Demand and Supply

In the market economy, it is a common phenomenon that, both the demand and supply of commodities changes frequently and thereby, prices of the commodities fluctuates. So, it leads to changes in the consumer's surplus.

(a) Increase or Decrease in Demand and Changes in Consumer's Surplus

An increase in demand for the commodity implies shift in the demand curve towards right, not due to change in price of the commodity, but due to other factors like increase in income, increase in size of family members, change in fashion towards that commodity etc. In Panel A of the Figure 13.7, due to increase in demand for the commodity, the demand curve DD shifts towards right as D_1D_1. Due to this, more quantity of commodity *i.e.,* OQ_1 is brought into the market and traded at the same price OP. But, the new equilibrium price is OP_1, where OQ_2 quantity was traded in the market, which is higher than the quantity traded at the initial equilibrium price OP. But, since more quantity is purchased and consumed even at higher price, the consumer's surplus increases from DEP to $D_1E_1P_1$. That means, $D_1E_1P_1$ is greater than DEP. Conversely, when there is decrease in demand for the commodity, the demand curve shifts towards left side and it leads to fall in quantity purchased in the market and hence, the consumer's surplus will decrease.

(b) Increase or Decrease in Supply and Changes in Consumer's Surplus

In Panel B of the Figure 13.7, because of adoption of cost-effective production technology, the supply of commodity into the market increases and thereby, the supply curve SS shifts towards right as S_1S_1. So, more quantity of commodity *i.e.,* OQ_1 is supplied into the market at the same price, OP. But, the new equilibrium price is OP_1, where OQ_2 quantity of commodity was traded in the market, which is higher than the quantity traded at the initial equilibrium price OP. Since, more quantity is purchased and consumed at lower price OP_1, it leads to increase in consumer's surplus from DEP to DE_1P_1. That means, DE_1P_1 is greater than DEP. Conversely, when the supply of the commodity is declined in the market due to adverse seasonal and climatic conditions, supply curve shifts left side, thereby, price of the commodity will increase in the market. As a result, consumer's surplus will decrease.

Consumer's Surplus for Related Commodities

Consumer's surplus of a commodity is also affected due to the changes in prices of substitute goods and complementary goods. In case of substitute goods, if the price of substitute good changes, the consumer's surplus for the good in question will change in the same direction. For example, rice and wheat are good substitutes. If price of

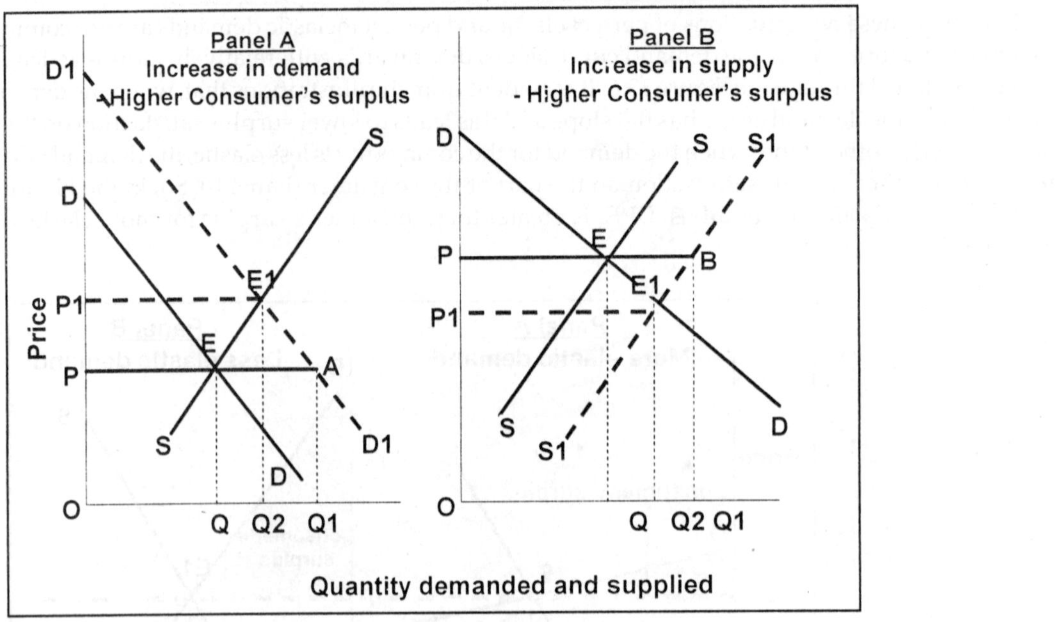

Figure 13.7: Relation between consumer's surplus and changes in demand and supply.

wheat is increased in the market, while the price of rice remains same, consumers prefer to have more of rice for wheat. So, more rice is demanded and its demand function will shift towards right, thereby, the area under its demand curve will rise, indicating rise in the consumer's surplus for rice. On the other hand, if price of wheat is decreased in the market, while the price of rice remains same, consumers prefer to have more of wheat for rice. So, less rice is demanded and its demand function will shift towards left, thereby, the area under its demand curve will decrease, indicating fall in the consumer's surplus for rice.

In case of good complements like bread and butter, if the price of complementary good changes, the consumer's surplus for the good in question will change in the opposite direction. For example, if price of butter is increased in the market, consumers buy less of both bread and butter. If less bread is demanded, then the demand function shifts downward *i.e.*, towards left side or towards origin, and hence, the area under its demand curve decreases, indicating fall in consumer's surplus for bread. On the other hand, if price of butter is decreased in the market, consumers buy more of both bread and butter. If more bread is demanded, then the demand function shifts towards right side or away from the origin, and hence, the area under its demand curve increases, indicating rise in consumer's surplus for bread.

2. Measurement of Consumer's Surplus through IDC Analysis

The Marshallian approach of measurement of consumer's surplus discussed above suffers from two important limitations *viz.*, cardinal measurement of utility and MU of money is assumed constant. Prof. Hicks and Prof. Allen criticized Marshallian view on these grounds, as they are not valid assumptions. So, they considered Marshallian approach of measurement of consumer's surplus as the most unsatisfactory method. According to them, 'utility' being subjective and a psychological phenomenon, it cannot be measured cardinally. They also opined that, MU of money will not remain constant, if the consumer goes on spending the money in consuming different units of the commodity or different commodities. If it is assumed constant, the income effect cannot be estimated. However, Marshall defended this criticism stating that, the expenditure of a consumer on a particular commodity accounts for only a meager share in his total income. But, according to Hicks and Allen, this is not true in case of all commodities and thereby, they rejected this assumption. So, both Hicks and Allen explained the concept of consumer's surplus through IDC technique, which was proposed to study utility analysis on ordinality grounds and without assuming the constancy of MU of money. So, this method of explaining the concept of consumer's surplus by using IDC approach is superior compared to Marshallian approach and therefore, it went a long way in establishing the validity of concept of consumer's surplus. Hicks measured the Marshallian consumer's surplus through applying IDC technique in two ways *viz.*, by assuming constant MU of money and by assuming diminishing MU of money.

(a) Hicksian Approach of Measuring the Marshallian Consumer's Surplus by Assuming Constant MU of Money

In the Figure 13.8, on X-axis, commodity X and on Y-axis, money are measured. Let us suppose, the consumer possess OP amount of money. Further, we suppose that, the consumer is not aware about the price of the commodity X at present. For the given level of income, the consumer wishes to purchase a desired quantity of commodity, which is

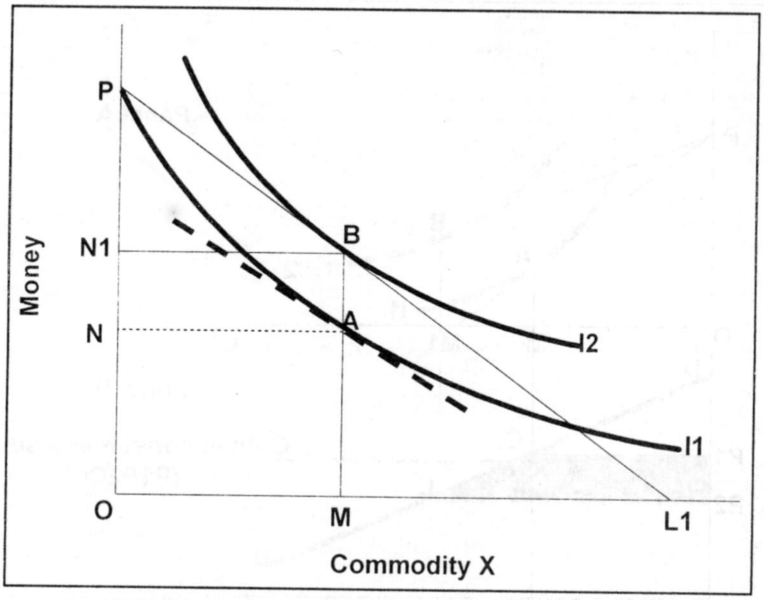

Figure 13.8: Hicksian approach of measuring the Marshallian consumer's surplus (Assuming constant MU of money).

indicated by point A on the IDC I_1 originating from point P. So, in choosing the point A on the IDC I_1, the consumer wishes or prefers to purchase OM quantity of commodity X and he is ready to sacrifice or spend PN amount of money. This point A on IDC I_1 infers the consumer's willingness to pay for the commodity, as this point A lies on the same IDC as point P, the initial position of IDC I_1, where he does not consume any amount or quantity of commodity X.

Now, let us suppose, the consumer is informed about the price of the commodity X. By knowing the price information about the commodity X, it helps to draw his price line to attain the consumer's equilibrium, as the slope of the price line indicates the price of the commodity (please, refer price line in IDC technique). As shown in the Figure 13.8, the price line is tangent with the higher IDC I_2 at point B, and this implies that, the consumer is at equilibrium by purchasing the same quantity OM, but by actually paying only PN_1 amount of money. So, finally, the analysis indicate that, to purchase OM quantity of commodity, the consumer is willing to pay PN amount of money, but he actually pays only PN_1 amount of money. So consumer's surplus is given by $PN-PN_1=N_1N$. It is important that, in explaining this concept, Hicks assumed vertically parallel IDCs *i.e.*, I_1 and I_2 and their slopes at point A and B are same indicating that MU of money is kept constant for analyzing the consumer's surplus.

The above explanation of consumer's surplus through IDCs can also be extended to a situation when the price of commodity X changes. Figure 13.9 shows the extension of Figure 13.8, when the price of commodity X changes and this also enables us to derive the demand curve for the commodity X. Panel A of the Figure 13.9 shows the indifference map. Here, notice that, the shift in price (budget) line away from the origin, is due to fall in price of commodity X from OP_1 to OP_2. As a result, the consumer's equilibrium was shifted from A to B, thereby, making the consumer to purchase more quantity of commodity X (OM_1>OM). To represent the influence of price change on the quantity demanded of commodity X, the equilibrium points of Panel A were extended to Panel B, wherein, on Y-axis, price of the commodity is taken and X-axis remain same *i.e.*, quantity of commodity X. This is because, the slope of the price line represent the price of the commodity and this price remains same through out the slope of the price line. So, the corresponding points of A and B of the Panel A will now become C and E in Panel B and if these points are joined, it gives a downward sloping demand curve DD in the lower Panel B. This demand curve now helps to measure the change in consumer's surplus due to change in price of the commodity. When price of the commodity X is declined to OP_2 from OP_1, the consumer's surplus increased to DP_2E from DP_1C, thereby, the net gain in consumer's surplus is P_1P_2CE.

(*b*) Hick's Approach of Measuring the Consumer's Surplus by Assuming Diminishing MU of Money

The common practical situation is that, the MU of money changes inversely with the amount of money income. Hicks applied this concept of diminishing MU of money to analyze the Marshallian consumer's surplus. In the Figure 13.10, on X-axis, commodity X and on Y-axis, money are measured. For the given level of money income *i.e.*, OP, when the consumer is not aware about the price of the commodity X at present, he express his preference to purchase a desired quantity of commodity, which is indicated by point A on the IDC I_1 originating from point P. So, in choosing the point A on the IDC I_1, the consumer wishes or prefers to purchase OM quantity of commodity X and he is ready to sacrifice or spend PN amount of money.

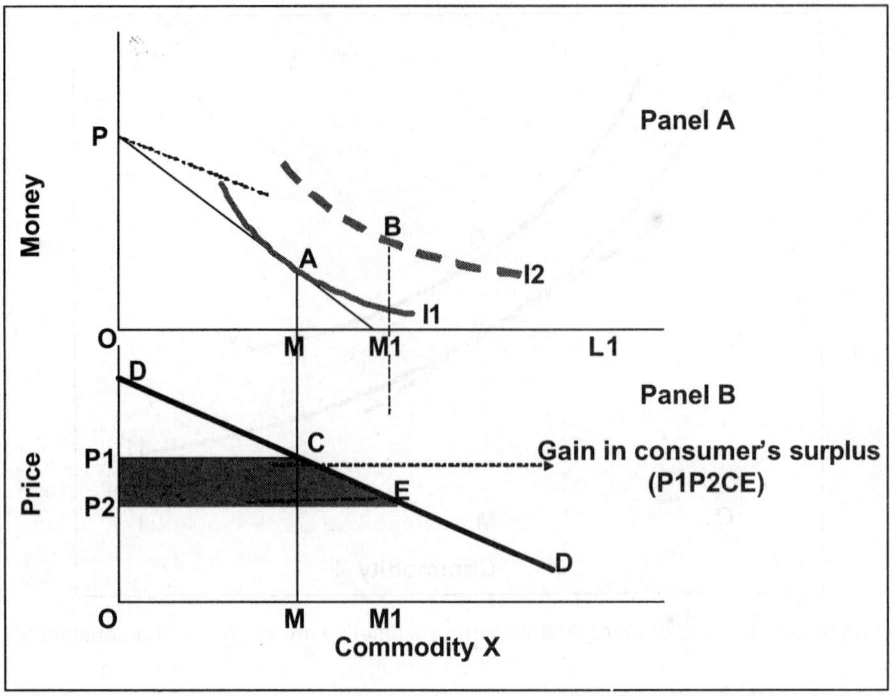

Figure 13.9: Hicksian approach of measuring the Marshallian consumer's surplus when price of the commodity changes.

Now, when the consumer is aware about the price of the commodity X, it helps him to draw his price line, so as to attain the consumer's equilibrium. This price line is found to be tangent to higher IDC I_2 at point B and this implies that, the consumer is in equilibrium by purchasing the same quantity OM, but by actually paying only PN_1 amount of money. It is important that, the slope of IDC I_2 is less flatter compared to the slope of IDC I_1, as here, we assumed diminishing MU of money. So, on IDC I_2, the consumer will be in equilibrium at point B, where he purchases the same OM quantity of commodity, but he actually pays only PN_1 amount of money. So, consumer's surplus is given by PN-PN_1= N_1N.

In order to differentiate between Hicksian approach and Marshallian approach of measurement of consumer's surplus, another IDC I_3 is drawn originating from point P. As discussed under Marshallian approach, which assumes constant MU of money, this new IDC I_3 lies vertically parallel to IDC I_2 and the slopes of these two IDCs *i.e.*, I_2 and I_3 are equal. This is clear from the dotted line drawn parallel to PL_1. The corresponding point of M on the IDC I_3 is 'C', indicating that, now to purchase the same OM quantity of commodity, the consumer is willing to spend PN_2 amount of

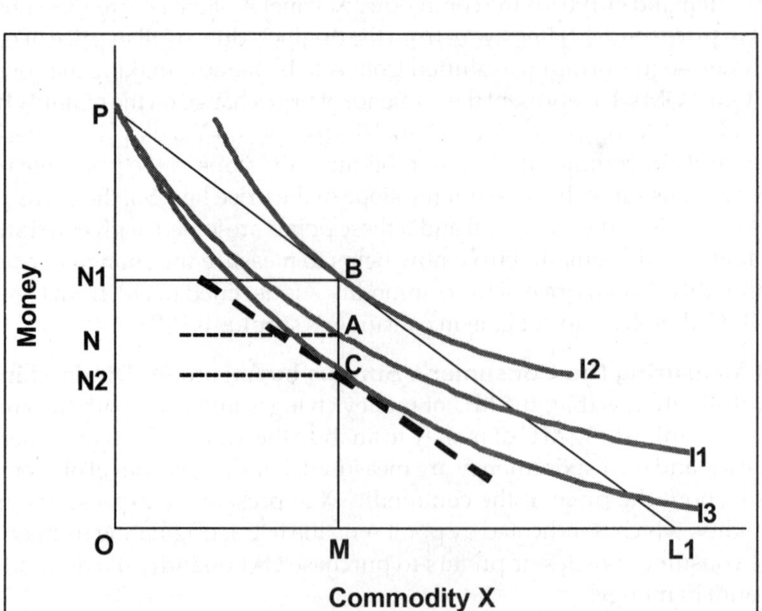

Figure 13.10: Hicksian approach of measuring the Marshallian consumer's surplus (Assuming diminishing MU of money).

money, when MU of money is assumed to be constant. So, IDC I_3 shows the consumer is willing to pay PN_2 amount of money for the commodity when MU of money is assumed constant and IDC I_2 shows the consumer actually pays PN_1 amount of money for purchasing same quantity of commodity X *i.e.*, OM. So, the consumer's surplus is $PN_2-PN_1=N_1N_2$.

Thus, the analysis indicate that, Marshallian consumer's surplus N_1N_2 is larger than Hicksian consumer's surplus N_1N, as Hicksian methodology assumed diminishing MU of money. But, this Hicksian's methodology of consumer's surplus is superior to Marshallian's approach, as it assumed diminishing MU of money.

(c) Hicks's Four Consumer's Surpluses when the Price of Commodity X Changes

After ordinalism was introduced to Economics, the concept of consumer's surplus is dealt in terms of interpersonal comparison of utility. Since, utility is not measurable, J.R. Hicks re-defined the whole concept of consumer's surplus with the help of IDC analysis. This concept of consumer's surplus received special attention from Hicks, as he laid useful foundations in the field of Welfare Economics. According to him, pure laissezfaire economies does not necessarily lead to an optimum structure of production and consumption. So, it is essential to formulate sound economic policies, which would be more effective and yield or maximize the net economic benefits. If the aggregate benefits in the economy are so large that can compensate the losses, thereby, leaving surplus benefits, it will contribute to a net social advantage. So, in this context, it is not essential to deal with aggregate utility, rather, we can simply focus for a size of compensation, the individuals would be willing to pay/obtain to cover their benefits/losses. In this context, Hicks defined consumer's surplus as, '*the measure of the compensation which consumers would need in order to maintain them at the same level of satisfaction as before, after the supply of the commodity had been withdrawn*'.

Hicks criticized that, the Marshallian measure of consumer's surplus is not a true measure of the benefit of a price change, as it holds income constant. But, according to Hicks, for a true measure of welfare change, it is welfare (utility) that needs to be held constant. The issue then is what the reference point is for holding welfare constant. Hicks then identified four different measures which could reasonably serve for evaluation of a market change and they include:

- Price compensating variation or Compensating variation
- Price equivalent variation or Equivalent variation
- Quantity compensating variation or Compensating surplus
- Quantity equivalent variation or Equivalent surplus

In a given situation (Figure 13.11), for the given level of income OP and for a given price of the commodity P_1, the price line is PL_1. The consumer will be in equilibrium at point A on IDC I_1 curve, where he purchases OM_1 quantity of commodity by spending PN amount of money. If price of the commodity is decreased to P_2 in the market, his price (budget) line shifts towards right as PL_2 and he moves to higher IDC I_2, and he is in equilibrium at B, where he purchases OM_2 quantity by spending PN_1 amount of money.

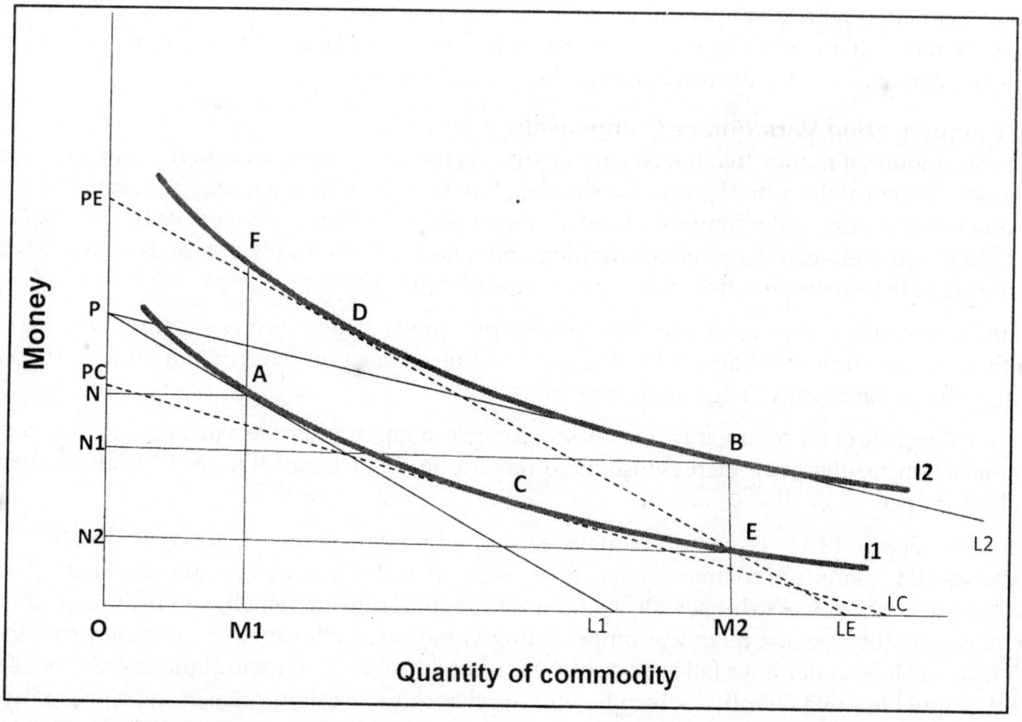

Figure 13.11: Hicks – Four Consumers' surpluses.

(i) Price Compensating Variation or Compensating Variation

When price of the commodity is decreased in the market then, the consumer feels he is better-off due to price decrease than without it. Then, compensating variation is the maximum sum that could be taken away from the consumer, so that, he is indifferent between the post-change (new) situation and the pre-change (original) situation. The reference point is the original level of welfare *i.e.*, on the original or initial IDC.

On the other hand, if price of the commodity is increased in the market, then the consumer becomes worse-off due to price increase than without it. Then, compensating variation is the compensation required by the individual to make him indifferent between the new and old (original) situations. The reference point is again the original level of welfare *i.e.*, on the original or initial IDC.

In the Figure 13.11, let us study, how much income could be extracted or removed or taken away from the consumer in order to compensate the price fall and leave the consumer to stay on the original IDC I_1. This is given by, PP_C. This is because, if we remove this income PP_C, the new budget line is P_CL_C and it touches the original IDC I_1 at C. So, with income P_C and at decreased price P_2, now the consumer is back onto the original IDC I_1 at C. This PP_C is called price compensating variation.

(ii) Price Equivalent Variation or Equivalent Variation

When price of the commodity is decreased in the market, then the consumer will be better-off with price fall than without it. So, equivalent variation refers to the sum of money that would have to be given to the consumer in the original situation, instead of change in price, so that, it makes him well off, as he would be in the new situation. The reference point is the level of welfare in the new situation or new IDC. So, Equivalent variation for a price fall is the minimum willingness to accept the money by the consumer to forego the price fall.

Suppose, price of the commodity is increased in the market. So, the Equivalent variation refers to the consumer's willingness to pay the money to avoid the price increase, *i.e.*, to avoid the decrease in welfare that would arise in the post-change situation. The reference point is the level of welfare in the new situation or new IDC. So, Equivalent variation for a price rise is the maximum willingness to pay by the consumer to avoid the price rise.

In the Figure 13.11, the minimum payment that is to be made to the consumer or required by the consumer in his original situation (without change in price), so that, it makes him well-off, as he would be in the new equilibrium situation on higher IDC I_2 is PP_E. That means, this amount PP_E should be added to the consumer's income, so that, the new budget line is P_EL_E and thereby, he will move to higher IDC I_2 and comes to equilibrium at D in the absence of change in price of the commodity. That means, the consumer is at new equilibrium D, when the price is same at P_1. Since, there is no change in price, he is in equilibrium at D instead of B on IDC I_2, where equilibrium B is due to price change from P_1 to P_2. So, the consumer is well-off now, as he is on high IDC I_2, but at old price P_1, and reaches equilibrium at D, instead of B.

So, the difference between Price compensating variation and Price equivalent variation is that, Price compensating variation measures relate to a context in which the price change takes place, whereas, Price equivalent variation measures relate to a context in which the price change does not take place.

(iii) Quantity Compensating Variation or Compensating Surplus

It refers to the amount of money that the consumer would be willing to pay to benefit from a fall in price of the commodity, so as to remain at the initial level of satisfaction. But, he is constrained to buy the same quantity which he will purchase due to fall in price of the commodity. So, Compensating surplus, is defined as the sum that would make the individual indifferent between the original situation and a new situation in which he is constrained to buy the quantity of commodity in the new situation (when price is decreased).

If price of the commodity is decreased, compensating surplus refers to the measure of the willingness to pay by the consumer, so that, the consumer will derive the original level of satisfaction (*i.e.*, remain on the original IDC), but purchase the quantity of commodity at declined price.

On contrary, when price of the commodity is increased, compensating surplus refers to the measure of the willingness to accept compensation for the price increase, so as to remain on the original IDC, but purchase the quantity of commodity at higher price.

As shown in the Figure 13.11, assume the consumer is at original position *i.e.*, at point A on IDC I_1, where the consumer purchases OM_1 quantity of commodity at price P_1. So, with the fall in price of commodity to P_2, the new budget line is PL_2, and the consumer moves to higher IDC I_2 and reaches equilibrium at point B, where the consumer purchases OM_2 quantity of commodity. So, the quantity compensating variation *i.e.*, the amount of money that the consumer would be willing to pay to benefit from a fall in price of the commodity, so as to remain at the initial level of satisfaction, but constrained to buy the same quantity, which he will purchase due to fall in price of the commodity is BE. That means, at B, the consumer can pay BE sum of money and return to original IDC I_1 at point E, but purchases OM_2 quantity of commodity *i.e.*, at declined price of the commodity.

This component BE *i.e.*, N_1N_2 also represents the Marshallian consumer's surplus. This is because, PN_2 is the total expenditure, which the consumer would be willing to pay rather than go without the thing (or, in this case, rather than he pay price P_1 and go to the point A), but he actually pays PN_1 at the point B, thereby, the Marshallian consumer's surplus is N_1N_2. But, according to Hicks, compensating surplus would not be of much importance to Welfare Economics, as it does not primarily concentrate on rationing.

(iv) Quantity Equivalent Variation or Equivalent Surplus

It is the sum that would make the consumer indifferent between the new situation (with the price change) and the old situation (original price), but the individual is constrained to buy the original quantity of commodity in the original or initial (price) situation. So, if the price of the commodity is decreased, quantity equivalent variation implies the minimum compensation that the consumer is willing to accept, in order to forgo the benefit or opportunity to purchase the commodity at lower price. But, he is constrained to purchase the initial quantity of the commodity at the original price or before the fall in price of the commodity. On contrary, if price of the commodity is increased, quantity equivalent variation implies the minimum compensation that the consumer is willing to pay to avoid purchasing the commodity at higher price and he will buy the original quantity at the original price or before the increase in price of the commodity.

In the Figure 13.11, AF is the quantity equivalent variation that the consumer is willing to accept, so that, he will forgo the benefit or opportunity to purchase the commodity at lower price and he is constrained to purchase the initial quantity of the commodity *i.e.*, OM_1 at the original price P_1 or before the fall in price in price of the commodity. So, this quantity equivalent variation will leave the consumer on a higher level of satisfaction (IDC I_2).

On the other hand, if we analyze the four consumer's surpluses due to increase in price of the commodity, the initial point is now B, the compensating measures becomes equivalent measures and *vice versa*. So, the following are the conclusions:

- PP_C now becomes Price equivalent variation.
- PP_E now becomes Price compensating variation.
- BE now becomes Quantity equivalent variation or Equivalent surplus.
- AF now becomes Quantity compensating variation or Compensating surplus.

13.6. Difficulties in the Measurement of Consumer's Surplus

The concept of consumer's surplus has subjected to lot of criticism than any other concept in Utility analysis. This concept was vehemently criticized by Economists like Nicholson, Cannon, Robinson, Davenport etc., on the grounds of unrealistic assumptions, which makes its measurement inaccurate. The main points of criticism are discussed here under.

- ***Senseless concept:*** The concept of consumer's surplus is based on LDMU. This LDMU concept is based on cardinal measurement utility, which is subjected to severe criticism, as measurement of utility is purely subjective and moreover related to psychological phenomenon of the individual. So, it is totally meaningless to study the concept of consumer's surplus, as the basic assumption on which this concept was framed was subjected to severe criticism. Further, it is totally senseless to measure the extra satisfaction derived by the consumer, even the realized satisfaction itself have no validity under cardinal terms.
- ***Purely hypothetical:*** Several Economists criticized this concept as hypothetical and imaginary. We know a consumer purchases different goods and services in the economy. If we summate the 'prices' at which the consumer is willing to pay for all the goods and services that he is intended to purchase, it may exceed the consumer's income. But, a consumer cannot afford to pay for the commodities more than his income. From this view point, there will not be any consumer's surplus. However, in a practical situation, the amount meant for expenditure on goods and services is limited to the extent of money income earned by the consumer.
- ***Commodities having prestige value:*** For example, commodity like diamond has high prestige value and they are purchased by very few rich people. If the diamonds becomes cheap in the market, the demand will not increase in the economy and even rich people may not prefer them, as they have lost the prestige value. So, when diamond price is reduced, consumer's surplus will not increase, but decreases.
- ***Not applicable to necessaries of life:*** In case of necessaries, the concept of consumer's surplus is not applicable and considered as infinity or immeasurable. This is because, for fulfilling necessaries of life, the people try to give up anything, if it is a must. For example, to save his life from a dreaded disease, a millionaire is ready pay any amount to the doctor. After his recovery from the disease, if the doctor charges Rs. 50000, it does not mean consumer's surplus is his total wealth minus doctor's fees worth of Rs. 50000. This sounds rather absurd. Since, the individuals will be prepared to pay any amount to fulfill the basic necessaries of life, it is not possible to compute consumer's surplus. Further, in case of necessaries, consumption only removes pain and there will not be any surplus satisfaction.

☆ ***Differences in consumer's circumstances:*** We know, rich people are capable of paying large amounts to have the commodity compared to poor people. So, the consumer's surplus will be high for rich people and low for poor people even for the same commodity. Moreover, the consumers differ in their tastes and sensibilities and hence, the consumer's surplus differs from person to person for the same commodity. So, this makes the concept absurd.

☆ ***'Independent utilities of commodities' is not valid:*** This assumption is not valid, as the utility derived from a commodity not only depends on its supply, but also on the supply of other related commodities. For example, there are two commodities *viz.*, rice and wheat. The MU derived from rice not only depends on its quantity supplied, but also on the quantity of wheat supplied in the market. Say, if the supply of rice is increased in the market, its MU decreases and hence, its price decreases. Consequently, the MU of wheat will rise in the market. So, no commodity can be considered as absolutely independent.

☆ ***Existence of substitutes***: The definition of consumer's surplus is purely misleading. It states the concept as, the consumer is 'willing to pay for a commodity rather than go without it'. But, this is not the case in practical situation, as there are many number of substitute commodities and hence, the concept 'rather than go without it' will not arise at all. Moreover, if there are substitute commodities, the concept of willingness to pay a high price for a particular commodity also will not arise. However, in the absence of substitutes for the commodity, the consumer's surplus will be very high. For example, if only rice is available in the market, the consumer is prepared to pay any amount for this commodity to fulfill his need. In such case, the consumer's surplus will be very high. However, if substitutes like wheat, jowar etc., are available in the market, the consumer would not have prepared to pay high price for rice, rather he switch over to wheat or jowar for consumption. So, the consumer's surplus for rice will decrease in the presence of substitutes. This implies, the magnitude of consumer's surplus derived from a commodity depends upon the availability of its substitutes. So, we cannot say that, consumer's surplus is a definite concept. To overcome this difficulty, Marshall clubbed all the commodities together and considered as one single commodity.

☆ ***Not applicable to Giffen goods:*** This concept of consumer's surplus is not applicable to Giffen goods, as they do not obey the law of demand.

☆ ***Potential prices are not known***: It is not possible to know the potential prices for each and every commodity brought and sold in the economy. So, it makes the concept of consumer's surplus purely imaginary. Further, the potential price varies with respect to person, commodity, time, place etc.

☆ ***MU of money is not constant:*** If the consumer goes on spending the money, MU of money will not remain constant. But, Marshall combated this argument saying that, in actual practice, an individual spends an insignificant portion of his income on the purchase of commodity and therefore, the change in MU of money is almost negligible.

☆ ***Zero consumer's surplus:*** Prof. Gobbi criticized the Marshall's concept of consumer's surplus. According to him, the consumer will decide to pay for the commodity after considering the utilities of all the units contained in that quantity of commodity. So, there will be no difference between 'willing to pay' and 'actual pay' for the commodity. So, the consumer's surplus will be zero and hence, this concept will not arise.

☆ ***Not useful to measure welfare:*** According to Prof. Nicholson, the concept of consumer's surplus is not useful to measure economic welfare of the people. Say, if prices of commodities in England are lower than in South Africa, it does not mean, economic welfare is greater in England compared to South Africa. This is because, income levels of the consumers across the two countries greatly differ. So, it is absolutely absurd to compare the welfares of people across the countries based on consumer's surplus.

☆ ***Utility of earlier units changes:*** It is a known fact that, if we go on consuming the commodity, different units of the commodity gives different levels of satisfaction. That means, MUs of different units of the commodity successively decrease, if the consumer continuously consumes the same commodity. Based on this concept, Marshall computed the consumer's surplus. At the margin of consumer's equilibrium, MU derived in terms of money of a unit of commodity is equal to the price of the commodity. So, for the previous units, intra–MU in terms of money is higher than the price and so on these intra–marginal units, the consumer obtains consumer's surplus. At this point, the concept of consumer's surplus is subjected to severe criticism, in the sense that, when a consumer goes on consuming a commodity, it is not only the utility of the marginal unit that declines, but also the MUs of all previous units of the commodity consumed also declines. This is because, all the units of a commodity are assumed alike and hence, all would have the same utility and behave alike. So, at the margin, when price is equal to MU derived in terms of money of the last unit purchased, the same will also be applicable to the earlier units of the commodity and therefore, the consumer will not get any consumer's surplus. This decrease in the utility of earlier units is not taken into account while calculating the consumer's surplus. So, the utility schedule will not be as before. It has to be changed every time when we make a purchase or consume all the units of a commodity. However, even this criticism is also not valid. This is because, the MUs of successive units of consumption of a commodity declines in accordance with LDMU concept. So, if the concept of consumer's surplus is criticized on the lines of equi-MU contributed each and every additional unit of the consumption of commodity, LDMU concept is also not valid. Since, the LDMU forms the base for the concept of consumer's surplus, its validity counts much in its application to the concept of consumer's surplus.

- ☆ *Neglects income effect:* According to Hicks, the concept of Marshallian consumer's surplus neglected the income effect of the change in price along the demand curve. But, Marshall justified his measurement stating that, the price changes are very small, so that, the income effect can be neglected. But, however, according to Hicks, income effect should be considered and this will lead to four types of consumers surpluses *viz.*, Price compensating variation, Price equivalent variation, Quantity compensating variation and Quantity equivalent variation.

Thus, the concept of consumer's surplus is based on unrealistic assumptions and hence, it is criticized as hypothetical, imaginary and illusory. However, it cannot be said that, it is totally incorrect and useless concept, as we cannot deny the extra surplus that we enjoy at times of price fall and better services availed in the economy. However, the exact measurement is not possible.

13.7. Practical Importance of the Concept of Consumer's Surplus

Consumer surplus is a concept that appears in many fields of economic analysis and its importance is discussed here under.

- ☆ *To compare economic welfare or conjectural advantages:* The concept of consumer's surplus helps us to compare the economic conditions and standard of living of the people living at different places. The greater the consumer's surplus, the greater the economic welfare. On this ground, the person living in a town or city enjoys more consumer's surplus than a person living in a village, as the person living in town enjoys more amenities at a very low cost.
- ☆ *Show advantages derived from falling prices:* If the price of the commodity decreases in the market without any change in the income of the consumer, consumer's surplus increases. The consumer then spends this 'saved' money on the same commodity and on other commodities and thereby, the sum total of his satisfaction from all the commodities will increase, atleast, to the extent of the increase in consumer's surplus on the commodity, whose price has declined. In this context, the concept of consumer's surplus serves as a useful tool in Welfare Economics.
- ☆ *Determination of monopoly price:* Monopolists often take advantage of consumer's surplus when setting prices for the products. If a monopolist can identify different groups of consumers in the market economy, who are willing and able to pay different prices for the same product, then he may engage in price discrimination. This price discrimination enables the monopolist to extract consumer's surplus as extra revenue. That means, if a consumer enjoys more surplus satisfaction on the commodity, it indicates the monopolist to raise the price of the commodity, such that, the actual price of the commodity matches the price at which consumer shows willingness to have the commodity, thereby, leaving no surplus to the consumer. However, in practice, the monopolist will not fix such a high price because, it may dissatisfy the consumers. But, if the commodity has inelastic demand in the market, the monopolist will quote a high price for the commodity, as it will not affect the quantity demanded by the consumers.
- ☆ *Importance in Public finance:* The concept of consumer's surplus is of great importance in the field of public finance, particularly in the field of taxation. A Finance Minister of a country will impose higher taxes on the commodities, for which the consumers enjoy more consumers' surplus. So, he carefully selects the commodities for imposing higher taxes, if the commodities have inelastic demand and at the same time yields higher consumers' surplus. Further, this concept is useful in deciding, whether imposing a direct tax or an indirect tax is better for the individuals in the economy. Based on the concept of consumer's surplus, we can prove that, imposing direct tax involves less sacrifisation on the part of consumers compared to indirect tax, thereby, imposing direct tax on the consumers is more beneficial to the consumers. This is because, imposition of direct tax involves loss of income on the part of the consumer and this loss of income will be distributed over various items of expenditure and hence, the consumer loses only a marginal unit on each commodity and consequently, he loses only a small amount of MU. On the other hand, due to imposition of indirect tax, the impact falls entirely on all the commodities, so that the consumer sacrifices not only the marginal unit but also some intra-marginal units. So, the total loss of satisfaction is greater in the case of an indirect tax than a direct tax. However, this concept is discussed earlier under IDC analysis and it is more valid, as IDC analysis assumed ordinality approach for utility measurement.

- *Granting subsidies:* This concept also guides the Government in granting subsidies to the firms or industries or individuals in the economy. We know, the Government sanctions subsidy to the beneficiaries, so as to promote extra satisfaction. This concept guides the Government in such a way that, the resulting gain to the beneficiary due to grant of subsidy should be greater than the loss to the Government on granting the subsidy. On the other hand, if the burden is greater than the gain, it implies loss to the Government in granting subsidy to the beneficiaries.
- *Evaluating the investment projects:* This concept helps in studying the cost-benefit analysis of project investments. If the benefits derived from the project investment leads to higher consumer's surplus, it implies the investment is fruitful and *vice versa.*
- *Importance in international trade:* The concept of consumer's surplus has more applicability in the field of international trade. In general, a country will export or import a commodity based on the principle of comparative advantage. Say, for example, a commodity X is produced in India at a cost of production of Rs. 1000/quintal and the same is produced in USA at Rs. 600/quintal. So, it is better for India to import the commodity from USA, as the people in India enjoy more consumers' surplus from importing the commodity than producing at home. The difference between the price that would have been paid if commodity is produced at home and price paid on that commodity if imported, measures the consumer's surplus arising out of international trade. So, the larger the difference, the larger the consumer's surplus arising out of international trade and more is the gain from international trade.
- *Consumer's surplus and E_D:* If the consumer's surplus on a commodity is high, it implies the demand for the commodity is less elastic or inelastic, as the people would be still enjoying some surplus even if the price rises. For instance, in the case of necessaries of life like wheat, rice etc., the consumer's surplus is supposed to be high and consequently these commodities have inelastic demand. Similarly, in the case of postal services, the people enjoy enormous consumer's surplus and consequently any increase in price will not affect the demand. However, in case of luxurious goods, the consumer's surplus will be either too low or absent, and hence, they will have elastic demand. So, E_D for a commodity or service depends on the consumer's surplus contained in it.
- *Measures the difference between value-in-use and value-in-exchange:* We know that, only economic goods will have value-in-exchange and free goods will have only value-in-use. So, economic goods alone will have market value. This market value (value-in-exchange) differs greatly from value-in-use because, a good (free good) may have value-in-use, but may not have value-in-exchange. But, a good (economic good), if possess value-in-exchange, will definitely possess value-in-use also. For example, goods like salt, news paper, soaps etc., have both value-in-use and value-in-exchange. But, the value-in-use outweighs the value-in-exchange. For such commodities, the consumers enjoy high consumer's surplus, as we are prepared to pay much more than we actually pay. So, it is important to note that, the concept of consumer's surplus depends on TU *i.e.*, value-in-use of the commodity, but the price we pay for the commodity depends upon value-in-exchange or MU of the commodity. So, the concept of consumer's surplus clearly measures the difference between value-in-use and value-in-exchange and it is evident from the above example, consumer's surplus is high for the commodities having more value-in-use, but less value-in-exchange.

From the above discussion, it infers that, the concept of consumer's surplus has both theoretical and practical uses. As Pigou pointed that, *'the engine which Marshall has devised, though limited in range, can often serve us'*.

13.8. Can Firms Reduce Consumer Surplus?

Firms try to reduce or exploit the consumer surplus in the following ways:

- Firms can reduce consumer surplus, if they enjoy market power. This enables them to raise the prices of the commodities above the competitive equilibrium.
- A monopoly firm aims at profit maximization by reducing consumer surplus. Monopolist generally practices price discrimination strategy to exploit consumer surplus. He charges higher prices in the market in which the demand for the product is more inelastic.
- To gain market power, a firm could advertise to create brand loyalty and this will make the demand for the product more inelastic.

Part III

Production Theory

14

Theory of Producer Behaviour

In the previous chapters, we have discussed about the behaviour of the consumers. In the ensuing Chapter 17, we discuss the 'supply' concept, which deals with the quantity of output offered for sale by the producer. So, before studying the 'supply' concept, it is essential to frame a background with the 'production' concept, where the farmer employs different resources and resource services in the production programme to produce an output. So, this chapter examines the behaviour of a producer regarding the selection of different combinations of resources, their allocation and efficiency in producing the optimum level of output. We know, a farmer employs several resources and services such as seed, manures, fertilizers, pesticides, irrigation, human labour, bullock labour, machinery labour etc., to produce a wide variety of crops. Similarly, raw cotton will be used as a factor along with other factors like labour, capital etc., to produce cloth. This process of conversion of resources and resource services into final output or product is called Production. In other words, production is defined as the transformation of inputs into output. This also infers that, production or output is the resultant of the factors employed in the production programme. According to Watson, '*production function is the name of relation between physical inputs and physical output of a firm*'.

For the use of resources and resource services in the production programme, the farmer incurs cost and the total cost incurred on per unit area of land resource constitutes Cost of cultivation. If we express the same in terms of per unit of output or product produced, it refers to cost of production. Once the output has been produced, the farmer sells the same in the market and earns revenue. The revenue earned by the farmer must cover the cost of cultivation, so as to continue the farm business. So, the objective of the production programme is to earn maximum profits. To attain maximum profits, the farmer will look at three basic components *viz.*, resource combination, resource allocation and resource use efficiency. These three components are considered as basic production problems *viz.*, 'how to produce', 'what to produce' and 'how much to produce' respectively. In the present chapter, we deal with all these three components in-detail.

14.1. Production Function

The production function is the relationship between the resources and resource services employed by the farmer and the output realized in the production programme. For various quantities of resources and services used, the farmer realizes different levels of output. Let us suppose, a farmer cultivates paddy with a high yielding variety seed and adopt package of practices regarding the use of resources (*i.e.*, seed treatment, organic manures, fertilizers, pesticides, irrigation etc), as per the scientific recommendations, thereby, he will optimize the output in the production programme.

On the other hand, if the farmer cultivates paddy with a local variety, without seed treatment, indiscriminate use of fertilizers, pesticides etc., the output he realizes is less compared to scientific management. This gives the impression that, when the resources are employed judiciously, as per the scientific recommended levels, the output will be optimized, but inefficient allocation of resources will not optimize the output. So, the concept of production function highlights the extent of influence of various resources and services on the output produced in the production programme and thereby, guides the farmer in employing the resources that execute significant positive influences on the output. So, the production function is defined as a technical and mathematical relationship between the resources employed and output realized in the production programme at a given level of technology and at a point of time. Mathematically,

$$Y = f(X_1, X_2, X_3, X_4 \text{———} X_n) \qquad \textit{Equation 14.1}$$

where,

$Y =$ Output produced

$X_1 =$ Seed cost

$X_2 =$ Fertilizer cost

$X_3 =$ Pesticide cost

$X_4 =$ Irrigation cost etc.

So, it is the technological knowledge that determines the optimization of output in a production programme from using different levels and combinations of resources and resource services. If the technology is effective and efficient, it ensures higher levels of output at cost-effectiveness. If a new production technology employed in the business ensures higher output, then, we will have a new production function.

14.2. Concept of Period in Production Function

We know a farmer will use different combinations of resources and resource services to ensure optimality in the production programme. Considering the 'variability' of these resources and resource services, the production function can be classified into two types *viz.*, short run production function and long run production function.

1. Short Run Production Function

Here, 'short run' refers to short-period of time. This is the period of time for which two conditions hold good *viz.*, there is at least one fixed input in the production programme and firms can neither enter nor exit an industry. So, in the short run, the farmer cannot vary all the factors. Some of the factors are fixed and some are variable. So, in order to vary the output level, the farmer has to vary only the variable factors, but the size or scale of the firm remains fixed. The factors that remains fixed are called the Fixed factors, whereas the other factors, which the farmer can vary are called the Variable factors. For example, in a short run production programme, the farmer cannot vary factors like land, machinery etc., and hence, they are fixed factors and other factors like seed, fertilizer, pesticides etc., are variable factors. Fixed factors are not related with the level of output. That means, irrespective of the level of output, the level of fixed factors remains same. So, when the output is increased or decreased or even zero, the level of fixed factors remains same. Conversely, with reference to variable factors, their level of usage is related to the output level produced in the production programme. That means, if the output level increases, the level of variable factors also increases and *vice versa*. So, when there is no production programme (*i.e.*, zero output), the level of variable factors is also zero. However, the distinction between fixed factors and variable factors arises only in the short run, as in the long run all the factors/resources are variable. Thus, 'short run' is a short period of time, in which there is no possibility for the farmer/firm to vary the plant size. So, in short run production function, since supply of output can only be slightly varied, still demand plays vital role in determining the price of the output. So, in short run production function, we will study the influence of these variable factors on the output of the production programme keeping the other (fixed) factors constant. Mathematically,

$$Y = f(X_1, X_2, X_3 / X_4, X_5, X_6 \ldots\ldots\ldots\ldots\ldots X_n) \qquad \textit{Equation 14.2}$$

where, Y= Output from the enterprise, X_1, X_2, X_3 are variable factors, $X_4, X_5, X_6 \ldots\ldots\ldots\ldots X_n$ are fixed factors, / denotes the separation of variable factors from fixed factors.

So, short run production function implies, changes in levels of few variable factors to analyze their influence on the output. The concept of LDR falls under short run production function, wherein, only one factor is continuously varied, keeping other factors constant, so as to study the influence of the variable factor on the output of the production programme.

2. Long Run Production Function

Here, 'long run' refers to long period of time. This is the period of time for which two conditions hold good *viz.*, there are no fixed factors of production (*i.e.*, firms can increase or decrease the amounts of all inputs and so can vary the scale of operation) and new firms can enter and existing firms can exit the industry. So, in the long run, the farmer can vary all the factors and no factor is fixed. For example, a farmer is having five acres of land. In the long run, the farmer can purchase more machinery, more land or sell away his machinery and land and may plan for any other business etc. So, no factor is fixed in the long run and all the factors are variable. Thus, 'long run' is the period of time that is sufficient enough to change the plant size of the firm to produce the desired level of output at lowest possible Long Run Average Total Cost (LRATC). This long period ensures the new firms to enter the industry, when the existing firms enjoy super normal profits in the short run and similarly makes some of the existing firms to leave the industry when they experience losses in the short run production programme. Hence, in the long run production programme, we will study the influence of all these variable factors on the output of the enterprise. A farmer, in order to produce different output levels in the long run, varies all the inputs simultaneously. Mathematically,

$$Y = f(X_1, X_2, X_3, X_4, X_5, X_6 \ldots\ldots\ldots\ldots X_n) \qquad \textit{Equation 14.3}$$

where, Y= Output from the enterprise, $X_1, X_2, X_3, X_4, X_5, X_6 \ldots\ldots\ldots\ldots X_n$ are variable factors

So, in the long run, the farmer will have ample time to vary all the resources and resource services, so that, he can adjust the level of output to suit the demand conditions in the market. So, the farmer tries out to achieve economies of large scale production. The equilibrium price of the output/product is influenced by both demand and supply forces. So, long run production function implies, changes in levels of all the factors to analyze their influence on the output. The concept of Returns to scale falls under long run production function, wherein, all the factors are increased or decreased simultaneously and in the same proportion, so as to analyze their influence on the output of the production programme. It is important that, the increase in levels of all the factors in the production programme, may increase the output in the same proportion, less than proportionately or more than proportionately and accordingly, the returns to scale is categorized as constant returns to scale or diminishing returns to scale or increasing returns to scale respectively.

It is important that, for different production processes, the long run periods may be different. Further, it is not advisable to define short run and long run in terms of days, months or years and we define a period as long run or short run simply by looking at whether all the factors can be varied or not.

14.3. Basic Concepts in Short Run Production Function

To understand the different stages of the short run production function, it is essential to understand certain basic concepts and they are discussed here under.

Total Physical Product (TPP)

It is otherwise called as Total product. It is the output realized by the farm-firm expressed in physical terms. It is the sum total of Marginal Physical Products (MPPs). For example, 40 quintals of output of paddy produced by the farmer infers TPP.

Average Physical Product (APP)

It is the ratio between total output and a variable input employed in the production programme. So, APP=Y/X and it indicates the technical efficiency of a variable resource.

Marginal Physical Product (MPP)

It is the additional output produced in a production programme due to additional usage of an input. It is, otherwise, expressed as the change in total output due to one more or one less unit of a variable input. So, the rate of change of APP indicates MPP. It is given by, MPP= $\Delta Y/\Delta X$ or MPP = $TPP_n - TPP_{n+1}$

Total Value Product (TVP)

It is the expression of output or TPP in monetary terms. It is given by, TVP=TPP*P_Y, where, P_Y is price per unit of output.

Average Value Product (AVP)

It is the product of APP and price per unit of output, given by

$AVP = APP{*}P_Y = (Y/X){*}P_Y = TVP/X$

Marginal Returns or Marginal Revenue (MR)

It refers to additional returns or revenue generated due to production of an additional unit of output. It is given by, MR= $(\Delta TR/\Delta Y) = \Delta TPP \times P_Y$

Marginal Cost (MC)

It is the additional cost incurred in producing an additional unit of an output. It is given by, MC= $\Delta TC/\Delta Y = \Delta TVC/\Delta Y$. (where, TC = Total cost, TVC = Total variable cost)

Marginal Input Cost (MIC_X)

It is the additional cost incurred in using an additional unit of an input in the production programme. It is given by,

MIC_X= Change in TC/Change in input usage = $\Delta TC/\Delta X = \Delta TVC/\Delta X$

Marginal Value Product (MVP_X)

It refers to additional returns or revenue derived from using an additional unit of an input in the production programme. It is given by,

MVP_X= Change in Total Returns/Change in input usage = $\Delta TR/\Delta X$

14.4. Production Function with One Variable Vactor (Returns to a variable factor)

Let us discuss the production function involving one variable factor influencing the output. Returns to a variable factor refers to the relation between a variable factor employed in the production programme and output realized, when all other factors are held constant. That means, when one factor is continuously varied in the production programme, keeping other factors constant, it exerts an influence on the output and this influence is known as Returns to a variable factor. There are two approaches to study the returns to a variable factor *viz.*, Traditional approach and Modern approach.

I. Traditional Approach

The Classical Economists are of the opinion that, there are three possible influences of varying a factor on the output (Laws of returns) *viz.*,

- ✰ Increase in output at an increasing rate (Increasing returns to a variable factor or Law of increasing returns)
- ✰ Increase in output at a constant rate (Constant returns to a variable factor or Law of constant returns)
- ✰ Increase in output at a diminishing rate (Diminishing returns to a variable factor or LDR)

This traditional explanation of the behaviour of returns *viz.*, at increasing rate, constant rate and diminishing rate was explained by Alfred Marshall.

1. Increasing Returns to a Variable Factor or Law of Increasing Returns

In law of increasing returns, each additional unit of a variable input when applied to a fixed level of other inputs, it will add more amount of additional output to the total output than the preceding unit of an additional input. Due to increase in total output (TPP) at an increasing rate, both MPP and APP also increases for every unit of an additional input. Inversely, from cost point of view, it is termed as 'Law of decreasing costs', as the MC goes on decreasing for every additional unit of an input. This is because, each and every additional unit of an input is accompanied by more than proportionate increase in output and thereby, the MC of each unit of an input goes on decreasing. In the words of Marshall, *'an increase of labour and capital leads generally to improved organization, which increases the efficiency of the work of labour and capital. Therefore, in those industries, which are not engaged in raising raw produce, an increase of labour and capital generally gives a return increase more than in proportion and further this improved organization tends to diminish or even override any increased resistance which nature may offer to raise increased amounts of raw produce'*. This definition given by Alfred Marshall highlights two important aspects *viz.*, increasing returns in the firm is due to better organization and the increasing returns is common in manufacturing industries, which do not engage in producing its own raw material. Benham stated the law *'as the production of one factor in the combination of factors is increased upto a point, the marginal product of the factor will increase'*. In the words of Joan Robinson, *'when an increased amount of any factor of production is devoted to certain use, it is often the case that improvement in organization can be introduced which will make natural units of the factor (men, acres or money capital) more efficient, so that an increase in output does not require a proportionate increase in the physical amount of the factor. This law or rather tendency, like the law of diminishing returns, may apply equally to all the factors of production, but, unlike the law of diminishing returns, it does not apply in every case. Sometimes an increase of the factors will lead to improvements in efficiency and sometimes it will not'*. The important features of this definition includes, increasing returns in the firm is due to better organization (as explained by Alfred Marshall), increasing returns can be seen irrespective of the production activity *i.e.*, either industry or agriculture, provided the level of the factor or factors are increased and the increasing returns is not 'inevitable' unlike diminishing returns. So, the definition given by Joan Robinson also highlights the significance of diminishing returns, as its operation is always inevitable in a production programme, but the law of increasing returns may be exhibited, but it is not a must always. The law of increasing returns is explained through the Table 14.1.

Table 14.1: Increasing (marginal) returns to a variable factor

Fixed Input, Land- (Units)	Variable Input Fertilizer, X (Units)	TPP (Output, Y) (Units)	APP (Y/X)	ΔX	ΔY	MPP (ΔY/ΔX)
2	5	2	0.40	–	–	–
2	10	5	0.50	5	3	0.60
2	15	9	0.60	5	4	0.80
2	20	14	0.70	5	5	1.00
2	25	20	0.80	5	6	1.20

It is clear from the Table 14.1 that, if the variable input, fertilizer is increased continuously keeping other inputs like land, fixed in the production programme, the output increases at an increasing rate, as indicated by the increase in MPP. Since, output is increased at an increasing rate, the APP also increased. The diagrammatic representation of the Table 14.1 infers that, the production function is convex to origin (Panel A of Figure 14.1) and the slope of production function *i.e.*, MPP is increasing (Panel B of Figure 14.1). However, this law is not applicable to agriculture and applies mostly to manufacturing industries that too, upto the point of optimality, beyond which, the constant and diminishing MRs will operate. The MPP for law of increasing returns is given by, $(\Delta_1 Y/\Delta_1 X) < (\Delta_2 Y/\Delta_2 X) < (\Delta_3 Y/\Delta_3 X) <$ ——— $< (\Delta_n Y/\Delta_n X)$.

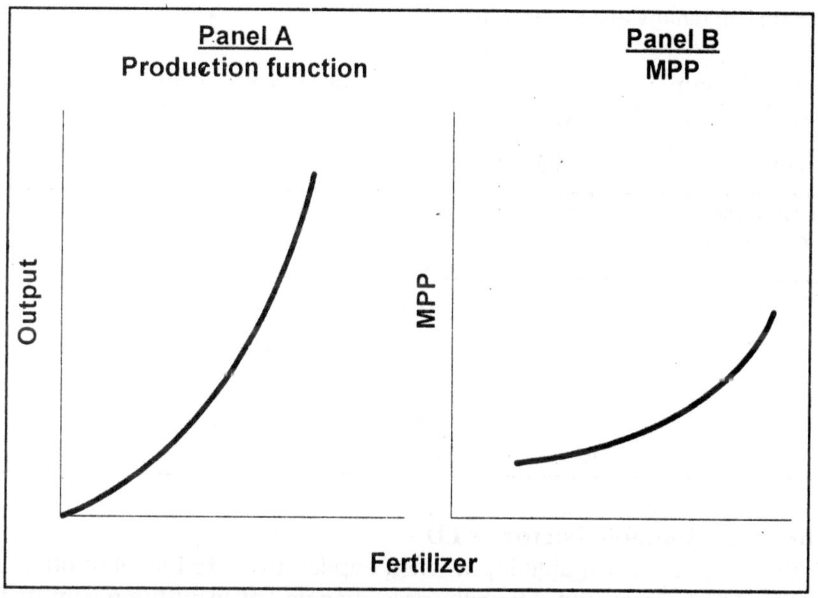

Figure 14.1: Increasing (marginal) returns to a variable factor.

2. Constant Returns to a Variable Factor or Law of Constant Returns

This law represents the transition from increasing returns to diminishing returns and *vice versa* in a production programme. In law of constant returns, each additional unit of a variable input when applied to a fixed level of other inputs, it will add same amount of additional output to the total output, as the preceding unit of an additional input. Due to increase in total output at constant rate, both MPP and APP remains constant for every additional unit of an input. Inversely, from cost point of view, it is termed as 'Law of constant costs', as the MC remains same for every additional unit of an input. This is because, each and every additional unit of an input is accompanied by same proportionate increase in output and thereby, the MC of each unit of an input remains same. In the words of Marshall, *'the stage of constant returns comes at that point, where the effects of increasing returns and diminishing returns balance each other'*. According to Prof. Stigler, *'when all of the productive services are increased in a given proportion, the product is increased in the same proportion'*. The law of constant returns is explained through the following Table 14.2.

It is clear from the Table 14.2 that, if the variable input, fertilizer is increased continuously keeping other inputs like land, fixed in the production programme, the output increases at constant rate, as indicated by the constancy of MPP. Since, output is increased at constant rate due to constancy of MPP, APP is also constant. The diagrammatic representation of the above Table 14.2 infers that, the production function is a straight line sloping upwards from the origin (Panel A of Figure 14.2) and the slope of production function *i.e.*, MPP is constant and equal to APP, thereby, lies

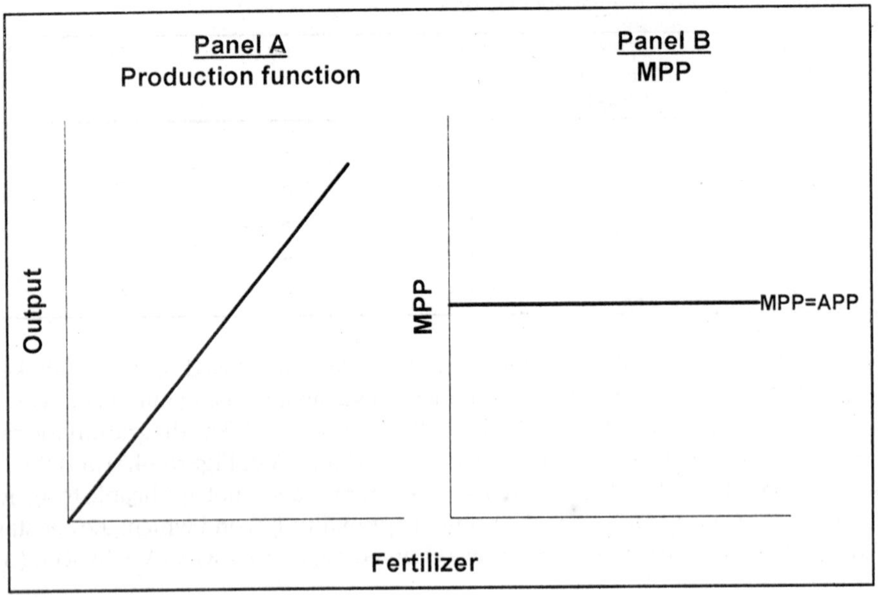

Figure 14.2: Constant (marginal) returns to a variable factor.

parallel to X-axis (Panel B of Figure 14.2). However, this law is also not applicable to agriculture. The MPP for law of constant returns is given by, $(\Delta_1Y/\Delta_1X) = (\Delta_2Y/\Delta_2X) = (\Delta_3Y/\Delta_3X) = \text{———} = (\Delta_nY/\Delta_nX)$.

Table 14.2: Constant (marginal) returns to a variable factor

Fixed Input, Land- (Units)	*Variable Input Fertilizer, X (Units)*	*TPP (Output, Y) (Units)*	*APP (Y/X)*	*ΔX*	*ΔY*	*MPP (ΔY/ΔX)*
2	5	2	0.40	–	–	–
2	10	4	0.40	5.00	2.00	0.40
2	15	6	0.40	5.00	2.00	0.40
2	20	8	0.40	5.00	2.00	0.40
2	25	10	0.40	5.00	2.00	0.40

3. Diminishing Returns to a Variable Factor or LDR

In LDR, each additional unit of a variable input when applied to a fixed level of other inputs, it will add less amount of additional output to the total output than the preceding unit of an additional input. Due to increase in total output at diminishing rate, both MPP and APP also decreases for every additional unit of an input. Inversely, from cost point of view, it is termed as 'Law of increasing costs', as the MC goes on increasing for every additional unit of an input. This is because, each and every additional unit of an input is accompanied by less than proportionate increase in output and thereby, the MC of each unit of an input goes on increasing. This concept of LDR was first proposed by Turgot, but Malthus and Ricardo have refined this concept. In the words of Marshall, '*keeping the fixed factors constant, when marginal product diminishes with the increase in the quantities of a variable factor, it is called law of diminishing returns*'. According to Boulding, '*as we increase the quantity of one input which is combined with a fixed quantity of other inputs, the marginal physical productivity of the variable input must eventually decline*'. In the words of Joan Robinson, '*the law of diminishing returns states that, with a fixed amount of any one factor of production, successive increases in the amount of other factors will, after a point, yield a diminishing increment of the product*'. The LDR is explained through the following Table 14.3.

It is clear from the Table 14.3 that, if the variable input, fertilizer is increased continuously keeping other inputs like land, fixed in the production programme, the output increases at diminishing rate, as indicated by the decline in MPP. Since, output increased at diminishing rate, the APP also decreased. The diagrammatic representation of the above Table 14.3 infers that, the production function is concave to origin (Panel A of Figure 14.3) and the slope of production function *i.e.*, MPP is diminishing (Panel B of Figure 14.3). This law is more common in agriculture and equally applies to other sectors *viz.*, industry, mining etc., as well. The MPP for LDR is given by, $(\Delta_1Y/\Delta_1X) > (\Delta_2Y/\Delta_2X) > (\Delta_3Y/\Delta_3X) > \text{———} > (\Delta_nY/\Delta_nX)$.

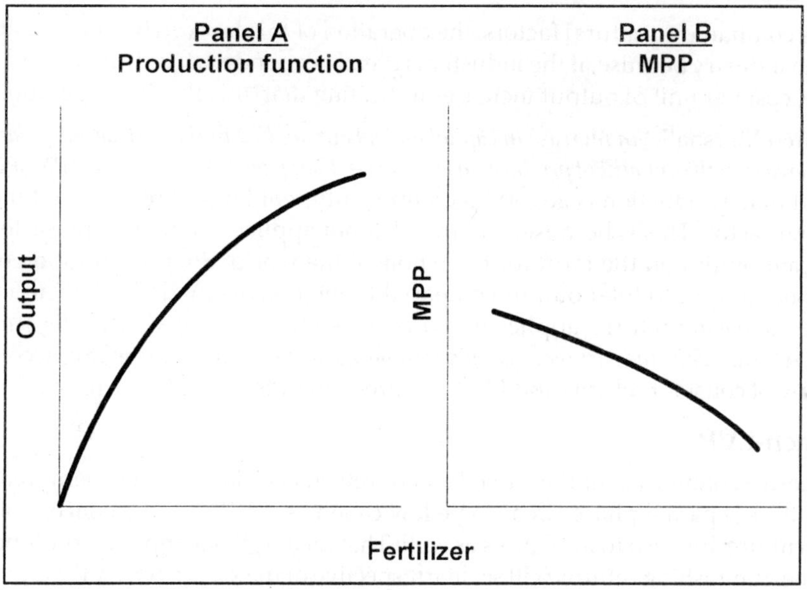

Figure 14.3: Diminishing (marginal) returns to a variable factor.

Table 14.3: Diminishing (marginal) returns to a variable factor

Fixed Input, Land- (Units)	*Variable Input Fertilizer, X (Units)*	*TPP (Output, Y) (Units)*	*APP (Y/X)*	*ΔX*	*ΔY*	*MPP (ΔY/ΔX)*
2	5	5.00	1.00	–	–	–
2	10	8.50	0.85	5	3.50	0.70
2	15	10.80	0.72	5	2.30	0.46
2	20	12.60	0.63	5	1.80	0.36
2	25	13.00	0.52	5	0.40	0.08

This concept of three different laws of returns *i.e.*, increasing, constant and diminishing returns is a classical or traditional approach. However, Modern Economists discarded this view and opined that, the returns to a variable factor *i.e.*, increasing, constant and diminishing returns are not three distinct laws, but they are different phases of a single law called 'Law of Variable Proportions (LVP)'. Further, Marshall opined that, the LDR applies to 'agriculture' sector only, because of the following special features:

- ✰ In agriculture, nature plays a very vital role in the production programme compared to human element. So, due to this excess dependence on natural and climatic factors, there is less control on the part of the farmer to execute the production programme.
- ✰ There is less scope for division of labour because, farmer himself has to play three important roles such as, farm manager, capitalist and labour.
- ✰ Less scope for mechanization due to sub-division and fragmentation of land holdings.
- ✰ Bringing inferior lands into cultivation is another important reason for the early operation of LDR. This is because, to meet the food security needs for mounting population, inferior lands will be brought into cultivation and for such lands, the productivity is very low.
- ✰ Due to continuous cultivation of crops on the same piece of land, soil fertility gets exhausted.
- ✰ There is less scope for efficient planning of the production programme because of excess dependence on natural factors on one side and limited capital availability to farmer on other side.
- ✰ The agricultural labour are less skilled and hence, their efficiency is low.
- ✰ Agricultural operations are spread over a larger area and hence, supervision may not be effective.
- ✰ Availability of a specialized machinery to perform a particular operation is limited.

This clearly infers that, in the production programme, where nature plays a crucial role or there is less control by the human element, this LDR sets in early. But in the sectors, where there is more human element involved in controlling production programme compared to natural factors, the operation of this law can be delayed. It is wrong to say that, this law will not apply in industry because, if the industry is greatly expanded, it is difficult to supervise on the part of the manager and hence, cost per unit of output increases indicating that, this law is in operation.

In the words of Alfred Marshall, *'an increase in capital and labour applied in the cultivation of land, causes, in general, a less than proportionate increase in the amount of produce raised, unless it happens to coincide with the improvement in the arts of agriculture'*. He believed that, production is governed by three different laws of returns and the LDR is particularly applicable to agricultural sector. This is because, had the LDR not applies to agricultural sector, the farmers would have produced significant output on the farm for the whole of the world. So, the operation of LDR in agriculture contributes less additional output to total output compared to increments made in the variable factor. He further opined that, the law of increasing returns applies to industrial sector and stated that, *'the part played by the nature conforms the diminishing returns, while the part that man plays conforms to increasing returns'*. So, according to Marshall, law of increasing returns, law of constant returns and LDR are three separate laws of returns to a variable factor.

II. Modern Approach-LVP

According to Modern Economists, all the three laws of returns *i.e.*, law of increasing returns, law of constant returns and LDR operate as separate phases in a single law called as 'LVP'. So, the concept of diminishing returns applies not only to agriculture, but also to industries as well. That means, if one input is continuously varied, keeping other inputs constant, the diminishing returns will set in irrespective of the sector. So, LVP is a universal law according to the Modern Economists. This law is so called because, the proportion among the inputs used is different *i.e.*, other inputs are held constant and only one input is continuously increased to study the influence on the output. Hence, LVP falls under short run production function. This law is also called as 'Law of non-proportional returns' because, returns are not proportionately increased, as the increase in input usage. This law is also called as, 'Law of Added cost and Added returns' because MC=MR is the output optimizing criterion in the production programme. This LVP is also considered as 'Fundamental law of Economics'. The following are the popular definitions of LVP:

'When total output or production of a commodity is increased by adding units of a variable unit, while the quantities of other inputs are held constant, the increase in total production becomes, after some point, smaller and smaller'

—(Watson)

'As the proportion of one factor in a combination of factors is increased, after a point, first the marginal and then the average product of that factor will diminish'.

—(Benham)

'As equal increments of one input are added, the inputs of other productive services being held constant, beyond a certain point the resulting increments of product will decrease i.e., the marginal products will diminish'.

—(G. Stigler)

'An increase in some inputs relative to the other fixed inputs will, in a given state of technology, causes output to increase, but after a point the extra output resulting from the same addition of extra inputs will become less'.

—(Paul Samuelson)

'The law of variable proportions states that, if the input of one resource is increased by equal increments per unit of time while the inputs of other resources are held constant, total product (output) will increase, but beyond some point the resulting output increases will become smaller and smaller'.

—(Leftwitch)

The definitions explain that, all the three laws of returns *viz.*, increasing returns, constant returns and diminishing returns will operate in the production programme, if we go on increasing one input continuously, keeping other inputs constant. That means, first the (marginal) returns will increase at increasing rate, then at constant rate, then at diminishing rate and finally decreases at increasing rate, if one factor is continuously increased, keeping other inputs constant in the production programme. This law can be better explained by studying the patterns of TPP, MPP and APP in different stages or zones of LVP.

Assumptions

This law holds good under the following assumptions:

- ☆ Quantity of input is given
- ☆ Only one input should be varied keeping other inputs constant, as this facilitates to study the effect of changes of one input on the output or production.
- ☆ All the units of a variable input should be homogeneous. For example, same type of nitrogen fertilizer should be employed as a continuous variable factor.

☆ There should not be any change in the state of technology or methods of production.

☆ It assumes short run production function, as in the long run, all the inputs are variable.

☆ The output *i.e.*, TPP, APP and MPP should be measured in terms of physical units only. They should not be considered in monetary terms because, even if the output falls due to operation of LDR, but total returns may increase, if price of the output increases in the market and *vice versa*.

☆ There should be possibility of varying the proportions of factors, so as to produce the output. That means, the factors involved in the production programme should not be perfect complements, as we vary only one factor keeping other factors constant.

☆ This law is applicable in terms of costs, only if the prices of variable inputs as well as output remain same.

Explanation of the Law

The concept of LVP explained through Table 14.4 and Figure 14.4 reveals that, if we go on increasing one input, keeping other inputs constant, first the output increases at increasing rate, then increases at constant rate, at diminishing rate and finally decreases at increasing rate. It is also evident that, as long as MPP is positive, TPP increases, when MPP is zero, TPP is maximum and when MPP is negative, TPP starts declining. Considering the behaviour of these three curves *viz.*, TPP, MPP and APP curves, the Figure 14.4 can be conveniently divided into three zones or stages and the features of these stages are discussed here under.

Table 14.4: Different stages of short run production function (LVP)

Fixed Input Land – (Acres)	*Variable Input Fertilizer (X)*	*Output (Y)*	*ΔX*	*ΔY*	*MPP (ΔY/ΔX)*	*APP (Y/X)*	*Returns to a Variable Factor*
1	1	2	1	2	2	2.00	Increasing (marginal) returns
1	2	6	1	4	4	3.00	
1	3	11	1	5	5	3.67	
1	4	17	1	6	6	4.25	Constant (marginal) returns
1	5	23	1	6	6	4.60	
1	6	28	1	5	5	4.67	Diminishing (marginal) returns
1	7	32	1	4	4	4.57	
1	8	33	1	1	1	4.13	
1	9	33	1	0	0	3.67	Zero (marginal) returns
1	10	32	1	-1	-1	3.20	Negative (marginal) returns
1	11	29	1	-3	-3	2.64	

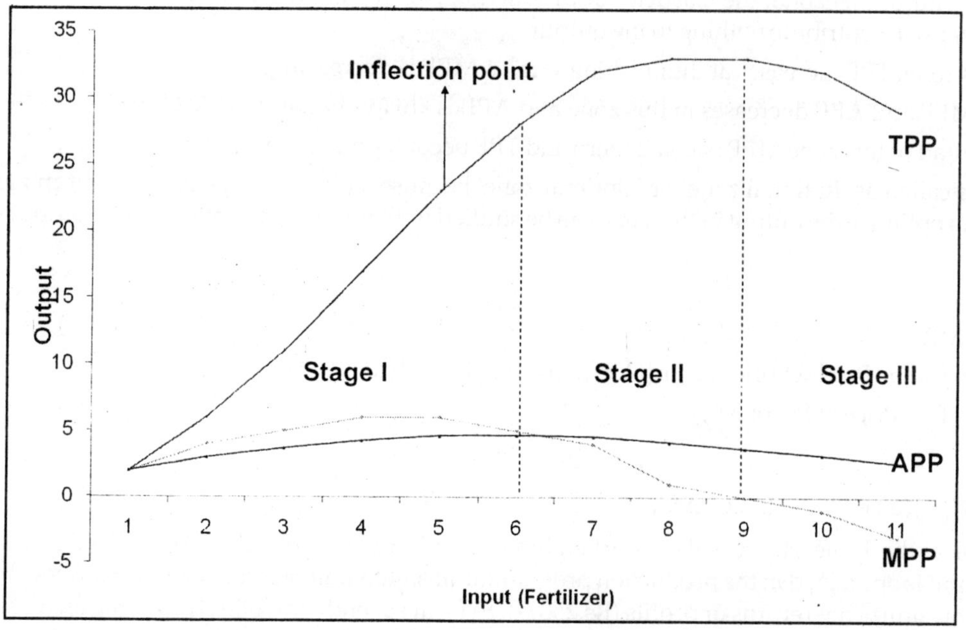

Figure 14.4: Three stages of LVP.

(a) First Zone or Stage

This zone starts from the point of origin upto the point, where MPP is equal to APP or APP is maximum. The important features of this stage include,

- ☆ In this zone, the elasticity of production (MPP/APP) is more than one because, MPP is higher than APP throughout the zone.
- ☆ At the end of first zone, MPP=APP. So elasticity of production is equal to one, implying that percentage change in output is equal to percentage change in input.
- ☆ When MPP is maximum, its corresponding point on TPP curve indicates 'Inflection point', which indicates that, TPP is now increasing at diminishing rate and there is change in curvature of TPP from convex to concave.
- ☆ In this zone, TPP increases at increasing rate upto the point of inflection, then at constant rate and after inflection point, TPP increases at diminishing rate because, MPP first increases, becomes constant and finally decreases in this zone. Hence, the curvature of TPP changes from convex to concave and this is indicated by the inflection point. This further indicates that, all the three laws of returns will be experienced in the first zone, thereby, the LDR starts operating in the first zone itself after employing five units of fertilizer *i.e.*, after inflection point on the TPP curve.
- ☆ APP continuously increases and reaches maximum at the end of this zone, where it equals to MPP.
- ☆ In this zone, though MPP falls, APP still rises between fifth and sixth units of fertilizer, as MPP is more than APP.
- ☆ In this zone, both APP and TPP will rise continuously, whereas, MPP rises, reaches maximum and then starts falling.
- ☆ The quantity of fixed factor in this zone is too much relative to the use of variable factor, such that, if we even withdraw some of the fixed factor, TPP will not fall. So, in this zone, the MPP of fixed factor is negative.

This zone is called as 'Irrational zone' or 'Sub-optimal zone' because, the farmer will not optimize the output in this zone, as TPP is still increasing in the second zone (diminishing rate) and in the second zone, though MPP and APP are falling, but they are positive. This implies that, if the farmer increases the input dosage, he can still increase the output and thereby, he definitely moves to second zone.

(b) Second Zone or Stage

This zone starts from the point, where MPP is equal to APP and ends, where MPP is equal to zero or TPP is maximum. The important features of this stage include,

- ☆ In this zone, elasticity of production is less than one, as MPP is less than APP implying that, the percentage change in output is less than the percentage change in input.
- ☆ At the end of this zone, the elasticity of production is zero, as MPP is zero indicating that, ninth unit of fertilizer will contribute nothing to the output.
- ☆ In this zone, TPP increases at diminishing rate, as MPP is decreasing.
- ☆ Both MPP and APP decreases in this zone and APP is always higher than MPP.
- ☆ At the end of this zone MPP becomes zero and TPP becomes maximum.

This zone is called as 'Rational zone' or 'Optimal zone' because, there is a possibility of optimization of output and the criteria to optimize the output in this zone can be studied with reference to both from the output side and input side.

Output Side

1. MC = MR
2. MPP = Factor-Product price ratio: $[(\Delta Y/\Delta X) = (P_X/P_Y)] = [(\Delta Y^*P_Y) = (\Delta X^*P_X)]$

   (*where*, P_X = Price of factor X)

Input Side

$MVP_X = MIC_X = [(\Delta TR/\Delta X) = (\Delta VC/\Delta X)]$

A close look at the Table 14.5 reveals that, at eighth unit of fertilizer, MR=MC, $MPP=(P_X/P_Y)$ and $MVP_X=MIC_X$, thereby, the output is optimized in the production programme at eighth unit of fertilizer. So, at this level of input usage, the farmer will maximize net returns or profits (Rs.250/-). Even at seventh unit of fertilizer, the farmer maximizes the net returns (Rs.250/-) as in eighth unit of fertilizer, but at seventh unit of fertilizer, the optimality conditions *viz.*, MC(Rs.2.50/-)$\neq\Delta$MR(Rs.10/-), MPP(4)$\neq\Delta P_X/P_Y$(1) and MVP_X(Rs.40)$\neq MIC_X$(Rs.10) are not fulfilled and more over, it is advisable for the farmer to increase the input usage upto eighth unit of fertilizer because, total output in quantity terms increases from 32 units to 33 units.

Table 14.5: Profit maximization in short run production function (LVP).

Input X (Fertilizer)	TPP Y	MPP ($\Delta Y/\Delta X$)	APP (Y/X)	ΔX	ΔY	TVC ($X * P_x$) (Given P_x= 10/-)	ΔTVC	MC ($\Delta TVC/\Delta Y$)	TR ($Y * P_y$) (Given P_y=10/-)	ΔTR	MR ($\Delta TR/\Delta Y$)	MICx ($\Delta TVC/\Delta X$)	MVPx ($\Delta TR/\Delta X$) or (MPP * Py)	NR* (TR-TVC)
1	2	2	2.00	1	2	10	10	5.00	20	20	10	10	20	10
2	6	4	3.00	1	4	20	10	2.50	60	40	10	10	40	40
3	11	5	3.67	1	5	30	10	2.00	110	50	10	10	50	80
4	17	6	4.25	1	6	40	10	1.67	170	60	10	10	60	130
5	23	6	4.60	1	6	50	10	1.67	230	60	10	10	60	180
6	28	5	4.67	1	5	60	10	2.00	280	50	10	10	50	220
7	32	4	4.57	1	4	70	10	2.50	320	40	10	10	40	250
8	33	1	4.13	1	1	80	10	10.00	330	10	10	10	10	250
9	33	0	3.67	1	0	90	10	0.00	330	0	0	10	0	240
10	32	-1	3.20	1	-1	100	10	-10.00	320	-10	10	10	-10	220
11	29	-3	2.64	1	-3	110	10	-3.33	290	-30	10	10	-30	180

*: Net returns/profits

Inequality between MR and MC or MVPx and MICx

Case1: MR>MC or $MVP_X > MIC_X$

This case is observed in the first zone of production function and the farmer is advised to increase the input usage, so that MC or MIC_X increases and becomes equal to MR or MVP_X respectively to attain optimality in the production programme.

Case 2: MR<MC or $MVP_X < MIC_X$

This case is observed in the third zone of production function and the farmer is advised to reduce the input usage, so that MC or MIC_X decreases and becomes equal to MR or MVP_X respectively to attain optimality in the production programme.

(c) Third Zone or Stage

This zone starts from the point, where MPP is equal to zero or TPP is maximum and continue till the end of production programme. The important features of this stage include,

- ☆ At the beginning of third zone, elasticity of production is zero because, MPP is zero and in the third zone, elasticity of production turns negative or less than zero because, MPP is negative.
- ☆ TPP is maximum at the beginning of this zone and later it falls at an increasing rate because, MPP turns negative and decrease at an increasing rate. Hence, this stage is also called 'stage of negative returns'.
- ☆ APP still falls when compared to second zone due to fall in TPP, but remains positive. It will not become zero or negative, as long as TPP is positive.
- ☆ The level of fixed factor is very low compared to level of variable factor. Since, the level of variable factor is too high relative to fixed factor, the MPP of variable factor is negative.

This zone is also called as 'Irrational zone' or Supra-optimal zone' because, farmer cannot optimize the output in this zone and more over, if he operates in this zone, he experiences 'double loss' in terms of a greater fall in output on one hand and increase in the cost expenditure on the resource on the other hand.

So, Stage I and Stage III are called 'Irrational zones', as the farmer cannot optimize the output in these two zones and hence, they are completely symmetrical. In stage I, the level of fixed factor is too high relative to variable factor, thereby, the MPP of fixed factor is negative. However, in stage III, the level of variable factor is too high relative to fixed factor, thereby, the MPP of variable factor is negative. So, a rational producer will never choose stage III to produce the output, as he experiences 'double loss' in this stage. Even if the input (fertilizer) is offered at free of cost, he will not enter the stage III and he will produce the output, where MPP is equal to zero or TPP is maximum. On the other hand, if the input is priced, he will try to optimize the output based on the criteria discussed earlier. It is interesting that, the farmer will not stop in stage I in producing the output, as in stage II, we see the TPP is increasing (at diminishing rate) and though APP and MPP are falling, but remain positive. So, the farmer should operate in second zone to optimize the output. Hence, second zone is otherwise called as rational zone or optimal zone. This law is also called as LDR because, in the stage II, the optimization of output is possible, when the TPP increases at diminishing rate. But, it is important to note that, LDR is only one phase of the more comprehensive LVP.

Shapes of TPP, MPP and APP Curves

In Figure 14.4, we measure units of input fertilizer along the X-axis and output along the Y-axis. The TPP in the input-output plane is a positively sloped curve. According to the LVP, the MPP of an input initially rises and then after a certain level of employment, it starts falling. The MPP curve in the input-output plane, therefore, looks like an inverse 'U'-shaped curve. Regarding APP curve, for the first unit of the variable input, both MPP and APP are same. If the variable input is continuously increased, the MPP rises initially in the Stage I of production function. APP being the average of summated MPPs, it also rises, but rises less than MPP. Then, after a point in the Stage I, the MPP starts falling. However, as long as the value of MPP remains higher than the value of the APP, the latter continues to rise. Once, MPP has fallen sufficiently, its value becomes less than the prevailing APP and the latter also starts falling. So APP curve is also inverse 'U'-shaped. As long as the APP increases, it must be the case that, MPP is greater than APP. Otherwise, APP cannot rise. Similarly, when APP falls, MPP has to be less than APP. It, therefore, follows the MPP curve and MPP curve cuts APP curve from above at its maximum.

Relationship between TPP, MPP and APP

From the three zones of LVP, we can formulate useful relationship between the TPP, MPP and APP.

(i) Relationship between TPP and MPP

- ☆ When MPP increases at increasing rate, TPP also increases at increasing rate and the TPP curve is convex to origin.
- ☆ When MPP increases at constant rate, TPP also increases at constant rate. Hence, TPP curve is a straight line.
- ☆ When MPP increases at diminishing rate, TPP also increases at diminishing rate. Hence, the TPP curve is concave to origin.
- ☆ When MPP is zero, TPP is maximum.
- ☆ When MPP becomes negative, TPP starts falling and it decreases at increasing rate.

(ii) Relationship between MPP and APP

- ☆ When MPP is above APP, APP also rises.
- ☆ When MPP is equal to APP, APP is maximum.
- ☆ When MPP is below APP, APP starts falling, but remains positive even when MPP turns negative.
- ☆ MPP rises rapidly compared to APP and also falls rapidly compared to APP.

(iii) Relationship between APP and TPP

- ☆ When APP increases, TPP increases at increasing rate, constant rate and diminishing rate.
- ☆ When APP decreases, TPP increases at diminishing rate and also decreases at increasing rate.

Reasons for Increasing MRs

The concept of increasing MRs is seen in the first stage of production function and the reasons for this increasing MRs are given below.

- ☆ *Effective supervision of less variable factor:* In first zone, the farmer employs more of fixed factor and less of variable factor, so that, he can effectively supervise the variable factor, thereby, the efficiency of variable factor will be increased (indicated by the rise in MPP).
- ☆ *Division of labour:* The farmer can go for specialization or division of labour with the variable factor he employs at the initial stages of production function and due to better organizational methods, increasing MRs can be experienced. However, this is not possible, if the scale of production is small. If the scale of production is considerably large, specialization or division of labour can be introduced in the production programme, leading to economies of scale, implying increasing MRs.
- ☆ *Efficient use of fixed factor:* When more and more of a variable factor is added to the fixed factor, the fixed factor can be effectively utilized and this leads to rise in production at rapid pace. This is because, at the initial stages of production, the large quantities of fixed factor remain unutilized (say, machinery) and when variable factor is increased, it ensures fuller utilization of fixed factor and this leads to rise in TPP.

- ☆ ***Indivisibility of fixed factor:*** It is difficult to employ the fixed factor in requisite quantity suiting the levels of variable factor employed in the production programme, because of 'indivisibility' of the fixed factor. This means that, the allocation of fixed factor cannot be divided in accordance with the level of variable factor and level of output to be produced and hence, a certain quantity of fixed factor has to be employed irrespective of the level of output. For example, processing machinery in a sugar factory is capable of processing say, 1000 units of sugar cane in a day. But, it is actually utilized to process only 700 units of sugar cane due to less availability of raw material. So, certain capacity of the processing machinery remains idle and this is due to indivisibility of the machinery. Similarly, certain costs are fixed in a business unit, whether the production is more or less or even nil, such as, rent of factory, wages or salaries of permanent employees etc., and this infers the indivisibility of the fixed factor in the production programme. So, by adding more and more of a variable factor, the fixed factor will be better utilized and the increase in production will lead to decline in fixed cost per unit of product produced from the factory.
- ☆ ***Economies of scale:*** Both internal and external economies of scale result in the increasing MRs in the production programme. Internal economies are those, which arise in a particular firm as a result of expansion of output and large scale operation. It includes indivisibility of machinery, employment of skilled labour (specialization of labour), efficient management etc. External economies arise for the industry as a whole, as a consequence of the expansion of that industry and are shared by all the firms in that industry.

Reasons for Constant MRs

The concept of constant MRs is seen in stage I and it is a transition phase from increasing MRs to diminishing MRs. The reasons for this constant MRs are given below.

- ☆ ***Optimum utilization of variable factor***: The allocation of variable factor in the production programme is at optimum level and this ensures its efficient utilization in proper combination with the fixed factor.
- ☆ ***Ideal factor ratio***: There is an ideal combination between the variable factor and fixed factor employed in the production programme and this combination leads to constancy of MRs in increasing the returns in the business.

The constancy of MRs is common in industries like sugar industry growing its own raw material (sugar cane), cotton industry growing its own raw material (cotton), jute industry growing its own raw material (jute) etc. In such industries, the production of raw materials experiences diminishing returns, whereas the manufacturing of products experiences increasing returns, thereby, one neutralizes the other. So, we can say, constant MRs is the simultaneous working of law of increasing MRs and law of diminishing MRs. So, the concept of constant MRs is the case of an optimum firm. Below the point of optimality, the firm experiences increasing MRs and beyond the point of optimality, the firm experiences diminishing MRs. So, the point of optimality is the ideal combination of factors of production and it is the equilibrium point of man's ingenuity and nature's niggardliness.

Reasons for Diminishing MRs

The concept of diminishing MRs starts in stage I and continues through out in stage II and the reasons for this diminishing MRs are given below.

- ☆ ***Less supervision:*** As farmer employs more of a variable factor compared to fixed factor in the production programme, the supervision of variable factor on the part of farmer will be difficult and hence, MPP starts falling, thereby, TPP increases at diminishing rate. However, there is scope for profit maximization even when the farm-firm experiences diminishing MRs because, even MPP is falling, it is still positive and TPP is still increasing, but at diminishing rate.
- ☆ ***Imbalance between variable and fixed factors:*** Continuous increase of a variable factor leads to total imbalance between the excess amount of variable factor and less amount of fixed factor. As a result, the technical efficiency of both these factors gets affected leading to diminishing MRs from the variable factor. This is because, the continuous increase of one variable factor makes the fixed factor inadequate relative to the quantity of variable factor and thereby, the additional units of variable factor will have less and less units of fixed factor to work with. So, the excess of variable factor receives less and less aid from the scarce fixed factor and this contributes to diminishing MRs.
- ☆ ***Certain factors remain fixed:*** In a production programme, some factors like land, machinery etc., are fixed in short run and some are variable, thereby, the adjustment and judicious combination of these two factors is not possible and hence, the MRs gets affected.
- ☆ ***Indivisibility of fixed factor:*** The indivisibility of fixed factor contributes to diminishing MRs like that of increasing MRs. When the variable factor is increased continuously, a stage will be reached in the production programme, such that, there will optimum combination of fixed factor and variable factor. If still the variable

factor is increased beyond that optimum combination, the MRs will decline because, the optimal combination of fixed factor and variable factor is disturbed and fixed factor is subjected to intensive usage, as it is indivisible. For example, if a machinery in a sugar factory is capable of processing 1000 units of sugar cane and if more raw material is available to the factory due to bumper harvest of sugar cane in that area, then the machinery (fixed factor) is subjected to intensive usage. Likewise, the fertility of land is fixed and if it is subjected to continuous cultivation, the MRs will decrease, as the fertility gets exhausted in due course of time. This indivisibility of fixed factor compared to variable factor leads to diseconomies in large scale production.

- ***Certain variable factors become scarce:*** In a production programme, even all the variable factors cannot be varied as desirable, because of their scarcity. For example, labour availability, especially skilled labour may not be available in plenty to boost the MRs. So, due to scarcity of certain factors, the combination of factors in the production programme gets disturbed and hence, the diminishing MRs sets in.
- ***Lack of perfect substitutes:*** The absence of perfect substitutes for the factors further aggravates the problem of judicious combination of factors in the production programme. For example, there is no perfect substitute for skilled labour to execute the technical works in a sugar factory. Similarly, if the electricity supply to a sugar factory is affected, it will affect the production, as there is no perfect substitute to this factor, thereby, leading to diminishing MRs from the factory. So, since the elasticity of substitution between the factors is not infinite, the diminishing MRs operate in the production programme. According to Joan Robinson, *'the law of diminishing returns operates due to imperfect substitutes'.*
- ***Wrong combination:*** Scarcity of some variable factors and imperfect substitutability of factors lead to wrong combination of factors. The fixed factor becomes less compared to variable factor and this wrong combination leads to diminishing MRs. Further, the units of a variable factor employed may differ in their efficiency.

Reasons for Negative MRs

The following are the reasons for the negative MRs:

- If still the variable factor is increased indiscriminately, the MRs will decline and turns negative, as it is relatively difficult on the part of the farmer to supervise the excess quantity of variable factor relative to fixed factor.
- The excessively large variable factor relative to fixed factor also impairs the efficiency of fixed factor due to over exploitation of fixed factor, thereby, the efficiency of both fixed and variable factors will be affected.

Had the fixed factor is perfectly divisible, there is no possibility of having either increasing MRs or diminishing MRs. This is because, even at the beginning of production programme, the fixed factor is employed in requisite quantity, such that, it will have optimal combination with the variable factor. So, the possibility of having large quantity of fixed factor against small quantity of variable factor will not arise at all, even at the beginning of production programme. So, at every output level, there will be optimum combination of fixed and variable factors to ensure constant MRs on sustained basis. In the words of Prof. Bober, *'let divisibility enter through the door, law of variable proportions rushes out through the window'.*

To sum up the above discussion, the operation of LVP is due to the fact that, as we continuously increase one variable factor, keeping the other factors fixed in the production programme, initially, the factor proportions become more and more suitable for the production and MRs increases. But, after a certain level of employment of resources, the production process becomes too crowded with the variable input and the factor proportions become less and less suitable for the production. It is from this point that, the MRs of the variable input starts falling.

Differences between Three Stages of LVP

The Table 14.6 highlights the important differences between three stages of LVP.

Limitations

This law will not be applicable,

- When new soils are brought into cultivation
- If there is change in technology adopted by the farmer. For example, if the farmer employs modern technology, the output will increase continuously and the operation of LDR will be delayed.
- If the farmer is having insufficient resources or capital.
- If the farmer goes for scientific management practices like selection of good quality seed, seed treatment, judicious use of manures and fertilizers, use of modern implements, deeper tillage, provision of good irrigation and drainage facilities etc., the operation of LDR can be delayed.

Table 14.6: Differences between three stages of LVP.

First Stage	*Second Stage*	*Third Stage*
Starts from the point of origin and ends when MPP=APP or when APP is maximum.	Starts from the point, where MPP=APP and ends where MPP is zero or TPP is maximum	Starts from the point, where MPP is zero or TPP is maximum and continue till the end of production programme
TPP increases at increasing rate up to inflection point, then at constant rate and later increases at diminishing rate after inflection point.	TPP increases at diminishing rate	TPP decreases at increasing rate
APP rises throughout first stage and reaches maximum at the end of the stage.	APP starts falling and decreases	APP still falls but remains positive
MPP rises, reaches maximum and starts falling.	MPP falls, but remains positive and becomes zero at the end of second stage.	MPP becomes negative and falls at an increasing rate
Elasticity of production is greater than one because, MPP>APP, Elasticity of production is equal to one at the end of first stage because, MPP=APP,	Elasticity of production is less than one because, MPP<APP, Elasticity of production is equal to zero at the end of second stage because, MPP=0	Elasticity of production is equal to zero at the beginning of third stage because, MPP=0 and in the third stage, Elasticity of production is less than zero because, MPP is negative.
The technical efficiency of both fixed and variable resources increases	The technical efficiency of fixed resource increases (rise in TPP), but technical efficiency of variable resource decreases (fall in MPP and APP).	The technical efficiency of both fixed and variable resources decreases (decline in TPP, APP and MPP)
Fixed factor is more than variable factor	Both fixed factor and variable factor are in optimal combination	Variable factor is more than fixed factor (indiscriminate use of variable factor)
Also called as 'irrational zone' or 'sub-optimal zone'.	Also called as 'rational zone' or 'optimal zone'.	Also called as 'irrational zone' or 'supra-optimal zone'.
MR>MC or $MVP_x > MIC_x$ and hence, farmer is advised to increase factor usage to achieve optimal output.	Scope for MR=MC or $MVP_x = MIC_x$ and it is the stage for optimization of output.	MR<MC or $MVP_x < MIC_x$ and hence, farmer is advised to reduce factor usage to ensure optimization of output.
Scope for re-organization of fixed and variable factors	No scope for re-organization of fixed and variable factors	Scope for re-organization of fixed and variable factors

Importance of LVP/LDR

This LDR has both theoretical as well practical applications and they are discussed here under.

- The LDR forms the basis for Malthusian Theory of Population. According to Thomas Robert Malthus, the growth of food production is at slow pace compared to population growth mainly because of the operation of the LDR in the food production process.
- This law will guide the farmer that, the point of optimality is crossed, once the operation of LDR is witnessed in the production programme. So, invariably at the point of equality of MC and MR from a variable factor, the producer must continue the production programme.
- The concept of LDR helps to explain the disguised unemployment in developing countries like India. We know, capital is a scarce factor in farming and hence, the farmer employs more labour-intensive methods in the production process. As the labour input is too intensive against the scarce capital factor, the MRs we derive from the labour is zero or almost equal to zero. This infers that, even if some labour are removed from the farming, the fall in output is zero. Such type of unemployment, where MPP of labour is zero refers to Disguised unemployment or Concealed employment. So, the concept of LDR guides to withdraw labour from the farming, when their MPP is equal to zero and to allocate these labour services in other productive works such as irrigation works, constructing roads etc., where they contribute higher MPP to the total output in different sectors.
- The LDR also explains the secular stagnation in developed countries, which leads to slow growth of the economy. This is because, the new investments made on the capital equipment will lead to diminishing MRs, thereby, the new investors in the economy are discouraged, till the new capital investments contribute to production.

- This LDR also forms the basis for LDMU in the theory of demand and diminishing marginal productivity theory of distribution.
- LDR forms the basis for Ricardian theory of rent. According to this theory, a superior (fertile) land will earn differential rent because, in the inferior lands, the concept of LDR will be in operation, as the MRs will decline with the increased applications of labour and capital.
- The LDR not only applies to land, building sites, mining, fisheries, plantations etc., but also to other types of production. For example, in the use of machinery, if we employ more and more of raw material (say, sugarcane in a sugar factory), initially the machinery may yield higher levels of output (sugar). But, sooner or later, the additional doses of raw material will increase the total output only at a diminishing rate, implying the operation of LDR. So, the concept of LDR is in operation in all types of production programmes irrespective of the sector. But, in the production programme, where more influence of natural factors is there compared to human element, the LDR operates rather quickly (agriculture). Conversely, when the influence of human element is more compared to natural factors, the operation of LDR is delayed in the production programme (manufacturing industries).

So, the LDR is a universal law and in the words of Wicksteed *'LDR is as universal as the law of life itself'*. So, the universal validity of this law has elevated the subject of Economics to the realm of pure science. Being the operation of LDR is inevitable in any production activity, it is essential to formulate the strategies to delay the operation of this law. It is understood from the limitations of the LDR that, new technological inventions and innovations is the only remedy to delay the operation of LDR in any production activity.

Differences between Traditional Approach and Modern Approach of Returns to a Variable Factor

- As per Traditional approach, the production programme is subjected to three different laws of returns due to a variable factor *i.e.*, law of increasing return, law of constant returns and LDR. But, according to Modern approach, there is one law called LVP in which, the above three laws of returns operate.
- The Traditional approach proposed that, the law of increasing returns operate in manufacturing industries and LDR operates in agriculture sector. But according to Modern approach, all the three laws of returns operate as separate phases in any production programme irrespective of the sector.
- In Traditional approach, land alone is considered as a fixed factor. But, according to Modern approach, any factor can be fixed in the production programme.

14.5. Production Function with Two Variable Factors

So far, we have discussed about the influence of one variable factor on the output of a production programme (short run production function). Now, let us analyze the long run production programme, wherein, all the factors are variable and no factor is fixed. Since, it is difficult to analyze the changes in output due to changes in all the variable factors, it is proposed to consider two variable factors, so as to study their variability and influences on output in the long run production programme. Before studying the production function with two variable factors, it is essential to study the factor combinations and the choice of optimal factor combination in the production programme.

14.5.1. Isoquant

An isoquant is a line or curve, which indicates the various possible combinations of two factors or resources (among which the farmer is indifferent) that yield same level of output to the farmer. Since, any combination of two resources on the same isoquant yield same level of output, isoquant curve is also called as Product-indifference curve or Constant-Product curve or Equal product curve or Iso-Product curve. The basic objective of studying the isoquant is how a rational producer chooses between the various combinations of two resources, so as to achieve cost minimization in the production programme.

We know, the farmer chooses his best combination of resources based on his income and prices of the resources. It is generally assumed that, the farmer has a well-defined preference to the set of all possible combinations of resources. This facilitates him to compare his preferred combination over other combinations of resources. It is further assumed that, the farmer can compare any two resources and between these two resources, he can have more of one to less of the other and he remains indifferent with respect to any combination of resources on the same isoquant. That means, an isoquant shows all possible combinations of two resources that yield same level of output to the farmer. This makes the farmer indifferent between any combinations of two resources on the same isoquant regarding the level of output. In other words, output is held constant along an isoquant. In the Table 14.7, different combinations of two resources *viz.*, manure and fertilizer are shown and any combination of these two resources will yield same level of output to the farmer and the table showing such information is called as Isoquant schedule. If the farmer prefers the first combination, he employs one unit of manure and 13 units of fertilizer to produce a given output of 100 units. In the second combination, he prefers to employ one more unit of manure and he is prepared to give up four units of fertilizer, in order to produce the same level of output. That means, for increasing the usage of one unit of manure, he sacrificed four units of fertilizer. Likewise, when the farmer reaches the fifth combination, he sacrifices one unit of fertilizer for increasing one unit of manure in the production programme.

Table 14.7: Isoquant schedule with reference to manure and fertilizer combinations

Combinations	*Manure (Units) (Added resource)*	*ΔManure (ΔM)*	*Fertilizer (Units) (Replaced resource)*	*ΔFertilizer (ΔF)*	*MRTS Manure for Fertilizer $MRTS_{MF}$ = (ΔF/ΔM)*	*Output (Units)*
A	1		13		–	100
B	2	1	9	–4	4	100
C	3	1	6	–3	3	100
D	4	1	4	–2	2	100
E	5	1	3	–1	1	100

The above discussion also infers that, the concept of isoquant in production theory is similar to the concept of IDC of consumer theory. But, the major difference is that, in IDC analysis, the level of satisfaction cannot be measured or quantified and can only be ranked from low to high, as it is based on ordinalism. But, in the concept of isoquant, the output level can be quantified or measured in terms of physical units, as the production of a commodity being a physical phenomenon. So, different isoquants indicate different levels of output and this facilitate to compare the output levels in physical terms, which is not the case in IDC analysis. So, in an isoquant map, we can compare the output levels in physical units *i.e.*, how much higher output the higher isoquant yields compared to lower isoquant and such comparative analysis in terms of quantification of satisfaction is not possible in IDC analysis.

Properties of Isoquants

The main attributes or properties of isoquants are discussed here under.

(a) Isoquants are Negatively Sloped

The isoquant slope down from left to right. That means, the slope of isoquant is convex to origin or negatively sloped. It slopes downward because, as the farmer employs more of manure, he has to give up certain units of fertilizer, so as to derive the same level of output (Table 14.7 and Figure 14.5). This also implies that, the two resources under consideration are imperfect substitutes. The rate of substitution of manure for fertilizer is in diminishing fashion to produce same level of output of 100 units and this gives the convex shape to isoquant. The Figure 14.5 indicates that, the farmer is indifferent among different combinations of manure and fertilizer, as any combination of these two resources on the same isoquant represents equal level of output. For example, at combination B on the isoquant, the

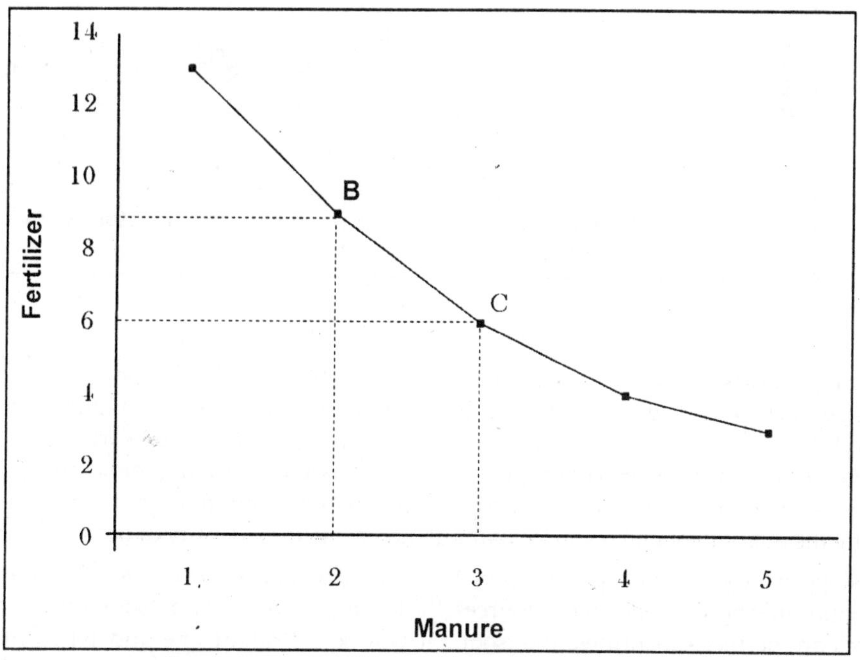

Figure 14.5: A negatively sloped isoquant.

farmer employs two units of manure and nine units of fertilizer to produce 100 units of output. The output is same, even at combination C *i.e.*, when the farmer employs three units of manure and six units of fertilizer on the same isoquant. So, this indicates that, any combination of manure and fertilizer on the same isoquant will yield same level of output to the farmer.

This Figure 14.5 also reveals that, it is only on the negatively sloped isoquant, the different combinations of resources (manure and fertilizer) will yield the same level of output and it makes the farmer indifferent. It is important that, every point on isoquant represent a different combination of both manure and fertilizer, but the farmer is indifferent between any two points on the same isoquant in terms of output. So, all the combinations are equally desirable to the farmer in terms of level of output. But, the farmer is different among the combinations of two resources taking into consideration his income level and prices of the two resources. Say, for example, if the farmer prefers to go for organic cultivation, he employs more of manure compared to fertilizer and thus, he prefers the combination comprising more units of manure and less units of fertilizer, though the output he derives at any point on the isoquant remains same. So, we can infer that, isoquant is a locus of different combinations of two resources, which yield the same level of output to the farmer.

Different Possible Shapes of Isoquants

The shape of isoquant discussed above (convex) illustrate the substitution relationship between the two resources under consideration. Likewise, different shapes of isoquants represent different types of relationships between the resources. They are discussed below.

(*i*) ***Perfect substitutes:*** In case of good substitutes like manure and fertilizer (discussed above), machinery and labour etc., the isoquant is convex in shape. However, in case of perfect substitutes like brand A of nitrogen fertilizer and brand B of nitrogen fertilizer, owned seed and purchased seed, owned human labour and hired human labour, owned bullock labour and hired bullock labour etc., the isoquant is a straight line connecting the two axes (Panel A of Figure 14.6). The parallel straight lines of isoquants for different levels of output represent that, the farmer would be willing to increase the output level at a fixed ratio.

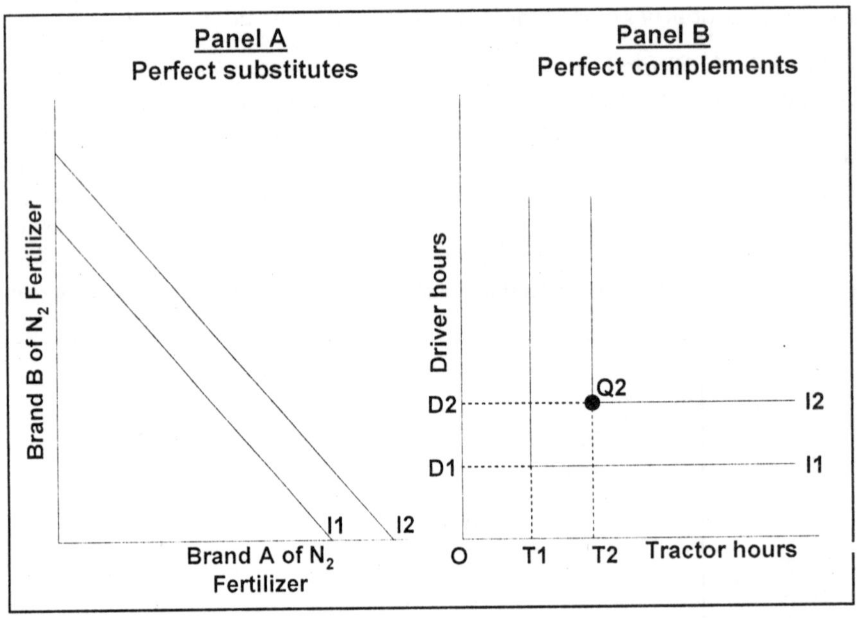

Figure 14.6: Different shapes of isoquants.

(*ii*) ***Perfect complements:*** These are the resources, which combine in a fixed proportion for deriving a given level of output. In such a case, the isoquant is right angle or 'L' shaped (Panel B of Figure 14.6). For example, tractor hours and driver hours, technical ingredients in a fertilizer or pesticide formulations combine in a fixed proportion. To increase the level of output, the farmer has to move to higher isoquant by employing both the resources in large quantities in a fixed proportion. No movement to higher isoquant is possible by simply increasing the quantity of one resource without increasing the quantity of other resource.

The extent of slope of the isoquant depends upon the extent of relationship between the two resources. The greater the substitution relationship between the two resources, the less is the convexity of isoquant and *vice versa*. That means, when the resources are perfect substitutes, the isoquant will be a straight line and when the resources are good substitutes, the isoquant will have a convex shape. On other hand, when the substitution between the resources is completely absent, (perfect complements), the isoquant will have a right angle or 'L' shape. So, we can say that, the greater the complementarity between the resources, the greater the convexity of the isoquants.

(*b*) Slope of Isoquant indicates Diminishing Marginal Rate of Technical Substitution (MRTS) of Resources

As studied through Table 14.7, if the farmer goes on employing more and more manure, he is sacrificing less and less of fertilizer in the production process. So, in order to maintain same level of output, manure replaces less and less of fertilizer. As shown through Table 14.7, as the farmer moves from combination A to E, the rate of substitution of manure (Added resource) for fertilizer (Replaced resource) goes down. In other words, as the farmer employs more and more units of manure, he is prepared to forego less and less of fertilizer in each successive case. This behaviour of compulsory fall in the use of one resource (fertilizer) due to increase in use of other resource (manure) is called Substitution. In the production process, for each and every additional unit of use of manure, it leads to compulsory sacrifisation of some additional units of fertilizer use and this rate of substitution is called MRTS. Since, the MRTS is at diminishing fashion, it is called as diminishing MRTS. It is otherwise defined as 'the rate at which the farmer must gives up one resource (fertilizer) in order to employ an additional unit other resource (manure), so as to maintain same level of output. So, the MRTS is calculated between two resources (manure and fertilizer) placed on an isoquant, which yields a frontier of equi-output for each combination of two resources. This concept of diminishing MRTS was parallel to the diminishing MRS of IDC analysis. The farmer being rational, tend to substitute one resource for the other, so as to derive the same level of output. The following are the two reasons for diminishing MRTS:

- If the farmer goes on employing one resource (manure) and in this process, the MPP derived from each and every unit of this resource will diminish continuously. So, the farmer will be prepared to sacrifice less and less amount of the other resource (fertilizer) in order to employ more and more amount of this added resource (manure), so as maintain same level of output.
- The resources are imperfect substitutes for one another. Since, they are not perfectly substitutable, every time, an increase in units of one resource will not lead to substitution of other resource in equal amounts, but substitute in decreasing amount due to decline in MPP of the added resource and thereby, leading to diminishing MRTS of resources.

The concept of diminishing MRTS between the two resources (manure and fertilizer) is illustrated through Figure 14.7. As the farmer moves from A to E, it indicates the willingness to substitute manure for fertilizer diminishes. This means that, as the amount of manure is increased by equal amounts, the use of fertilizer diminishes by smaller amounts. Thus, the MRTS of manure for fertilizer $MRTS_{MF}$ is the quantity of fertilizer that the farmer is willing to give up to gain a marginal increase of manure in the production programme, so as to maintain same level of output. This diminishing MRTS contributes negative (convex) slope to the isoquant. The greater the fall in MRTS, the greater is the convexity of the isoquant.

In the Figure 14.7, initially, at point A, the farmer is prepared to have more of fertilizer compared to manure. Since, manure factor is scarce here, he will give more value to manure factor than fertilizer factor and hence, he will come

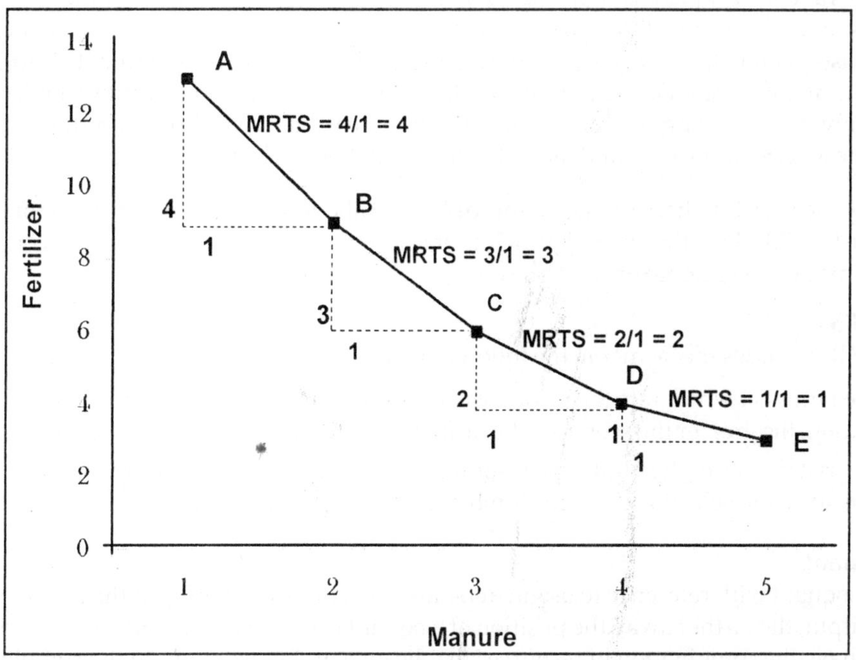

Figure 14.7: Isoquant showing diminishing MRTS of resources.

down the isoquant, thereby, substituting manure for fertilizer. That means, when the farmer is at higher or top most position of isoquant, the economic significance is more for manure. So, the farmer wishes to substitute manure for fertilizer. On the other hand, when the farmer is down the isoquant, he will employ more of manure and less of fertilizer. So, the economic significance of fertilizer will increase now and hence, he will substitute less quantity of fertilizer due to increase in manure. That means, the rate of substitution of fertilizer by employing more of manure will gradually decrease, as the farmer moves down the isoquant because, while coming down the isoquant, the economic significance of fertilizer increases gradually. This phenomenon of changing economic significance between the resources at the top and lower positions of isoquant also leads to diminishing MRTS.

The concept of MRTS is explained by the following formula:

$$MRTS_{MF} = \frac{\text{Change in number of units of replaced resource (fertilizer)}}{\text{Change in number of units of added resource (manure)}}$$

So, MRTS of manure for fertilizer is given by, $MRTS_{MF} = (\Delta F)/(\Delta M)$ *Equation 14.4*

The same concept is explained through Table 14.7 and Figure 14.7. Please note, the convention is to ignore 'minus' sign. It is important to note that, the $MRTS_{MF}$ *i.e.*, $(\Delta F)/(\Delta M)$ is equal to the inverse ratio of MPP of two factors *i.e.*, (MPP_M/MPP_F). This is because, if we move down the isoquant, the loss in physical output due to fertilizer is equal to the gain in physical output due to manure, as at any point on the same isoquant, the output level remains same. The loss in output due to decline in fertilizer usage is given by $\Delta F \times MPP_F$. Similarly, the gain in output due to increase in manure usage is given by $\Delta M \times MPP_M$.

So, we can write, $(\Delta F \times MPP_F) + (\Delta M \times MPP_M) = 0$ or $\Delta F \times MPP_F = \Delta M \times MPP_M$.

So, $(\Delta F)/(\Delta M) = (MPP_M/MPP_F)$ or $MRTS_{MF} = (MPP_M/MPP_F)$ *Equation 14.5*

MRTS of Resources and Shapes of the Isoquants

It is essential to know that, the MRTS of resources also influences the shape of isoquants.

- ☆ ***MRTS is diminishing:*** If the MRTS is diminishing, it implies, both the resources are good substitutes and the shape of isoquant is convex to origin (Figure 14.7). This diminishing rate of factor substitution is more common in agriculture.
- ☆ ***MRTS is increasing:*** If the MRTS between resources is increasing, the shape of isoquant is concave to origin. This is not common in agriculture.
- ☆ ***MRTS is constant:*** If the MRTS between resources is constant or infinite at all points on the isoquant, it implies the resources are perfect substitutes and the shape of isoquant is a straight line connecting the axes (Panel A of Figure 14.6).
- ☆ ***MRTS is zero:*** If the MRTS between two resources is zero, then the two resources are said to be perfect complements and the isoquant is right angled or 'L' shaped. (Panel B of Figure 14.6). To produce more output on higher isoquant I_2, we need to employ more of both the resources. For example, to increase the output to point Q_2 on isoquant I_2, the levels of both the resources should be increased from OT_1 to OT_2 with respect to tractor hours and OD_1 to OD_2 with respect to driver hours. The higher output Q_2 on isoquant I_2 cannot be achieved only by increasing tractor hours without increasing driver hours. In the vertical portion of the isoquant, the $MRTS_{TD}$ is infinity and along the horizontal portion of the isoquant, the $MRTS_{TD}$ is zero.

However, among the possible shapes of isoquants discussed above, convex shaped isoquant is more common in agriculture, as diminishing MRTS is the general principle of isoquant and other shapes such as concavity, straightness and L-shaped isoquants are exceptional or rare cases.

Importance of MRTS

The concept of MRTS guides the farmer in the following aspects:

- ☆ It guides the farmer at which rate the resources under consideration are substituting each other. The greater the substitution, the greater the scope for the farmer to increase the allocation of added (cheaper) resource.
- ☆ It is relative, as it measures the utility of a resource in relation to another resource and not in isolation. Hence, it guides the farmer in selecting the right combination of related resources.

(c) Position of Isoquant

The position of isoquant with reference to origin depends upon the level of output the farmer wishes to produce. Higher the level of output, the farther away the position of isoquant from the origin and *vice versa*. So, it implies that, a farmer will prefer to have the combination of resources that lie on a higher isoquant to the combination that lies on a lower isoquant, as higher isoquant indicates larger level of output. So, greater is the distance of an isoquant from the point of origin, higher it will be in the farmer's preferential order, as output level increases. In other words, an isoquant that lies above and to the right of another isoquant denotes preferred level of higher output. In the Figure 14.8, the farmer would prefer to lie on isoquant I_5 rather than isoquant I_1, as higher isoquant denotes higher output from the two resources. So, the order of preference by the farmer considering the output level is denoted as $I_5>I_4>I_3>I_2>I_1$. For example, points 'B' or 'C' or any point between them on I_2 have more of atleast one resource without having less of the other resource compared to point 'A' on isoquant I_1. So, these points on isoquant I_2 represent higher output level compared to any point (combination of resources) on isoquant I_1. Since, any point on the same isoquant represent same level of output, so all points on isoquant I_2 imply higher output as compared to all points on isoquant I_1. Thus, farther the distance of isoquant away from the origin, suggests higher level of output. However, the output level to be produced by the farmer will be based on the outlay with the farmer and prices of the resources.

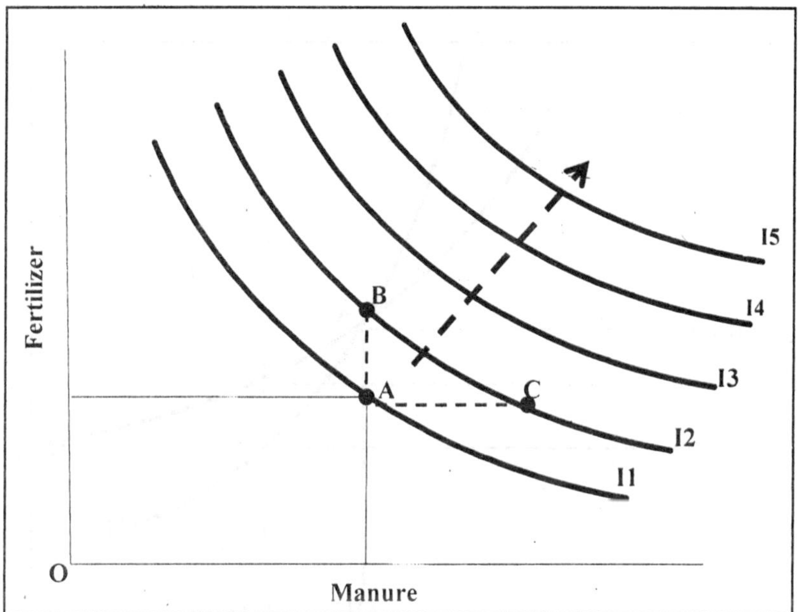

Figure 14.8: Isoquant map and Position of isoquant indicates level of output.

Isoquant Map

We can draw any number of isoquants by considering different quantities of both the resources. But, the farmer remains indifferent among different combinations of the two resources on each isoquant. So, we can say, an isoquant map is a geometrical expression of a number of isoquants at different resources levels indicating different output levels on the same graph. So, an isoquant map comprises of a set of isoquants of different levels of output. The concept of isoquant map further signifies that, in between two isoquants, there can be any number of isoquants indicating different levels of output for different levels of resources combinations.

Isoquant Map vis-à-vis IDC Map

The merit of the isoquant map over IDC map is that, each isoquant indicate the output in physical units and output levels of different isoquants can be compared in physical terms, unlike IDC map. Isoquant map indicates production technology, whereas, IDC map depicts consumer psychology. That means, different isoquants on the isoquant map indicate technologically-decided combinations of different levels of factors to produce different levels of output, but an IDC map indicates subjective ordering of different combinations of two commodities yielding different levels of satisfaction to the consumer. In case of isoquant map, the choice for optimal factor combination is between two ridge lines, where the MPP of factors is diminishing, but positive. However, in case of IDC analysis, the consumer can derive maximum satisfaction at any point on the IDC, provided the three essential conditions are fulfilled *viz.*, tangency between IDC and price line, slope of IDC is equal to slope of price line and at the point of tangency, IDC should be convex to origin.

(*d*) The Isoquants does not Intersect or Tangent to each other

The isoquants does not intersect each other, as different isoquants will indicate different levels of output. As shown in the Figure 14.9, at the point of the intersection 'A' of isoquant I_1 and isoquant I_2, it implies that, the level of output indicated by the two isoquants are equal, which is absurd and impossible. As discussed earlier, higher isoquant I_2 indicates combinations of larger quantities of two resources and lower isoquant I_1 indicates combinations of smaller quantities of the same two resources. So, different combinations of two resources in smaller quantities (I_1) and different combinations of two resources in larger quantities (I_2) will not yield same level of output. But, at point A, it implies, both the isoquants are yielding same level of output, which is absolutely impossible.

As shown in the Figure 14.9 will reference to isoquant I_2, it gives the impression that, at combinations 'A' and 'C', the farmer derives same level of output, as they both lie on the same isoquant I_2. Similarly, at combinations 'A' and 'B', the farmer derives same level of output, as both points lie on the same isoquant I_1. So, it gives the impression that, the output derived at point 'A' on isoquant I_2 is equal to isoquant I_1. But, this is not at all possible, as higher isoquant I_2 represent higher level of output than lower isoquant I_1. Moreover, if combination C is equal to combination A in terms of output and combination B is equal to combination A in terms of output, it follows that, the combination B will give same output as combination C, which is no where possible. This is because, combination C contains more of fertilizer compared to combination B and hence, output level will not remain same. Further, it can be proved from the equations given below.

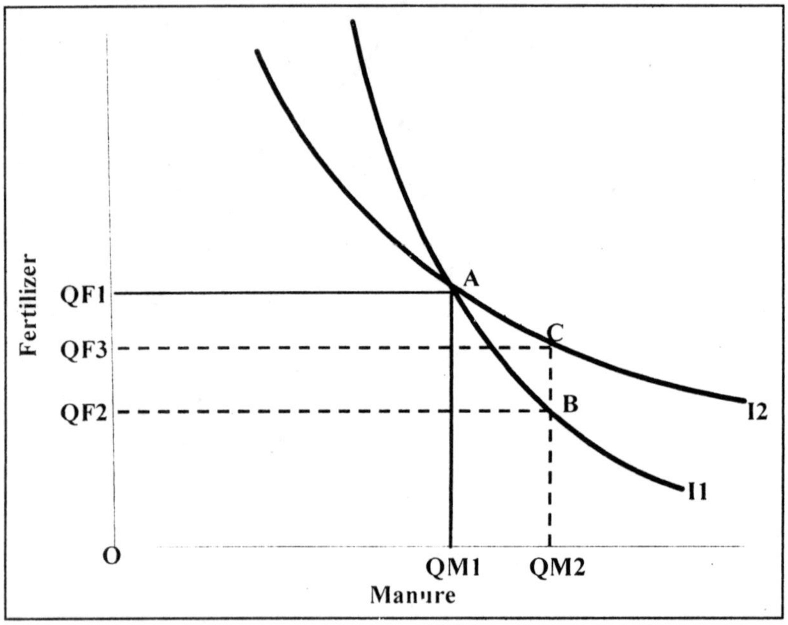

Figure 14.9: Isoquants do not intersect each other.

As combinations A and B lies on the same isoquant I_1, the farmer is indifferent between them. So, we can write,

OQ_{M1} of manure + OQ_{F1} of fertilizer = OQ_{M2} of manure + OQ_{F2} of fertilizer *Equation 14.6*

Similarly, as the combinations A and C lie on the same isoquant I_2, the farmer is indifferent between them. So, we can write,

OQ_{M1} of manure + OQ_{F1} of fertilizer = OQ_{M2} of manure + OQ_{F3} of fertilizer *Equation 14.7*

Since, in the above two equations 14.6 and 14.7, the terms on the left hand side *i.e.*, OQ_{M1} of manure + OQ_{F1} of fertilizer are same and hence, we can write the above two equations 14.6 and 14.7 as,

OQ_{M2} of manure + OQ_{F2} of fertilizer = OQ_{M2} of manure + OQ_{F3} of fertilizer *Equation 14.8*

From the above equation 14.8, we can conclude that, OQ_{F2} of fertilizer is equal to OQ_{F3} of fertilizer, which is impossible. So, no two isoquants will intersect or tangent to each other.

(e) The Isoquant is Asymptotic to Both the Axes

It implies, the isoquant will not touch either of the two axes. If isoquant touches one of the axes, it means, the farmer prefers only one of the two resources. In the Figure 14.10, it is shown that, the isoquant (broken line) touches both the axes *i.e.*, at point Q_M on X-axis and at point Q_F on Y-axis. So, at point Q_M it gives the impression that, the farmer purchase only OQ_M quantity of manure and no quantity of fertilizer. Similarly, at point Q_F, the farmer purchases only OQ_F quantity of fertilizer and no quantity of manure. Such isoquants are not possible in producer analysis, as isoquant concept assumes resources combinations should be employed to produce the output.

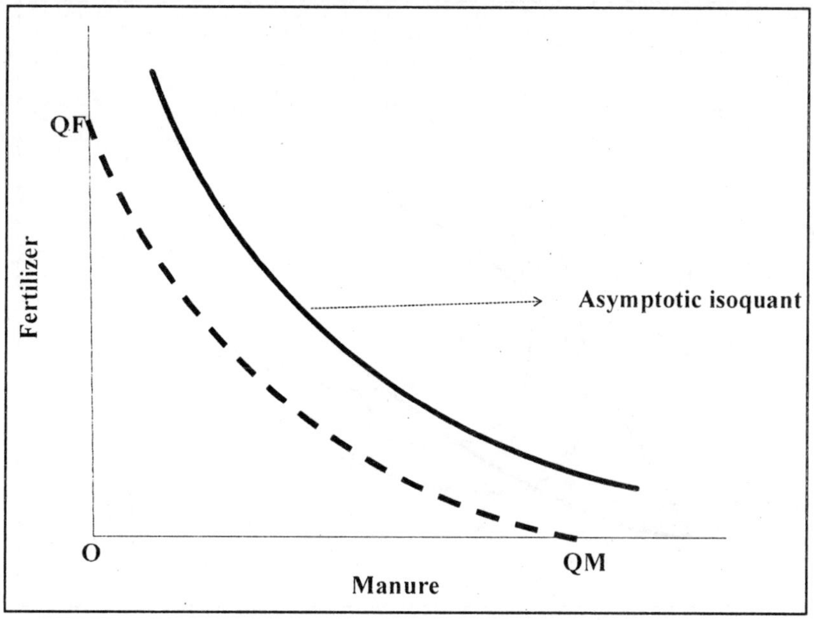

Figure 14.10: Isoquant is asymptotic to both the axes.

(f) Isoquants not Necessarily Parallel to Each Other

Since the resources under consideration are related resources (imperfect substitutes), isoquants will not be parallel to each other. On the other hand, if the resources under considerations are perfect substitutes and perfect complements, the isoquants remain parallel to each other in an isoquant map. But, since the resources are imperfect substitutes and the MRTS differs with reference to quantities of resources employed, the isoquants will slope downward from left to right, but may not remain parallel to each other. The same is explained through Figure 14.11 and the three isoquants I_1, I_2 and I_3 are not parallel to each other. The steeper the slope of the isoquant, the greater the MRTS of two resources and hence, the isoquants may not be parallel to each other.

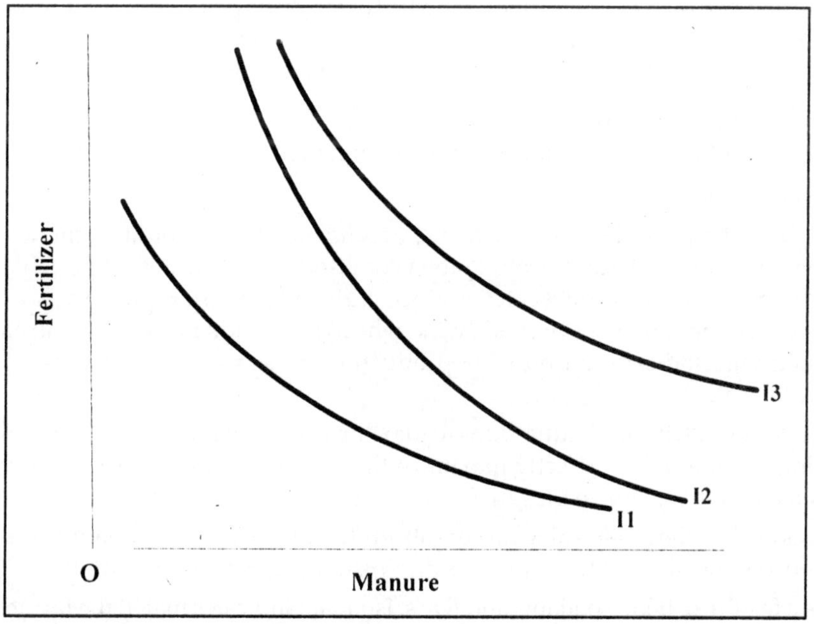

Figure 14.11: Isoquants are not parallel to each other.

(g) Isoquants are Oval Shaped

In reality, the isoquant is oval shaped, which means that, at some level of resources usage, the isoquant begins to recede from each of the axes. In the Figure 14.12, above OA line (ridge line), the isoquants bends away from the Y-axis and below the OB line (ridgė line), the isoquants bends away from the X-axis. This bending of isoquants away from the axes implies that, both the resources should be increased to maintain same level of output, if the farmer wishes to produce the output beyond the ridge lines. But, between the ridge lines, the resources substitute each other. That means,

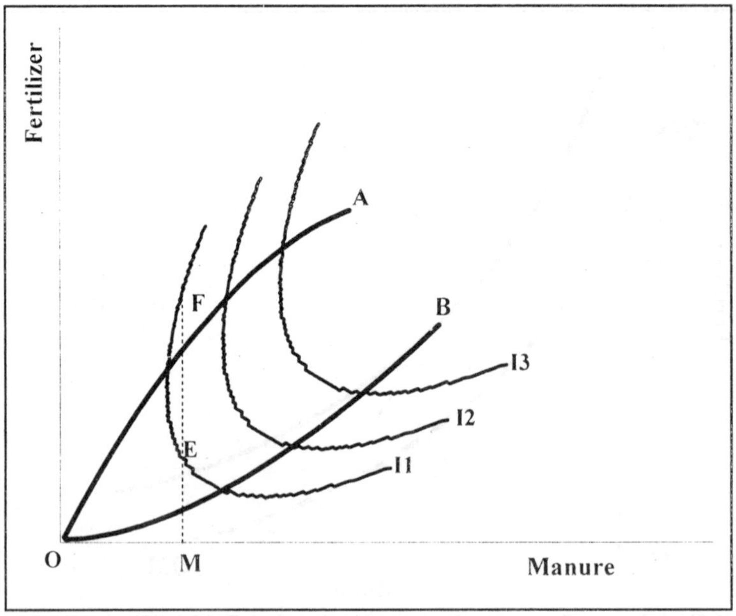

Figure 14.12: Isoquants are oval in shape.

beyond OA ridge line, the MPP of fertilizer becomes negative, but beyond OB ridge line, the MPP of manure becomes negative. This is explained through drawing a straight line MEF with respect to isoquant I_1 producing 100 units of output (100Y). At point E on the isoquant I_1, to produce 100Y, the farmer employs OM units of manure and ME units of fertilizer. Since the point E lies between the two ridge lines, the two resources substitute each other. If the farmer wishes to produce 100Y at point F on I_1, he has to employ the same OM units of manure, but MF units of fertilizer. This infers that, even by reducing fertilizer dosage to ME from MF and by employing same OM units of manure, the farmer can maintain the output at 100Y. That means, beyond the OA ridge line, the MPP of fertilizer is negative. Similarly, producing output at any point outside the ridge line OB, the MPP of manure is negative. On the ridge lines, the MPP of one of the factors is zero and between the ridge lines, the MPP of factors is positive, but diminishing. So, the segments of isoquant beyond the ridge lines are called 'Waste bearing segments' or 'Uneconomic regions of production'. The portion of isoquant between the ridge lines is called 'Region of economic production'.

Similaritie between Isoquant and IDC

An isoquant is equivalent to an IDC in a number of ways and they are discussed here under.

- ☆ Both the curves are convex to origin.
- ☆ Both curves show the possible choices when spending a certain amount of money. That is, an business graphs image by Chad McDermott from Fotolia.com isoquant indicates all possible combinations of two resources (say, manure and fertilizer) to produce a given level of output. An IDC indicates all possible combinations of two commodities (say, paddy and wheat) that yield same level of satisfaction to the consumer. So, an isoquant map indicates all possible production choices and an IDC map indicates all available consumption choices.
- ☆ Both curves assume that, the individual aims to maximize the benefits in the activity. An isoquant guides the farmer to minimize the cost (indirectly maximize the benefits) in the production programme and an IDC guides the consumer to maximize the satisfaction in the consumption behaviour.
- ☆ Both the curves explain the diminishing rate of substitution. The slope of isoquant curve indicates diminishing MRTS of resources and slope of IDC indicates diminishing MRS of commodities.
- ☆ Tradeoffs are a feature of both isoquants and IDCs. Isoquant indicates that, if the farmer employs more of one resource, he has to reduce the level of other resource, so as to maintain same level of output. In case of IDC, if the consumer consumes more of one commodity, he has to reduce the consumption of other commodity, so as to derive same level of satisfaction.

Dissimilarities between Isoquant and IDC

As discussed above, the isoquant of production theory plays much the similar kind of role that an IDC plays in consumer theory. However, there are definite differences amidst the two and they are discussed here under.

- ☆ An IDC shows various combinations of two commodities that yield same level of satisfaction to the consumer. But, an isoquant shows various combinations of two resources that yield same level of output. That means, on an IDC, satisfaction level remains same, but on isoquant, output level remains same.
- ☆ The output level indicated by an isoquant is measurable, but the satisfaction level indicated by the IDC cannot be measured, as IDC concept is based on ordinality approach.
- ☆ IDC shows various combinations of two commodities in the consumption behaviour of the consumer, where as an isoquant shows the possible combinations of two resources in the production programme taken up by the famer.
- ☆ An isoquant provides information pertaining to economic and non-economic regions of production, but an IDC will not provide such information regarding the combinations of consumption of commodities.
- ☆ Slope of an isoquant diminishing MRTS of resources, whereas, slope of an IDC indicates diminishing MRS of commodities.
- ☆ On an IDC map, one can only say that, a higher IDC gives more satisfaction than a lower one. However, it cannot be said how much more or less satisfaction is being derived from one IDC corresponding to the other. But, one can easily judge by how much output is greater on a higher isoquant relating to a lower isoquant. Thus, IDC approach is subjective and isoquant approach is objective.
- ☆ IDCs are strictly convex with a continuously diminishing MRS, but isoquants may have linear segments with a sectionally declining MRTS.
- ☆ An isoquant shows the choices of a farmer in the allocation of two resources in different combinations that yield same level of output and the IDC shows the choices of a consumer in the consumption of two commodities in different possible combinations that yield same level of satisfaction.

14.5.2. Iso-cost Line

From the isoquant map alone, it is not possible to compute the Least Cost Combination (LCC) of resources, as it simply indicate the various quantities of two resources under consideration by the farmer for deriving different levels of output. On the same isoquant, different combinations of two resources will yield same level of output. But, the best choice of combination of resources the farmer will make, however, depend on his cash outlay and relative prices of the two resources. So, to ascertain the point of LCC, we should have information about the cash outlay of the farmer and prices of the two resources, which helps to derive the iso-cost line of the farmer. So, this concept of studying the iso-cost line of the farmer is essential for knowing the point of LCC. An iso-cost line represents the various combinations of two resources, which can be purchased by the farmer with a given cash outlay and at given prices of two resources. This iso-cost line is drawn as a continuous line and it identifies the options from which the farmer can choose the combination of resources. It also indicates the limitations of farmer's choices for different combinations of resources. The iso-cost line is also called as Price line, Outlay line, Input price line and Factor cost line.

For example, a farmer is having an outlay of Rs.60 and he prefers to have two resources *viz.*, manure and fertilizer. The price of manure is Rs. 5/unit and the price of fertilizer is Rs. 6/unit. Given the outlay and the prices of the two resources, the farmer prefers to have 13 possible combinations of two resources as shown through Table 14.8 and Figure 14.13.

The Table 14.8 shows that, if the farmer spends all his money on fertilizer (Combination B), he purchases 10 units of fertilizer at Rs. 6/unit thereby, nothing is left to purchase manure. Similarly, if the farmer allocates all his money on manure (Combination M), he purchases 12 units of manure at Rs. 5/unit and nothing is left to purchase fertilizer. However, the intermediate purchases of the combinations of manure and fertilizer from B to L shows the mixes of both manure and fertilizer resources purchased worth of his cash outlay Rs. 60/-. For example, at combination B, the farmer purchases one unit of manure and 9.2 units of fertilizer by allocating all his money (Rs. 60) on both the resources. The same can be explained through Figure 14.13.

Table 14.8: Attainable combinations of manure and fertilizer by the farmer for a given cash outlay and prices of the resources.

Combinations	*Manure (Units)*	*Fertilizer (Units)*
A	0.0	10.0
B	1.0	9.2
C	2.0	8.3
D	3.0	7.5
E	4.0	6.7
F	5.0	5.8
G	6.0	5.0
H	7.0	4.2
I	8.0	3.3
J	9.0	2.5
K	10.0	1.7
L	11.0	0.8
M	12.0	0.0

In the Figure 14.13, the line PL indicates the iso-cost line of the farmer, indicating various combinations of

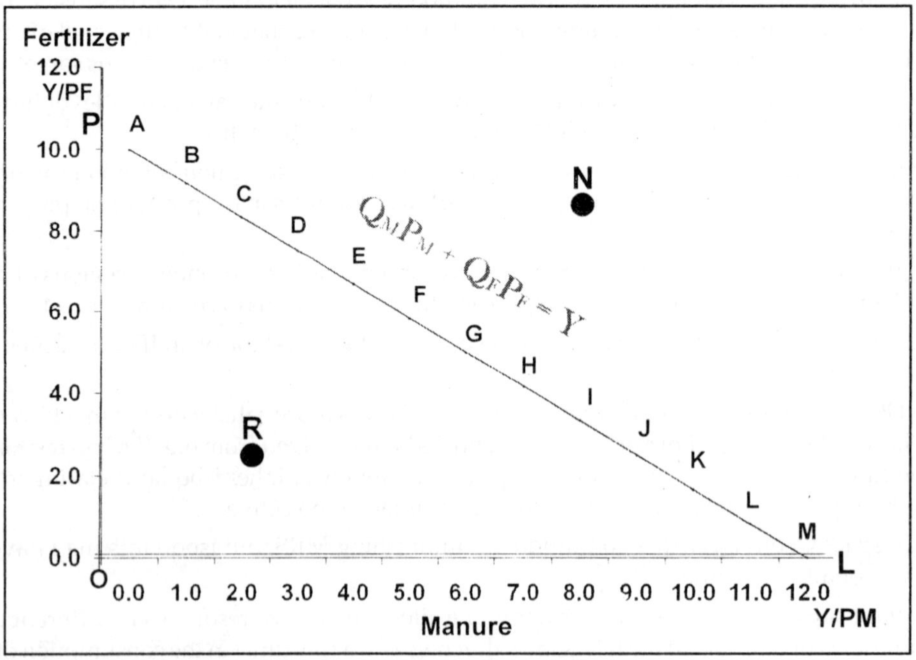

Figure 14.13: Iso-cost line of the farmer for manure and fertilizer combinations.

manure and fertilizer resources purchased by the farmer at a given level of outlay and at given prices of two resources. It shows 13 possible combinations of manure and fertilizer purchased by the farmer from A to M. Combination A indicates that, the farmer has purchased 10 units of fertilizer by allocating all his money *i.e.*, Rs. 60 on fertilizer only and nothing of manure is purchased. Similarly, point M shows the farmer has purchased 12 units of manure by allocating all his money *i.e.*, Rs. 60 on manure only and nothing of fertilizer is purchased. The remaining points on the iso-cost line *i.e.*, from B to L indicates various combinations of both manure and fertilizer the farmer can purchase at given outlay and prices of the two resources.

Another interesting point we can ascertain from the iso-cost line is that, the farmer can purchase any combination of resources that lies inside the iso-cost line. In the Figure 14.13, combination R of manure and fertilizer resources is feasible to purchase by the farmer, but some money will be saved on the part of the farmer. This is because, purchase of manure and fertilizer resources at combination R involves outlay of money less than Rs.60. On the contrary, to purchase any combination of resources of manure and fertilizer outside the iso-cost line, such as, N requires an outlay larger than the farmer's income of Rs.60, thereby, such combination is not attainable. So, total spending for the resources can fall short of money with the farmer, but should not exceed it.

Let us assume that, the income of the farmer is Y and the prices of two resources *viz.*, manure and fertilizer are P_M and P_F respectively. If the farmer wishes to purchase Q_M and Q_F quantities of manure and fertilizer respectively as the preferred combination, he has to spend Q_MP_M amount of money on manure and Q_FP_F money on fertilizer respectively. So, the farmer to purchase Q_M and Q_F quantities of both manure and fertilizer, he will have to spend $Q_MP_M + Q_FP_F$ amount of money. That means, the farmer can purchase this combination of resources, if he possess at least $Q_MP_M + Q_FP_F$ amount of money. Of course, the farmer can prefer any combination of manure and fertilizer resources at given prices and given money of the farmer, provided the costs incurred on them is less than or equal to the money he possess. So, we can infer that, the farmer can select any combination of resources such that, $Q_MP_M + Q_FP_F = Y$. The points below the iso-cost line represent the combinations of manure and fertilizer resources, which cost strictly less than income Y. The iso-cost line, $Q_MP_M + Q_FP_F = Y$ can also be expressed with respect to Q_M as $(Y/P_M) - (Q_FP_F/P_M)$. Similarly, the iso-cost line, $Q_MP_M + Q_FP_F = Y$ can also be expressed with respect to Q_F as $(Y/P_F) - (Q_MP_M/P_F)$. So, the iso-cost line is a straight line with horizontal intercept Y/P_M and with vertical intercept Y/P_F. The horizontal intercept (Y/P_M) represents the combination of resources that the farmer purchases, if he spends all his money on manure *i.e.*, he purchases only manure and no amount of fertilizer. Similarly, the vertical intercept (Y/P_F) represents the combination of resources that the farmer purchases, if he spends all his money on fertilizer, *i.e.*, he purchases only fertilizer and no amount of manure (Figure 14.13).

Slope of the iso-cost line

The slope of the iso-cost line indicates the price ratio of the two resources. The slope of iso-cost line can be studied through the following three methods:

Method 1

The slope of iso-cost line can be studied through the Figure 14.13. In the Figure 14.13, OL units of manure can be exchanged with OP units of fertilizer, as the money value with respect to OL units of manure and OP units of fertilizer is same. So, one unit of manure is equal to (OP/OL) units of fertilizer. This is the same thing as the tangent of angle PLO. So, the rate of exchange of the manure and fertilizer is, thus, given by tangent PL. This PL is the iso-cost line. The slope of the iso-cost line can be derived mathematically, as given below:

- Along X-axis, the total quantity of manure purchased by the farmer is OL at a given price of P_M, if the whole money of the farmer is allocated on the manure. This is given by Y = OL x P_M. So, OL = Y/P_M
- Similarly, along Y-axis, the total quantity of fertilizer purchased by the farmer is OP at a given price of P_F, if the whole money of the farmer is allocated on the fertilizer. This is given by Y = OP x P_F. So, OP = Y/P_F
- As discussed earlier and from the Figure 14.13, slope of the iso-cost line is given by OP/OL. So, from the equation 14.9, the slope of iso-cost line indicates the price ratio of two resources.

  $OP/OL = (Y/P_F)/(Y/P_M) = (P_M/P_F)$ *Equation 14.9*

In the given example, as the price per unit of manure is Rs.5 and price per unit of fertilizer is Rs.6, the slope of iso-cost line is 0.83. Since, the prices are assumed constant at a given time, the iso-cost line is a straight line.

Method 2

We know, the slope of the iso-cost line measures the amount of change in one resource (say, fertilizer) due to per unit change in other resource (say, manure) along the iso-cost line. In the Figure 14.14, let us consider two points A and B on the iso-cost line, where point A represents Q_M and Q_F combinations of two resources and point B represents the movement down the iso-cost line given by $Q_M+\Delta Q_M$, $Q_F-\Delta Q_F$. So, the iso-cost line with reference to these two points A and B is given as follows:

At point A: $(P_M \times Q_M) + (P_F \times Q_F) = Y$ *Equation 14.10*

At point B: $(P_M \times (Q_M + \Delta Q_M)) + (P_F \times (Q_F - \Delta Q_F)) = Y$ *Equation 14.11*

Subtracting Equation 14.10 from Equation 14.11, we get,

$P_M \,\Delta Q_M - P_F \,\Delta Q_F = 0$ *Equation 14.12*

By rearranging the terms in Equation 14.12, the slope of the iso-cost line is $(\Delta Q_F/\Delta Q_M) = (P_M/P_F)$.

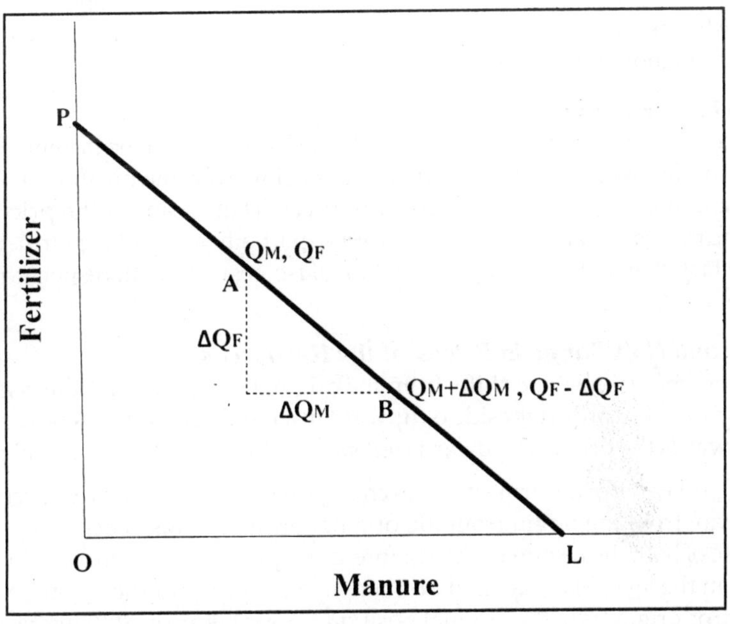

Figure 14.14: Method 2 – Estimation of slope of the iso-cost line.

Method 3

The slope of the iso-cost line can be simply estimated through computing the ratio between vertical intercept and horizontal intercept of the iso-cost line. As discussed earlier, we know, iso-cost line can be denoted as $(P_M \times Q_M) + (P_F \times Q_F) = Y$.

☆ ***Horizontal intercept of iso-cost line :*** The above iso-cost line can be expressed with respect to Q_M as
$(P_M \times Q_M) = Y - (P_F \times Q_F)$.
So, $Q_M = (Y/P_M) - (Q_F P_F/P_M)$ ***Equation 14.13***
In the above equation 14.13, (Y/P_M) is the horizontal intercept of the iso-cost line with respect to manure factor.

☆ ***Vertical intercept of iso-cost line :*** Similarly, we can compute the vertical intercept of iso-cost line. The iso-cost line can be expressed with respect to Q_F as
$(P_F \times Q_F) = Y - (P_M \times Q_M)$.
So, $Q_F = (Y/P_F) - (Q_M P_M/P_F)$ ***Equation 14.14***
In the above equation 14.14, (Y/P_F) is the vertical intercept of the iso-cost line with respect to fertilizer factor.

☆ So, slope of iso-cost line = Vertical intercept/Horizontal intercept = $(Y/P_F)/(Y/P_M) = (P_M/P_F)$.

Notice that, the slope of the iso-cost line depends only on the ratio of the prices of two resources and not on their absolute values. In this example, the price of manure is Rs. 5/unit and price of fertilizer is Rs. 6/unit. So, the ratio of the prices of manure and fertilizer is 1.0:1.2 and thereby, if the farmer wants to increase the allocation of manure by one more unit, he has to sacrifice 1.2 units of fertilizer. On the other hand, if the farmer increases the fertilizer allocation by one more unit, he has to sacrifice 0.83 units of manure. Similarly, if we consider prices of manure and fertilizer as Rs. 10/unit and Rs. 12/unit respectively, the farmer must still forego 1.2 units of fertilizer for increasing the allocation of one more unit of manure and 0.83 units of manure for increasing the allocation of one more unit of fertilizer. So, as long as the price of fertilizer is 1.2 times the price of manure, the farmer must forgo the same amounts, if he increases the allocations of either manure or fertilizer by one more unit. This guides that, the slope of the iso-cost line is guided by the price ratio of the resources rather than their absolute values. So, this concept also illustrates the opportunity cost of manure allocation vis-à-vis fertilizer allocation and it may be written as:

(P_M/P_F) = Opportunity cost of manure in terms of fertilizer or Opportunity cost of fertilizer in terms of manure.

So, the above explanation infers that, any change in cash outlay and changes in the prices of both manure and fertilizer in the same proportion will not affect the price ratio, thereby, the slope of iso-cost line will remain unaffected. This discussion further helps to clarify the distinction between money prices and relative prices. Here, both P_M and P_F are the money prices of manure and fertilizer respectively, while the ratio P_M/P_F is a relative price.

It is important to note that, both isoquant and iso-cost line have negative slopes, but the MRTS indicated by the slope of isoquant diminishes as we move down the curve, whereas, the price ratio of factors indicated by slope of iso-cost line remains same throughout the iso-cost line.

Position and Shifts in Iso-cost Line

The position of iso-cost line from the origin indicates the cash outlay with the farmer. The closer the iso-cost line to the origin, the lesser the outlay with the farmer and *vice versa*. However, the position of iso-cost line with respect to origin is inversely influenced by the prices of the two resources. That means, if the prices of both the resources are increased, the iso-cost line moves towards origin and *vice versa*. So, both outlay of the farmer and prices of the resources are the major influential factors in determining the position of iso-cost line with respect to origin and their influences are discussed here under.

(a) Changes in Outlay and No Change in Prices of the Resources

In this given situation, the iso-cost line shifts from the original position. With the rise in outlay, the iso-cost line shifts away from the origin *i.e.*, towards right side or upward in a parallel position. Conversely, with the fall in outlay, the iso-cost line shifts towards the origin *i.e.*, towards left side or downward in a parallel position.

A rise in the level of outlay with the farmer and no change in the prices of the resources, will shift the iso-cost line towards right side or away from the origin from the original position. This is because, with increase in outlay, the farmer can purchase more of both the resources than before, as the cost of one resource in terms of the other remains the same. Conversely, a fall in the level of outlay and no change in the prices of the resources, will shift the iso-cost line towards left side or towards origin from the original position. This is because, with decrease in outlay, the farmer can purchase less of both the resources than before. The same explanation shown through Figure 14.15 reveals that, due to rise in cash outlay with the farmer, the iso-cost line PL shifts away from the origin or parallel upwards as $P_I L_I$. So, the farmer is able to purchase more of both the resources *viz.*, manure and fertilizer. Conversely, due to fall in outlay, the iso-cost line PL shifts towards the origin or parallel downwards as $P_D L_D$. So, the farmer will purchase less of both the resources.

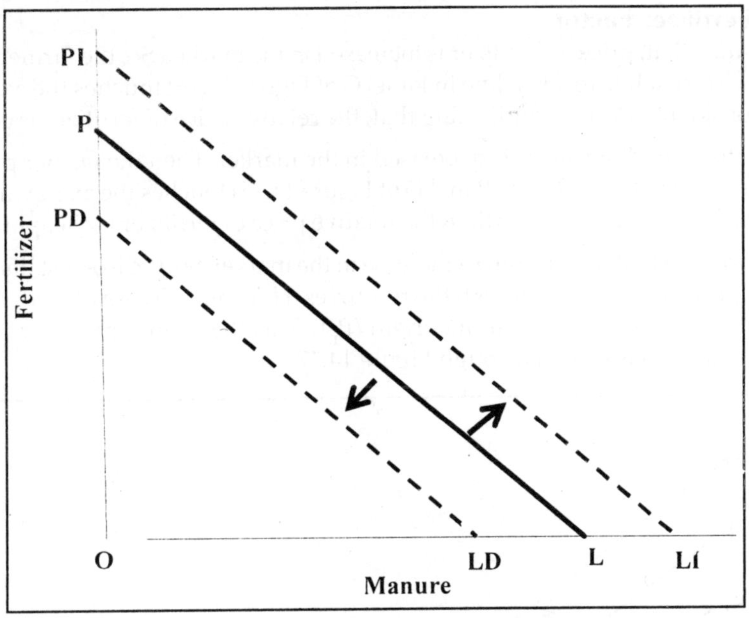

Figure 14.15: Shifts in iso-cost line due to changes in money income with the farmer.

(b) Changes in Prices and No Change in Outlay with the Farmer

Now, let us assume, the prices of the resources under consideration *viz.*, manure and fertilizer fluctuate in the market, whereas, the outlay with the farmer remains same.

Changes in Price of Manure Factor

In the first case, assume that, price of manure is increased in the market. Then the farmer purchases less of manure than before. So, the new iso-cost line (broken line in Panel A of Figure 14.16) becomes more steeper than before and touches the axis closer towards origin on which the manure is considered (X-axis), indicating that, the relative price of manure is costlier.

Now assume that, the price of manure is decreased in the market. Then the farmer purchases more of manure than before. So, the new iso-cost line (broken line in Panel B of Figure 14.16) becomes more flatter than before and touches the axis away from the origin on which the manure is considered (X-axis), indicating that, the relative price of manure is cheaper.

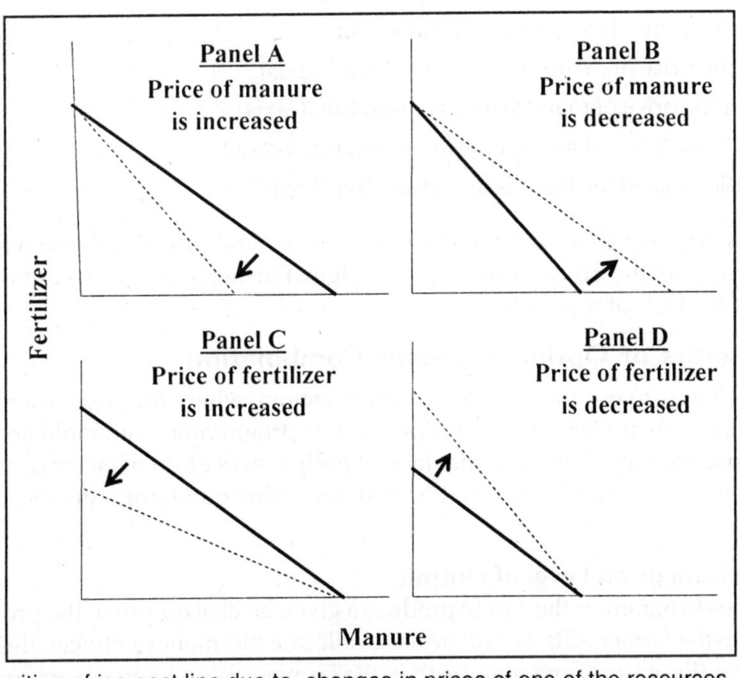

Figure 14.16: Shifts in the position of iso-cost line due to changes in prices of one of the resources.

Changes in Price of Fertilizer Factor

In the first case, assume that, price of fertilizer is increased in the market. So, the farmer purchases less of fertilizer than before. So, the new iso-cost line (broken line in Panel C of Figure 14.16) touches the axis closer towards origin on which the fertilizer is considered (Y-axis), indicating that, the relative price of fertilizer is costlier.

Now assume that, the price of fertilizer is decreased in the market. Then the farmer purchases more of fertilizer than before. So, the iso-cost line (broken line in Panel D of Figure 14.16) touches the axis away from the origin on which the fertilizer is considered (Y-axis), indicating that, the relative price of fertilizer is cheaper.

In some cases, the prices of both the resources increase in the market. So, the iso-cost line moves towards the origin (P_IL_I), as the farmer can purchase only less of both the resources. However, in case of price fall of both the resources in the market, the iso-cost line shifts away from the origin (P_DL_D), as the farmer can now purchase more of both the resources. The same explanation is shown through Figure 14.17.

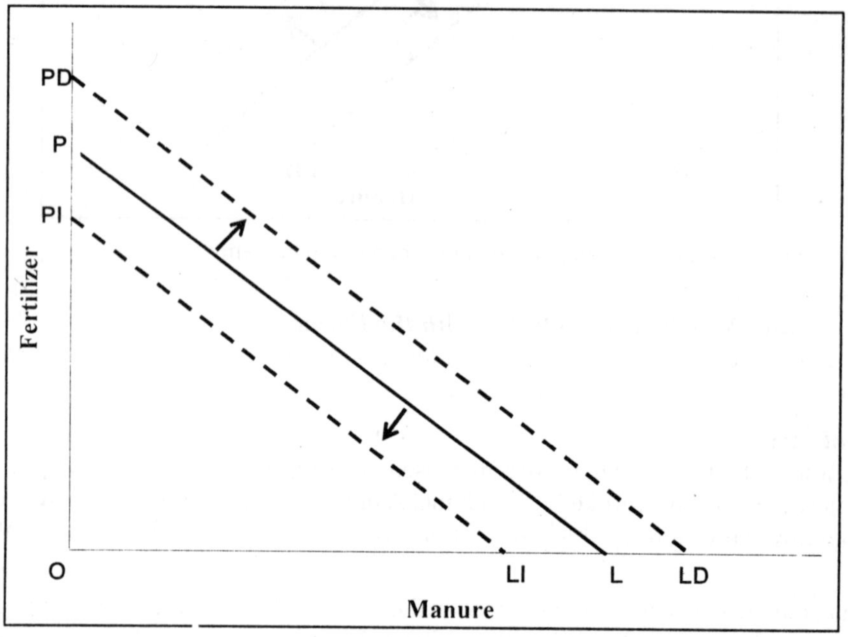

Figure 14.17: Shifts in iso-cost line due to fluctuations in prices of both the resources.

To summarize, the following are the six possible ways that iso-cost line can move:

- ☆ Shift left: If the farmer has less money to spend
- ☆ Shift right: If the farmer has more money to spend
- ☆ Swing left: If the price of manure is increased (on X-axis)
- ☆ Swing right: If the price of manure is decreased (on X-axis)
- ☆ Swing down: If the price of fertilizer is increased (on Y-axis)
- ☆ Swing up: If the price of fertilizer is decreased (on Y-axis)

The above analysis indicates that, both income of the farmer and prices of the two resources are the two major determinants of the position of the iso-cost line from the origin. Further, the analysis advocates the importance of iso-cost line in determining the LCC of resources.

14.5.3. LCC of Factors or Optimum Factor Combination

The LCC of factors refers to optimum combination of factors, where the costs are minimum in the production programme. To ensure the costs are minimum in the production programme, we should be aware of two basic concepts *viz.*, isoquant and iso-cost line. Generally, a farmer faces two choices of optimal factor combinations or LCC of factors *viz.*, cost minimization for a given level of output or output maximization for a given cash outlay and prices of the resources.

1. Cost Minimization for a given Level of Output

Here, the farmer tries to minimize the cost to produce a given level of output in the production programme. Let us suppose, the money with the farmer is Rs. 60 and he has to allocate this money between the two resources *viz.*, manure (Rs. 5/unit) and fertilizer (Rs. 6/unit) and assume that, the farmer wishes to produce 100 units of output (100Y). The earlier discussion on isoquant and iso-cost line reveals two important aspects *viz.*, the isoquant infers 'what combination of resources the farmer is willing to buy' and the iso-cost line guides the farmer, 'what the farmer is able to buy', so as to attain the LCC of resources. So, when the isoquant and the iso-cost line are combined, we find the quantities of each resource the farmer is both willing and able to buy, so as to produce a desired level of output. So, the isoquant and iso-cost line for a given level of output helps the farmer to find out the combination of resources that will lead to LCC of resources. So, the LCC refers to the point of cost minimization in the production programme at a particular combination of resources, given outlay with the farmer and prices of the resources in the market.

From the Figure 14.18, it is clear that, the iso-cost line PL is tangent to the isoquant of 100Y at point 'A' and this indicates that, by allocating OQ_M units of manure and OQ_F units of fertilizer, the cost is minimum in the production programme. It is also important that, the farmer can allocate any combination of resources below PL iso-cost line, but 100Y cannot be achieved. Similarly, the iso-cost line P_1L_1 touches the isoquant at B and C, but to produce same 100Y, the farmer needs more cash outlay, as these points B and C are with reference to higher iso-cost line. But, the objective of the farmer is to minimize the cost in the production programme to produce a given output of 100Y and this is feasible at point A only. This discussion infers that, three important conditions must be fulfilled to arrive at the point of LCC of resources. They include,

(*a*) The iso-cost line should be tangent to the isoquant from below

(*b*) Slope of the iso-cost line should be equal to the slope of isoquant

(*c*) Isoquant should be convex to the origin

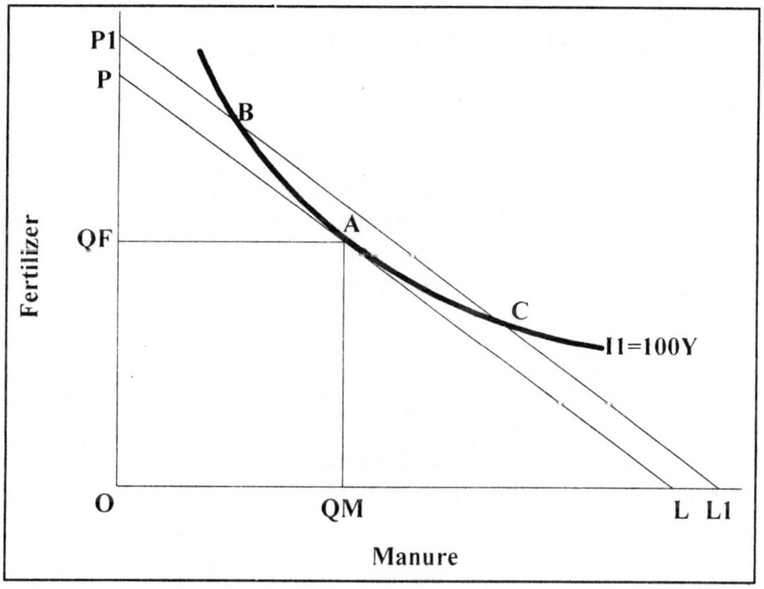

Figure 14.18: Farmer's equilibrium – Iso-cost line tangent to isoquant (Cost minimization for a given output).

(a) The Iso-cost Line should be Tangent to the Isoquant from below

The farmer's equilibrium position is only at a point, where the iso-cost line is tangent to the isoquant from below. This is explained through the Figure 14.18, where an isoquant I_1 and PL iso-cost line are tangent to each other at point A to produce 100Y. That means, the farmer attains LCC of resources by allocating OQ_M quantity of manure and OQ_F quantity of fertilizer. Even though, the points B and C on the same isoquant yields 100Y, they are not the points of cost-minimization, as these points B and C are with reference to higher iso-cost line P_1L_1. To put it simply, cost-minimization cannot be ensured to produce a given level of output of 100Y either at B or C. So, we can infer that, at point A, where iso-cost line is tangent to isoquant I_1 from below, it is the point of LCC of resources *i.e.*, by employing OQ_M quantity of manure and OQ_F quantity of fertilizer resources.

(b) Slope of the Iso-cost Line should be Equal to the Slope of Isoquant

The second condition essential for arriving at the point of LCC is that, the slope of the iso-cost line should be equal to the slope of isoquant from below of the desired output level (100Y). As explained through Figure 14.19, the iso-cost line PL of the farmer is tangent to the isoquant I_1 of 100Y from below at point A. So, at point A on isoquant I_1, the slope of the iso-cost line is equal to the slope of the isoquant I_1, where the farmer employs OQ_M and OQ_F quantities of manure and fertilizer respectively. We know, the slope of isoquant indicates MRTS of manure for fertilizer and slope of iso-cost line indicates price ratio of two resources and the point of LCC is explained below.

Slope of the isoquant = $MRTS_{MF} = (\Delta F)/(\Delta M)$

Slope of the iso-cost line = (P_M/P_F)

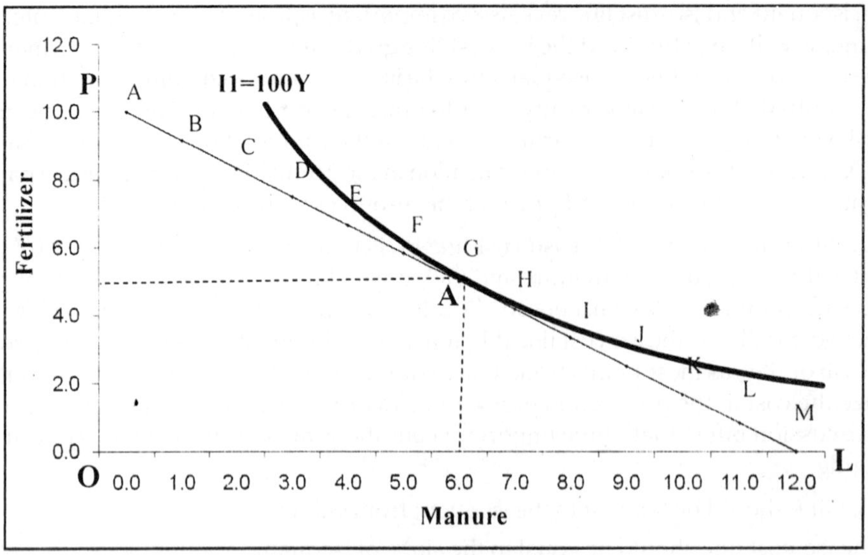

Figure 14.19: Farmer's equilibrium for manure and fertilizer combinations (Slope of isoquant = Slope of iso-cost line).

So, the point of farmer's equilibrium is when the slope of the isoquant is equal to the slope of the iso-cost line and this is given by,

$(\Delta F)/(\Delta M) = (P_M/P_F)$ or $MRTS_{MF} = (P_M/P_F)$ *Equation 14.15*

The above Equation 14.15 infers that, at point A, what the farmer is willing to pay *i.e.*, his personal exchange rate between manure and fertilizer resources ($MRTS_{MF}$) is equal to what he actually pays *i.e.*, the market exchange rate P_M/P_F. So, at the point of LCC, the slope of isoquant is equal to the slope of iso-cost line or the $MRTS_{MF}$ is inversely equal to the price ratio of the manure and fertilizer resources. The point of LCC can also be stated that, the rate at which the farmer is willing to substitute manure for fertilizer is inversely equal to the price ratio between manure and fertilizer resources.

(c) Isoquant should be Convex to the Origin

This is an essential condition to be fulfilled for arriving at the point of producer's equilibrium. When the iso-cost line is tangent to the isoquant, at the point of tangency, the isoquant should be convex to origin and this is a necessary condition for farmer's equilibrium. In other words, the MRTS of manure for fertilizer must be diminishing at the point of equilibrium. The same is shown through the Figure 14.20, where at the point of consumer's equilibrium *i.e.*, at A, the shape of the isoquant I_2 is convex to origin.

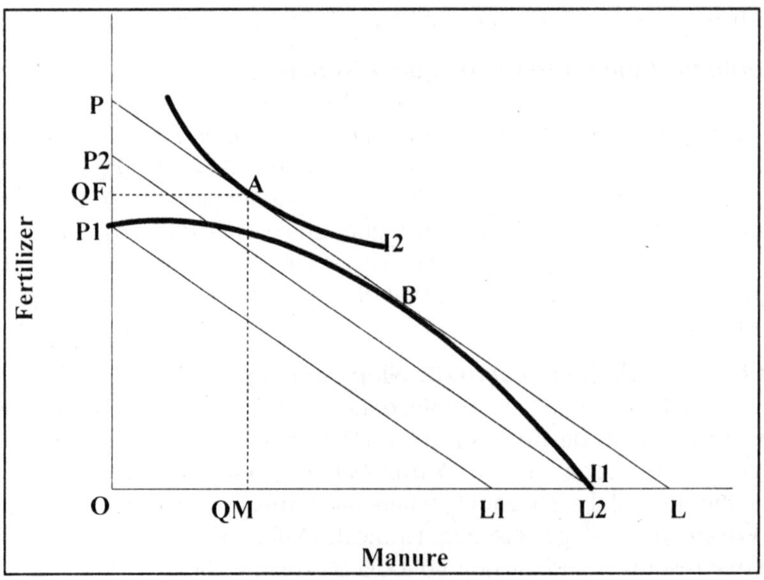

Figure 14.20: Farmer's equilibrium – Iso-cost line is tangent to 'convex' isoquant from below.

In the Figure 14.20, the iso-cost line PL touches isoquant I_1 at point B and isoquant I_2 at point A. A close examination of the Figure 14.20 shows that, at the point of tangency of iso-cost line PL and isoquant I_1, at point B, the isoquant I_1 is concave to the origin, where it implies $MRTS_{MF}$ is increasing rather than diminishing. So, definitely the farmer moves away from point B and reaches the isoquant I_2 and attains LCC at point A, where he employs OQ_M and OQ_F quantities of manure and fertilizer respectively. If the farmer moves along the same isoquant I_1, he can obtain the same output either at points P_1 or L_2, that too at lower cost. But, to attain LCC at P_1 or L_2, the farmer will have a corner solution *i.e.*, the farmer has to employ only one resource in the production programme (*i.e.*, at P_1 he employs only fertilizer and at L_2 he employs only manure). But, producing same output at P_1 or at L_2 on par at point B by employing only one of the resources is absolutely impossible. So, this gives the information that, the farmer will be in equilibrium, when iso-cost line is tangent to the isoquant I_2 and that too, when that isoquant is convex to origin (*i.e.*, at point A), where it shows diminishing MRTS of resources.

The above explanation clearly infers that, at point A on the isoquant I_2 (Figure 14.20), all the above three conditions are fulfilled or satisfied and hence, we can conclude that, the farmer is in equilibrium at point A (LCC), through employing OQ_M and OQ_F quantities of manure and fertilizer respectively. So, we can define the farmer's equilibrium as a situation, where the farmer purchases a particular combination of two resources with a given level of cash outlay and at given prices of the resources, such that, he is not willing to move away from the attained equilibrium situation or show no tendency to re-arrange his purchases for the resources.

Inequality between MRTS and Price Ratio of Resources

Case 1: When $MRTS_{MF}$ is Greater than Price Ratio of Resources

This is given by, $(\Delta F)/(\Delta M) > (P_M/P_F)$. So, to attain the point of LCC of resources, the farmer should employ more of added resource (manure) or less of replaced resource (fertilizer), such that, the equality between MRTS and price ratio of resources is ensured and thereby, the farmer is at the point of equilibrium (LCC).

Case 2: When $MRTS_{MF}$ is less than Price Ratio of Resources

This is given by, $(\Delta F)/(\Delta M) < (P_M/P_F)$. So, to attain the point of LCC of resources, the farmer should employ less of added resource (manure) or more of replaced resource (fertilizer), such that, the equality between MRTS and price ratio of resources is ensured and thereby, the farmer is at the point of equilibrium (LCC).

2. Output Maximization for a given Cash Outlay and Prices of the Two Resources

The other way of attaining the optimal combination of factors is maximizing the output for a given level of outlay with the farmer and for the given prices of the two resources. Considering the given cash outlay with the farmer and given prices of the two resources, we can draw the iso-cost line PL (Figure 14.21). This PL line is tangent to the isoquant I_2 of 200Y at point A, which implies the farmer allocates OQ_M quantity of manure and OQ_F quantity of fertilizer to maximize the output of 200Y. The iso-cost line PL also touches the isoquant I_1 at points B and C, but the isoquant I_1 represent less level of output (100Y). So, the farmer prefers to allocate OQ_M quantity of manure and OQ_F quantity of fertilizer to maximize the output of 200Y. Further, at the point of tangency of iso-cost line and isoquant I_2 at point A, all the three necessary conditions discussed earlier are fulfilled.

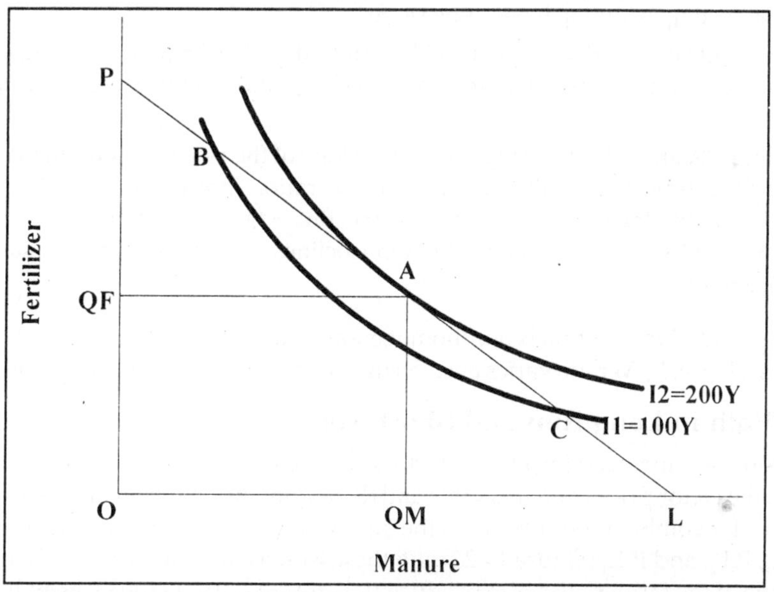

Figure 14.21: Farmer's equilibrium – output maximization for a given cash outlay and prices of the resources.

14.5.4. Isocline

Isocline is a line, which connects the points of equal MRTS of isoquants on an isoquant map. The isocline may be in the form of a straight line or a curve depending upon the nature of the production function. If the production function is linear and homogenous, the isocline is a straight line and if the production is non-homogenous, the isocline takes an irregular shape and the same is explained through Panels A and B of the Figure 14.22.

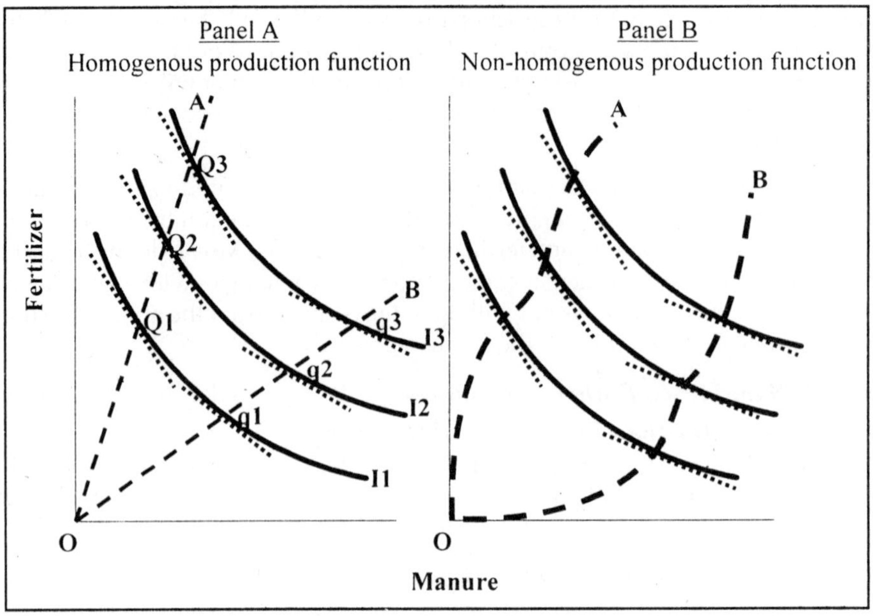

Figure 14.22: Type of production function and shape of isocline.

In Panel A of the Figure 14.22, isocline OA connects the points Q_1, Q_2 and Q_3, where the MRTS (slope of isoquants) is equal for different isoquants I_1, I_2 and I_3, as indicated by the same slope of the three tangents drawn at these points. Similarly, isocline OB connects the points q_1, q_2 and q_3 where the MRTS is equal for different isoquants I_1, I_2 and I_3 as indicated by the same slope of the three tangents drawn at these points. The following are the inferences we can draw from Panel A of the Figure 14.22:

- Since the production function is linear and homogenous, the isoclines OA and OB are straight lines.
- Along OA isocline, the $MRTS_{MF}$ at points Q_1, Q_2 and Q_3 is same on different isoquants I_1, I_2 and I_3, as the tangents drawn at Q_1, Q_2 and Q_3 have same slope.
- Similarly, along OB isocline; the $MRTS_{MF}$ at points q_1, q_2 and q_3 is same on different isoquants I_1, I_2 and I_3, as the tangents drawn at q_1, q_2 and q_3 have same slope.
- The $MRTS_{MF}$ at Q_1, Q_2 and Q_3 along isocline OA is more than $MRTS_{MF}$ at q_1, q_2 and q_3 along isocline OB, as indicated by the steep slope of the tangents drawn at Q_1, Q_2 and Q_3 compared to the tangents drawn at q_1, q_2 and q_3.
- Since, the isoclines OA and OB are straight lines, the slope of the isoclines remains same throughout and the slope of the isocline indicate fertilizer/manure (F/M) ratio. That means, along a particular straight line isoclines (say, OA or OB), the F/M ratio remains same. But, F/M ratio differs between the isoclines and F/M ratio is higher along OA isocline compared to OB isoclines, as indicated by the steepness of the tangents drawn at Q_1, Q_2 and Q_3.

However, when the production function is non-homogenous (Panel B of the Figure 14.22), isoclines OA and OB will have irregular shapes. Hence, F/M ratio varies even on the same isocline and of course, between the two isoclines.

14.5.5. Expansion Path in Long Run and Short Run

In the earlier discussion, we analyzed the LCC of factors at which the farmer wishes to produce the desired level of output. Let us assume, the factor prices remains same and depending upon the level of output the farmer wishes to produce, he changes his factor combinations provided the outlay is given. So, considering the given outlay, we draw several iso-cost lines P_1L_1, P_2L_2, and P_3L_3 (Figure 14.23) and these iso-cost lines are parallel to each other, as the prices of the factors remains same. If the farmer wishes to produce 100 units of output (100Y), he will choose the iso-cost line P_1L_1 considering the prices of the factors and he will attain LCC at the point Q_1, where iso-cost line P_1L_1 is tangent to the

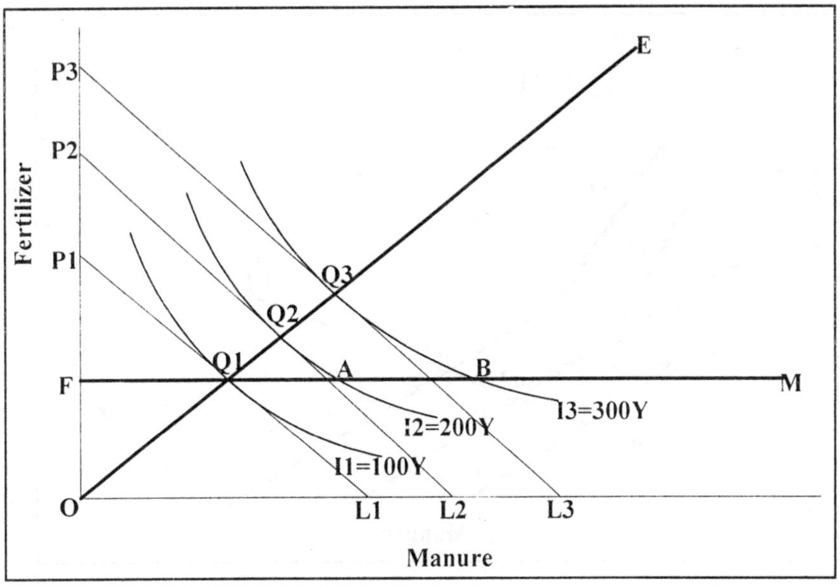

Figure 14.23: Expansion path in long run and short run.

isoquant I_1 (100Y). If the farmer wishes to produce a higher level of output say, 200Y indicated by isoqunt I_2, his new iso-cost line P_2L_2 lies tangent to the isoquant I_2 at Q_2, where he attains LCC of factors. Likewise, if the farmer wishes to increase the output to 300Y, he will attain LCC at point Q_3 *i.e.*, at the point of tangency of isoquant I_3 with iso-cost line P_3L_3. If we connect all the points Q_1, Q_2 and Q_3 *i.e.*, all the points of LCC of factors at different levels of output and at different levels of cash outlay of the farmer, the line OE represents the to Expansion path. The name 'expansion path' is given because, if the farmer wishes to increase the output in his production programme, he has to move on this line OE and it also fetches him to attain the higher levels of output at LCC of factors. So, the expansion path is also called as 'scale line', as it guides the farmer in attaining the higher levels of output through increasing the scale of two factors. So, either a given outlay or a given output level will guide the farmer to operate on this expansion path, so as to ensure LCC of factors. Like the isocline, the slope of the expansion path also indicates F/M ratio.

From the above analysis, we can compare expansion path and isoclines. An isocline is a line connecting the points of equal $MRTS_{MF}$ of different isoquants and it is drawn independently of the prices of the factors under consideration. So, the shape of isocline do not depend upon the prices of the factors. However, expansion path is a line connecting the points of LCC of factors on successive isoquants and iso-cost lines and the shape of expansion path depends upon the relative prices of the factors under consideration. As in the above case (Figure 14.23), if the factor prices are constant, the expansion path (OE) will be a straight line sloping upward. So, an expansion path indicates the actual combination of factors employed by the farmer in his production programme, as it connects the points of LCC of factors at different levels of output, unlike isocline. However, the shape and slope of the expansion path depends upon the shapes of the isoquants (extent of substitutability between the two factors) and relative prices of the two factors.

If the ratio of factor-prices changes, the direction of expansion path also changes, as iso-cost line changes its slope. The same is discussed through Figure 14.24. Let us assume, the ratio of factor prices *i.e.*, P_M/P_F decreases, thereby, the iso-cost lines become more flatter and lies tangent with isoquants at points q_1, q_2 and q_3, thereby, the new expansion path is OE_1.

In the above cases, the straight line expansion paths OE and OE_1 implies a homogenous production function or constant returns to scale. This is so because, as discussed above, the expansion path connects the points of LCC of two factors, where $MRTS_{MF}$ is inversely equal to the ratio of factor prices. (P_M/P_F) *i.e.*, $\Delta F/\Delta M = P_M/P_F$. Since, factor-prices remains constant along the expansion path, $MRTS_{MF}$ also remains constant along the expansion path. Since, expansion path is a straight line, factors ratio (F/M) also remains same. To prove expansion path is a straight line for a linear and homogenous production function (constant returns to scale), let us consider the function in the form of Cobb-Douglas equation $Y = AM^{1/2}F^{1/2}$

Computing the MPP of manure = $(\Delta Y/\Delta M) = ½\, AM^{-1/2}F^{1/2}$ *Equation 14.16*

Computing the MPP of fertilizer = $(\Delta Y/\Delta F) = ½\, AM^{1/2}F^{-1/2}$ *Equation 14.17*

Dividing Equation 14.16 with Equation 14.17, we get

$(\Delta Y/\Delta M)/(\Delta Y/\Delta F) = (½\, AM^{-1/2}F^{1/2})/(½\, AM^{1/2}F^{-1/2}) = F/M$ *Equation 14.18*

So, $(\Delta F/\Delta M) = F/M$ *Equation 14.19*

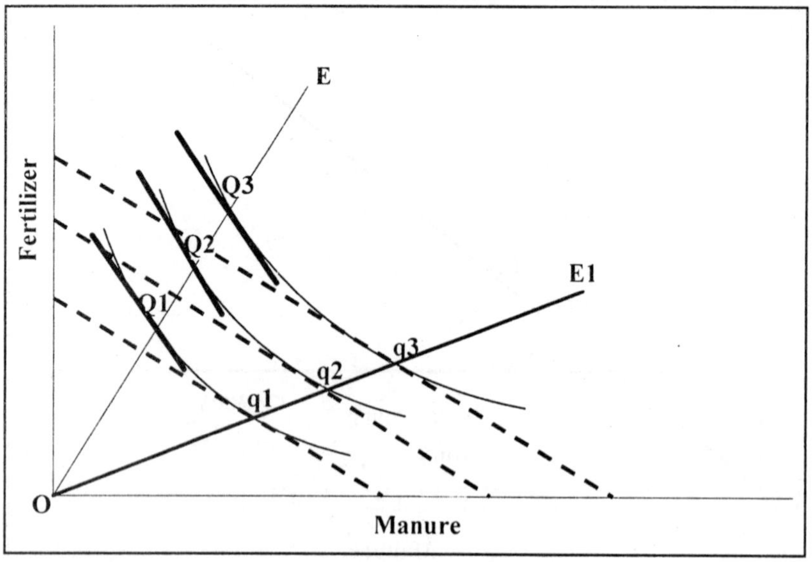

Figure 14.24: Changes in direction of expansion path due to changes in factor prices.

But, $(\Delta Y/\Delta M)/(\Delta Y/\Delta F) = \Delta F/\Delta M = MPP_M/MPP_F$ As per the ***Equation 14.15***

But, $\Delta F/\Delta M = MRTS_{MF} = P_M/P_F$ As per the ***Equation 14.15***

So, $\Delta F/\Delta M = MRTS_{MF} = P_M/P_F = MPP_M/MPP_F = F/M$ ***Equation 14.20***

So, from the above Equation 14.20, for a given linear homogenous Cobb-Douglas production function, $MRTS_{MF}$ is equal to F/M. We know, at the point of LCC of factors, the $MRTS_{MF}$ is equal to P_M/P_F and for the given prices of two factors, $MRTS_{MF}$ is also constant along the expansion path. Since $MRTS_{MF}$ is constant, F/M is also constant throughout the expansion path. So, constant factor ratio *i.e.*, F/M implies expansion path is a straight line. However, in case of non-homogenous production function, the expansion path will have an irregular shape, as discussed under the concept of isocline.

The above discussion infers the optimal expansion path in the long run, as both the factors are variable. However, in short run, the farmer can vary only few factors and not all the factors. Let us assume, the farmer, due to limited capital can apply only OF quantity of fertilizer and he can apply more of manure to increase the output in the production programme. So, FM is the product line (Figure 14.23), as it shows the changes in total output with reference to changes in manure dosage only, but keeping the fertilizer dosage constant at OF. So, the expansion path is not OE, but the farmer can expand output along FM and it is not the optimum expansion path, as the points Q_1, A and B do not lie on the points of LCC.

14.5.6. Ridge Line

According to Traditional economic theory, the farmer can minimize the cost in the production programme, when the factors are used efficiently and when the MPP of factors are diminishing, but positive. We discussed earlier, isoquants have oval shape and the farmer will produce only in those segments, where isoquant is convex to origin and backward bending segments of the isoquant (positive slope) are not suitable for production to ensure cost minimization.

Figure 14.25 shows that, between OA and OB lines, the isoquants are convex to origin, but above OA line and below OB line, the isoquants have positive slope, which means that, above OA line and below OB line, we need more of both the factors (manure and fertilizer) to produce a given level of output. So, in the context of cost minimization, the portions of the isoquants above OA line and below OB line are technically inefficient and these two portions are called 'waste bearing segments' of the isoquant. Above the ridge line OA, the MPP of fertilizer is negative, as output can be maintained by using less fertilizer and by keeping the manure factor constant. Similarly, below OB ridge line, the MPP of manure is negative, as output can be maintained by using less manure and by keeping fertilizer factor constant. However, between OA and OB lines, the factors are technically efficient, as the LCC of factors can be achieved between these lines and MPP of both the factors are diminishing, but positive. So, these two lines OA and OB, which separate the waste bearing segments of isoquant from technically efficient portion of the isoquants in an isoquant map are called ridge lines. They are also called as Border lines. So, a ridge line indicates the maximum output derived from each factor given a fixed level of other factor. So, it is clear that, the line OA connects those points on the isoquants, when MPP of fertilizer is zero. The line OB connects those points on the isoquants, when MPP of manure is zero. So, the lines OA and OB connects the points on the isoquants, when MPP of one of the factors is zero. So, no farmer will operate outside the

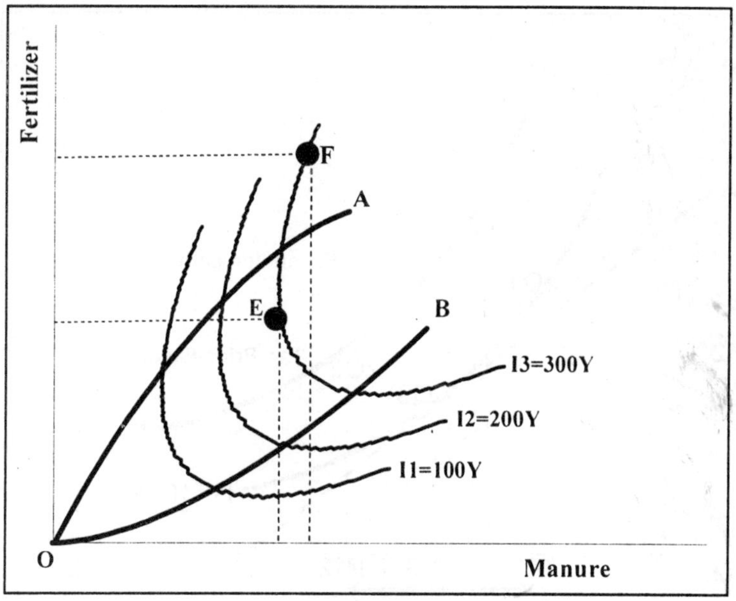

Figure 14.25: Ridge lines – Region of economic production.

ridge line, as the production is technically inefficient and MPP of one of the factors is negative. This is because, if the farmer operates between the ridge lines, he can produce same level of output but with less cost. For example, consider two points F and E on the isoquant I_3 (Figure 14.25). So, to produce 300Y, the farmer needs more of both the factors at point F compared to point E. That means, at point E, the farmer can achieve the same level of output, but employing less of both the factors. So, a rational farmer, will operate between OA and OB ridgelines, where production is technically efficient, MPP is diminishing, but positive and isoquants are negatively sloping. In this context, the portion of isoquants between ridge lines is called Region of economic production and the portions of isoquants above (OA) or below (OB) ridgelines are called Region of economic non-sense. However, it is important that, exactly at what point the farmer operates between the ridge lines depends upon the level of cash outlay with the farmer and prices of the two factors.

14.5.7. Effect of Changes in Factor Price on Factor Usage

In the concepts of iso-cost line and computation of LCC of factors, we made an assumption that, factor prices remain constant. Bur, in practical situation, the factor prices fluctuate frequently and this will influence the extent of their allocation in the production programme. We already discussed there are two types of relationships between the factors *viz.*, substitutes and complements. Let us discuss the influence of change in a factor price on its usage in the production programme for the above two cases of factor relationships.

1. Substitution Relationship Between the Factors

Let us assume, the farmer wishes to employ two resources *viz.*, manure and fertilizer in cultivating paddy. For the given outlay of the farmer and the prices of two resources, PL_1 is the iso-cost line drawn (Figure 14.26). This iso-cost line PL_1 is tangent to isoquant I_1 at Q_1, where the farmer employs OM_1 quantity of manure and OF_1 quantity of fertilizer to ensure LCC of resources. Now, if the price of manure falls, the iso-cost line PL_1 pivots to the right as PL_2 and it is tangent to isoquant I_2 at Q_2, where the farmer employs OM_2 quantity of manure and OF_2 quantity of fertilizer to ensure LCC of resources. That means, with the fall in price of manure, the farmer employs more of manure ($OM_2>OM_1$) and less of fertilizer ($OF_2<OF_1$) than in original situation. This movement from combination Q_1 on isoquant I_1 to combination Q_2 on isoquant I_2 refers to price effect, which is due to fall in price of manure. The line connecting these two combinations Q_1 and Q_2 refers to price factor curve, as it shows the factor-price effect. The downward falling price factor curve shows the increased allocation of manure (due to fall in its price) and less allocation of fertilizer to derive higher output. This factor price effect is the resultant of two effects called Substitution effect or Technical substitution effect and Expansion effect or Output effect.

Substitution Effect or Technical Substitution Effect

When the price of manure is decreased, the farmer psychologically feels his purchasing power is increased and thereby, he allocate his outlay on both the resources. So, to trace out the true substitution effect, we have to remove this indirect increase in outlay of the farmer and as a result, the new iso-cost line AB is drawn such that, it is parallel to PL_2 iso-cost line and it is tangent to the original isoquant I_1 at Q_3, where the farmer employs OM_3 quantity of manure and OF_3 quantity of fertilizer to ensure LCC of resources. This implies that, before the fall in price of manure, the farmer

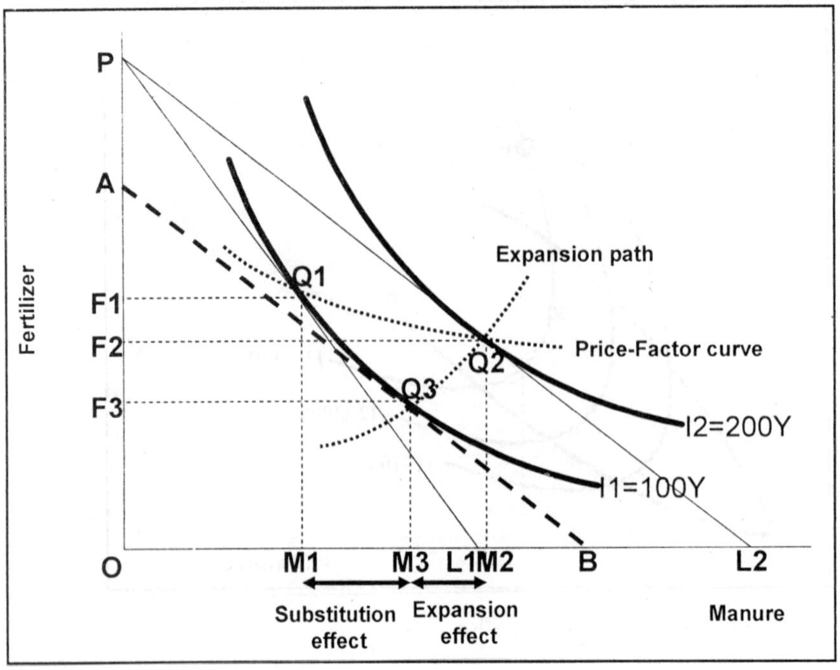

Figure 14.26: Factor substitutes (Price effect = Substitution effect + Expansion effect).

employed OM_1 quantity of manure and OF_1 quantity of fertilizer to produce output on isoquant I_1. But, with the fall in price of manure, the farmer employed OM_3 quantity of manure and OF_3 quantity of fertilizer to produce same output on isoquant I_1. That means, with the fall in price of manure, the farmer employed more of manure ($OM_3 > OM_1$) and less of fertilizer ($OF_3 < OF_1$), so as to produce same amount of output on isoquant I_1. This movement on the same isoquant from Q_1 to Q_3 (M_1M_3) is called Substitution effect.

Expansion Effect or Output Effect

As explained earlier, when the price of manure falls, the farmer psychologically feels his purchasing power is increased and thereby, he allocate his outlay on both the resources, thereby, he can produce more output. So, with reference to new iso-cost line PL_2, the farmer produces higher output on isoquant I_2 at point Q_2 by employing OM_2 quantity of manure and OF_2 quantity of fertilizer. So, under new outlay situation (PL_2), the farmer employed more of both the resources *i.e.*, $OM_2 > OM_3$ and $OF_2 > OF_3$ and thereby, total output increases (shift from isoquant I_1 to isoquant I_2) and this movement from Q_3 to Q_2 is called Expansion effect or Output effect. So, due to expansion effect, the farmer moves from Q_3 of lower output (isoquant I_1) to Q_2 of higher output (isoquant I_2) along the expansion path.

So, the total price effect due to fall in price of one of the factors (if two factors are good substitutes) is a combination of two effects *viz.*, substitution effect and expansion effect. That means, the total price effect in terms of fall in price of manure is given by combination of substitution effect (M_1M_3) and expansion effect (M_3M_2). Thus, $M_1M_2 = M_1M_3 + M_3M_2$. It is also clear that, if the two factors are good substitutes, then substitution effect (M_1M_3) is greater than expansion effect (M_3M_2) *i.e.*, $M_1M_3 > M_3M_2$.

2. Complementarity Relationship between the Factors

Suppose, two resources X and Y are good complements. When a fall in price of resource X, it also leads to increase in quantity demanded for resource Y. Let us assume, the farmer wishes to employ these two resources X and Y in a production programme. For the given cash outlay of the farmer and the prices of two resources, PL_1 is the iso-cost line drawn (Figure 14.27). This iso-cost line PL_1 is tangent to isoquant I_1 at Q_1, where the farmer employs OM_1 quantity of resource X and OF_1 quantity of resource Y to ensure LCC of resources. Now, if the price of resource X falls, the iso-cost line PL_1 pivots to the right as PL_2 and it is tangent to isoquant I_2 at Q_2, where the farmer employs OM_2 quantity of resource X and OF_2 quantity of resource Y to ensure LCC of resources. The line connecting these two combinations Q_1 and Q_2 refers to price factor curve, as it shows the factor-price effect. This upward sloping price factor curve indicates that, with the fall in price of resource X, the quantities employed of both the resources increases *i.e.*, $OM_2 > OM_1$ with respect to resource X and $OF_2 > OF_1$ with respect to resource Y. This movement from combination Q_1 on isoquant I_1 to combination Q_2 on isoquant I_2 refers to price effect, which is due to fall in price of resource X. This factor price effect is the resultant of two effects called Substitution effect or Technical substitution effect and Expansion effect or Output effect.

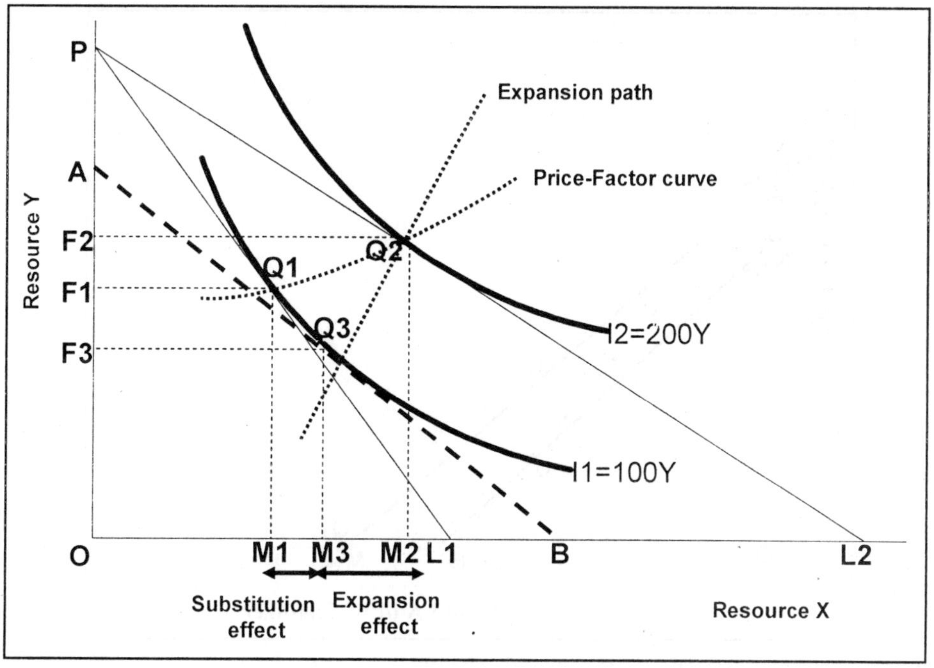

Figure 14.27: Factor complements (Price effect = Substitution effect + Expansion effect).

Substitution Effect or Technical Substitution Effect

When the price of resource X has fallen, the farmer psychologically feels his purchasing power is increased and thereby, he allocate his outlay on both the resources. So, to trace out the true substitution effect, we have to remove this indirect increase in outlay of the farmer and as a result, the new iso-cost line AB is drawn such that, it is it is parallel to PL_2 iso-cost line and tangent to the original isoquant I_1 at Q_3, where the farmer employs OM_3 quantity of resource X and OF_3 quantity of resource Y to ensure LCC of resources. This implies that, before the fall in price of resource X, the farmer employed OM_1 quantity of resource X and OF_1 quantity of resource Y to produce output on isoquant I_1. But, with the fall in price of resource X, the farmer employed OM_3 quantity of resource X and OF_3 quantity of resource Y to produce same output on isoquant I_1. That means, with the fall in price of resource X, the farmer employed more of resource X ($OM_3>OM_1$) and less of resource Y ($OF_3<OF_1$), so as to produce same amount of output on isoquant I_1. This movement on the same isoquant from Q_1 to Q_3 (M_1M_3) is called Substitution effect.

Expansion Effect or Output Effect

As explained earlier, when the price of resource X has fallen, the farmer psychologically feels his purchasing power is increased and thereby, he allocate his outlay on both the resources, thereby, he can produce more output. So, with reference to new iso-cost line PL_2, the farmer produces higher output on isoquant I_2 at point Q_2 by employing OM_2 quantity of resource X and OF_2 quantity of resource Y. So, under new outlay situation (PL_2), the farmer employed more of both the resources *i.e.*, $OM_2>OM_3$ and $OF_2>OF_3$ and thereby, total output increases (shift from isoquant I_1 to isoquant I_2) and this movement from Q_3 to Q_2 is called Expansion effect or Output effect. So, due to expansion effect, the farmer moves from Q_3 of lower output (isoquant I_1) to Q_2 of higher output (isoquant I_2) along the expansion path. Note the expansion path has upward steep slope and this indicates increased allocation of both the resources X and Y to produce higher output due to fall in price of resource X.

So, the total price effect due to fall in price of one of the factors (if two factors are good complements) is a combination of two effects *viz.*, substitution effect and expansion effect. That means, the total price effect in terms of fall in price of resource X is given by combination of substitution effect (M_1M_3) and expansion effect (M_3M_2). Thus, $M_1M_2=M_1M_3+M_3M_2$. It is also clear that, if the two factors are good complements, the expansion effect (M_3M_2) is greater than substitution effect (M_1M_3) *i.e.*, $M_3M_2>M_1M_3$.

3. Factors are Perfect Substitutes

We know, two resources Brand A of nitrogen fertilizer and Brand B of nitrogen fertilizer are perfect substitutes, thereby, the farmer employs only one resource (cheaper resource) in the production programme. We know, in case of perfect substitutes, the isoquants are parallel straight lines (Figure 14.28) for which the MRTS remains constant. In such cases, the farmer attains equilibrium (LCC) at one corner of the isoquant, implying corner solution. The analysis shown through the Figure 14.28 reveals that, for the given level of income of the farmer and for the prices of two resources PL_1 is the price line and it touches the isoquant I_1 at Q_1 (LCC or farmer's equilibrium), where the farmer

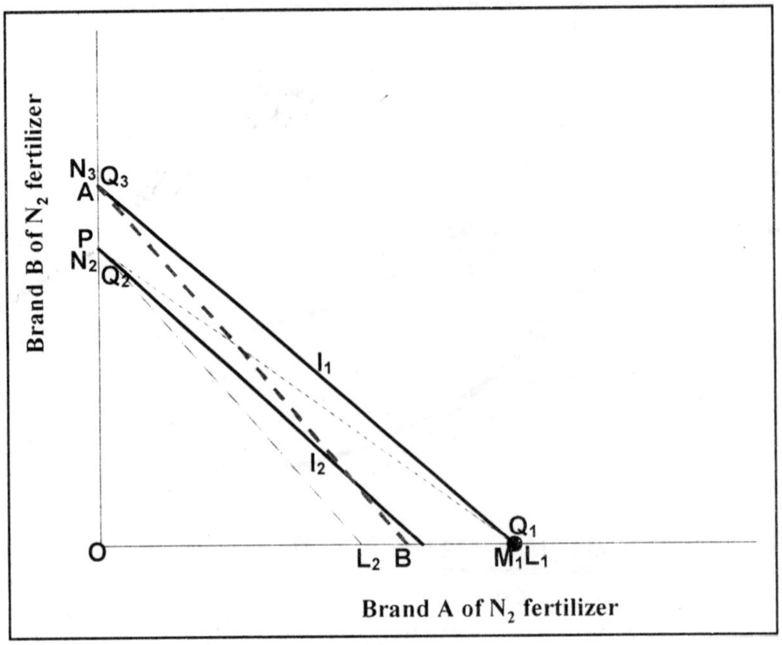

Figure 14.28: Perfect substitutes (Price effect = Substitution effect).

employs only OM_1 quantity of Brand A of nitrogen fertilizer (because of corner equilibrium). Assume, the price of Brand A of nitrogen fertilizer is increased in the market and thereby, the new price line is PL_2 and it touches the new isoquant I_2 at Q_2 (LCC or farmer's equilibrium), where the farmer employs only ON_2 quantity of Brand B of nitrogen fertilizer (because of corner equilibrium).

In order to eliminate the indirect decrease in real income of the farmer due to increase in price of Brand A of nitrogen fertilizer, compensatory variation in income has been made, so as to maintain his real income constant. So, BL_2 is the compensatory variation in income with reference to Brand A of nitrogen fertilizer and thereby, AB is the compensated price line and it touches the original isoquant I_1 at Q_3 (LCC or farmer's equilibrium), where the farmer purchases only ON_3 amount of Brand B of nitrogen fertilizer (due to rise in price of Brand A of nitrogen fertilizer and they are prefect substitutes). So, the substitution effect in case of perfect substitutes is very large *i.e.*, from Q_1 to Q_3. So, substitution effect is maximum and price effect includes only substitution effect.

4. Factors are Perfect Complements

We know, two resources tractor hours and driver hours are perfect complements in a production programme. When tractor hiring rate is decreased, the farmer employs more of both the resources. Let us assume, the farmer wishes to employ tractor in his farm operations. For the given outlay of the farmer and the prices of two resources, PL_1 is the iso-cost line drawn (Figure 14.29). This iso-cost line PL_1 is tangent to isoquant I_1 at Q_1, where the farmer employs OM_1 tractor hours and OF_1 driver hours to ensure LCC of resources. Now, if the tractor hiring rate is decreased, the iso-cost line PL_1 pivots to the right as PL_2 and it is tangent to isoquant I_2 at Q_2, where the farmer employs OM_2 tractor hours and OF_2 driver hours to ensure LCC of resources. The line connecting these two combinations Q_1 and Q_2 refers to price factor curve, as it shows the factor-price effect. This upward sloping price factor curve indicates that, with the fall in tractor hiring rate, the farmer employs more of both the resources *i.e.*, OM_2>OM_1 with respect to tractor hours and OF_2>OF_1 with respect to driver hours. This movement from combination Q_1 on isoquant I_1 to combination Q_2 on isoquant I_2 refers to price effect, which is due to fall in tractor hiring rate. As the two resources are perfect complements, the substitution effect is zero and the new iso-cost line AB (after compensating the indirect increase in purchasing power of the farmer due to decline in tractor hiring rate) is drawn parallel to PL_2, such that, it is tangent to the original isoquant I_1 at Q_1, where the farmer employs original OM_1 number of tractor hours and OF_1 number of driver hours to ensure LCC of resources. So, substitution effect is zero and price effect includes only output effect or expansion effect. That means, with the fall in tractor hiring rate, the farmer employs more of both the resources *i.e.*, OM_2>OM_1 with respect to tractor hours and OF_2>OF_1 with respect to driver hours.

Difference between Expansion Path and Price-Factor Curve

As discussed above, expansion path connects the points of LCC of two resources at different output levels. That means, there will be change (increase) in cash outlay of the farmer and change (increase) in output of the production programme. But, the iso-cost lines are parallel to each other, indicating that, prices of the two resources are constant (Figure 14.23).

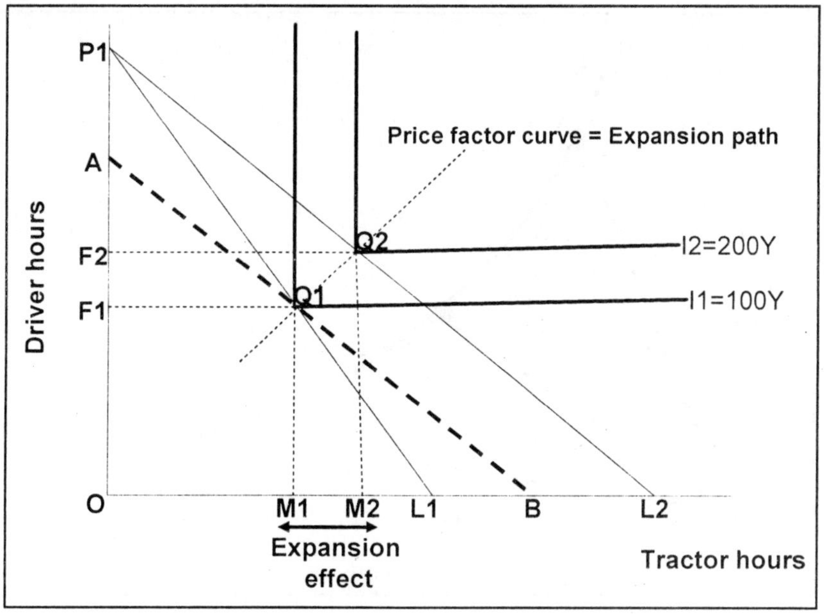

Figure 14.29: Perfect complements (Price effect = Expansion effect).

As shown through Figures 14.26 and 14.27, the line joining the LCC of resources at different levels of output also represent price-factor curve. But, the major difference between price-factor curve and expansion path is that, in case of price factor curve, there is change in price of one factor relative to other factor, thereby, the iso-cost lines are not parallel to each other.

14.5.8. Elasticity of Factor Substitution (E_{LC})

In Chapter 12, we dealt with E_S and here, we study the elasticity of substitution between the two factors. This concept was proposed by J.R. Hicks and it refers to the measure of ease with which the two factors are substituted in the production programme. In case of perfect substitutes, where we completely substitute one factor for the other (say, family labour and hired labour), the E_{LC} is infinity. In case of perfect complements like tractor hours and driver hours, substitution between the factors is no where possible, as they combine in fixed proportion and hence, the E_{LC} is zero. These two cases are the extremes of the concept of E_{LC}.

The E_{LC} is measured by computing the ratio between percentage change in the ratio of the amounts of two factors employed to percentage change in the ratio of the MRTS of the two factors. Assuming, labour (L) and capital (C) are the two factors employed in the production programme, the E_{LC} is given by

$$E_{LC} = \frac{\text{\% change in the ratio of amounts of (C/L) employed in the production}}{\text{\% change in the MRTS of labour for capital}} \quad \textit{Equation 14.21}$$

This is expressed as:

$$E_{LC} = [\Delta(C/L)/(C/L)]/[\Delta(MRTS_{LC})/MRTS_{LC}] \quad \textit{Equation 14.22}$$

$$E_{LC} = [\Delta(C/L)/(C/L)] \times [(MRTS_{LC}/\Delta(MRTS_{LC})]$$

$$E_{LC} = [\Delta(C/L)/\Delta(MRTS_{LC})] \times [(MRTS_{LC})/(C/L)] \quad \textit{Equation 14.23}$$

We can also express the above equation 14.22 in terms of MPP as given below (as per the Equation 14.5)

$$E_{LC} = [\Delta(C/L)/(C/L)]/[\Delta(MPP_L/MPP_C)/(MPP_L/MPP_C)] \quad \textit{Equation 14.24}$$

As shown in the Figure 14.30, if we move down the isoquant from left to right, the $MRTS_{LC}$ decreases gradually. That means, at different points on the isoquant, the $MRTS_{LC}$ differs. This responsiveness of the changes in the $MRTS_{LC}$ due to changes in the use (combination) of two factors refers to E_{LC}. The same concept explained through Figure 14.30 and as per the equation 14.23 reveals that,

Between Points A and B

Between points A and B on the substitution (isoquant) curve, the percentage change in the ratio of amounts of (C/L) employed in the production is less than percentage change in the MRTS of labour for capital, thereby E_{LC} is low. That means, the range over which the substitution curve A_1C_1 is steep *i.e.*, between A and B portion of the curve, the E_{LC} is low

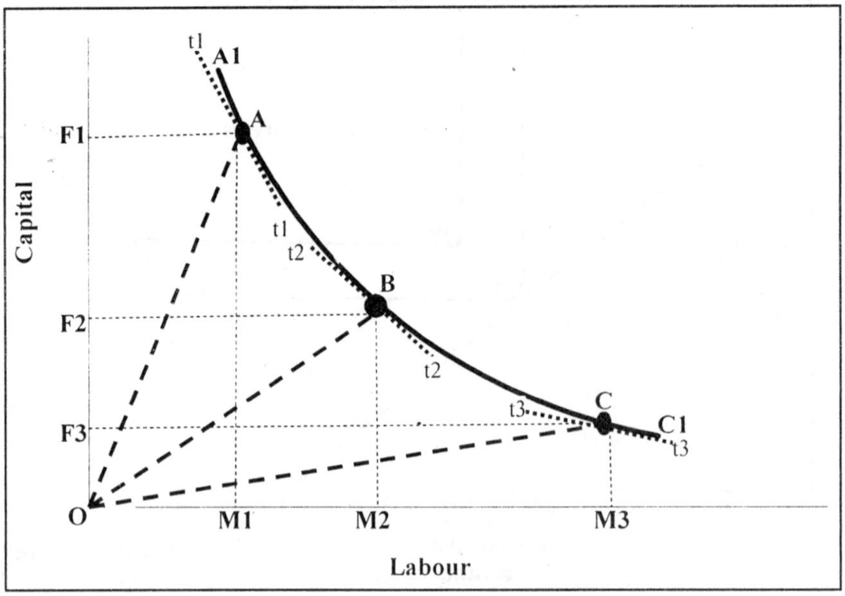

Figure 14.30: E_{LC} along the isoquant.

Between Points B and C

Between points B and C on the substitution (isoquant) curve, the percentage change in the ratio of amounts of (C/L) employed in the production is more than percentage change in the MRTS of labour for capital, thereby E_{LC} is high. That means, the range over which the substitution curve A_1C_1 is flat *i.e.*, between B and C portion of the curve, the E_{LC} is high.

The above analysis further infers that:

- If we move down the isoquant, we substitute labour for capital, thereby the labour-capital ratio increases or capital-labour ratio decreases.
- The points on the isoquant, where the slope is less steep or more flat, the E_{LC} increases compared to the points on the isoquant, where the slope is more steep or less flat. So, in the Figure 14.30, at point C, both labour are capital are good substitutes compared to at point A.
- On the curve, at point A_1, where the [Δ(C/L)] is zero, (as the isoquant is almost parallel to Y-axis) the E_{LC} is zero.
- On the curve at point C_1, where the $\Delta MRTS_{LC}$ is zero, (as the isoquant is almost parallel to X-axis), the E_{LC} is infinity.
- We can easily infer that, at mid point B, on the isoquant, the E_{LC} is unity, because a given change in L/C ratio, brings about an equal proportionate change in the $MRTS_{LC}$.
- So, in case of good substitutes, the proportionate change in C/L ratio is more compared to the proportionate change in $MRTS_{LC}$, thereby, E_{LC} is high.
- In case of perfect substitutes, where $MRTS_{LC}$ is same at all points on the isoquant and hence, its proportionate change is zero (*i.e.*, $\Delta MRTS_{LC} = 0$), thereby, E_{LC} between the factors is infinity.
- In case of perfect complements, where there is no possibility of substitution, as the factors combine in a fixed proportion, Δ(C/L) is zero, thereby E_{LC} is zero.
- Considering the values of E_{LC}, we can say that, when E_{LC} is unity, the production obeys law of constant returns, when $E_{LC}>1$, the production obeys law of increasing returns and when $E_{LC}<1$, the production obeys law of diminishing returns.

We know, the point of LCC is when $MRTS_{LC}$ is equal to the inverse of price ratio of factors. So, in the above Equation 14.21, we can substitute the proportionate change in the price ratio of factors in place of proportionate change in the $MRTS_{LC}$. So, we can write the equation 14.16 as

$$E_{LC} = \frac{\text{\% change in the ratio of amounts of (C/L) employed in the production}}{\text{\% change in the price ratio of factors}} \qquad \textit{Equation 14.25}$$

$E_{LC} = [\Delta(C/L)/(C/L)]/(\Delta(P_L/P_C)/(P_L/P_C)]$ *(with reference to Equation 14.22)*

$E_{LC} = [\Delta(C/L)/(C/L)] \times [(P_L/P_C)/\Delta(P_L/P_C)]$

$E_{LC} = [\Delta(C/L)/\Delta(P_L/P_C)] \times [(P_L/P_C)/(C/L)]$ *Equation 14.26*

14.5.9. Returns to Scale

It implies, long run production function, where no factor is fixed and all the factors are variable. Moreover, the proportion of change among the factors remains same *i.e.*, all the factors will be increased or decreased in the same proportion. That means, all the factors are increased by two or three or four times or decreased by two or three or four times and such variations in the factors will be studied regarding their influence on the output. This increase or decrease in the levels of all the factors in the same proportion refers to increase in scale and decrease in scale respectively. Before studying the concept of Returns to scale, it is essential to draw the distinction between the 'changes in scale' and 'changes in factor proportion' and this is explained through Figure 14.31.

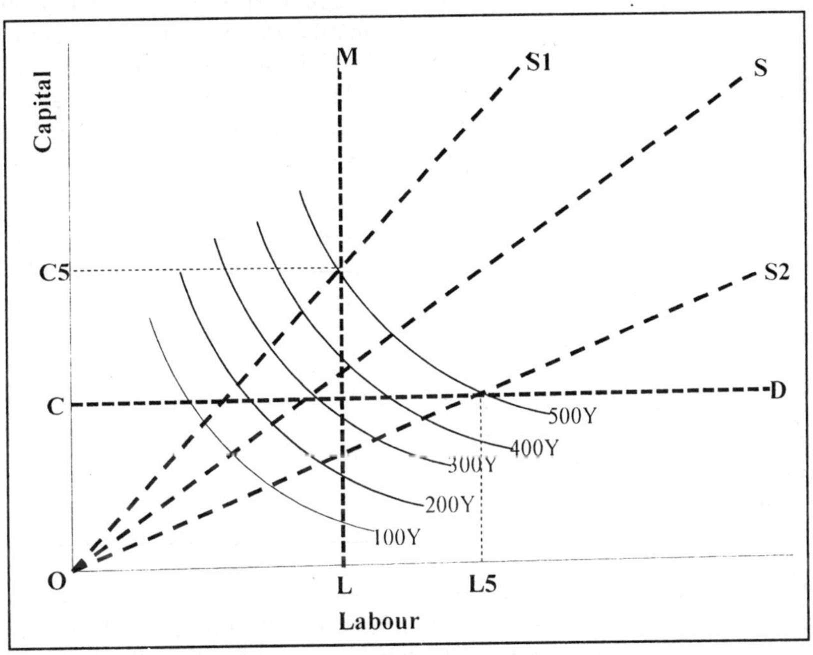

Figure 14.31: Scale vis-à-vis Proportions of factor employed in a production programme.

Assume that, both labour and capital factors are required together to produce a commodity (paddy) and the output levels of a commodity are indicated by different isoquants. Let us take a point 'C' on Y-axis and draw a line parallel to X-axis *i.e.*, CD. From the CD line, it is clear that, to produce 500 units of output (500Y), we have to employ OC level of capital and OL_5 level of labour factor. Further, if the output is increased in production programme from 100Y to 500Y, capital factor is kept constant and only labour factor is varied.

Let us take another point L on X-axis and draw a line parallel to Y-axis *i.e.*, LM. From LM line, it is clear that, to produce 500 units of output, we need OC_5 level of capital and OL level of labour. So, to increase the output level from 100Y to 500Y, we will employ more and more of capital and same number of labour. These two cases represent that, the factor proportions are altered by keeping one factor constant and other factor is varied, so as to increase the output. That means, different combinations of fixed and variable factors are employed to study their influence on the output, which is nothing but the concept of LVP. So, when we increase only one factor keeping other factors constant, it implies 'change in factor proportion', which is seen in LVP.

On the other hand, we can increase the output in the production programme, through varying both the factors *i.e.*, capital and labour. So, when we vary all the factors to study their influence on the output, it implies 'change in scale of factors' (increase in scale means, all the factors are increased in the same proportion) and the returns we derive through changing the scale of factors represent Returns to scale. The factors in the production programme can be employed in different combinations, but the change in factor proportions may remain same and this is explained through drawing OS scale line from the origin. This OS line is a 'scale line' because, the farmer has to increase both capital and labour in absolute amounts to increase output from 100Y to 500Y. Further, the change in proportions between the factors remains same through out the scale line.

Let us draw another line OS_1 from the origin and this is also a scale line because, the farmer has to increase both capital and labour in absolute amounts to increase output from 100Y to 500Y. But, to increase the output from 100Y to

500Y, both the factors are employed in different combinations *i.e.*, capital should be employed more compared to labour, but the change in proportions between the factors remains same through out the scale line.

Reverse is the case for OS_2 scale line. This is because, the farmer has to increase both capital and labour in absolute amounts to increase output from 100Y to 500Y. But, to increase the output from 100Y to 500Y, both the factors should be increased in different combinations *i.e.*, labour should be employed more compared to capital, but the change in proportions between the factors remains same through out the scale line.

So, various straight lines drawn through origin indicates that, different combinations of both the factors (scale) are used to increase the output, but the change in proportions between the factors remains same through out the scale line. So, considering the influences of different scales of factors employed, but with equal changes in factor proportions on the production of output, there are three types of returns to scale *viz.*, Increasing returns to scale, Constant returns to scale and Diminishing returns to scale. So, just like in LVP, here also, the returns due to increase in scale may be either more or equal to or less than proportionate to the change in proportion of factors. That means, if we increase the scale of factors, we may get either increasing returns to scale or constant returns to scale or diminishing returns to scale. If the increase in returns in the production programme is more than proportionate to the changes in all the factors (*i.e.*, scale), the returns to scale is said to be Increasing returns to scale. If the increase in returns is in equal proportion to the changes in all the factors (*i.e.*, scale), the returns to scale is said to be Constant returns to scale. If the increase in returns is less than proportionate to the changes in all the factors (*i.e.*, scale), the returns to scale is said to be Diminishing returns to scale. This concept is explained through Table 14.9.

Table 14.9: Changes in scale of factors and their influence on output in the long run production programme.

S.No	*Scale*	*TPP*	*MPP*	*Returns to Scale*
1.	1 labour + 4 capital	10	10	Increasing
2.	2 labour + 8 capital	25	15	
3.	3 labour + 12 capital	45	20	
4.	10 labour + 40 capital	100	80	Constant
5.	18 labour + 72 capital	180	80	
6.	30 labour + 120 capital	250	70	Diminishing
7.	40 labour + 160 capital	310	60	

It is clear from the Table 14.9 that, both the factors *viz.*, labour and capital are increased simultaneously in the same proportion in the production programme to study their influence on the output. In the first case, when one labour and four units of capital are employed, the total output (TPP) is 10 units. When both the factors are doubled, the output is more than doubled, as indicated by the rise in MPP by 15 units and this is the case of Increasing returns to scale. If the factors are trebled, the total output is more than trebled, as indicated by the rise in TPP from 10 units to 45 units or increase in MPP by 20 units, indicating Increasing returns to scale. So, upto this stage, we have increasing returns to scale in the production programme and this is the Stage I of returns to scale. If both the factors are increased ten times, the total output also increases by ten times *i.e.*, to 100 units and this is the case of Constant returns to scale. Even if both the factors are increased by 18 times, TPP also increased by 18 times *i.e.*, to 180 units and this is again the case of Constant returns to scale. This is further indicated by the constancy of MPP at 80 units. So, upto this stage, we have constant returns to scale and this is the Stage II of returns to scale. If we still increase both the factors in the same proportion beyond this stage of constant returns to scale, the MPP declines by 70 units and 60 units, indicating Diminishing returns to scale and this is the Stage III of returns to scale.

The data of the Table 14.9 is presented graphically through Figure 14.32 considering scale on X-axis and MPP on Y-axis. In the Figure 14.32, AD is the returns to scale curve, where between A and B portion, there is increasing returns to scale, between B and C portion, there is constant returns to scale and between C and D portion, there is diminishing returns to scale. Let us discuss each stage of returns to scale one by one here under.

1. Increasing Returns to Scale

If the increase in all the factors in the same proportion (say, by 10 per cent) leads to more than proportionate increase in the amount of output (say, 20 per cent), then the returns to scale is called as Increasing returns to scale. That means, for less and less usage of factors, we will derive more and more of output. This implies, the factors are used more efficiently in the production programme. This is exemplified below:

100Y=3C+3L

200Y=5C+5L

300Y=6C+6L

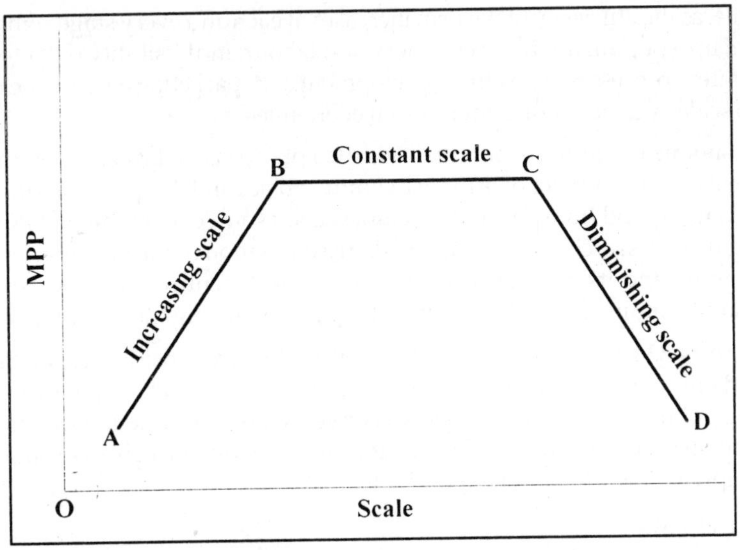

Figure 14.32: Three phases of returns to scale in a long run production programme.

From Panel A of the Figure 14.32, it is clear that, since the factors are used efficiently, the space between the successive isoquants is decreased. That means, to produce higher output *i.e.*, extra 100 units of output, we need less and less of factors at successive levels, as factors are used more efficiently in the production programme. This infers that, the output increases more than proportionately than the increase in the scale of factors. So, the space between successive isoquants decreases and it is given by, $I_1I_2>I_2I_3>I_3I_4>I_4I_5>I_5I_6$. So, in case of increasing returns to scale, the production function is non-homogenous of degree more than one.

According to Joan Robinson, Lerner and Knight, the firms enjoy increasing returns to scale mainly because of internal economies of scale like increased use of machinery, indivisibility of machinery, employment of skilled labour and their efficient management. Chamberlin opined that, even if the factors are divisible, the firm enjoys increasing returns to scale because, the specialization of machinery and labour contributes much to increase in the output and this will lead to higher returns. This higher output is accompanied with good quality and it fetches higher prices and profits to the firm and again it will lead to installation of modern capital equipments in the firm. Further, the integration among different processes of product development makes the firm to enjoy economics of large scale production. For example, if a fruit processing firm is also engaged in taking up of packing of processed products, labelling, distant transport, good distribution network for marketing etc., it will enjoy economies of large scale and it contributes to

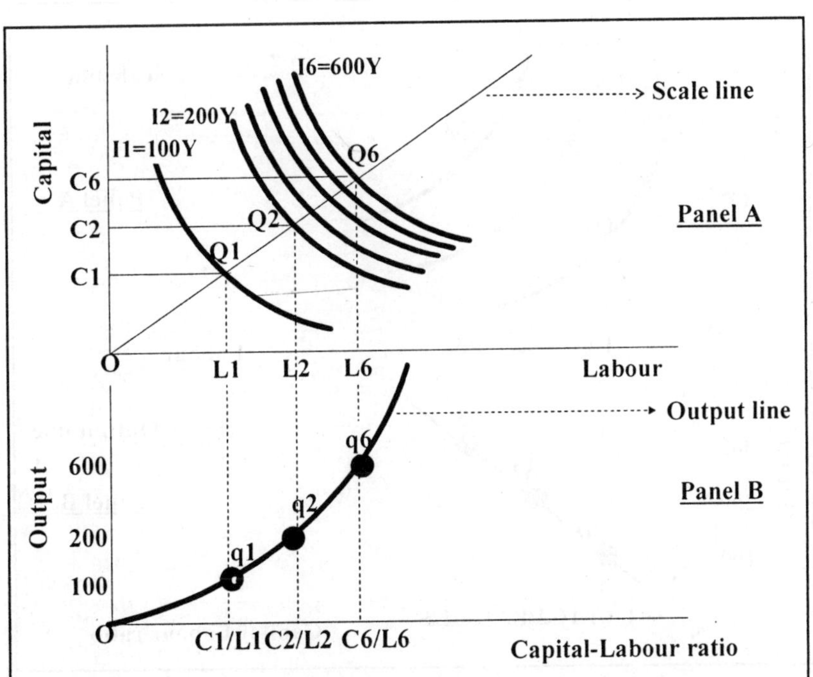

Figure 14.33: Increasing returns to scale and output line.

increasing returns to scale, as the efficiency of the firm increases at each and every stage right from product development to marketing. So, to sum up, specialization of machinery and labour, indivisibility of factors, economies of large scale due to integration of different processes in product development and marketing etc., are the major factors that contribute to increasing returns to scale and they represent internal economies.

Besides internal economies, external economies will also play a crucial rule. When the entire industry prospers, there will be establishment of more number of firms at a common place and this promotes more raw material production, easy raw material availability, good transportation facilities, good sources of credit facilities, good marketing facilities etc. There will be setting up of research centers and quick dissemination of the information about the present status of the industry in the economy and its future prospects and when the entire industry prospers, the individual firms follows the same suit mainly aiming for higher returns through enhancing the resource use efficiency.

Panel B of the Figure 14.33 shows the derivation of output line by considering the factors ratio (*i.e.*, capital and labour) on the X-axis and output realized on the Y-axis. The points Q_1 to Q_6 of Panel A are extended downward to Panel B and the corresponding points q_1 to q_6, if joined gives a convex shaped production function with respect to origin (or concave shaped from above) indicating that, if all the factors are increased in a given proportion, output increases more than proportionately implying increasing returns to scale.

2. Constant Returns to Scale

If the increase in all the factors in the same proportion (say, by 10 per cent) leads to same proportionate increase in the amount of output (*i.e.*, 10 per cent), then the returns to scale is said to be Constant returns to scale. This is exemplified below:

100Y=3C+3L

200Y=6C+6L

300Y=9C+9L

From Panel A of the Figure 14.34, it is clear that, since the increase in the output is in the same proportion as increase in the factors, the space between the successive isoquants remains same. That means, to produce higher output *i.e.*, say, by 10 per cent, we need same proportionate increase in factors at successive levels. This infers that, the output increases in the same proportion as the increase in scale of the factors. So, the space between successive isoquants remains same and it is given by, $I_1I_2=I_2I_3$. So, in case of constant returns to scale, the production function is linear homogenous of degree equal to one. The following are the major reasons for the constant returns to scale:

- When the internal economies of scale are neutralized by the internal diseconomies of scale.
- When external economies of scale are neutralized by external diseconomies of scale.
- When the factors of production are perfectly divisible and homogenous.

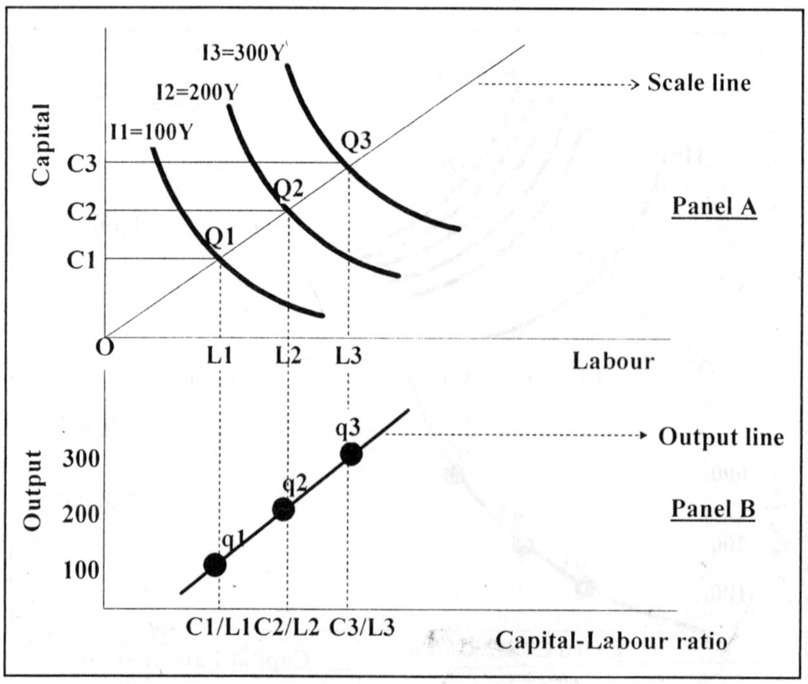

Figure 14.34: Constant returns to scale and output line.

There are two different views to explain the concept of constant returns to scale. Joan Robinson, Frank Knight and Lerner are of the opinion that, if all the factors are perfectly divisible and can be increased or decreased in equal proportion, output can also be increased or decreased in the same proportion, implying constant returns to scale. But, they argued that, the factors employed in the production programme cannot be increased in the same proportion because of two reasons *viz.*, scarcity of some factors and indivisibility of some factors. Had these two constrains are absent, the firm enjoys constant returns to scale. According to Prof. Chamberlin, even if the factors are perfectly divisible and can be varied in small quantities, constant returns to scale cannot prevail, instead increasing returns to scale is more common. This is because, the firm employs modern technology or machinery in the production programme and this in association with skilled labour enable the firm to specialize the business activities, thereby, leading to economics of large scale production.

Panel B of the Figure 14.34 shows the derivation of output line by considering the factors ratio (*i.e.*, capital and labour) on the X-axis and output realized on the Y-axis. The points Q_1, Q_2 and Q_3 of Panel A are extended downward to Panel B and the corresponding points q_1, q_2 and q_3, if joined gives a straight line or linear production function indicating that, if all the factors are increased in a given proportion, output also increases in the same proportion (constant returns to scale).

3. Diminishing Returns to Scale

If the increase in all the factors in the same proportion (say, by 10 per cent) leads to less than proportionate increase in the amount of output (say, 5 per cent), then the returns to scale is called as Diminishing returns to scale. That means, for more and more usage of factors, we will derive less and less of output. This implies, the factors are used less efficiently in the production programme. This is exemplified below,

100Y=3C+3L

150Y=6C+6L

200Y=9C+9L

From Panel A of the Figure 14.35, it is clear that, since the factors are used less efficiently, the space between the successive isoquants is increased. That means, to produce higher output *i.e.*, extra 100 units of output, we need more and more of factors at successive levels. This infers that, the output increases less than proportionately than the increase in the scale of factors. So, the space between successive isoquants increases and it is given by, $I_1I_2<I_2I_3$. So, in case of diminishing returns to scale, the production function is non-homogenous of degree less than one.

The reasons for diminishing returns to scale are due to both internal diseconomies and external diseconomies. Internal diseconomies includes lack of proper supervision on the part of farmer in case of expanding business and more over due to indivisibility of fixed factors like machinery make them less productive or less efficient (due to over

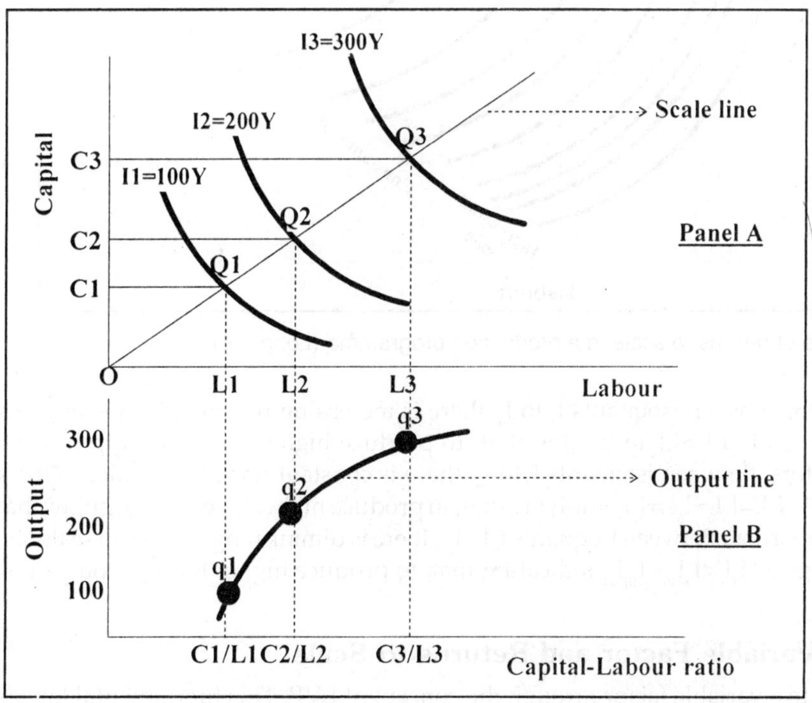

Figure 14.35: Diminishing returns to scale and output line.

exploitation) in producing the output. External diseconomies include rise in prices of factors of production such as rent, wages, interest etc. The raw material prices may also shoot up, thereby, escalating the cost of production of commodity. Besides these, high transportation costs and marketing problems may also arise and all these factors lead to diminishing returns to scale even though the factors are increased at higher scale.

However, there are different opinions regarding the operation of diminishing returns to scale in the production programme. When one input is increased continuously keeping other inputs constant as in LVP, due to improper factor combinations, the diminishing returns may arise. But, when all the factors are varied (increased), there is no constraint that leads to diminishing returns to scale, but it will be experienced in the production programme. One view is that, when all the factors are increased in the production programme, it is not possible to increase the number of entrepreneurs. That means, one entrepreneur has to handle all the other factors increased in the same proportion and this improper combination leads to diminishing returns to scale. So, considering entrepreneur as a fixed factor, we can infer that, diminishing returns to scale is a special case of LVP. The other view is that, with continuous increase in all the factors in the same proportion, it aggravates the managerial problems (planning, direction, control, coordination etc) in handling the large variable factors and this leads to diminishing returns to scale.

Panel B of the Figure 14.35 shows the derivation of output line by considering the factor ratio (*i.e.*, capital and labour) on the X-axis and the output realized on the Y-axis. The points Q_1, Q_2 and Q_3 of Panel A are extended downward to Panel B and the corresponding points q_1, q_2 and q_3, if joined gives a concave shaped production function with respect to origin indicating that, if all the factors are increased in a given proportion, output increases less than proportionately (diminishing returns to scale).

Representation of all the Three Types of Returns to Scale in a Production Programme

Just like three laws of returns *i.e.*, increasing, constant and diminishing MRs operate in LVP, the three types of returns to scale, *i.e.*, increasing returns to scale, constant returns to scale and diminishing returns to scale also operate in a single production programme, when the farmer varies all the factors in the same proportion. This is represented through Figure 14.36, where a number of isoquants (I_1 to I_{11}) are drawn.

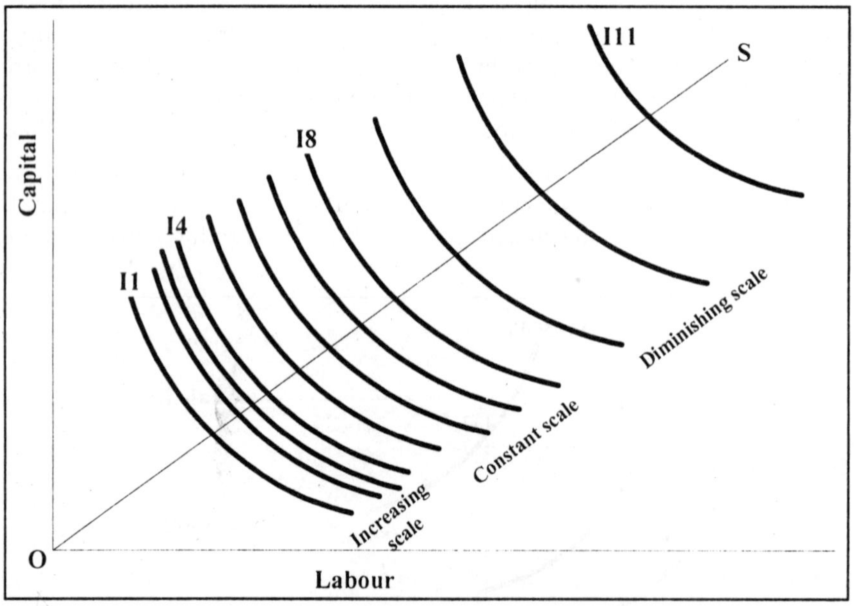

Figure 14.36: Three phases of returns to scale in a production programme (Long run).

From the Figure 14.36, between isoquants I_1 to I_4, there is increasing returns to scale, as the space between the isoquants is decreasing *i.e.*, $I_1I_2>I_2I_3>I_3I_4$ indicating that, to produce higher levels of output, we need less and less quantities of both the factors. Between isoquants I_4 to I_8, there is constant returns to scale, as the space between the isoquants remains same *i.e.*, $I_4I_5=I_5I_6=I_6I_7=I_7I_8$ implying that, to produce higher levels of output we have to increase both the factors in the same proportion. Between isoquants I_8 to I_{11}, there is diminishing returns to scale, as the space between the isoquants is increasing *i.e.*, $I_8I_9<I_9I_{10}<I_{10}I_{11}$ indicating that, to produce higher levels of output, we have to use still higher levels of factors.

14.5.10. MRs to a Variable Factor and Returns to Scale

We studied the MRs to a variable factor through the concept of LVP. This fundamental law explains that, when one input is continuously varied, keeping other inputs constant, the TPP first increases at increasing rate, then at constant rate, then at diminishing rate and finally decreases at increasing rate. So, a rational producer will try to optimize the output, when the output increase at a diminishing rate (or when MR is diminishing, but positive) *i.e.*, in stage II of the production function. So, the farmer optimizes the output, when the returns to a variable factor is diminishing (but, positive) in the production programme.

However, in long run production function, we study the responsiveness of output for a given proportionate change in the quantities of all the factors. That means, we study in what proportion the output changes, when there is equal proportionate change in the quantities of all the factors. So, by changing all the factors in the long run in equal proportion, the output may increase in more than or equal to or less than proportionately and this is referred as increasing returns to scale, constant returns to scale and diminishing returns to scale respectively. As per the Traditional economic theory, if all the factors are varied in equal proportion in a production programme, the firm first shows increasing returns to scale, then constant returns to scale and if still the scale of factors are increased, then the firm exhibit diminishing returns to scale. So, this concept of returns to scale is called as Varying returns to scale.

So, the two concepts *viz.*, LVP and returns to scale will guide us in analyzing the MRs of a variable factor. We have already discussed earlier, the MRs derived from a variable factor under LVP concept. Now, let us analyze the MRs to a variable factor, when the firm experiences increasing returns to scale, constant returns to scale and diminishing returns to scale. The same is discussed below through the Figures 14.37 to 14.39.

1. MRs to a Variable Factor and Increasing Returns to Scale

We call increasing returns to scale as 'Non-homogenous production function of a degree greater than one'. As explained earlier, in case of increasing returns to scale, the distance between successive isoquants will decrease with increase in output in the (long run) production programme. This is because, for less and less usage of factors, we will derive more and more of additional output. In the Figure 14.37, three isoquants I_1, I_2 and I_3 are drawn indicating different levels of output *i.e.*, 100Y, 200Y and 300Y respectively. A straight line OS is drawn from the origin connecting these three isoquants and this represents scale line or returns to scale, as it represents different scales of two factors employed in producing different levels of output. Since, the returns to scale is increasing, the space between the successive isoquants is decreased *i.e.*, $Q_1Q_2>Q_2Q_3$. To study the MRs to a variable factor, a horizontal straight line CL is drawn with reference to point C on the Y-axis and along this straight line, we can study the behaviour of the output when only labour factor is changed, keeping the other factor, capital fixed. Hence, this line CL is called Product line. Along this CL line, we can easily figure out that, $AQ_2>Q_2B$ and this infers that, the MRs of a variable (labour) factor increases to produce an additional output. This is because, to increase the output from 100Y to 200Y, we have to increase the labour factor from CA to CQ_2 *i.e.*, by AQ_2, but to increase the output by the same proportion *i.e.*, by 100Y from 200Y to 300Y, we employ less number of labour factor than earlier *i.e.*, only Q_2B. Since, $AQ_2>Q_2B$, we can infer that, the MRs of a variable (labour) factor increases, when the farm experiences increasing returns to scale. It is important that, this increase in MRs of a variable factor will be experienced only when the returns to scale strongly increases. But, when the returns to scale increases slightly, the MRs of a variable factor will diminish.

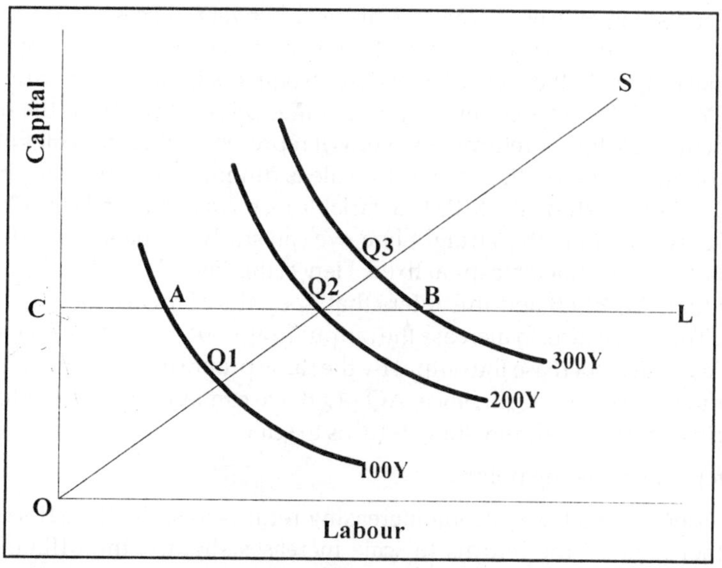

Figure 14.37: Increasing returns to scale and Marginal Returns to a variable factor.

2. MRs to a Variable Factor and Constant Returns to Scale

We call constant returns to scale as 'Linear and homogenous production function or Production function of the first degree'. As explained earlier, in case of constant returns to scale, the distance between successive isoquants will remain same with increase in output in the (long run) production programme. This is because, the increase in output is in the same proportion as increase in the factors. In the Figure 14.38, three isoquants I_1, I_2 and I_3 are drawn indicating different levels of output *i.e.*, 100Y, 200Y and 300Y respectively. A straight line OS is drawn from the origin connecting these three isoquants and this represents scale line or returns to scale, as it represents different scales of two factors employed in producing different levels of output. Since, the returns to scale is constant, the space between the successive isoquants is same *i.e.*, $Q_1Q_2=Q_2Q_3$. To study the MRs to a variable factor, a horizontal straight line CL is drawn with reference to point C on the Y-axis and along this straight line, we can study the behaviour of the output when only labour factor is changed, keeping the other factor, capital fixed. Hence, this line CL is called Product line. Along this CL line, we can easily figure out that, $AQ_2<Q_2B$ and this infers that, the MRs of a variable (variable) factor diminishes to produce an additional output. This is because, to increase the output from 100Y to 200Y, we have to increase the labour factor from CA to CQ_2 *i.e.*, by AQ_2, but to increase the output by the same proportion *i.e.*, by 100Y from 200Y to 300Y, we employ still more number of labour factor than earlier *i.e.*, Q_2B. Since, $AQ_2<Q_2B$, we can infer that, the MRs of a variable (labour) factor decreases, when the farm experiences constant returns to scale.

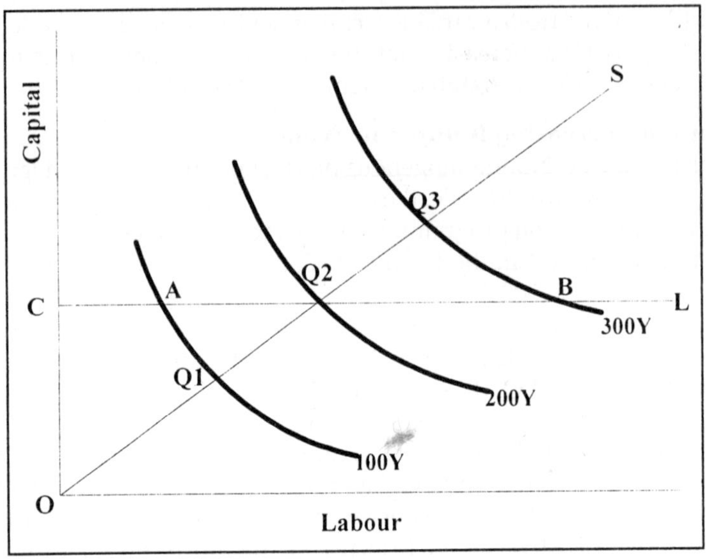

Figure 14.38: Constant returns to scale and Marginal Returns to a variable factor.

3. MRs to a Variable Factor and Diminishing Returns to Scale

We call diminishing returns to scale as 'Non-homogenous production function of a degree less than one'. As explained earlier, in case of diminishing returns to scale, the distance between successive isoquants will increase with increase in output in the (long run) production programme. This is because, for more and more usage of factors, we will derive less and less of additional output. In the Figure 14.39, three isoquants I_1, I_2 and I_3 are drawn indicating different levels of output *i.e.*, 100Y, 200Y and 300Y respectively. A straight line OS is drawn from the origin connecting these three isoquants and this represents scale line or returns to scale, as it represents different scales of two factors employed in producing different levels of output. Since, the returns to scale is diminishing, the space between the successive isoquants is increased *i.e.*, $Q_1Q_2<Q_2Q_3$. To study the MRs to a variable factor, a horizontal straight line CL is drawn with reference to point C on the Y-axis and along this straight line, we can study the behaviour of the output when only labour factor is changed, keeping the other factor, capital fixed. Hence, this line CL is called Product line. Along this CL line, we can easily figure out that, $AQ_2<Q_2B$ and this infers that, the MRs of a variable (labour) factor decreases to produce an additional output. This is because, to increase the output from 100Y to 200Y, we have to increase the labour factor from CA to CQ_2 *i.e.*, by AQ_2, but to increase the output by the same proportion *i.e.*, by 100Y from 200Y to 300Y, we employ more of labour factor than earlier *i.e.*, Q_2B. Since, $AQ_2<Q_2B$, we can infer that, the MRs of a variable (labour) factor decreases, when the farm experiences diminishing returns to scale.

The above discussion infers the following points:

- ☆ When the farm experiences greater or strong increasing returns to scale, the MRs of a variable factor will increase. On the other hand, if the returns to scale increases slightly, the MRs of a variable factor will diminish.
- ☆ When the farm experiences constant returns to scale, the MRs of a variable factor will diminish.
- ☆ When the farm experiences diminishing returns to scale, the MRs of a variable factor will diminish.

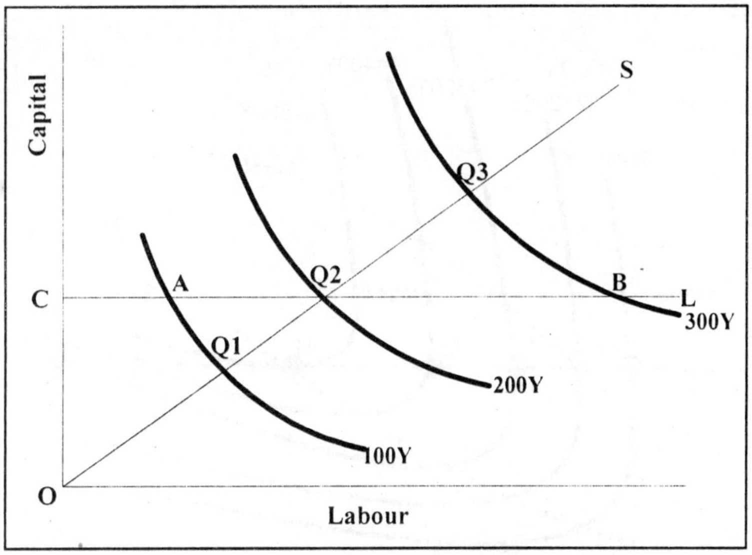

Figure 14.39: Diminishing returns to scale and Marginal Returns to a variable factor.

14.5.11. Differences between LVP and Returns to Scale

The important differences between LVP and Returns to scale are depicted in Table 14.10.

Table 14.10: Differences between LVP and Returns to scale.

LVP	*Returns to scale*
It implies short run production function	It implies long run production function
In LVP, only one factor Is varied and other factors are kept constant to study the influence on the output.	In returns to scale, all the factors are varied to study the influence on the output
The proportion among the factors is different hence, the name LVP.	All the factors are increased or decreased simultaneously in the same proportion. Hence, the proportion among the factors remains same.
It has more practical application because, farmer can increase one factor keeping other factors constant in the production programme.	Returns to scale will have limited application because, all the factors cannot be increased or decreased simultaneously, that too in the same proportion.
Mathematically, $Y=f(X_1/X_2,X_3,X_4..........X_n)$	Mathematically, $Y=f(X_1,X_2,X_3,X_4..........X_n)$
LVP is a reality in agriculture	Returns to scale is hypothetical or myth because, it is difficult on the part of farmer to increase or decrease all the factors simultaneously that too in the same proportion.
The diminishing MRs in case of LVP is due to total imbalance between fixed and variable factors and due to over exploitation of fixed factor.	The diminishing returns to scale is mainly due to both internal and external diseconomies of scale.
Optimization of output is due to proper combination of fixed and variable factors.	Optimization of output is due to management of optimum size of farm and proper combination of factors.

14.5.12. Explanation of LVP in Terms of Isoquant Approach

Like the concept of returns to scale explained in terms of isoquants, the concept of LVP can also be explained in terms of isoquant analysis. The concept of LVP explains that, if one factor is continuously varied, keeping other factors constant, we experience three distinct stages, where in Stage I, increasing MRs is seen, in Stage II diminishing MRs is seen and in Stage III, we experience, negative MRs. The same is explained through Figure 14.40 in terms of isoquants approach.

In the Figure 14.40, I_1, I_2, I_3, I_4 and I_5 are the isoquants drawn indicating different levels of output *viz.*, 100Y, 200Y, 300Y, 400Y and 500Y respectively for different combinations of two factors *viz.*, labour and capital. Since, we know in LVP, only one factor is continuously varied and other factors are held constant to study the influence on the output, let us assume only labour factor is varied, keeping capital factor fixed in the production programme. So, assume OC level

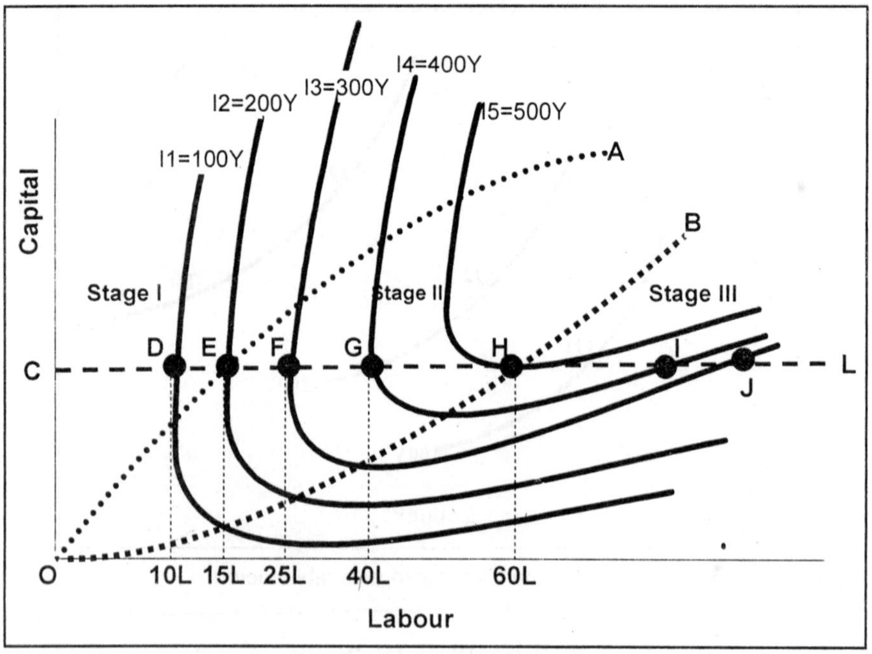

Figure 14.40: LVP in terms of isoquants approach.

of capital factor is fixed in the production programme and we draw a horizontal line CL parallel to X-axis indicating that, only labour factor is varied in the production programme.

We know OA and OB are the two ridge lines drawn and between these two ridge lines the MPP is diminishing, but positive indicating that, economically feasible units of labour and capital can be employed to increase the output. On the ridge line OA, the MPP of capital is zero, on the ridge line OB, MPP of labour is zero and outside the ridge lines, the MPP of both the factors are negative. That means, on the points F and G, the MPP of the factors is positive. On the point D, the MPP of capital is negative and on the points I and J the MPP of labour is negative. On the point E, the MPP of capital is zero and on the point H, the MPP of labour is zero. Let us consider the movement from point D to point E on the horizontal line CL. It reveals that, to increase the output from 100Y to 200Y, we employed OC level of capital and we increased only labour factor. It is interesting to note that, the amount of output is doubled from 100Y to 200Y, but the amount of labour factor employed is not increased in the same proportion, but by less proportion (from 10L to 15L only). That means, for less increase in employment of labour factor, we derived more increase in amount of output indicating that, MRs of labour factor is increasing. This is further indicated by the decline in space between the isoquants I_1 and I_2 *i.e.*, CD > DE. Since, MRs of labour factor is increasing, the area between points C to E represents stage I of the production function.

Now, let us consider the movement between E to H on the horizontal line CL. It reveals that, to increase the output from 200Y to 500Y, we employed OC level of capital and we increased only labour factor. It is clear that, the amount of output increased is less than proportionately compared to the increase in labour factor. That means, for more and more employment of labour factor (15L to 60L), output increased less than proportionately (200Y to 500Y) indicating that, MRs of labour factor is diminishing. This is further indicated by the increase in gap between the isoquants I_2 to I_5. So, the area between two ridge lines *i.e.*, between E to H represents the stage II of production function, as the MRs of labour factor is diminishing.

If we further move along the horizontal straight line beyond H, it means we are going beyond the ridge line OB indicating that, MPP of labour factor turns negative. This is indicated by the fall in output from 500 Y to 400 Y (at I) and to 300Y (at J) and this is due to improper combination of excessive variable (labour) factor relative to fixed (capital) factor and hence, it is yielding negative MRs. So, this stage of production function represents stage III.

So, isoquant approach also gives the same explanation of LVP and it further guides the farmer to operate in stage II of the production function (*i.e.*, between the ridge lines) to optimize the output. Further, it also explains the regions outside the ridge lines (*i.e.*, left side of OA ridge line represents stage I and right side of OB ridge line represents stage III) are not economical to optimize the output.

14.6. Multi-product firm

So far, we discussed about the production programme, where output is influenced by one factor or more than one factor. But, sometimes the farmer practices more than one enterprise In the farm. This concept of production problem refers to 'what to produce' by the farmer with the given level of resources, so as to maximize the profits. That means, it implies, analyzing the product-product relationship *i.e.*, different enterprise combinations and the farmer has to select the right combination of enterprises that maximizes the profits. We know, production function of enterprise Y1 is given by, Y1 = f(X1, X2, X3/X4, X5, X6 ——— Xn). Similarly, production function of enterprise Y2 is given by, Y2 = f(X1, X2, X3/X4, X5, X6 ——— Xn). So, we can write, Y1 = f(Y2). If the farmer deals with more than two enterprises, we can write the functional relationship as Y1 = f(Y2. Y3, ——— Yn). The enterprise combination, so selected by the farmer at a given level of resources should ensure maximization of profits in the production programme. Thus, in product-product relationship, resource level at the farmer is assumed constant and with that level of resources, he has to maximize the profits in the production programme through selecting and practicing optimum enterprise combinations. The following are the concepts involved in product-product relationship:

14.6.1. Production Possibility Curve or PPF

It is a line or curve, which indicates the various possible combinations of two products that can be produced with the same level of resources. Since, any combination of two products on the same production possibility curve indicate same level of resources, it is also called as Iso-resource curve or Iso-factor curve or Iso-outlay curve. It is also called as Opportunity curve, as it indicates the possible opportunities available with the farmer in producing two products with the given level of resources. It is other wise called as Transformation curve, at it studies the rate of transformation from one product to another product by using same level of resources. It is also called as Frontier curve, as it indicates the limits of production of two products with the given level of resources.

We know, a farmer chooses the best combination of products based on his available resources. It is generally assumed that, the farmer has a well-defined preference to the set of all possible combinations of enterprises. This facilitates him to compare his preferred combination over other combinations of enterprises. It is further assumed that, the farmer can compare any two enterprises and between these two enterprises, he can have more of one to less of the other and he remains indifferent with respect to any combination of enterprises on the same production possibility curve regarding the resource level. That means, production possibility curve shows all possible combinations of two enterprises that can be produced with the given level of resources with the farmer. This makes the farmer indifferent between any combinations of two enterprises on the same production possibility curve for the given level of resources. In other words, resource level is held constant along a production possibility curve. In the Table 14.11, different combinations of two enterprises *viz.*, paddy and maize are shown and any combination of these two enterprises can be produced with the given level of resources with the farmer and such information is called as Production possibility schedule. It is evident that, all the five combinations of paddy and maize enterprises can be produced by the farmer with reference to the land resource under consideration. If the farmer prefers the first combination, he cultivates only maize (Y_2) in his total acreage of four acres of land and derives an output of 75 units and nothing of paddy, as no land resource was allocated under paddy. In the second combination, he allocates one acre of land to paddy and thereby, he can allocate only three acres of land for maize production. That means, by allocating one acre of land to paddy, he is prepared to give up one acre of land under maize cultivation. That means, for increasing the usage of land resource on one enterprise, he has to sacrifice the same on the other enterprise. Likewise, when the farmer reaches the fifth combination, he allocates all his land resource (four acres) for paddy cultivation alone. The above discussion also infers that, both the enterprises substitute each other for land resource.

Properties of Production Possibility Curve

The main attributes or properties of production possibility curve are discussed here under.

(a) Production Possibility Curve is Positively Sloped

The production possibility curve is concave to origin (positive slope), as the products substitute each other at an increasing rate. As shown in Table 14.11, to produce different combinations of two enterprises *viz.*, paddy and maize, four acres of land was shared between the two. As more and more of land is allocated under paddy, the rate of substitution of paddy for maize is increasing and this gives the concavity to the production possibility curve (Figure 14.41). On the same production possibility curve, at any point or any combination of two enterprises, the resource level employed by the farmer remains same. For example, at combination B (Table 14.11), the farmer employed four acres of land for crops cultivation *i.e.*, one acre under paddy and three acres under maize. Similarly, the land resource allocated remains same even at combination C (four acres) *i.e.*, the farmer allocates two acres each under paddy and maize cultivation.

(b) Slope of Production Possibility Curve Indicates Increasing MRPS of Enterprises

As studied through Table 14.11, if the farmer goes on increasing the added product (paddy), he is sacrificing more and more of replaced product (maize) in the production programme. This behaviour of compulsory fall in the output of one product (maize) due to increase in the output of other product (paddy), because of allocation of more land resource under added product (paddy) is called Substitution. In the production programme, for each and every additional

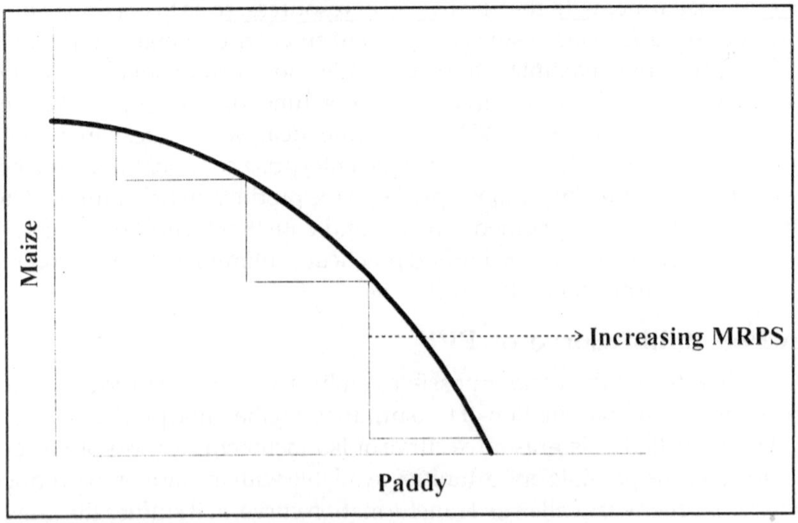

Figure 14.41: Positively sloped Production possibility curve showing increasing MRPS.

output of paddy, it leads to compulsory sacrifisation of some additional units of maize output and this rate of substitution is called MRPS. Since, the MRPS is at increasing rate, it is called as increasing MRPS. (The reasons for this increasing rate of MRPS of concave shaped production possibility curve are already discussed under section 4.4.3.). This increasing rate of product substitution is more common in agriculture. It is, otherwise, defined as 'the rate at which the farmer must gives up output of one enterprise (maize) in order to produce more of an additional output of other enterprise (paddy) for a given level of resources.

Table 14.11: Production possibility curve schedule with reference to paddy and maize enterprise combinations.

Combinations	*Land Area (Acres)*		*Output (Units)*				*Marginal Rate of Product Substitution (MRPS) of Paddy for Maize*
	Y1 (Paddy)	*Y2 (Maize)*	*Y1 (Paddy)*	*ΔY1*	*Y2 (Maize)*	*ΔY2*	$MRPS_{PM}=(\Delta M/\Delta P)$
A	0	4	0	—	75	—	—
B	1	3	8	8	60	15	1.88
C	2	2	16	8	44	16	2.00
D	3	1	24	8	26	18	2.25
E	4	0	32	8	0	26	3.25

The concept of increasing MRPS between the two products (paddy and maize) is illustrated through the Figure 14.41. As the farmer moves down the production possibility curve, it indicates the willingness to substitute paddy for maize increases. This means that, as the amount of paddy output is increased by equal amounts, the output of maize diminishes by higher amounts. Thus, the MRPS of paddy for maize $MRPS_{PM}$ is the output of maize that the farmer is willing to give up to gain a marginal increase of output of paddy in the production programme for the given level of resources. This increasing MRPS contributes positive (concave) slope to the production possibility curve.

The concept of MRPS is explained by the following formula:

$$MRPS_{PM} = \frac{\text{Change in number of units of replaced product (maize)}}{\text{Change in number of units of added product (paddy)}}$$

So, MRPS of paddy for maize is given by, $MRPS_{PM} = (\Delta M)/(\Delta P)$ *Equation 14.27.*

(c) Position of Production Possibility Curve

The position of production possibility curve with reference to origin depends upon the level of resources with the farmer. Higher the level of resources, the farther away the position of production possibility curve from the origin and *vice versa*.

(d) The Production Possibility Curves does not Intersect or Tangent to Each Other

The production possibility curves does not intersect each other, as different production possibility curves represent different levels of resources with the farmer.

Production Possibility Curve vis-a-vis Isoquant

The following are the differences between production possibility curve and isoquant:

- Production possibility curve indicates various combinations of two enterprises or products (say paddy and maize) for a given outlay of resources with the farmer. But, an isoquant indicates various combinations of two resources (say manure and fertilizer) that yield same level of output.
- On a production possibility curve, resource level is constant and on isoquant, output level is constant.
- Production possibility curve is concave to origin. Isoquant is convex to origin.
- The slope of production possibility curve indicates increasing MRPS. The slope of isoquant indicates diminishing MRTS.
- Tradeoffs are a feature of both production possibility curve and isoquant. Production possibility curve indicates the trade off between two products, whereas, isoquant indicates the trade off between two resources.
- The position of production possibility curve from the origin depends upon the level of resources with the farmer, whereas the position of isoquant from the origin depends upon the output level.
- Production possibility curve guides the farmer in the maximization of profits, whereas isoquant guides the farmer in the minimization of costs.

14.6.2. Iso-Revenue Line

From the production possibility curve alone, it is not possible to compute the Optimum Product combination, as it simply indicate the various combinations of two products under consideration by the farmer for the given level of resources. Among the different combinations of products, the best choice the farmer will make, depends on the profit maximization in the production programme. So, to ascertain the point of optimum product combination, we should have information about the prices of the products and revenue the farmer wishes to earn from the production programme. So, this concept of studying the iso-revenue line for the farmer is essential for knowing the point of optimum product combination. An iso-revenue line represents the various combinations of two products that are capable of yielding same amount of revenue to the farmer for the given prices of the products. The iso-revenue line is also called as Iso-income line.

For example, a farmer produces two commodities *viz.*, paddy and maize. The price of paddy is Rs. 5/unit and the price of maize is Rs. 6/unit. If the farmer desires to earn the income of Rs. 60, he has to sell 12 units of paddy or 10 units of maize. So, the line connecting 12 units of paddy and 10 units of maize, which yields an income of Rs.60 is called an iso-revenue line.

Table 14.42 shows that, if the farmer produces only maize, he has to sell 10 units of maize to derive the income of Rs. 60. Similarly, if the farmer produces only rice, he has to sell 12 units of paddy to derive the income of Rs. 60. If the farmer allocates the land resource between paddy and maize, he has to sell the combinations of paddy and maize (any combination between B and L) to earn Rs. 60.

Table 14.12: Different combinations of paddy and maize capable of yielding same revenue to the farmer.

Combinations	*Paddy Output (Units)*	*Maize Output (Units)*
A	0.0	10.0
B	1.0	9.2
C	2.0	8.3
D	3.0	7.5
E	4.0	6.7
F	5.0	5.8
G	6.0	5.0
H	7.0	4.2
I	8.0	3.3
J	9.0	2.5
K	10.0	1.7
L	11.0	0.8
M	12.0	0.0

In the Figure 14.42, the line PL indicates the iso-revenue line of the farmer, indicating various combinations of paddy and maize that can be offered for sale by the farmer, so as to earn Rs. 60. It shows 13 possible combinations of paddy and maize can be offered for sale by the farmer from A to M. Combination A indicates that, the farmer has to sell 10 units of maize to earn Rs. 60, by allocating all his resources on maize crop only and nothing on paddy. Similarly, point M shows the farmer has to sell 12 units of paddy to earn Rs. 60, by allocating all his resources on paddy only and nothing on maize. The remaining points on the iso-revenue line *i.e.*, from B to L indicates various possible combinations of both paddy and maize the farmer has to sell, so as to derive the income of Rs. 60.

Let us assume that, the farmer wishes to derive income Y and the prices of two products *viz.*, paddy and maize are P_P and P_M respectively. If the farmer wishes to sell Q_P and Q_M quantities of paddy and maize respectively as the preferred choice, he will obtain Q_PP_P amount of revenue from paddy and Q_MP_M amount of revenue from maize respectively. So, the farmer by selling Q_P and Q_M quantities of both paddy and maize, he will get a TR of $Q_PP_P + Q_MP_M$.

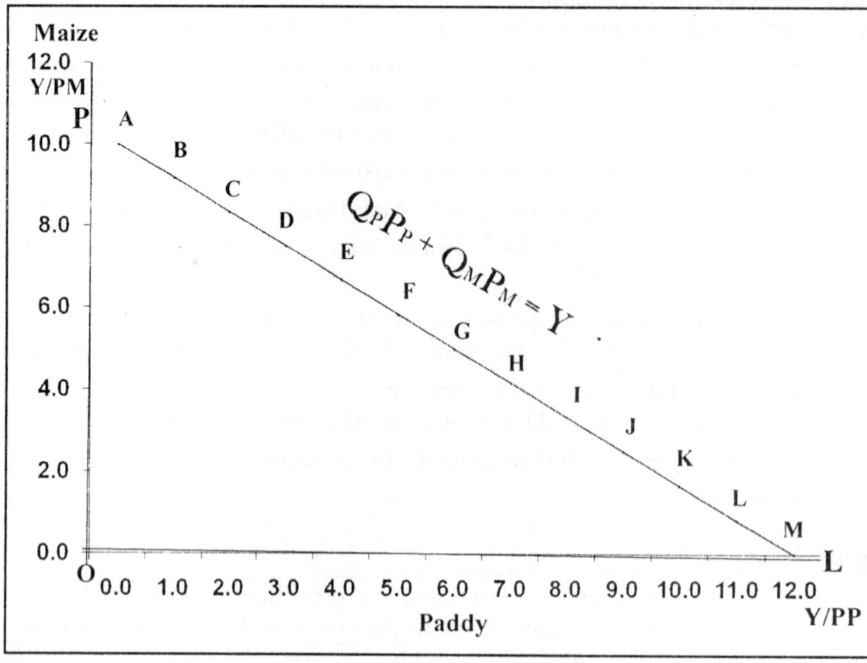

Figure 14.42: Iso-revenue line of the farmer for paddy and maize combinations.

That means, the farmer has to sell this combination of products, if he desires to earn at least $Q_PP_P + Q_MP_M$ amount of revenue. So, we can infer that, the farmer has to sell the above combination of products such that, $Q_PP_P + Q_MP_M = Y$. The iso-revenue line, $Q_PP_P + Q_MP_M = Y$ can also be expressed with respect to Q_P as $(Y/P_P) - (Q_MP_M/P_P)$. Similarly, the iso-revenue line, $Q_PP_P + Q_MP_M = Y$ can also be expressed with respect to Q_M as $(Y/P_M) - (Q_PP_P/P_M)$. So, the iso-revenue line is a straight line with horizontal intercept Y/P_P and with vertical intercept Y/P_M. The horizontal intercept (Y/P_P) represents the quantity of paddy sold, if the farmer allocates all the resources on paddy alone. Similarly, the vertical intercept (Y/P_M) represents the quantity of maize sold, if the farmer allocates all the resources on maize alone. (Figure 14.42).

Slope of the Iso-Revenue Line

The slope of the iso-revenue line indicates the price ratio of the two products. The slope of iso-revenue line can be studied through the following three methods:

Method 1

The slope of iso-revenue line can be studied through the Figure 14.42. In the Figure 14.42, OL units of paddy can be exchanged with OP units of maize, as the revenue with respect to OL units of paddy and OP units of maize is same. So, one unit of paddy is equal to (OP/OL) units of maize. This is the same thing as the tangent of angle PLO. So, the rate of exchange of the paddy and maize is thus given by tangent PL. This PL is the iso-revenue line. The slope of the iso-revenue line can be derived mathematically, as given below:

- Along X-axis, the total quantity of paddy sold by the farmer is OL at a given price of P_P, if all the resources are allocated on the paddy. This is given by $Y = OL \times P_P$. So, $OL = Y/P_P$
- Similarly, along Y-axis, the total quantity of maize sold by the farmer is OP at a given price of P_M, if all the resources are allocated on the maize. This is given by $Y = OP \times P_M$. So, $OP = Y/P_M$.
- As discussed earlier and from the Figure 14.42, slope of the iso-revenue line is given by OP/OL. So, from the equation 14.28 given below, the slope of iso-revenue line indicates the price ratio of two products.

$$(OP/OL) = (Y/P_M)/(Y/P_P) = P_P/P_M \qquad \textit{Equation 14.28}$$

In the given example, as the price per unit of paddy is Rs.5 and price per unit of maize is Rs.6, the slope of iso-revenue line is 0.83. Since, the prices are assumed constant at a given time, the iso-revenue line is a straight line.

Method 2

We know, the slope of the iso-revenue line measures the amount of change in one product (say, maize) due to per unit change in other product (say, paddy) along the iso-revenue line. In the Figure 14.43, let us consider two points A and B on the iso-revenue line, where point A represents Q_P and Q_M combinations of two products and point B represents

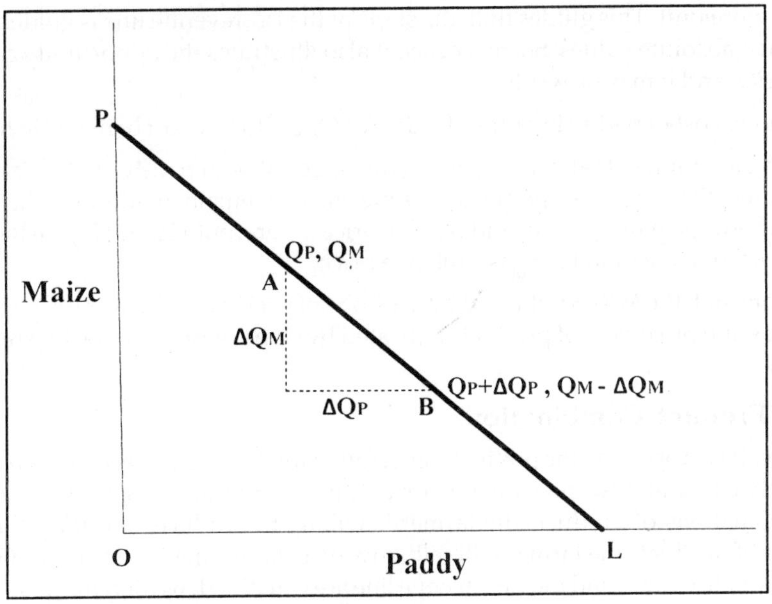

Figure 14.43: Method 2 – Estimation of slope of the iso-revenue line.

the movement down the iso-revenue line given by $Q_P+\Delta Q_P$, $Q_M-\Delta Q_M$. So, the iso-revenue line with reference to these two points A and B is given as follows:

At point A: $(P_P \times Q_P) + (P_M \times Q_M) = Y$ *Equation 14.29*

At point B: $(P_P \times (Q_P + \Delta Q_P)) + (P_M \times (Q_M - \Delta Q_M)) = Y$ *Equation 14.30*

Subtracting Equation 14.29 from Equation 14.30, we get,

$P_P \, \Delta Q_P - P_M \, \Delta Q_M = 0$ *Equation 14.31*

By rearranging terms in Equation 14.31, we get the slope of the iso-revenue line as,

$(\Delta Q_M/\Delta Q_P) = (P_P/P_M)$.

Method 3

The slope of the iso-revenue line can be simply estimated through computing the ratio between vertical intercept and horizontal intercept of the iso-revenue line. As discussed earlier, we know, iso-revenue line can be denoted as $(P_P \times Q_P) + (P_M \times Q_M) = Y$.

- ***Horizontal intercept of iso-revenue line:*** The above iso-revenue line can be expressed with respect to Q_P as $(P_P \times Q_P) = Y - (P_M \times Q_M)$.
  So, $Q_P = (Y/P_P) - (Q_M P_M/P_P)$ *Equation 14.32*
  In the above equation 14.32, (Y/P_P) is the horizontal intercept of the iso-revenue line with respect to paddy.
- ***Vertical intercept of iso-revenue line:*** Similarly, we can compute the vertical intercept of iso-revenue line. The iso-revenue line can be expressed with respect to Q_M as:
  $(P_M \times Q_M) = Y - (P_P \times Q_P)$.
  So, $Q_M = (Y/P_M) - (Q_P P_P/P_M)$ *Equation 14.33*
  In the above equation 14.33, (Y/P_M) is the vertical intercept of the iso-revenue line with respect to maize.
- So, Slope = Vertical intercept/Horizontal intercept = $(Y/P_M)/(Y/P_P) = (P_P/P_M)$.

Notice that, the slope of the iso-revenue line depends only on the ratio of the prices of two products and not on their absolute values. In this example, the price of paddy is Rs. 5/unit and price of maize is Rs. 6/unit. So, the ratio of the prices of paddy and maize is 1.0:1.2 and thereby, if the farmer wants to increase the production of paddy by one more unit, he has to sacrifice 1.2 units of maize. On the other hand, if the farmer increases the production of maize by one more unit, he has to sacrifice 0.83 units of paddy. Similarly, if we consider prices of paddy and maize as Rs. 10/unit and Rs. 12/unit respectively, the farmer must still forgo 1.2 units of maize for increasing the production of one more unit of paddy and 0.83 units of paddy for increasing the production of one more unit of maize. So, as long as the price of maize is 1.2 times the price of paddy, the farmer must forgo the same amounts, if he increases the production of either paddy or maize by one more unit. This guides that, the slope of the iso-revenue line is guided by the price ratio of the products rather than their absolute values. So, this concept also illustrates the opportunity cost of paddy production vis-à-vis maize production and it may be written as:

(P_P/P_M) = Opportunity cost of paddy in terms of maize or Opportunity cost of maize in terms of paddy.

So, the above explanation infers that, any change in the prices of both paddy and maize in the same proportion will not affect the price ratio, thereby, the slope of iso-revenue line will remain unaffected. This discussion further helps to clarify the distinction between money prices and relative prices. Here, both P_P and P_M are the money prices of paddy and maize resepectively, while the ratio P_P/P_M is a relative price.

It is important to note that, the MRPS indicated by the slope of production possibility curve increases as we move down the curve, whereas, the price ratio of products indicated by slope of iso-revenue line remains same throughout the iso-revenue line.

14.6.3. Optimum Product Combination

Generally, the farmer has to optimize the product combination for the given level of resources. The earlier discussion on production possibility curve and iso-revenue line reveals two important aspects *viz.*, the production possibility curve infers 'what combinations of products the farmer is willing to produce' and the iso-revenue line guides the farmer, 'what quantities of products the farmer will sell', so as to produce the optimum combination of products. So, when the production possibility curve and the iso-revenue line are combined, we find the quantities of each product the farmer is both willing and able to sell for the given level of resources. So, the production possibility curve and iso-revenue line helps the farmer to find out the optimum combination of products, where he can maximize the profits. So, the optimum product combination refers to the point of profit maximization in the production programme at a particular combination of products for the given level of resources. Three important conditions must be fulfilled to arrive at the point of optimum combination of products. They include,

(*a*) The iso-revenue line should be tangent to the production possibility curve from above

(*b*) Slope of the iso-revenue line should be equal to the slope of production possibility curve

(*c*) Production possibility curve should be concave to the origin.

(a) The Iso-Revenue Line should be Tangent to the Production Possibility Curve from Above

The point of optimum product combination is where, the iso-revenue line is tangent to the production possibility curve from above. Figure 14.44 shows that, iso-revenue line PL is tangent to the production possibility curve at point A, where the farmer attains optimum product combination by selling OQ_P quantity of paddy and OQ_M quantity of maize. Any point other than A *i.e.*, at M_1 or N_1 will not contribute to maximum profits, as these points touch the lower iso-revenue line P_1L_1. Moreover, at these points, the iso-revenue line is not tangent to the production possibility curve.

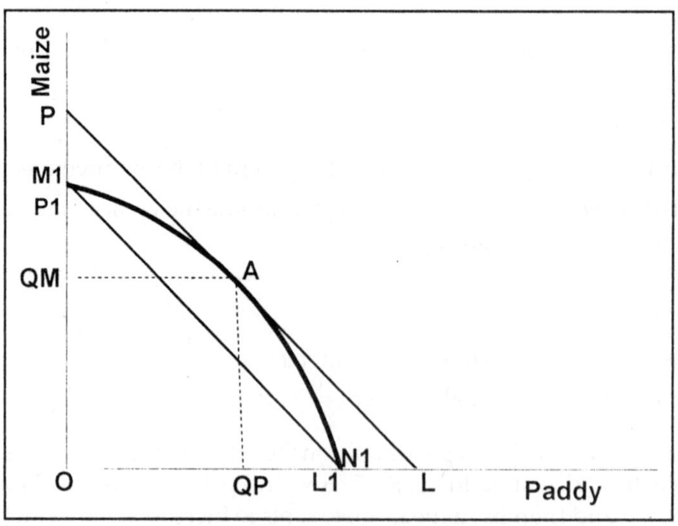

Figure 14.44: Farmer's equilibrium – Iso-revenue line tangent to Production possibility curve form above.

(b) Slope of the Iso-Revenue Line should be Equal to the Slope of Production Possibility Curve

The second condition essential for arriving at the point of optimum product combination is that, the slope of the iso-revenue line should be equal to the slope of production possibility curve from above for the desired combination of products. As explained through Figure 14.44, the iso-revenue line PL of the farmer is tangent to the production possibility curve from above at point A. So, at point A on production possibility curve, the slope of the iso-revenue line is equal to the slope of the production possibility curve, where the farmer produces OQ_P and OQ_M quantities of paddy and maize respectively. We know, the slope of production possibility curve indicates MRPS of paddy for maize and slope of iso-revenue line indicates price ratio of two products and the point of optimum product combination is explained below.

Slope of the production possibility curve = $MRTS_{PM} = (\Delta M)/(\Delta P)$

Slope of the iso-revenue line = (P_P/P_M)

So, the point of profit maximization is when the slope of the production possibility curve is equal to the slope of the iso-revenue line and this is given by,

$(\Delta M)/(\Delta P) = (P_P/P_M)$ or $MRTS_{PM} = (P_P/P_M)$ *Equation 14.34*

So, at the point of optimum product combination, the slope of production possibility curve is equal to the slope of iso-revenue line or the $MRPS_{PM}$ is inversely equal to the price ratio of paddy and maize. The point of optimum product combination can also be stated that, the rate at which the farmer is willing to substitute paddy for maize is inversely equal to the price ratio between paddy and maize.

(c) Production Possibility Curve should be Concave to the Origin

This is an essential condition to be fulfilled for arriving at the point of optimum product combination. When the iso-revenue line is tangent to the production possibility curve, at the point of tangency, the production possibility curve should be concave to the origin. In other words, the MRPS of paddy for maize must be increasing at the point of equilibrium. The same is shown through the Figure 14.44, where at the point of optimum product combination *i.e.*, at A, the shape of the production possibility curve is concave to origin. If at the point of tangency, where production possibility curve is convex to origin (Figure 14.45), it is not the point of profit maximization, as the movement from B towards either M_1 or N_1 will yield more revenue. In such a situation, the farmer will produce either at M_1 or N_1 rather than at B. But, the farmer will definitely operate at M_1, where production possibility curve touches higher iso-revenue line P_1L_1.

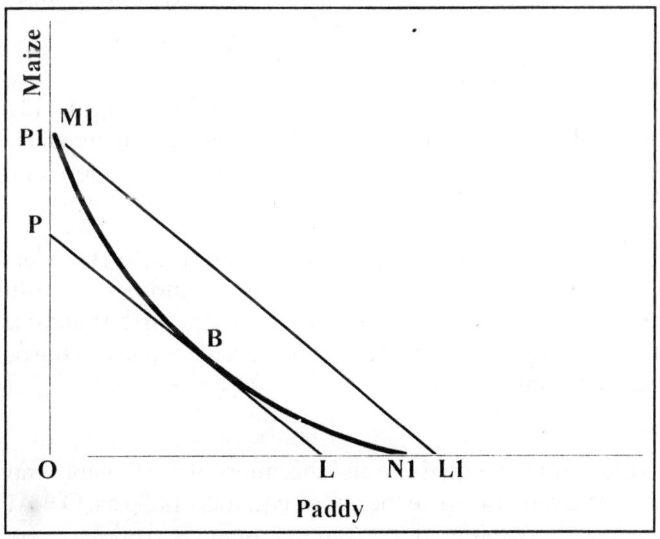

Figure 14.45: Farmer's equilibrium – Production possibility curve should not be convex at the point of tangency.

So, this gives the information that, the farmer will attain maximum profits, when the iso-revenue line is tangent to the production possibility curve from above and that too, when that production possibility curve is concave to origin, where it shows increasing MRPS.

The above explanation clearly infers that, at point A on the production possibility curve (Figure 14.44), all the above three conditions are fulfilled or satisfied and hence, we can conclude that, the farmer attains maximum profits at point A, through selling OQ_P and OQ_M quantities of paddy and maize respectively.

Inequality between MRPS and Price Ratio of Products

Case 1: When $MRPS_{PM}$ is Greater than Price Ratio of Products

This is given by, $(\Delta M)/(\Delta P) > (P_P/P_M)$. So, to attain the point of optimum product combination, the farmer should produce more of replaced product (maize) or less of added product (paddy).

Case 2: When $MRPS_{PM}$ is Less than Price Ratio of Products

This is given by, $(\Delta M)/(\Delta P) < (P_P/P_M)$. So, to attain the point of optimum product combination, the farmer should produce less of replaced product (maize) or more of added product (paddy).

14.6.4. Types of Production Functions

There are two types of production functions and they are discussed here under.

1. Linear Homogenous Production Function

If the quantities of all the factors are increased by 'm' times, output Q also increases by 'm' times. It is represented by Qm = f(mL, mC), where 'm' stands for any real number, Q stands for total output, L represents labour input, C represents capital input. So, if all the factors are increased in a given proportion (10 per cent), output also increases in the same proportion (10 per cent). It implies constant returns to scale.

2. Non-homogenous Production Function of 'k' Degree

It is given by $Qm^K = f(mL, mC)$, where 'm' stands for any real number, k is constant, Q stands for total output, L represents labour input, C represents capital input. This infers that, if the quantities of all the factors are increased by 'm' times, the output may increase by 'm' times or more than 'm' times or less than m times. If 'k' is equal to one, it implies, 'linear homogenous production function of the first degree' showing constant returns to scale. In this case, the output increases in the same proportion as the change in factors. For example, if all the factors are increased by 10 per cent, output also increases by 10 per cent. If 'k' is greater than one, it is 'Non-homogenous production function of degree more than one' showing increasing returns to scale. In this case, the output increases in more proportion compared to the change in factors. For example, if all the factors are increased by 10 per cent, output increases by 20 per cent. If 'k' is less than one, it is 'Non-homogenous production function of degree less than one' showing diminishing returns to scale. In this case, the output increases in less proportion compared to the change in factors. For example, if all the factors are increased by 10 per cent, output increases by only five per cent. Among the different production functions, linear homogenous production function can be easily employed in conducting empirical studies without complications. Hence, it is also called as 'well behaved production function'. It is widely used in input-output analysis and linear programming.

14.6.4.1. Cobb-Douglas Production Function

While conducting empirical study on American manufacturing industry, Prof. H.Douglas and Prof. C.W. Cobb developed a linear homogenous production function of degree one considering two factors *viz.*, labour and capital to determine the output of the manufacturing industry. The Cobb-Douglas production function is given by

$$Q = AL^aC^b$$ ***Equation 14.35***

where, a and b are the parameters of the function, A = intercept, L = labour factor, C = Capital factor, Q = Output in the industry. This equation clearly shows that, the output produced in the industry is influenced by two major factors labour and capital. This functional analysis revealed that, the output increase in the manufacturing industry is influenced by the labour factor to the extent of 75 per cent and remaining 25 per cent increase in the output is influenced by capital input. So, the above equation can be written as

$$Q = AL^{3/4}C^{1/4}$$ ***Equation 14.36***

So, the total value of labour and capital is equal one and this infers, it is a linear homogenous production function showing constant returns to scale. We can also write the above equation 14.36 as, $Q = AL^aC^{1-a}$

Being a non-linear production function (Equation 14.35), it will be made linear by converting it into log form as given below,

$$\log Y = \log A + a \log L + b \log C$$ ***Equation 14.37***

Features/Properties of Cobb-Douglas Production Function

☆ The Cobb-Douglas production function is a linear homogenous production function, which exhibit constant returns to scale. That means, when both the factors are increased 'n' times, the output is also increased by 'n' times.

Proof: We know, $Q = AL^aC^b$

When both the factors are increased by 'n' times, the output Q* is given by

$Q^* = A(nL)^a (nC)^b$

$= n^{a+b} (AL^aC^b)$

$= n^{a+b}(Q)$, where $Q = AL^aC^b$ as per the ***Equation 14.35***

But, a+b = 1 as per the constant returns to scale.

So, $Q^* = n(Q)$.

This proves that, Cobb-Douglas production function is a linear and homogenous production function of degree one (a+b=1). In the above Cobb-Douglas function, if a+b>1, then it infers increasing returns to scale, if a+b<1, it refers to diminishing returns to scale.

☆ From Cobb-Douglas production function, we can compute the MPP and APP of both labour and capital factors.

We know, Cobb-Douglas production function is $Q = AL^aC^b$

The MPP of labour factor is given by,

$MPP_L = \Delta Q/\Delta L = a(AL^{a-1}C^b)$

$= a(AL^aC^b)L^{-1}$

$= a(Q)L^{-1}$, where, $Q = AL^aC^b$ as per *Equation* **14.35**

$MPP_L = a(Q/L) = a\,(APP_L) = f(C/L)$ ***Equation 14.38***

where, APP_L is the APP of labour

Similarly, MPP of capital is given by

$MPP_C = \Delta Q/\Delta C = b(AL^aC^{b-1})$

$= b(AL^aC^b)C^{-1}$

$= b(Q)C^{-1}$ where, $Q = AL^aC^b$ as per the *Equation* **14.35**

$MPP_C = b(Q/C) = b(APP_C) = f(C/L)$ ***Equation 14.39***

So, from the above Equations 14.38 and 14.39, we can infer that, the MPP and APP of the factors is a function of ratio of the factors. The MPP of both the factors depends upon the quantities of both the factors used in the production programme.

☆ The exponents 'a' and 'b' of the two factors *viz.*, labour and capital respectively measures the elasticity of production with reference to these factors. Elasticity of production refers to the proportionate change in output resulting from a proportionate change in the factor, keeping other factors constant.

So, the elasticity of production with reference to labour factor is given by,

Elasticity of production of labour $= (\Delta Q/Q)/(\Delta L/L)$

$= (\Delta Q/Q) \times (L/\Delta L)$

$= (\Delta Q/\Delta L) \times (L/Q)$ ***Equation 14.40***

We know, $\Delta Q/\Delta L$ is the MPP_L

Substituting MPP_L of Equation 14.38 in the above Equation 14.40, we get

Elasticity of production of labour $= a.(Q/L)(L/Q) = a$

So, the exponent 'a' of labour factor measures the elasticity of production for labour factor

Similarly, we can compute the elasticity of production with reference to capital factor and it is given by

Elasticity of production of capital $= (\Delta Q/Q)/(\Delta C/C)$

$= (\Delta Q/Q) \times (C/\Delta C) = (\Delta Q/\Delta C) \times (C/Q)$ ***Equation 14.41***

We know $\Delta Q/\Delta C$ is the MPP_C

Substituting the MPP_C as given in equation 14.39 in the above equation 14.41, we get,

Elasticity of production of capital $= b.(Q/C)(C/Q) = b$

So, the exponent 'b' of capital factor measures the elasticity of production of capital factor.

These exponents indicate the percentage change in output for every one per cent change in the factor. For example, if exponent 'a' of labour, L is 0.8626, it indicates that, for every one per cent increase in labour factor, the output increases by 0.8626 per cent. So, the interpretation of exponents will be in terms of 'percentages'.

☆ The sum of elasticities of production *i.e.*, exponents of factors indicate the returns to scale. That means, if a+b = 1, it implies constant returns to scale, if a+b>1, it implies increasing returns to scale and if a+b<1, it implies diminishing returns to scale.

☆ We can derive the MRTS between labour and capital as below,

$MRTS_{LC} = \Delta C/\Delta L = MPP_L/MPP_C$ As per the ***Equation 14.5***

This is because, the $MRTS_{LC}$ is equal to the ratio of MPP of the two factors. If we move down along an isoquant, the loss in physical output of one factor (Figure 14.46), say capital will be equal to the gain in

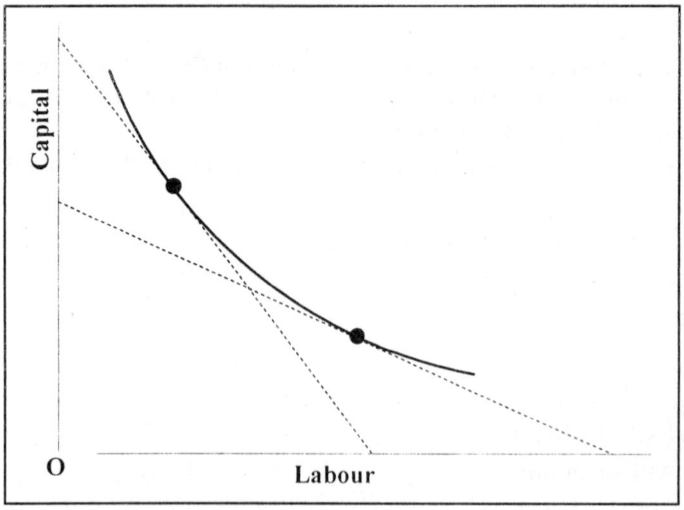

Figure 14.46: $MRTS_{LC}$ of isoquant indicating ratio of MPP of two factors.

physical output of another factor *i.e.*, labour. The loss in physical output of capital factor is given by the MPP_C multiplied by the change or reduction in capital factor, *i.e.*, $MPP_C \times \Delta C$. Similarly, the gain in physical output of labour factor is given by the MPP_L multiplied by the change or increase in labour factor, *i.e.*, $MPP_L \times \Delta L$. So,

Movement along the isoquant from left to right is given by,

$(MPP_C \times \Delta C) = (MPP_L \times \Delta L)$ or $(MPP_C \times \Delta C) + (MPP_L \times \Delta L) = 0$

$\Delta C/\Delta L = MPP_L/MPP_C$ *Equation 14.42*

As per the Equation 14.38, $MPP_L = a(Q/L)$

As per the Equation 14.39, $MPP_C = b(Q/C)$

Substituting Equations 14.38 and 14.39 above, in equation 14.42, we get

$\Delta C/\Delta L = MPP_L/MPP_C = [a(Q/L)]/[b(Q/C)] = (aC/(bL) = (a/b)(C/L)$

So, $MRTS_{LC} = \Delta C/\Delta L = MPP_L/MPP_C = (a/b)(C/L)$ *Equation 14.43*

The component (a/b) in Equation 14.43 measures the factor intensity employed in the production programme. That means, if the value of a/b is more than one, it implies the labour intensive techniques are employed in the production programme and if the value of a/b is less than one, it implies, the capital intensive techniques (say, machinery) are employed in producing the output.

- ☆ The E_{LC} in Cobb–Douglas production function is equal to unity. We know, E_{LC} is the ratio between proportionate change in the ratio between capital and labour factors employed in the production programme to the proportionate change in the MRTS of labour for capital. This is given by

  $E_{LC} = [\Delta(C/L)/(C/L)]/[(\Delta MRTS_{LC})/(MRTS_{LC})]$ As per the ***Equation 14.22***

  Substituting the $MRTS_{LC}$ of Equation 14.22 with the terms in Equation 14.43, we get,

  $E_{LC} = [\Delta(C/L)/(C/L)]/[\Delta(a/b)(C/L)/(a/b)(C/L)]$ *Equation 14.44*

  Since a/b is constant, we can write the above Equation 14.44 as

  $E_{LC} = [\Delta(C/L)/(C/L]/[\Delta(C/L)/(C/L)] = 1$

  Since, the E_{LC} is equal to one, the isoquants have convex shape, their production function is homogenous of degree one and it shows constant returns to scale. The same is explained through Figure 14.30 at point 'B'.

- ☆ As per the Equation 14.35, $Q = AL^aC^b$, we can say, the output of the production programme is influenced by the factors considered in the function and it is in the multiplicative form. That means, if the level of one factor is zero, the output will be zero. So, no factor level should be zero in the production function, implying that, the considered factors are all important in influencing the output level.

Criticisms of Cobb-Douglas Production Function

The following are the important criticisms laid out against the Cobb-Douglas production function:

- ☆ This function shows the constant returns to scale. But, in reality, the industry experiences either increasing returns to scale or diminishing returns to scale.

- ☆ When the same production function is applied to each firm and the industry as a whole, problems will arise, as the function applied to each firm may not give the same result as that of the industry.
- ☆ It is based on the assumption that, factors are substitutable and excludes complementarity of factors. But, in the short run, complementarity among the factors is also possible. Hence, we can say, Cobb-Douglas production function is more suitable for long run to the short run.
- ☆ It assumes perfect competition in the factor market, but this is unrealistic.

Importance of Cobb-Douglas Production Function

In spite of the above criticisms, the Cobb-Douglas production function has wider applicability. They include

- ☆ Its parameters 'a' and 'b' are used for inter-sectoral comparisons.
- ☆ It helps to estimate elasticity of production directly and thereby, returns to scale of the firm.
- ☆ It helps to understand the changes of factors of production on the output
- ☆ It helps to understand he nature of the costs
- ☆ It is used to prove Euler's theorem